The Auditory System and Human Sound-Localization Behavior

The Auditory System and Human Sound-Localization Behavior

John van Opstal
Department of Biophysics
Donders Centre for Neuroscience
Radboud University Nijmegen
The Netherlands

ELSEVIER

AMSTERDAM • BOSTON • HEIDELBERG • LONDON
NEW YORK • OXFORD • PARIS • SAN DIEGO
SAN FRANCISCO • SINGAPORE • SYDNEY • TOKYO
Academic Press is an Imprint of Elsevier

Academic Press is an imprint of Elsevier
125 London Wall, London EC2Y 5AS, UK
525 B Street, Suite 1800, San Diego, CA 92101-4495, USA
50 Hampshire Street, 5th Floor, Cambridge, MA 02139, USA
The Boulevard, Langford Lane, Kidlington, Oxford OX5 1GB, UK

British Library Cataloguing-in-Publication Data
A catalogue record for this book is available from the British Library

Library of Congress Cataloging-in-Publication Data
A catalog record for this book is available from the Library of Congress

ISBN: 978-0-12-801529-2

For information on all Academic Press publications
visit our website at https://www.elsevier.com/

Typeset by Thomson Digital

Dedication

To Dr. Dick Donker (Dec 21, 1934–Oct 1, 2014):
Neurologist, life artist, lover of music,
and of auditory science.

Contents

List of Abbreviations

2AFC	Two-alternative forced choice
AC	Auditory cortex
AM	Amplitude modulation
AMP	Auditory median plane
AN	Auditory nerve
ANN	Artificial neural network
AV	Audiovisual
AVCN	Antero-ventral part of Cochlear Nucleus
BB	Black box
BCD	Bone-conduction device
BM	Basilar membrane
BP	Band-pass
BS	Band-stop
CI	Cochlear implant
CN	Cochlear nucleus
CNS	Central nervous system
DCN	Dorsal part of Cochlear Nucleus
DTF	Directional transfer function
DMF	Dynamic movement field
FEF	Frontal eye fields
FM	Frequency modulation
FOV	Field of view
FRF	Future receptive field
FT	Fourier transform
GBC	Globular bushy cell
GME	Gaze-motor error
GWN	Gaussian white noise
HA	Hearing aid
HP	High-pass
HRTF	Head-related transfer function
HSE	Head-shadow effect
IC	Inferior colliculus
ICc	Central nucleus of IC
ICx	External nucleus of IC
IHC	Inner hair cell
ILD	Interaural level difference
INC	Interstitial nucleus of Cajal
IOR	Inhibition of return
IPD	Interaural phase difference

ISI	Intersaccadic interval
ITD	Interaural time difference
LP	Low-pass
LS	Linear system
LSO	Lateral superior olive
LT	Laplace transform
MAA	Minimum audible angle
MAP	Maximum a-posteriori
MDT	Medial-dorsal thalamus
MF	Movement field
MGB	Medial geniculate body
MI	Multisensory Index
MLB	Medium-lead burst neuron
MLE	Maximum Likelihood Estimate
MNTB	Medial nucleus of the trapezoid body
MSO	Medial superior olive
MVN	Medial vestibular nucleus
NI	Neural integrator
NLS	Nonlinear system
NPH	Nucleus prepositus hypoglossi
OAE	Oto-acoustic emission
OHC	Outer hair cell
OMR	Oculomotor range
OPN	Omni-pause neuron
PG	Pulse generator
PPC	Posterior parietal cortex
PPRF	Para-pontine reticular formation
PSG	Pulse-step generator
PVCN	Postero-ventral part of Cochlear Nucleus
riMLF	rostral interstitial nucleus of the medial longitudinal fasciculus
RM	Reissner's membrane
ROC	Receiver operator characteristic
RT	Reaction time
SBC	Spherical bushy cell
SC	Superior colliculus
SDT	Signal-detection theory
SNR	Signal to noise ratio
SPL	Sound-pressure level
SR	Spontaneous firing rate
SRT	Saccade reaction time
SSD	Single-sided deaf(ness)
STT	Spatial-to-temporal transformation
TM	Tectorial membrane
TOT	Tonotopic to oculocentric transformation
UCHL	Unilateral Conductive Hearing Loss
VOR	Vestibular-ocular reflex
VPG	Vectorial pulse generator
VS	Vector strength

Chapter 1

A Brief Introduction to the Topic

1.1 TWO TASKS FOR THE AUDITORY SYSTEM

Whenever a detectable sound wave reaches our ears, the brain will try to assign meaning to the acoustic event. In a split second, the auditory system succeeds in *identifying* the nature of the sound source out of a virtually unlimited number of possibilities: is it noise (the wind, the rain, the sea, a sigh)? Was it perhaps a familiar or an unfamiliar human voice? Was it an animal vocalization? Maybe it was a car, some other man-made machine, a musical instrument, or an orchestra? Perhaps, the sound was caused by the ticking of an object (a fork?) against another object (a dinner plate?), etc.

At the same time, the auditory system *localizes* the sound source. But just like source identification, the seemingly simple localization task could refer to multiple possibilities: where is the sound source located in "external space," that is, in the world around us, through which we navigate? Or: where is the sound relative to the ears or head? Where is it relative to other landmarks in the environment? Surprisingly, as we will see later, the brain seems to be particularly interested in determining where the sound source is located *relative to your eyes*! Thus, the auditory system has evolved to perform the following major tasks on the acoustic input:

Auditory task	It answers
Identification	What?
Localization	Where?

Obviously, the ability to rapidly identify and localize sound sources is vital for survival. In any case, it was crucial when in a not too distant past, we as hominids, had to struggle fiercely to stay alive, as food sources (good for us) and predators (very bad for us) had to be identified and localized as fast as possible. It is therefore not surprising that throughout evolution, the auditory systems of virtually all the animal species have developed dedicated neural circuits to efficiently and accurately solve identification and localization tasks.

Although simply formulated, these tasks are in fact astonishingly difficult to perform. Current technological advances, despite the tremendous increase in computer speed and memory storage over the last decades, are still not able to execute these tasks with the same accuracy, speed, flexibility, and efficiency as biological

The Auditory System and Human Sound-Localization Behavior. http://dx.doi.org/10.1016/B978-0-12-801529-2.00001-5

auditory systems. Indeed, the auditory system seems to carry a bag loaded with sophisticated tricks (neural algorithms) in order to do what it is supposed to.

How do we know all this, and how do we study, understand, and model the different aspects of sound processing in human and animal brains? Can we learn something essential from this, and perhaps implement this knowledge in future sound-recognition technologies, including healthcare applications, such as improved hearing aids and implants? This monograph forms my personal account of an exciting line of research on sound localization behavior in humans and nonhuman primates, which has kept me busy for well over 20 years, and is likely to keep me busy for the next decade as well. Some of the questions raised here form the central topic of this book.

To persuade the reader that sound processing in the brain is an interesting topic, wholly worthy of study, and above all, intellectually challenging and rewarding, here I would like to briefly illustrate, in a very general way, the sound–source identification problem as it presents itself to the auditory system.

1.2 AN ILL-POSED PROBLEM

The fundamental problem faced by audition has been particularly nicely formulated and illustrated by Albert Bregman (1990) in his "*man-at-the-lake*" analogy (Fig. 1.1). Imagine this guy, lying at the shore of a large lake. Although the story

FIGURE 1.1 **Our hero at the lake is allowed to only look at the movements of the two thin sheets to identify the sources that caused the water waves on the lake.** *(Courtesy: Carmela Espadinha.)*

doesn't tell, we may assume that the man is either deaf, or deafened by earplugs, so he can't hear. He just dug two narrow, parallel channels that fill themselves with water from the lake, and has then draped and fastened two thin plastic sheets onto the water surface of each channel.

Then the game starts: the man is allowed to only look at the up and down movements of the two sheets. Meanwhile, the lake's water surface is continuously perturbed by all kinds of objects (toy boats, swimming and splattering kids, dropped stones, landing ducks and geese, wind, etc.) that each cause their own specific water waves to travel at some fixed speed, from some direction, along the surface of the lake. Of course, the man doesn't know all this, since he is not allowed to look at the lake, and he can't hear either.

However, a small part of these traveling waves will at some moment, enter each of his two channels. The challenge for our guy is to decide, only on the basis of the motion patterns of the two sheets (which he may assume to oscillate without any energy loss with the water motion, and which he may analyze in every possible way), what exactly is happening on the lake.

Clearly, the two channels symbolize our ear canals, while the lake is the air, set in vibration by different sound sources; the two plastic sheets represent our eardrums. The "guy" in this story represents the homunculus within our auditory system that can only "look" at the one-dimensional temporal vibrations of the two eardrums to analyze acoustic input.

One doesn't have to be very imaginative to recognize that this is a formidable, if not an *unsolvable*, challenge for the auditory system! Indeed, mathematics tells us that such a problem is in fact, *ill posed* (Kabanikhin, 2008). This means there is no unique solution to the problem as in reality there are infinitely many mathematically valid solutions! How can we appreciate the severity of this problem?

Recall the "Hitchhiker's Guide to the Galaxy" by Douglas Adams, in which the earthlings have been told the answer (which is "42"), but have to guess *the correct question* in order to save Planet Earth from total destruction. Was the question: how old is your wife's sister? How many hairs does your adolescent son have on his chest? How much is 6×7, or $43.5 \times (42/43.5)$? How many presidents had governed the USA when G.W. Bush took office in 2001? Clearly, infinitely many questions can be formulated, all with the same correct answer: "42."

A similar mind-boggling problem bugs the auditory system. Think about it: the sound wave that reaches the ears is a linear superposition of all the sound sources in the environment. Suppose there are N such sources, and that the time-varying sound pressure for each source is $s_n(t-\{\tau_n\})$, for $t \in \{\tau_n\}$, where $\{\tau_n\}$ are the on- and offset timings of sound n. Each sound occupies a certain location in space, which we here measure in relation to the ears. For simplicity, we denote a location by its two directional angles (ignoring distance), here indicated by H (for horizontal), and V (for vertical). As we will see in later chapters of this book, different locations/directions of given sound sources lead to systematically different acoustic patterns at the two ears. As a result, a sound source is uniquely described by its temporal variations (and hence, its spectral content, which is given by the signal's set of constituent frequencies, $\{\omega_n\}$) and by its directional information (the *what* and *where*): $s_n(t-\{\tau_n\}, \{\omega_n\}, H_n, V_n)$. The resulting sound–pressure waves seen at the left (L) and right (R) eardrums are then given by:

$$S^{L,R}(t) = \sum_{n=1}^{N} s_n(t-\{\tau_n\},\{\omega_n\}, H_n, V_n) \tag{1.1}$$

We will see later that the left- and right-ear sound patterns for a given sound, differ from each other in a systematic way, whenever the horizontal position of the sound source moves away from the midsagittal plane.[a] The acoustic problem for our auditory homunculus therefore boils down to:

Find $\{\tau_n\}$, $\{\omega_n\}$, and H_n, V_n for all n *sources in the acoustic scene.*

For natural acoustic environments we can, however, construct the following truth table for the general situation of the problem described by Eq. (1.1).

Acoustic feature	A priori knowledge
Number of sources, N	Unknown
Spectral content of individual sources, $\{\omega_n\}$	Unknown
On- and offset times of individual sources, $\{\tau_n\}$	Unknown
Locations of individual sources, H_n, V_n	Unknown

That is, when the auditory system has no knowledge about the constituent sources that make up the total sound wave, there is absolutely *no way* that it can uniquely segregate individual sources from the acoustic mixture described by Eq. (1.1). The problem is even more severe, since many sound sources in the environment will have considerable spectral- and temporal overlap. They may start and end in roughly similar time windows, and their spectral bandwidths may be quite similar too. The problem even remains ill posed if we would know for sure that there are only two sources in the scene (ie, if $N = 2$)! Indeed: $42 = 21 + 21 = 43.5–1.5 = 3 + 39 = \ldots$ In other words, there is no hope for us, earthlings, to prevent ultimate disaster or is there?

[a]Later we will see that a sound-source displaced in the horizontal plane yields binaural differences in the timings of sound on- and offsets: $\{\tau_n\}^{L,R}$, and in the high-frequency content of the acoustic spectrum: $\{\omega_n\}^{L,R}$. In this general introduction, we will not delve into this matter further.

1.3 DEALING WITH ILL-POSED PROBLEMS

It turns out that in fact the auditory system does a remarkably good job at segregating, identifying, and following particular sound sources in the environment, and localizing them with astonishing accuracy. Yet, the problem of Eq. (1.1) remains ill posed and doesn't just disappear. To understand its successful performance, the auditory system must somehow rely on more practical, useful strategies, than on trying to solve unsolvable problems.

> *Strategy: Since the sound-identification problem cannot be solved, deal with it to the best of your abilities!*

A useful strategy to deal with ill-posed problems like Eq. (1.1) is to make clever assumptions regarding potential solutions, and to use these assumptions as efficiently as possible. In modern neuroscience theories (known under the collective name of Bayesian models) such assumptions are called "*priors*," as they refer to learned and stored probabilities of stimulus–response properties that the system may use as prior information, that is, advance knowledge, to deal (not "solve") with perceptual tasks. The idea behind these theories, which are essentially probabilistic, rather than deterministic in nature, is that the brain (ie, the auditory system) generates a response that can be considered its best *statistical estimate* for the current solution. In a (statistically) *optimal* system, such a response will, on average, have the smallest systematic error (highest accuracy) and variability (highest precision), given the uncertain and ambiguous sensory evidence of Eq. (1.1), and the potential advance prior knowledge (estimate, expectation) stored in the system. Now that is quite a mouth full, but we will come back to this topic in more detail later in this book. It here suffices to state that there is good evidence that our sensory and motor systems, the auditory system included, seem to operate according to such statistical (nearly) optimal principles.

In the case of sound sources, these assumptions could be based on certain relevant properties of the physical world, which have been learned through experience by interacting with the environment in many different ways: navigating through the environment, perceiving the same objects in the environment through our different sensory systems, orienting to objects in the environment, particular properties of familiar objects, etc. For example, our brains may have learned that in natural *physical* environments there can be only one object (read: sound source) at any given point in space at a certain time. Note that this statement excludes the possibility of artificial, technical devices, such as loudspeakers, or headphones, which can clearly violate this natural requirement! I nonetheless believe that all auditory systems have evolved to use this kind of natural logic in order to make sense of unknown acoustic environments.

This is a very important point, which immediately touches on the central topic of this book, namely that sound processing and sound localization are *active* processes that involve planned behaviors! The idea is that by doing so, the system becomes better and better at dealing with ill-posed problems like formulated by Eq. (1.1).

1.4 SPECTRAL REPRESENTATION AND SOURCE PRIORS

The vibrations of the tympanic membrane, as described by Eq. (1.1), are transmitted via the middle-ear bones to the stapes at the entrance of the inner ear (the cochlea) (Purves and Augustine, 2012). Fig. 1.2 illustrates a typical, complex sound–source signal, produced by a male human voice, uttering the sentence "*Your test starts now*" (in phonetic script: "jʊə tɛst sta:rts naʊ"), which took about 2 s to complete.

In the cochlea, the organ of Corti performs the first nontrivial sensory transformation on this sound–pressure wave: from a pure *temporal* in- and outward movement of the oval window to a combined *spectral–temporal* representation along the length (coordinate x, in mm) of the cochlea. We will see in chapter: The Cochlea, that as a result of the intriguing micromechanics in the cochlea, specifically the position-dependent resonance properties of the basilar membrane (BM), in combination with local nonlinear feedback through the function of outer hair cells (OHC), the temporal sound–pressure signal $S(t)$, induces a transverse traveling wave along the BM, the amplitude of which reaches a sharply defined maximum at a frequency-specific location. Through this biomechanical mechanism, sound frequencies are topographically represented along the BM. In this *tonotopic representation*, frequency f, is mapped roughly logarithmically along the BM: $x = -4\log_2 (f/f_{max})$ mm, with f_{max} the highest audible frequency at $x = 0$ (see chapter: The Cochlea). In cases where the sound is a single pure tone, that is, $S(t) = S_0\sin(2\pi f_0 t)$, high-frequency tones (in normal-hearing humans up to about 16–20 kHz) yield their maximum near the base ($x < 10$ mm) of the BM, low-frequency tones (between 40 and 100 Hz) vibrate maximally at the BM apex ($x \sim 30$–35 mm), while midrange frequencies (2–6 kHz) are encountered at more central locations.

Thus, evolution has figured out a nice way to extend the temporal representation of the sound mixture of Eq. (1.1) with an additional, now explicitly

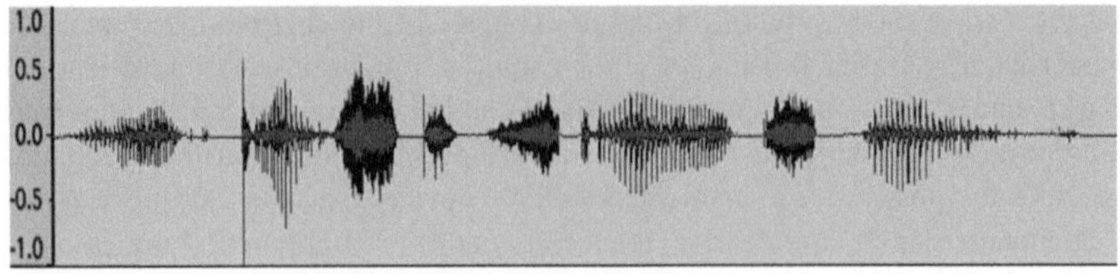

FIGURE 1.2 **Temporal vibration pattern of the complex sound–pressure wave, when a male human voice utters the sentence: "Your Test Starts Now."**

represented dimension (the sound's spectrum), which could further help the auditory system in its analysis of the acoustic scene.

Indeed, how the auditory system may be utilizing this additional representation is illustrated in the spectral–temporal representation (spectrogram) of the male utterance in Fig. 1.3. The features that make up the spectrogram suggest that the auditory system could make some useful prior assumptions that relate to the spectral–temporal structure of sound sources themselves. For instance, many potentially interesting sounds, like the vocalizations from fellow humans, or from a prey or a predator, are caused by mechanical vibrations of vocal chords or strings, and therefore unavoidably contain regular discrete harmonic complexes in their spectra. The frequencies in such vibrational spectra are thus related by: $\{\omega_n\} = \{n\omega_0\}$ with $n = 1, 2, 3, \cdots$, and where ω_0 is the fundamental (lowest) frequency in the sound spectrum.

Thus, the sonogram of Fig. 1.3 would allow rapid identification of some unique acoustic patterns, which are not obvious at all in the temporal signal of Fig. 1.2. Indeed, the distinct harmonic complexes in the voice (containing between 7 and 10 tones with a spectral interval of approximately 100 Hz, and a fundamental frequency of about 100 Hz) can be easily identified by visual

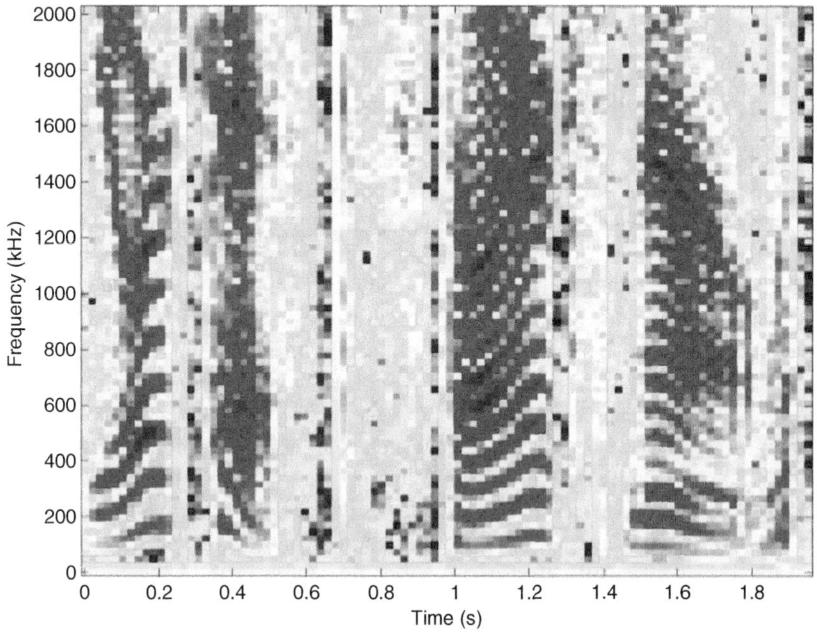

FIGURE 1.3 Spectral–temporal representation (or sonogram) of the sound in Fig. 1.2. Time runs along the abscissa (from 0 to 2.0 s), frequency along the ordinate (represented here on a linear scale from 0 to 2000 Hz to highlight the harmonic regularities better). Color represents amplitude (red: positive; blue: negative) of the frequency component. Any time point (a vertical line) relates to the instantaneous vibrational pattern of the BM.

inspection in the time windows 0–0.2 s ("ʊə"), 1.0–1.25 s ("*a*:") and 1.45–1.75 s ("aʊ") in the 0–1000 Hz range. A second interesting feature to note is that all tones in the harmonic complexes start and end at the same time, and sweep synchronously upward and downward in frequency, as the voice pronounces the "ʊə" "*a*:" and "aʊ" vowels.

It is easy to imagine that joint synchronous movements in the spectral–temporal domain will provide strong cues to the auditory system for wanting to group them as a single acoustic object. Why? Obviously, because of mere statistics: it is extremely *unlikely* that multiple, independent acoustic sources will be precisely synchronized in their on- and offsets (time), and follow joint complex movements in the frequency domain at fixed spectral intervals. On the other hand, for a single sound source it is extremely *likely* (and even a prerequisite!) that tight synchrony and spectral harmony both occur together.

So, even though there is no unique exact solution for the general auditory problem of Eq. (1.1), sensible prior assumptions may readily discard many potential mathematical solutions, for the simple reason that they are nonphysical, extremely unlikely, or do not fit task-related expectations

> *The auditory system deals with ill-posed problems by making sensible statistical assumptions about the world, and about potential target sounds.*

1.5 TOP–DOWN SELECTIVE FILTERING

From the above, it could follow that the auditory system may not be interested at all in finding "the" solution of the *N*-sources problem, since it is impossible to identify them anyway. In fact, the truth table for the general auditory identification problem has in reality become irrelevant: all acoustic features that turn out to be "unknown" could possibly be replaced by: *Don't know, and don't care.*

Instead, we could assume that the auditory system would prefer to make educated prior guesses about the most important, task-relevant, regions within the acoustic scene. Such a mechanism would call for *selection, based on task-specific prior information*. It invokes top–down decision mechanisms that set priorities on the acoustic task at hand. Which sources are we interested in at this moment, and for this particular task? What do we know about their properties (prior information)? What do we consider as "target," and what as "distracter," or "background"? Once the system has defined and set its priority list (the "task"), it can efficiently start its probabilistic search in the acoustic input of Eq. (1.1). The task constraints may thus act as clever "filters" on the spectral–temporal input, which suppress those spectral–temporal regions in the input that are considered too remote from the target.

There may exist an interesting analogy to the way in which *vision* is thought to process information (Purves and Augustine, 2012). The visual system is thought to use an attentional "filter," that in combination with a "covert"

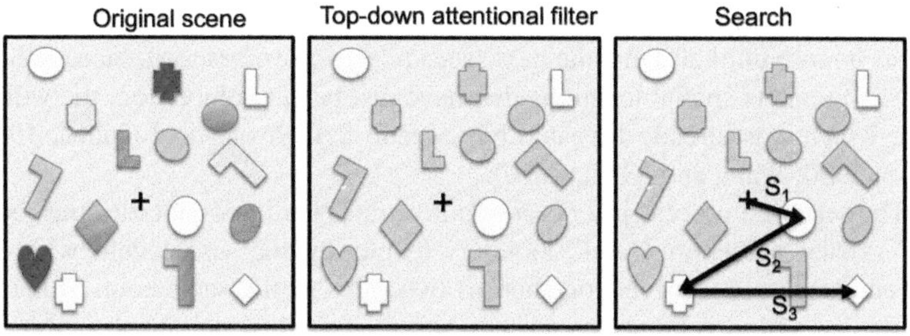

FIGURE 1.4 Search in a cluttered scene with many competing objects. The task is to find the yellow diamond. The fovea fixates the small cross. A top–down attentional filter enhances visual responses to yellow features in the scene, creating a "yellowness" saliency map. All different-colored objects become less salient. Saccadic eye movements are only directed to the conspicuous yellow targets in this saliency map. (A) Original scene; (B) top–down attentional filter; and (C) search.

(ie, attentional, mental), "spotlight," and "overt" (observable) fovea[b] efficiently deals with complex visual scenes. This strategy allows vision to quickly identify and localize a particular yellow object (suppose that this is the task) in a complex urban scene, like a busy street in which many advertisement boards, cars, and people all compete for attention.

Fig. 1.4 illustrates this idea for a simpler laboratory test, where the task is to find the yellow diamond among a set of different shapes with different colors. The attentional filter first "highlights/boosts" everything yellow in the scene (Fig. 1.4B), while the covert spotlight selects the next goal for foveation with the eyes (Fig. 1.4C). The sequence of fast (overt) saccadic eye movements across the scene (the arrows), will then quickly direct the fovea from one conspicuous yellow object to the other, until the visual system has identified the target (Koch and Ullman, 1985; Itty and Koch, 2000).

Saccades have to be executed one after the other, which is typically done with an intersaccadic interval of about 150–250 ms, so that our brains program and generate about three to four saccades per second. As a result, this visual search strategy may appear to be a time-consuming "serial" search, but because of the preprocessing by the attentional filter, which effectively acts as "parallel filtering," it actually becomes a "clever" serial search! The fovea will hardly ever land, for example, on irrelevant red or blue objects, so that the system avoids wasting valuable visual processing time. Moreover, use of additional forms of prior information (not only "yellowness," but perhaps other visual features, like orientation, relative size, shape, etc.) can in principle render the system even more efficient.

A second interesting property of saccadic eye movements, which could further increase the efficiency of goal-directed visual search, is the use of

[b]The fovea is the small area on the retina with a high spatial resolution (spanning less than a degree of visual angle), which is crucial for perceiving color and for analyzing fine details for visual object recognition. A saccade directs the fovea as fast as possible (at speeds up to 600–700 degrees/s) to a selected target in the periphery.

a mechanism called "*inhibition of return*" (IOR). This process prevents (or makes it more unlikely) that the next saccadic eye movement (eg, saccade S_3 in Fig. 1.4) revisits spatial locations that have just been explored (eg, the yellow oval, which was already foveated by saccade S_1) (Posner and Cohen, 1984; Klein, 2000; Wang and Klein, 2010).

The selection process just described illustrates the important idea that "seeing" (visual perception) and "looking" (exploring the visual field with eye movements) need to occur together, as two sides of the same coin. This idea entails that vision is an active process (*active vision*), which involves dynamic, adaptive filtering of the input, the use of prior, task-relevant information to select potential targets from a myriad of possibilities, and subsequent active exploration by the visuomotor system, which in turn involves clever decision making, and speed-accuracy trade off through goal-directed and fast orienting responses across the selected environment.

We will see that in the same realm the auditory system may employ a strategy of "*active hearing*" to perform the sound localization and identification task. By programming and generating combined eye–head movements to target sounds in the environment, results show that it can find the sound source within a few hundred milliseconds, with a reaction time, accuracy, and precision that often surpasses that of vision.

An important reason for the use of gaze control in both systems is that the spatial resolutions and spatial ranges for vision and audition are markedly

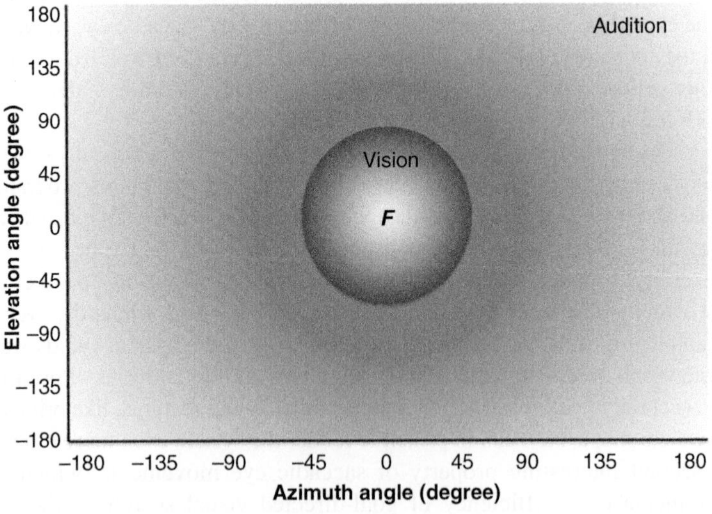

FIGURE 1.5 Vision and audition have very different spatial resolutions and spatial perceptual ranges. Whereas vision is restricted to a high resolution within a narrow (<10 degree) range around the fovea (*F*), audition has a lower, yet roughly constant spatial resolution across the entire directional space (−180 to +180 degree in both azimuth and elevation). Azimuth: angle in horizontal plane; elevation: angle in vertical plane. [*A*,*E*] = [0,0] = straight ahead; [180,180] = back; [0,90] = zenith.

different (Fig. 1.5). Whereas vision has a very high spatial resolution over a very limited viewing angle (ie, a parafoveal range of a few degrees), which decreases rapidly toward the visual periphery of about 45 degree and beyond, audition has a lower spatial resolution than vision, but an omnidirectional perceptual range (it covers the full sphere in azimuth and elevation). As a consequence, any peripheral target beyond the visual perceptual range, or in a range with very low spatial resolution, has to be acquired by the auditory system whenever possible! Fig. 1.5 suggests that the auditory system may in fact be the primary sensory system-driving eye–head orienting responses.

1.6 AUDIOVISUAL INTEGRATION

Vision and audition both use gaze shifts (gaze is the direction in which the fovea points in external space) as a natural response to rapidly look at (ie, localize) targets that were perceived in the periphery of sensory space. However, as in natural environments, objects tend to emit both visual and auditory signals it is of crucial importance that the two sensory systems collaborate intensively to perform the localization task as fast and as accurately as possible. There are many interesting and nontrivial issues involved in this integration, all of which will be dealt with in this book. Here, I will only briefly illustrate one of them, which is the need to bring the two sensory systems into "*register*." As the eyes can rotate freely within the head, and the head can independently rotate on the neck and trunk, the typical situation will be that the head (your nose) will point in a different direction in space than the fovea (gaze line) when a sudden audiovisual target appears in the periphery.

Fig. 1.6 describes the situation of a single target, perceived by the auditory and visual systems, when the nose is pointing upward, while the eyes are

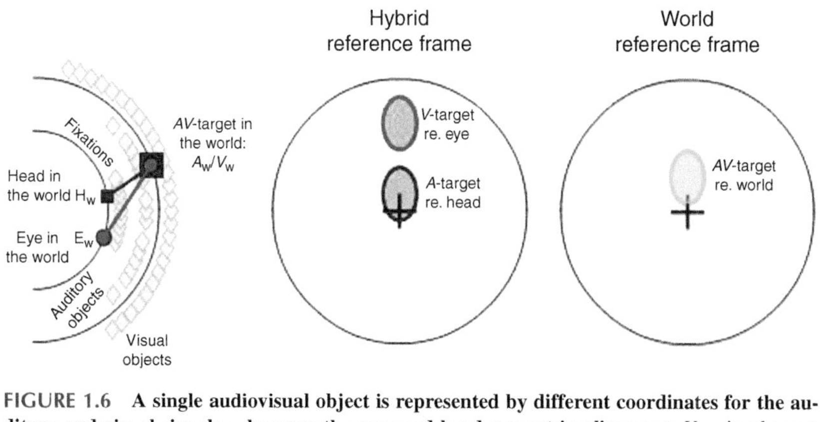

FIGURE 1.6 A single audiovisual object is represented by different coordinates for the auditory and visual signals, whenever the eyes and head are not in alignment. V_E, visual target relative to the eye; and A_H, auditory target relative to the head. To bring the disjoint sensory representations into a common reference frame, the brain should perform a sensorimotor coordinate transformation that uses motor signals about the misalignment of eyes and head.

looking straight ahead. The sound-localization system perceives the acoustic input from the target slightly upward with respect to the nose, while the visual location of the same target is projected more upward on the retina. Clearly, the two sensory locations are not in register, and could in principle have arisen from two different objects. So how can the brain decide whether or not the two sensory inputs came from the same location in space, and hence, according to the prior about natural spaces, from a single object? To that end, it should use the difference in alignment between the two sensory systems (ie, either the head orientation on the trunk, the orientation of the eyes in the head, or both) to recalibrate the sensory signals. For example, if the target is to be referenced with respect to the head, the visual retinal signal should be combined with the motor signal of the eye-in-head orientation, E_H: $V_H = V_E + E_H$. Alternatively, the auditory head-referenced signal could be combined with the motor signal about eye-in-head orientation, like: $A_E = A_H - E_H$. Note that both sensory signals could also be represented in a more general spatial reference frame, for example, relative to the body, which would require the use of motor information from both the head and the eyes. In that case:

$$A_S = A_H + aE_H + bH_S \text{ and } V_S = V_E + cE_H + dH_S. \tag{1.2}$$

As an exercise, the reader may verify that in the case of a single audiovisual target, the coefficients [*a*,*b*] and [*c*,*d*] of Eq. (1.2) are different for the visual and auditory modalities. In other words, these coordinate transformations are highly context and modality specific. Yet, audiovisual integration is crucial for adequate identification and localization of stimuli in the environment, and hence they need to be carried out fast with accuracy and precision.

1.7 HOW TO USE THIS BOOK

This monograph may serve as a full-semester course on the application of experimental and theoretical psychophysics to the human auditory and audiomotor systems. Much of the material covered in this book has been used in my own university courses, which were aimed at neurobiology, cognitive neuroscience, and physics bachelor students with a keen interest in quantitative understanding of the mechanisms of perception and sensorimotor integration. We believe this book will also appeal to biomedical engineering students, and to any natural scientist or engineer (biomedical, robotics, virtual reality, etc.) interested in the workings of the human brain. A basic understanding of elementary mathematics (calculus, standard differential equations, basic integration, linear algebra, and some vector calculus) and physics (mechanics) is needed, but sections that would require more advanced technical skills will be marked (*) as such. These sections can be skipped without losing touch with the narrative of the book.

Divided over 14 chapters, the reader will encounter a variety of related topics that range from the physical acoustic input stimulus of the system (see chapter: The Nature of Sound), to top–down neural mechanisms that may underlie audiovisual integration in goal directed–gaze orienting behavior (see chapter: Multisensory Integration), and in hearing impairments (see chapter: Impaired Hearing and Sound Localization).

The book covers topics ranging from the auditory periphery to central mechanisms of sound encoding and decoding, as well as mechanisms that underlie sound-localization encoding and the planning, and generation of coordinated eye- and head movements to sound sources. The book thus consists of roughly two equal-sized parts that could in principle be taught and studied to a large extent independently as two half-semester or trimester courses. In Chapters 2–7 (on *acoustics, theoretical concepts, and the neurobiology of audition*), the reader encounters relevant topics in acoustics, from the physical principles underlying the propagation of sound waves, to the physical–mathematical modeling of cochlear hydrodynamics, including active and nonlinear properties of cochlear OHC (see chapters: The Nature of Sound; The Cochlea). As a general background, the first part also provides some in-depth coverage of systems theory, linear as well as nonlinear, and Fourier analysis (including Laplace transforms, see chapters Linear Systems Analysis and Nonlinear Systems). The concepts from these two chapters are also needed in case one decides to study only the second half of the book. The first seven chapters also present neurobiological underpinnings of the auditory system, by describing some important neural encoding principles (and their analyses) observed in the activity of neurons at different stages in the ascending auditory pathway of mammals, like spike timing, rate coding, phase locking, spectral tuning, nonlinear interactions, monaural and binaural interactions, spatial sensitivity, and spectral–temporal receptive fields (see chapters: The Auditory Nerve; Sound-Localization Cues; and Assessing Auditory Spatial Performance).

The second half of the book (see chapters 8–14, on "*gaze control, sound-localization behavior, spatial updating and plasticity, and audiovisual integration*") deals with the behavioral aspects of sound-evoked orienting. These chapters are built on the idea that sound localization is an active process, in which the generation of rapid orienting responses of the eyes and head are needed to close the action-perception cycle. Additional theoretical background is given on psychometric methods and signal-detection theory (see chapter: The Gaze Orienting System), including the use of eye and eye–head movements as accurate, absolute, and fast sound-localization probes. We will also describe the neurophysiological underpinnings of eye–head gaze control in more detail by focusing on the involvement and modeling of the midbrain (see chapter: The Midbrain Colliculus). From the basic (static) sound-localization responses to single sound sources in the typical silent laboratory environment, we then proceed to the need for dynamic sensorimotor feedback control under more realistic auditory localization tasks, and how the required coordinate transformations may be represented in the activity patterns of neural populations (see chapter: Coordinate Transformations). In Chapter: Sound-Localization Plasticity, we describe the experimental evidence for some remarkable plasticity in the human auditory localization system, which extends well into adulthood. To some extent this topic returns in the chapter on impaired hearing (see chapter: Impaired Hearing and Sound Localization). In Chapter: Multisensory Integration, we will deal with the interesting interactions between the visual and auditory systems, and highlight some of the modern theoretical concepts of Bayesian inference that could underlie many of the phenomena observed in sound-localization behavior, and in audiovisual integration.

To the student: depending on your mathematics and physics background, this book can be studied either in its entirety, or by skipping the sections marked (*), which require advanced understanding of physics and/or mathematics. The same holds for the exercises that are given at the end of each chapter. Additional material, like computer scripts needed to run some of the simulations (in Matlab), or some extra hints that may help solve some of these exercises, are provided on the website, http://www.mbfys.ru.nl/~johnvo/LocalizationBook.html

To the lecturer: as the book is roughly divided into two equal parts that cover complementary fields of study (the auditory system and sound-localization behavior, respectively), it can be used as a full-semester (14 weeks) textbook for a course on human auditory psychophysics, in which each chapter can be covered in a typical week of 2 h of lectures, supplemented by two to three practical hours for the exercises and computer simulations. Alternatively, the book can be used in two separate semisemester (7–8-week quarters) or two trimester courses, which deal with the first eight chapters and with Chapters 9–14, respectively. Different combinations may be considered too. For example, the chapters on sound-localization plasticity (see chapter: Sound Localization Behavior and Plasticity) and on impaired hearing (see chapter: The Auditory System and Human Sound-Localization Behavior) could be included in the first course.

Physics students at the bachelor level (ie, in the second or third year of their curriculum), or engineering and informatics students with an equivalent math background, should be able to study all the material in this book. Students from the life sciences (eg, neurobiology, medical biology), or cognitive neuroscience and functional psychology, may want to skip the sections (*) that require some more advanced technical knowledge of math and/or physics.

Students are particularly encouraged to make the exercises and run the computer simulations (in Matlab) that are provided at the end of each chapter. Lecturers can obtain the fully worked-out problems of the exercises from the author upon request (j.vanopstal@donders.ru.nl).

All figures in the book are available for your PowerPoint presentations on the book's website, at http://www.mbfys.ru.nl/~johnvo/LocalizationBook.html

As a final remark, the author will be extremely grateful for any constructive feedback, interesting additional exercises, errors, etc. that could help to improve future editions of this book. Your contributions will of course be fully acknowledged.

1.8 OVERVIEW

Chapter: The Nature of Sound provides the physical basis of the sensory input stimulus to the auditory system: the sound wave, which is a longitudinal mechanical perturbation of the vibration of molecules that propagates at high velocity through the medium. From first principles of thermodynamics we deduce the propagation speed of monochromatic sound waves through air. We then look at the mechanical wave equation (both in homogeneous and in inhomogeneous media), the dispersion relation between the sound frequency and its wavelength, and at the transmission of acoustic power through the medium. We discuss reflection and transmission at the transition boundary of different media, which leads to the concept of acoustic impedance. We then introduce Fourier analysis of periodic signals, and finally define the phase velocity and group velocity of modulated acoustic signals.

Chapter: Linear Systems Analysis gives a thorough introduction into the field of linear black box analysis, a mathematical modeling technique that has become quite popular in the engineering sciences, but has also seen many useful applications in computational neuroscience and psychophysics, and in auditory science in particular. Starting from the superposition principle it is shown that this simple idea leads to a series of profound consequences: the concept of the system's impulse response as the system's memory, and the convolution integral to predict a system's response to arbitrary inputs. We apply Fourier analysis to extend the time-domain analysis of linear systems to the frequency domain, meanwhile introducing the system's transfer characteristic, and the Bode plot as a useful graphical tool to analyze its properties. We then introduce the Laplace transform as a convenient general method for dealing with a broad range of linear systems and signals in a semiintuitive way. We end the chapter with the introduction of the Gaussian white noise signal as a stimulus, and the auto- and cross-correlation functions of signals.

Chapter: Nonlinear Systems extends the linear black box theory of Chapter 3 to smooth nonlinear systems by introducing the Volterra and Wiener functional approaches to systems identification. In these nonlinear descriptions the system's response is described as a series representation of higher-order (ie, nonlinear) contributions. The core descriptors in these approaches are the system *kernels*, which are a natural extension of the impulse–response kernel for linear systems. Although the Volterra approach is a general systems analysis technique that may be applied to arbitrary input stimuli, it is not mathematically feasible to identify the underlying system kernels in a black box input–output measurement approach, because they are mutually dependent. This property has prompted researchers to develop a method that allows one to independently extract the nonlinear system kernels. The Wiener series, which is based on the autocorrelation properties of Gaussian White Noise as the system's input, is the most commonly used technique. Interestingly, a three-layered feed-forward artificial neural network with an appropriate input signal representation turns out to be mathematically equivalent to the complete Volterra series. It thus provides an alternative model description for the nonlinear system at hand. As a consequence, it is possible to independently extract the corresponding Volterra kernels from the trained synaptic weights in the network.

Chapter: *The Cochlea* applies several of the concepts described in the previous chapters to build and discuss a physical and physiologically realistic model of the cochlea. The chapter starts with the idea of tonotopy in the cochlea, and elaborates on the interesting analogy with the physics of water waves to the traveling wave along the cochlea's basilar membrane. It then describes the linear model based on the Nobel prize-winning work of Géorg von Békésy, as further elaborated by Joseph Zwislocki, and proceeds with the modern work on modeling the role of the electromotile response of OHC in combination with the tectorial membrane to explain the high sensitivity and huge dynamic range of the auditory system.

Chapter: *The Auditory Nerve* discusses the first neural stages in the auditory processing chain. Auditory nerve recordings indicate that responses faithfully reflect the motion patterns of the basilar membrane. Reverse correlation analysis links the tuning of low-frequency auditory nerve fibers to sharply tuned gammatone filters. Using a clever stimulation method the nerve recordings of single units can be used to fully reconstruct the amplitude and phase characteristics of the BM motion. The chapter concludes with a pragmatic model of the auditory periphery.

Chapter: *Sound-Localization Cues* describes the implicit localization cues that are available to the human auditory system in the horizontal plane (azimuth angle), the medial plane (elevation angle, and front/back), as well as sound–source distance. Binaural differences in time and intensity encode the sound–source azimuth angle for low- and high frequency sounds, respectively. The elevation angle is determined on the basis of complex spectral-shape cues that arise by acoustic reflections and diffraction within the pinna. Models that explain these computational processes are discussed. The chapter also describes the neural mechanisms for these cues that have been identified in animal studies.

Chapter: *Assessing Auditory Spatial Performance* discusses different methods that are used to study spatial hearing. The methods range from the mere detection of auditory stimuli, to left/right lateralization or discrimination of stimuli with respect to the center of the head to the localization of sounds in absolute external space. As the latter is measured with continuous pointing methods, the former measurements typically yield binary responses, and rely on the concepts of signal detection theory. We argue that the visual system is a natural ally of the sound-localization system, and therefore eye movements (gaze shifts) provide a natural, fast, and accurate pointer to study sound-localization behavior. We finally dwell on the method to virtual acoustics to study naturalistic sound-localization behavior with headphones, and which gives the flexibility to selectively manipulate the acoustic localization cues.

Chapter: The Gaze Orienting System describes the saccadic eye-movement system and the control of combined eye–head gaze-orienting movements. The latter is implicated in natural sound-localization behavior. Because of their inhomogeneous retina, foveate animals optimize their orienting behavior through rapid, accurate, and precise saccadic eye movements. The saccadic system is characterized by a prominent kinematic nonlinearity, which is implemented as a central vectorial pulse generator that provides a common eye-velocity drive to the horizontal and vertical motor plants. We argue that this same vectorial motor command drives the eye–head motor systems, albeit that it is modified by appropriate coordinate transformations that let the eyes and head move in the direction of auditory and visual targets. We discuss several quantitative models that explain the underlying circuitry of eye- and eye–head gaze shifts.

Chapter: The Midbrain Colliculus discusses how a localized population of saccade-related cells in the superior colliculus (SC) encodes the vector for the upcoming gaze shift. We model the mechanism underlying the afferent complex-logarithmic SC motor map, and how the efferent projections of its cells to the brainstem contribute to encode the saccade, and the tuning characteristics of SC movement fields. We then present evidence that the firing rates of SC cells determine the saccade kinematics, such that the population effectively acts as the nonlinear vectorial pulse generator, hypothesized for the common source control of eye–head gaze shifts. We show that a spatial gradient of peak firing rates along the rostral–caudal axis of the motor map, together with a fixed number of spikes in the burst, provide the nonlinear mechanism of the main sequence. Finally, we discuss the potential role of the midbrain inferior colliculus (IC) in spatial hearing, and how the tonotopic to spatial transformation in the auditory system might be understood.

Chapter: Coordinate Transformations describes the different egocentric reference frames that are relevant for the control of orienting gaze shifts to brief sounds and lights: oculocentric, craniocentric, and world-centered coordinates. We then discuss how experiments and quantitative models of eye–head gaze orienting can dissociate the different coordinate systems. Neurophysiological experiments have identified two different mechanisms that could potentially cope with coordinate transformations: gain fields and predictive remapping. We describe the static and dynamic double-step paradigms as prime examples of studying the nature of spatial target updating, and discuss the limits of spatial updating performance in situations where the system lacks crucial information about stimulus motion. Finally, we show how eye- and head-position signals interact within the auditory system, and postulate that the sound-localization system represents acoustic targets in a world-centered reference frame.

Chapter: Sound-Localization Plasticity discusses plasticity of the human sound-localization system in response to a variety of acoustic manipulations. We distinguish explicit perceptual learning from implicit sensory–motor feedback learning, and describe experiments that illustrate both types of learning. We introduce the phenomenal plasticity of the barn owl's auditory system, and describe the immediate localization effects of unilateral ear–canal plugging of human listeners. We discuss why spectral cues are essential to cope with this binaural perturbation. We then describe the adaptive response of humans to long-term unilateral plugging. When the pinna cues are perturbed by molds that change the pinna geometry, subjects relearn to map the new spectral cues to target elevation within a few weeks. Interestingly, these manipulations do not interfere with the original spectral cues, and there are no aftereffects. We argue that pattern-recognition learning (no aftereffects) differs in essential ways from parametric learning (with aftereffects). We demonstrate that the visual system is needed to fine tune the spatial mapping of spectral cues, by comparing sound localization of congenital blind listeners with normal sighted listeners. Finally, we demonstrate that perturbed spatial vision will perturb the sound-localization mappings in a similar way.

Chapter: Multisensory Integration discusses the mechanisms that guide multisensory integration. In particular, we focus on audiovisual and audiovestibular interactions that influence spatial perception. We describe the spatial–temporal factors that modulate the strength of multisensory integration, and introduce the phenomenon of inverse effectiveness. We forward three types of models to account for multisensory interactions on reaction times and response accuracy: race models, interactions within the superior collicular motor map, and Bayesian inference. We discuss the latter framework as a mathematical–statistical model to understand two prominent effects on response trajectories to multisensory stimuli: response averaging and decreased response variability. We show that audiovisual integration in cluttered audiovisual environments follows the spatial–temporal rules of multisensory integration, and provides strong support for inverse effectiveness at the behavioral level. We show that the sensory–motor system integrates stimuli only when they are spatially and temporally congruent, and keeps track of the stimulus statistics in the environment to assess the probability of audiovisual alignment. Finally, we apply the Bayesian framework to explain the considerable mislocalization of sounds during head tilts with respect to gravity.

Chapter: Impaired Hearing and Sound Localization describes the sound-localization abilities of hearing-impaired listeners. We distinguish listeners with a conductive hearing loss from sensory–neural impairments, and describe current technologies to restore impaired hearing: hearing aids, bone-conduction devices, middle-ear implants, and cochlear implants. Single-sided deaf listeners lack binaural hearing and may only use spectral cues from their hearing ear, and the head–shadow effect. Despite the fact that the latter cue is ambiguous, all SSD patients rely heavily on this cue. We argue that this behavior reflects a Bayesian strategy in which prior information is weighted more heavily than less reliable spectral cues. Listeners with a unilateral conductive hearing loss adopt a flexible localization strategy in which they weigh potentially remnant binaural cues, spectral cues, and the head–shadow effect to estimate the azimuth of a sound source under varying acoustic conditions. Finally, we propose that modest age-related, high-frequency hearing loss may be partially compensated by the acoustic effects of pinna growth. For certain sounds the elderly may thus outperform young listeners in localizing source elevation.

1.9 EXERCISES

Problem 1.1: Coordinates and coordinate transformations of sound locations.

The azimuth (A)–elevation (E) system is a *double-pole* coordinate system, in which the azimuth angle in the horizontal plane is specified by a rotation about a head-vertical rotation axis from the midsagittal plane to the source position, while the elevation angle in the vertical plane is described by a rotation about the inter-aural axis from the horizontal plane to the source (Fig. 1.7A).

(a) Suppose a target sound is described by azimuth–elevation angles (A_S,E_S). You wish to foveate the sound with your eyes, which initially fixate at straight-ahead, that is, $(A,E) = (0,0)$ degree. The sound location has to be transformed into oculocentric polar coordinates, expressing the rotation amplitude, given by eccentricity angle, R, and direction, Φ (Fig. 1.7B). Do the transformation, that is, calculate $R(A_S,E_S)$ and $\Phi(A_S,E_S)$.

(b) Show that for all sound locations in the frontal hemifield: $A + E \leq \pi/2$.

Problem 1.2: Audiovisual integration is useful only when sound S and visual target V are both at the *same* spatial location (Fig. 1.6)! Often, sensory coordinates may differ substantially, but when they emerge from a single object they

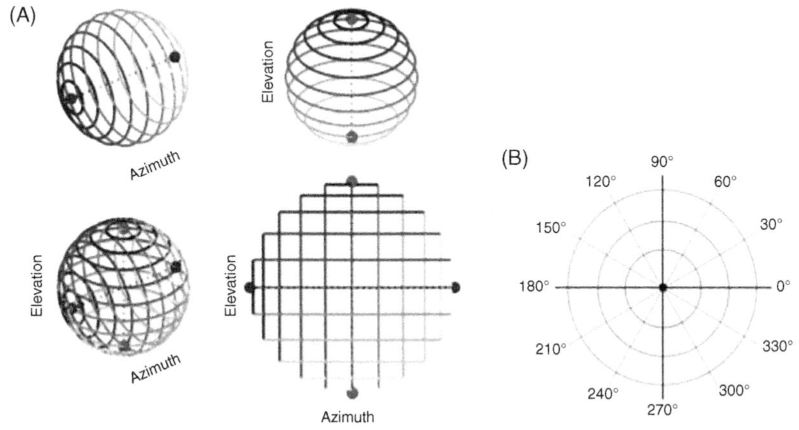

FIGURE 1.7 (A) Double pole–azimuth elevation coordinate system to specify sound locations with respect to the head of the subject. Top plots show the isoazimuth (left) and isoelevation (right) contour lines. Bottom plots show the full sphere (left) and the projection of (A,E) contour lines as seen from straightahead. (B) polar coordinates, showing iso eccentricity lines (circles) and iso direction lines (spokes).

should be integrated. From Eq. (1.2), determine the coefficients $[a,b,c,d]$ for adequate audiovisual integration.

Often, however, the *sensory* coordinates of S and V may be identical $(A_S,E_S) = (R_V,\Phi_V)$, yet originate from *different* objects. In such cases the stimuli should not be integrated.

Draw that situation. What happens if the coefficients that you just determined for integration are applied by the sensorimotor system? Give an argument as to why this may or may not help in the identification and localization tasks.

Problem 1.3: Inverse problems are often ill posed. A good example is given by the problem of perception: on the basis of a limited number of measurements (sensory observations, eg, foveation of points in the visual scene through saccades), the brain has to make an estimate (inference) about the environment and the stimuli that caused the percept. Mathematically, a problem is well posed if there exists a unique and stable solution to the problem. A solution is stable if it resists (small) perturbations of the starting values. If solutions are not stable, or unique, the problem is ill posed.

The latter may even happen for seemingly trivial calculations such as taking a derivative. As a numerical exercise we look at the following example: suppose that we have to determine the derivative of a function, $f(x)$:

$$q(x) = \frac{df}{dx}$$

However, it could occur that instead of $f(x)$, we have to deal with a slightly perturbed measurement of this function, say:

$$f_n(x) = f(x) + \frac{\sin(nx)}{\sqrt{n}}$$

Clearly, for $n \to \infty$ the difference between perturbed and original function, $\|f - f_n\|$, approaches zero. Show that for the derivative, however, this difference becomes arbitrarily high. This means that the operation is unstable, and hence calculating the derivative is ill posed.

Problem 1.4: Fig. 1.8 shows a simple two-layered (input–output) neural network consisting of N linear neurons in each layer. The activity of the input layer is drawn above the neurons. Only neurons $k-2$, k, $k+2$, and $k+4$ receive input strengths of 1, 3, 2, and 1 units, respectively.

The neurons of the network have repetitive connection patterns; only the connections of input neuron k are drawn for clarity. Each neuron in the input layer excites its corresponding output neuron with synaptic strength +1, and inhibits all the other neurons of the network with synaptic strengths $-1/(N-1)$.

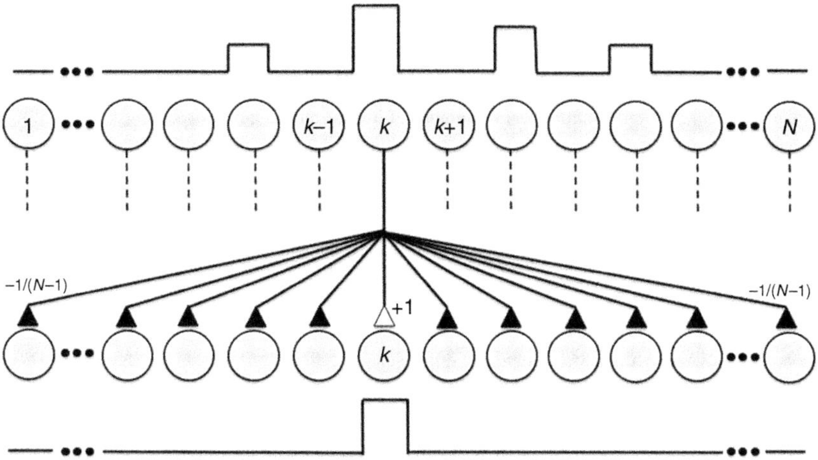

FIGURE 1.8 **A WTA two layer–feed forward network of linear neurons.** The connection scheme (highlighted here for input neuron k only) is identical for all neurons.

In this way, the total synaptic weight from each input neuron sums to zero. The activity of an output neuron is determined by:

$$y_k = \sum_{n=1}^{N} w_{kn} x_n$$

with w_{kn} the synaptic connection from input neuron n to output neuron k, and x_n the activity of input neuron n. Take $N = 11$ and $k = 5$. Show that the network indeed operates as a WTA network by calculating the activities of all N output neurons. In modeling saliency maps, WTA networks play an important role, as they weed out the contributions from all competing active neurons except from the one neuron with the strongest activation.

ACKNOWLEDGMENT

This book is dedicated to Dr Dick Donker (Dec 21, 1934–Oct 1, 2014): neurologist, life artist, a lover of music, and auditory science. I thank my colleagues, technicians, PhD students, post-docs, and graduate students, who greatly enriched my lab for more than 25 years. In particular, I am indebted to Maarten Frens, Jeroen Goossens, Marcel Zwiers, Paul Hofman, Denise van Barneveld, Tom van Grootel, Joyce Vliegen, Peter Bremen, Rob van der Willigen, Marc van Wanrooij, and Martijn Agterberg, who designed and completed the many tedious experiments and sophisticated data analyses that are presented in Chapters 7–14 of this monograph.

REFERENCES

General Readings

Bregman, A.S., 1990. Auditory Scene Analysis. MIT Press, Cambridge, MA.

Purves, D., Augustine, G.J. (Eds.), 2012. Vision: the eye, Central visual pathways, and The auditory system. Neuroscience, fifth ed. Sinauer Associates Inc., Sunderland, MA (Chapters 10, 11, and 12), pp. 229–314.

Purves, D., Augustine, G.J. (Eds.), 2012. Eye movements and sensory motor integration. Neuroscience, fifth ed. Sinauer Associates Inc., Sunderland, MA, pp. 453–468.

On Ill-Posed Problems

Kabanikhin, S.I., 2008. Definitions and examples of inverse and ill-posed problems. J. Inv. and Ill. Probs. 16, 317–357.

On Saliency Maps

Koch, C., Ullman, S., 1985. Shifts in selective visual attention. Hum. Neurobiol. 4, 219–227.

Itty, L., Koch, C., 2000. A saliency-based search mechanism for overt and covert shifts of visual attention. Vis. Res. 40, 1489–1506.

On Inhibition of Return

Posner, M.I., Cohen, Y., 1984. Components of visual orienting. In: Bouma, H., Bouwhuis, D.G. (Eds.), Attention and Performance. Hillsdale, NJ, Erlbaum (Chapter 32), pp. 531–556.

Klein, R.M., 2000. Inhibition of return. Trends Cogn. Sci. 4, 138–147.

Wang, Z., Klein, R.M., 2010. Searching for inhibition of return in visual search: a review. Vis. Res. 50, 220–228.

Chapter 2

The Nature of Sound

2.1 LONGITUDINAL PRESSURE WAVES IN A MEDIUM

Sound is a pressure perturbation in a medium like air or water, which is caused by a vibratory source that oscillates in the 50–20,000 Hz range (Serway and Jewett, 2013). The perturbation propagates as a longitudinal traveling wave through the medium, in which the molecular movements associated with the perturbation are along the same direction as the propagating traveling wave. Think of a vibrating membrane (a loudspeaker) at position $x = 0$, moving inward and outward in the x-direction at frequency f_0 Hz. When the membrane moves, it pushes or drags the air molecules at $x = 0$ in the same direction, thereby creating a local increase or decrease in the density of air molecules, which consequently increases or decreases the local pressure. These pressure changes propagate in the x-direction through the medium. The wave is thus described by a succession of compressions (regions of high pressure, c) and rarefactions (regions of low pressure, r), which propagate through the medium at velocity v_{sound}. In air, at sea level at a temperature of 20°C, and at a mean pressure of 10^5 N/m^2 (1 atm), the speed of sound is approximately $v_{\text{sound}} = 342$ m/s across the frequency range of hearing.

We first discuss a simple physical model for sound waves and their properties, to provide a basic and quantitative understanding of the input for the auditory system.

Such a model is provided by a long spring, with rest length L_0 m, mass density ρ_0 kg/m, and an elasticity K N/m (Fig. 2.1). The spring is attached to a vibrating source on one end (left), at $x = 0$, which can only move rightward or leftward. As the source moves to the right, it causes a local compression, c, in the spring; when it returns to the resting position, it causes a rarefaction, r. If the source moves in an oscillatory manner, the compressions and rarefactions are mechanically transmitted across the length of the spring. The phase velocity of the longitudinal wave through the spring is then determined by:

$$v_\varphi = \sqrt{\frac{KL_0}{\rho_0}} \tag{2.1}$$

According to Hooke's law, the compressive force needed to change the spring's length to $L < L_0$ is given by

$$F = K(L_0 - L) \tag{2.2}$$

The Auditory System and Human Sound-Localization Behavior. http://dx.doi.org/10.1016/B978-0-12-801529-2.00002-7

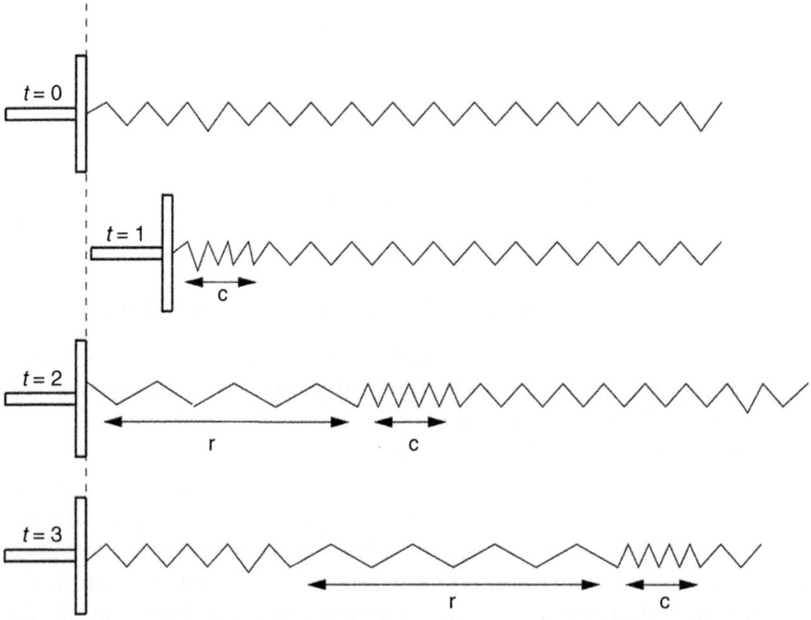

FIGURE 2.1 A simple physical model for longitudinal sound waves in one dimension, is a spring (L_0, rest length; and K, elasticity); a membrane at $x = 0$ causes compressions (*c*) and rarefactions (*r*) that transport along the length of the spring at a fixed velocity.

Isaac Newton was the first to apply the spring model to an ideal gas, to predict the sound velocity in air. The gas, at temperature T, is confined to a cylinder of rest length L_0, total rest volume V_0, and a freely moveable piston at $x = 0$ can compress and decompress the gas (Fig. 2.2). As the gas is isolated from the environment, it cannot exchange heat with it. This suggests that the gas compresses and expands under *adiabatic* conditions: when compressed, the temperature will increase, when decompressed it will decrease (explained in the further section).

When the length of a spring changes by ΔL (in case of the gas, through the piston's movement), the force exerted by the gas on the piston changes accordingly, see Newton's third law:

$$\Delta F = -K\Delta L \tag{2.3}$$

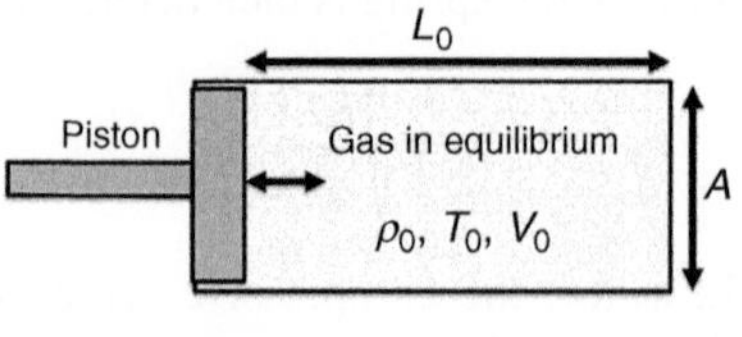

FIGURE 2.2 A thermodynamic model for an isolated, ideal gas, confined to a cylinder with a freely moveable piston. No heat can escape from the system. The column of gas can be modeled as a spring because it exerts an elastic force on the piston.

Let us find an expression for the equivalent elasticity of the gas, so that we can apply Eq. (2.1) to predict the speed of sound in the gas. The force exerted by the gas on the piston is determined by its pressure, p_0, and the cross-section of the piston, A:

$$F = p_0 A \tag{2.4}$$

When the piston moves over a distance ΔL the cylinder's volume will change by:

$$\Delta V = A \Delta L \tag{2.5}$$

and the force on the piston changes by:

$$\Delta F = A \Delta p \tag{2.6}$$

The change in pressure and (relative) volume are related through an extended version of Hooke's law:

$$\Delta p \equiv -B \frac{\Delta V}{V_0} \tag{2.7}$$

with the parameter $B > 0$, the *bulk modulus* of the gas. A decrease in relative volume is thus associated with a proportional increase in pressure. As a result, the force on the piston changes as:

$$\Delta F = -AB \frac{\Delta V}{V_0} = -B \frac{A^2}{V_0} \Delta L \tag{2.8}$$

and this is indeed a similar formulation as Hooke's law for a spring Eq. (2.4)! The equivalent elasticity of the gas is therefore given by:

$$K_{\text{gas}} = B \frac{A^2}{V_0} \tag{2.9}$$

Inserting this result in Eq. (2.2) yields an expression for the speed of sound in a gas. Note, however, that the mechanical spring equation contains the spring's 1D length density in kilogram per cubic meter, while the gas lives within a 3D volume; we thus have the 3D volume mass density, ρ, expressed in kilogram per cubic meter. To obey the requirements for the SI units of Eq. (2.2), we need to convert the gas *volume* density to the equivalent spring *length* density, ρ_0. This is done as follows, noting that the masses of spring and gas are defined by:

$$M_{\text{gas}} \equiv \rho V = \rho A L_0 \equiv \rho_0 L_0 \tag{2.10}$$

from which we obtain

$$\rho_{0,\text{gas}} \equiv \rho A \tag{2.11}$$

The speed of sound in a gas is thus found by the following, general equation:

$$v_{\varphi,\text{gas}} = \sqrt{\frac{BA^2L_0}{\rho_0}} = \sqrt{\frac{BA}{\rho_0}} = \sqrt{\frac{B}{\rho}} \tag{2.12}$$

Note that the sound velocity is determined by only two physical (macroscopic and observable) properties of the medium: its *bulk modulus* and its *volume mass density*. To calculate the speed we thus need to determine its bulk modulus, whereas the air density at sea level and 20°C is 1.3 kg/m^3. We only consider small changes in the volume of the gas, so that we may take the value for B at the equilibrium volume, V_0. From Eq. (2.7) we can define the bulk modulus by:

$$B \equiv -V_0 \left.\frac{\partial p}{\partial V}\right]_{V_0} \tag{2.13}$$

We need to find the value of the partial derivative of $p(n,V,T)$ (Crawford, 1968) with respect to volume, with n the number of moles in the gas, and T the temperature, at the equilibrium volume. Interestingly, it was at this point that Newton made a wrong assumption about the gas! He was well aware that pressure and volume in an ideal gas are related through the ideal gas law:

$$pV = nRT \tag{2.14}$$

and he used the fact that at constant temperature, the gas law reduces to Boyle's law:

$$pV = \text{constant} \equiv p_0V_0 \tag{2.15}$$

With this constraint, the derivative in Eq. (2.14) yields:

$$\left.\frac{\partial p}{\partial V}\right]_{V_0} = -\left.\frac{p_0V_0}{V^2}\right]_{V_0} = -\frac{p_0}{V_0} \tag{2.16}$$

from which Newton found that $B = p_0$, and hence a sound velocity of:

$$v_{\text{Newton}} = \sqrt{\frac{p_0}{\rho}} = 277\,\text{m/s} \tag{2.17}$$

However, this result is almost 20% below the real value of 342 m/s! (Crawford, 1968) Apparently, the error resulted from the assumption of constant temperature. Indeed, since the (isolated) gas does not exchange heat with the environment, the compressions and rarefactions may be adiabatic. Under adiabatic conditions, the temperature will not remain constant when work is exerted on, or taken from, the gas. As a result, in compression zones the temperature will be higher, while in rarefactions it will be lower than the average temperature in the cylinder. Instead of Boyle's equilibrium condition at constant T, the adiabatic relation reads:

$$p^{\gamma}V^{\gamma} = \text{constant} \equiv p_0^{\gamma}V_0^{\gamma} \text{ with } \gamma \equiv \frac{c_P}{c_V} = 1.4 \,(\text{adiabatic index}) \qquad (2.18)$$

Here, c_p and c_V are the specific heats of the gas (specifying how much energy is needed to raise the temperature of 1 mole of gas by 1K) at constant pressure and volume, respectively. For the $p(V)$ function we now obtain:

$$p = p_0 V_0^{\gamma} V^{-\gamma} \qquad (2.19)$$

As a result of the increased temperature differences between compression and rarefaction zones, when compared to the equilibrium condition, the pressure difference, which is the driving force for propagating the acoustic wave, is also larger than expected from Boyle's law. Therefore, the predicted speed of sound in the adiabatic case will be higher too. The reader can readily verify that in that case:

$$v_{\text{Adiabatic}} = \sqrt{\frac{\gamma p_0}{\rho}} = 332 \text{ m/s at sea level} \qquad (2.20)$$

Since this is a much better approximation for the measured speed of sound than predicted by Boyle's model, we should verify whether the adiabatic model is indeed applicable to the situation of acoustic waves. The adiabatic condition should be applied when heat cannot escape from the system. The question therefore boils down to whether or not there is enough time for heat to flow from a region of increased temperature (a compression zone) to decreased temperature (a rarefaction zone). If heat could do this sufficiently fast, the temperature would remain constant across the entire cylinder, and Boyle's law would still apply! In that case, the heat would have to flow over half a wavelength ($\lambda/2$; the distance between a compression and rarefaction zone) well within the time of half an oscillation period ($P/2$) of the piston motion (when c and r are going to change places). Therefore, the minimal requirement for effective heat transport would be:

$$v_{\text{heat}} \gg \frac{(1/2)\lambda}{(1/2)P} = v_{\text{sound}} \qquad (2.21)$$

Heat transport is due to thermal *conduction* (the transfer of molecular kinetic energy to other molecules, through collisions), and not to *convection* (which is the active transport of molecules from one area in space to another, as achieved by a ventilator). So, what is the thermal velocity of air molecules? The kinetic gas theory learns that the average molecular velocity:

$$< v >_{\text{heat}} = \sqrt{\frac{k_B T}{m}} \qquad (2.22)$$

where $k_B = 1.38 \times 10^{-23}$ J/K is Boltzmann's constant; and m is the molecular mass. From the ideal gas law we may verify (Problem 2.1) that for the speed of sound:

$$v_{\text{sound}} = \sqrt{\frac{\gamma k_B T}{m}} \tag{2.23}$$

Thus, apart from the adiabatic index, $\gamma = 1.4$, the thermal and sound velocities result seems to be quite comparable. So, *if* an air molecule could move along a straight line in the x-direction over a distance $\lambda/2$ before colliding with another air molecule (the free path length), heat would arrive just in time at the rarefaction zone to keep the temperature constant. However, this scenario is highly unlikely, as molecules will move in random directions instead of along straight lines. More realistically, therefore, the free path length for molecules under 1 atm pressure is in the order of 10^{-5} m. In other words, as long as the sound's wavelength is large compared to the free path length of 10 μm, the adiabatic condition will be fulfilled. Note that at the highest audible frequency of about 20 kHz the wavelength $\lambda = 1.6$ cm, which is still several orders of magnitude larger than the adiabatic requirement! Only for very high ultrasonic frequencies will the adiabatic approximation. This analysis also clarifies why the speed of sound is independent of sound frequency over a very large range.

As a final note, in a more advanced model for gases, the sound velocity will also depend on the temperature, so that a more exact equation for the speed of sound waves eventually becomes (Crawford, 1968):

$$v_{\text{sound}} = \sqrt{\frac{\gamma p_0[1+(T/273)]}{\rho}} = 344 \text{ m/s at sea level and } T = 20°\text{C} \tag{2.24}$$

This equation is quite a reasonable approximation of the measured value.

2.2 THE HOMOGENEOUS WAVE EQUATION

A longitudinal, traveling sound wave, which is propagating at a constant speed, v, in the positive or negative x-direction through a homogeneous medium with constant density and bulk modulus, can be described by the general function, $s(x,t) = s(x \pm vt)$, and is a solution (Problem 2.2a) of the following *homogeneous wave equation*:

$$\frac{\partial^2 s}{\partial x^2} = \frac{1}{v^2}\frac{\partial^2 s}{\partial t^2} \tag{2.25}$$

The wave equation is a linear, second-order, partial differential equation (for simplicity, we restrict the description of the wave to 1D, ie, x). Linearity implies that if the equation would have more than one solution, say $s_1(x,t)$ and $s_2(x,t)$, then

$$s(x,t) = as_1(x,t) + bs_2(x,t) \tag{2.26}$$

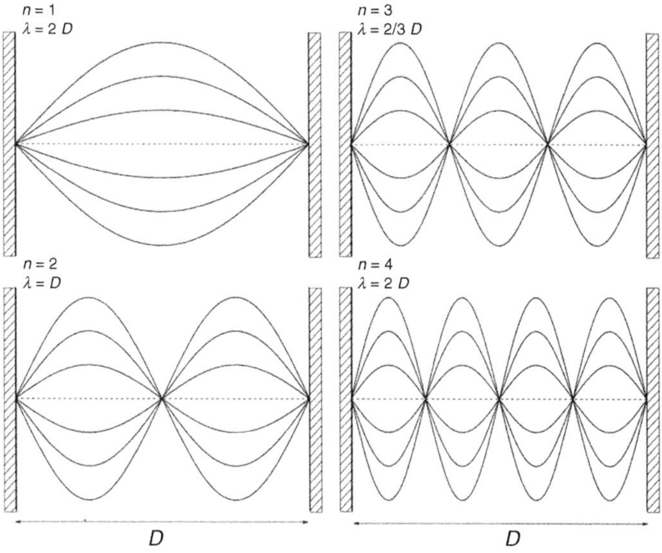

FIGURE 2.3 **Examples of solutions of the wave equation with fixed boundary conditions (modes: $n = 1, 2, 3$, and 4), shown at six different times.**

is also a solution, for all a,b∈R [(2.26) expresses the *superposition principle*; see Problem 2.2b, and also chapter: Linear Systems Analysis for an extensive discussion of linearity].

In particular, suppose that two *harmonic* traveling waves, one propagating leftward and the other rightward, are a solution of the wave equation:

$$s_1(x,t) = \cos(kx + \omega t) \text{ and } s_2(x,t) = \cos(kx - \omega t) \tag{2.27}$$

Then their sum is also a solution, which takes the following form (check!):

$$s(x,t) = \cos(kx + \omega t) + \cos(kx - \omega t) = 2\cos(kx)\cos(\omega t) \tag{2.28}$$

This solution represents a *standing wave* (there are points in the medium where the amplitude of the vibration is always zero. These so-called *nodes* are found at $x = (2n + 1) \times \pi/2k$, with $n = 1,2,3$, etc. (eg, Fig. 2.3). Note that the standing wave, Eq. (2.28), is a space–time *separable* function. We show here that the separability property of the wave equation is unique to harmonic solutions, and it is directly related to Fourier analysis.

Thus, apart from traveling waves of the general form $s(x \pm vt)$, the wave equation can have (superpositions of) harmonic solutions that are fully specified by the boundary and starting conditions of the medium. This idea can be appreciated from the following analysis, in which we propose potential solutions to Eq. (2.25) as separable in space and time:

$$s(x,t) = X(x)T(t) \tag{2.29}$$

Substitution of Eq. (2.29) into the homogeneous wave equation, Eq. (2.25) yields the following requirements for $X(x)$ and $T(t)$ (Problem 2.3):

$$\frac{d^2X}{dx^2} = -k^2X \text{ and } \frac{1}{v^2}\frac{d^2T}{dt^2} = -k^2T \text{ with } k > 0 \tag{2.30}$$

Solutions to Eq. (2.30) are given by the following general harmonic functions:

$$\begin{aligned} X_k(x) &= A_k \cos(kx) + B_k \sin(kx) \text{ and} \\ T_k(t) &= C_k \cos(vkt) + D_k \sin(vkt) \end{aligned} \tag{2.31}$$

The general standing wave solution for the wave equation is the superposition of all possible values for the amplitudes $[A_k, B_k, C_k, D_k]$ and the constant k (the *wave number*, which is defined by $k = 2\pi/\lambda$, where λ is the wave length of the standing wave). As discussed further, these parameters are fully determined by the boundary and starting conditions of the problem (see Exercises):

$$s(x,t) = \sum_k [\{A_k \cos(kx) + B_k \sin(kx)\}][C_k \cos(vkt) + D_k \sin(vkt)] \tag{2.32}$$

The spatial boundary conditions can be one of the following types:

1. Fixed boundary conditions: the medium is closed at both ends (at $x = 0$ and $x = L_0$), so that the amplitude at $s(0,t) = s(L_0,t) = 0$ for all t (Fig. 2.3).
2. Free boundary conditions: the medium is open at both ends, so that the amplitude at $s(0,t) = s(L_0,t) = maximal$ for all t.
3. Mixed boundary conditions: closed at one end (amplitude zero) and open at the other end (amplitude maximal).
4. Periodic boundary conditions: in that case, $s(0,t) = s(L_0,t)$, and also the spatial derivatives at $x = 0$ and $x = L_0$ are equal.

Each of these boundary conditions leads to a different set of solutions. In Problem 2.5, the reader is invited to analyze the different scenarios. Suffice it here to work out the solution for fixed boundary conditions, which is illustrated in Fig. 2.3 for four vibrational modes: $n = 1, 2, 3, 4$.

Under fixed boundary conditions, the constraints for the standing waves of Eq. (2.29) are determined by $s(0,t) = s(L_0,t) = 0$. We apply these constraints to the solutions of the spatial function, $X_k(x)$, of Eq. (2.28). The reader should verify (Problem 2.4) that in that case all $A_k = 0$, and that $B_k \sin(kL_0) = 0$, and that therefore the wave numbers, k, should obey the following constraints:

$$kL_0 = n\pi \quad \text{for} \quad n = 1,2,3,\cdots \tag{2.33}$$

Thus, only specific standing waves are admitted as a solution for this boundary condition; the wave numbers (ie, the associated wavelengths) will have to obey:

$$k_n = \frac{n\pi}{L_0} \quad \text{and thus,} \quad \lambda_n = \frac{2L_0}{n} \quad \text{for } n = 1,2,3,\cdots \tag{2.34}$$

where the natural number n identifies the vibrational mode of the sound wave (Fig. 2.3, which illustrates the case for n = 1,2,3,4).

Each vibrational mode (or *eigenmode*) is also associated with its particular frequency, through the *dispersion relation*:

$$f_n = \frac{vk_n}{2\pi} = \frac{nv}{2L_0} \quad \text{for } n = 1,2,3,\cdots \tag{2.35}$$

Taken together, the total solution of the wave equation with fixed boundary conditions is described by the following formidable equation:

$$s_{\text{fixed}}(x,t) = \sum_{n=1}^{\infty} \sin\left(\frac{n\pi x}{L_0}\right)\left[C_n \cos\left(\frac{n\pi vt}{L_0}\right) + D_n \sin\left(\frac{n\pi vt}{L_0}\right)\right] \tag{2.36}$$

The solution of the wave equation is a linear superposition of vibrational eigenmodes, each with their own amplitude, wavelength, and frequency. The amplitudes C_n and D_n are determined by the starting condition of the problem, and can be found by Fourier analysis (see Section 2.6). Moreover, wavelength and frequency are related through the *dispersion relation*:

$$\omega_n(k_n) = v_{\text{sound}} k_n \tag{2.37}$$

which for *nondispersive (ie, linear, homogeneous) media* is a simple linear relationship, as the propagation velocity of the waves is independent of their frequency (see Section 2.1, for a physical justification of this point).

A similar analysis follows for the other three boundary conditions (left as an exercise in Problem 2.5).

2.2.1 Example: Interference and Binaural Beats

The dispersion relation is a property of the medium in which the perturbation is inflicted, and acts as a hard constraint on the physically allowed harmonic solutions of the linear wave equation. Indeed, we saw that the propagation velocity of the wave is only determined by two medium-specific quantities: the bulk modulus, and the density [Eq. (2.12)]. Wave number, k, and angular frequency, ω, are thus strictly related through Eq. (2.37).

Suppose that two harmonic traveling waves with different wave numbers, k_1 and k_2, respectively, start to travel simultaneously in the positive x-direction. What is the resulting wave pattern? For simplicity, we assume that both waves have the same amplitude, A_0:

$$s_{\text{sum}}(x,t) = A_0 \cos(k_1 x - \omega_1 t) + A_0 \cos(k_2 x - \omega_2 t) \quad \text{with } \frac{\omega_1}{k_1} = \frac{\omega_2}{k_2} = v_{\text{sound}} \tag{2.38}$$

Let's look at this solution, by explicitly calculating the wave modulations that pass through a fixed point in space, say $x = 0$:

$$\begin{aligned} s_{\text{sum}}(0,t) &= A_0 \cos(\omega_1 t) + A_0 \cos(\omega_2 t) \quad \text{with } \omega_2 = k_2 v_{\text{sound}} \\ s_{\text{sum}}(0,t) &= 2A_0 \cos\left(\frac{\omega_1 - \omega_2}{2} t\right) \cos\left(\frac{\omega_1 + \omega_2}{2} t\right) \\ s_{\text{sum}}(0,t) &= 2A_0 \cos\left[\frac{1}{2}\omega_1\left(1 - \frac{k_2}{k_1}\right)t\right] \cos\left[\frac{1}{2}\omega_1\left(1 + \frac{k_2}{k_1}\right)t\right] \end{aligned} \tag{2.39}$$

In case the two wave numbers are close together, eg, $k_2/k_1 = 0.9$, the wave modulations follow the pattern shown in Fig. 2.4. We observe a slow amplitude modulation at $0.05 \cdot \omega_1$, which is called *beating*, at twice the amplitude of the constituent waves (resulting in constructive ($A = 2A_0$) and destructive ($A = 0$) interference). The fast oscillations occur at a frequency of $0.95 \cdot \omega_1$.

It is important to realize that despite the apparent presence of quite different frequencies in the signal of Fig. 2.4, the actual spectrum of the acoustic signal in Eq. (2.39) has not changed, and still contains only the two frequencies, ω_1 and $(k_2/k_1) \cdot \omega_1$!

The traveling wave through the medium is represented by Eq. (2.38), and has a more awkward analytical expression than Eq. (2.39). Fig. 2.5 (left-hand side) shows a graphical space–time representation of the total traveling wave, in which the wave amplitude is shown in gray shading (white is positive and black is negative

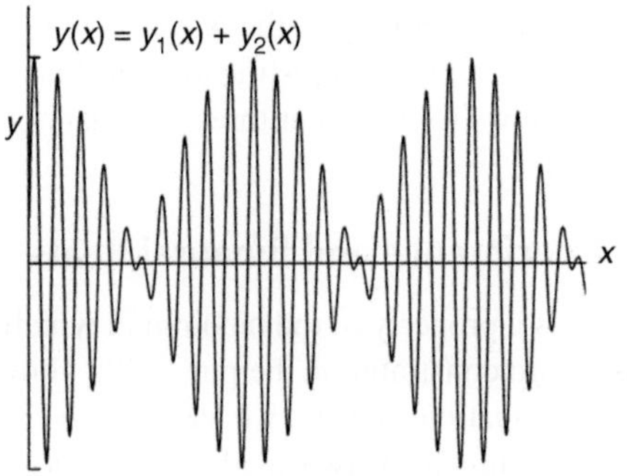

FIGURE 2.4 Interference of two traveling waves at $x = 0$, with a slightly different wave number: $k_2/k_1 = 0.9$. The resulting pattern has a slow amplitude beat.

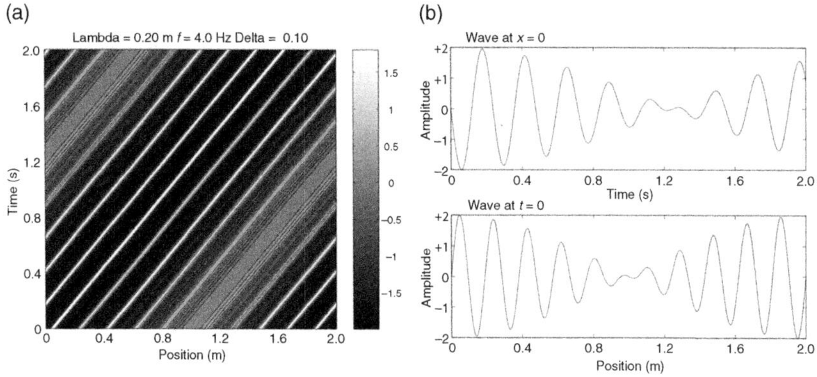

FIGURE 2.5 (a) Interference of two traveling waves in space and time through some arbitrary medium, showing the interference pattern emerging in both dimensions according to Eq. (2.38). (b) Cross-sections at $x = 0$ (time modulation) and $t = 0$ (space modulation), show the clear amplitude beats.

amplitude). In this case we took the wavelength of the first wave, $\lambda_1 = 0.2$ m and frequency, $f_1 = 4$ Hz. The slopes of the lines therefore correspond to the propagation velocity of the wave, which is found to be $v = f_1\lambda_1 = 0.8$ m/s (the medium in this example is obviously not air...). The right-hand side shows two cross-sections at a fixed position (top) and time (bottom), respectively, in which the harmonic beating patterns are immediately obvious. From these patterns it is also possible to estimate the wavelength (and frequency) of the second wave.

Binaural Beats: Interestingly, although acoustic beating [Eq. (2.39)] results from a clear physical phenomenon of constructive and destructive interference of two waves with comparable frequencies in space and time, it also has a nice *psychoacoustic* equivalent, which is known as *binaural beats*. To perceive binaural beats the listener wears earphones to prevent physical interference of the acoustic signals (dichotic listening). If we then present a pure tone, of say 500 Hz to one ear, and 502 Hz to the other ear, the listener will perceive a single tone of (about) 501 Hz with a clear amplitude beating at approximately 2 Hz. Clearly, as in this case there is no physical interference of the sound waves, the beating *must* be generated somewhere in the auditory system! Indeed, at the level of the cochlea and the auditory nerve of either ear, the two frequencies are still clearly separated, representing 500 and 502 Hz, respectively, and their activity patterns do not show any evidence for the presence of the other frequency. However, at *binaural* processing stages in the brainstem and midbrain inferior colliculus, the perceived beating is clearly observed in the firing patterns of single neurons (eg, Kuwada et al., 1979)!

Since in natural environments, acoustic stimuli that arise from a single object will never produce such a dichotic acoustic pattern, the brain is faced with an unusual problem. As a result, it "creates" a percept that is the most likely explanation for this odd acoustic input. As discussed in chapter: Cues for Human

Sound Localization, the beating pattern, that occurs because action potentials from left and right ear arrive at MSO cells with small, but changing time delays, is interpreted as a slow harmonic shift in the horizontal location of a single sound source inside the head. The listener therefore perceives the beats as a left–right movement of the sound source inside the head.

2.3 HARMONIC SOUND WAVES AND ACOUSTIC IMPEDANCE

When the piston of Fig. 2.2 moves harmonically, the local density changes of the air will follow the same harmonic behavior. The resulting traveling wave is then described by:

$$s(x,t) = s_{\max}\cos(kx - \omega t) \text{ with } k = \frac{\omega}{v_{\text{sound}}} \tag{2.40}$$

The traveling wave of Eq. (2.27) is due to a harmonic modulation of the local pressure, Δp, which we calculate from the definition of the bulk modulus [Eq. (2.13); Fig. 2.6]. This short derivation leads to the important concept of acoustic impedance.

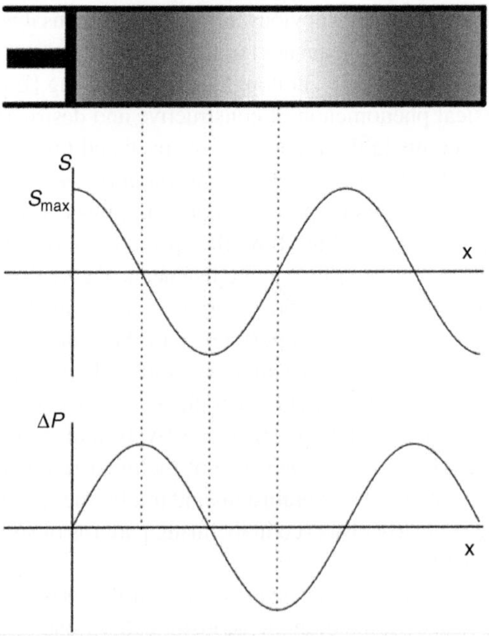

FIGURE 2.6 Harmonically driven acoustic wave in a gas. Note that the pressure modulation is exactly 90 degrees ($\pi/2$ rad) out of phase with the amplitude modulations of the vibration.

Imagine a small local cylindrical volume of gas, V_i, with length, Δx, and cross-section A, so that $V_i = A \times \Delta x$. The pressure change causes a change in volume, $\Delta V_i = A \times \Delta s$, in which Δs $s(x+\Delta x) - s(x)$. Using Eq. (2.13), we obtain

$$\Delta p = -B\frac{\Delta V}{V_i} = -B\frac{A\Delta s}{A\Delta x} \text{ which in } \lim_{\Delta x \downarrow 0} \text{ becomes } \Delta p = -B\frac{\partial s}{\partial x} \quad (2.41)$$

Upon substituting the harmonic wave function:

$$\Delta p = Bks_{\max}\sin(kx - \omega t) = \rho v_{\text{sound}}\omega s_{\max}\sin(kx - \omega t) \quad (2.42)$$

where on the right-hand side we inserted the dispersion relation Eq. (2.37), and the general expression (2.12) for the sound velocity. We write

$$\Delta p = p_{\max}\sin(kx - \omega t) \quad (2.43)$$

and note that the pressure is precisely $\pi/2$ rad out of phase with the local vibration, $s(x,t)$. In Eq. (2.42) we defined:

$$\Delta p_{\max} = \rho v_{\text{sound}}\omega s_{\max} \equiv Z\omega s_{\max} \quad (2.44)$$

where we introduced Z as the *acoustic impedance* of the medium (unit: $\text{kg/m}^2\text{s}$). Note that [using Eq. (2.20)] the acoustic impedance of a gas can also be written as:

$$Z = \rho v_{\text{sound}} = \sqrt{\gamma\rho p_0} \quad (2.45)$$

Finally, from Eqs. (2.43) and (2.45) one may verify that the change in pressure can also be formulated as:

$$\Delta p = Z\frac{\partial s}{\partial t} \quad (2.46)$$

which relates the pressure to the volume velocity of the acoustic perturbation (which is not the same as the propagation velocity of the wave!).

**Inhomogeneous media*: The homogeneous wave equation [Eq. (2.25)] refers to the situation of a constant propagation velocity. We have seen that the speed of sound depends on the bulk modulus and the density of the medium, but also on the temperature [Eq. (2.24)]. When these quantities are constant, so is the propagation velocity of the acoustic wave. However, a more general problem is one in which the medium is inhomogeneous, whereby both the volume density and the bulk modulus may depend on location (eg, the height in the atmosphere, or when there is a temperature gradient in the air). To appreciate the consequences of such more general situations, it merits a brief excursion into this problem, although we will not treat it in great detail here.

For the general case, both volume density, ρ, and bulk modulus, B, depend on x, that is, $\rho = \rho(x)$, and $B = B(x)$. As a result, the propagation speed of the acoustic wave will depend on x as well, and there will be no simple linear dispersion relation, $\omega(k)$ anymore. Let's consider a small compartment in the air cylinder of length Δx, so that its volume is $\Delta V = A\Delta x$, and its mass is $\rho(x)\Delta V$.

The driving force for the acoustic wave is the pressure difference across the compartment, ie, $p(x + \Delta x,t) - p(x,t)$, and according to Newton's second law [$F = md^2s/dt^2$, but take the force per area to revert to pressure, and use Eq. (2.41)]:

$$\rho(x)\Delta x\frac{\partial^2 s}{\partial t^2} = B(x+\Delta x)\left(\frac{\partial s}{\partial x}\right)_{x+\Delta x} - B(x)\left(\frac{\partial s}{\partial x}\right)_x \quad (2.47)$$

We now apply a first-order Taylor expansion (ie, ignore terms in $(\Delta x)^2$ and higher):

$$B(x+\Delta x) = B(x) + \Delta x\frac{\partial B}{\partial x} + \cdots \text{ and } \left(\frac{\partial s}{\partial x}\right)_{x+\Delta x} = \left(\frac{\partial s}{\partial x}\right)_x + \Delta x\left(\frac{\partial^2 s}{\partial x^2}\right)_x + \cdots \quad (2.48)$$

Multiplying the terms, and only including terms up to Δx yields:

$$\rho(x)\frac{\partial^2 s}{\partial t^2} = \left(\frac{\partial B}{\partial x}\right)\left(\frac{\partial s}{\partial x}\right) + B(x)\left(\frac{\partial^2 s}{\partial x^2}\right) \quad (2.49)$$

which is succinctly written as the *inhomogeneous wave equation*:

$$\rho(x)\frac{\partial^2 s}{\partial t^2} = \frac{\partial}{\partial x}\left[B(x)\left(\frac{\partial s}{\partial x}\right)\right] \quad (2.50)$$

Note that Eq. (2.50) is still a linear second order differential equation, but the behavior of its solutions now depends critically on the properties of the medium. For standing waves (closed boundary conditions), Eq. (2.50) still has harmonic solutions for the temporal modulations, $T(t)$, but in general the spatial component of the wave solution, $X(x)$, is no longer harmonic (Problem 2.6). This also means that the concept of *wavelength* is no longer defined, as it becomes a function of x itself, and there is not a simple linear dispersion relation anymore, because the propagation velocity is not constant either. For simple example cases Eq. (2.50) can be analytically solved. In the Exercises, the reader can look into two illustrative Problems (2.7 and 2.8, respectively), for which the density is a function of x, but the bulk modulus is assumed to be a constant.

For example, in the case of fixed boundary conditions at positions $x = a$ and $x = 2a$, and density and bulk modulus: $\rho(x) = \frac{m}{x^2}$, $B(x) = B_0$, the solution of Eq. (2.50) is found to be (Problem 2.7):

$$s(x,t) = A\sqrt{\frac{x}{a}}\sin\left[k\ln\frac{x}{a}\right]\cos(\omega t + \varphi) \text{ where } \omega^2 = \frac{B_0}{m}\left(k^2 + \frac{1}{4}\right) \quad (2.51)$$

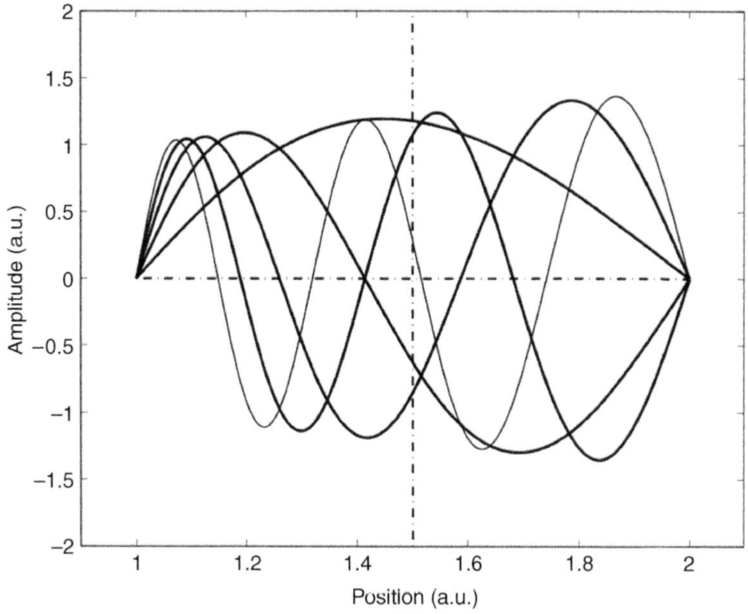

FIGURE 2.7 **The first five spatial vibration modes of the inhomogeneous wave equation [Eqs. (2.50) and (2.51)], with $a = 1$.** Note that the wavelength (distance between peaks) varies with x.

The spatial modulations are clearly nonharmonic (Fig. 2.7), as the wavelength depends on x. Thus, the "dispersion relation" is nonlinear too Eq. (2.51), right-hand side). The speed of sound in such a medium will therefore vary strongly with location. According to Eq. (2.12), in this medium:

$$v_{\text{inhom}} = x\sqrt{\frac{B_0}{m}}$$

2.4 ACOUSTIC ENERGY, INTENSITY, DECIBEL

A traveling sound wave transports an amount of energy that is equal to the work exerted by the piston (loudspeaker, voice) on the gas. The total kinetic energy of a harmonic sound wave in the gas cylinder, taken over a full wavelength, λ, is calculated to be (Problem 2.9):

$$K_\lambda = \frac{\rho}{4} A\left(\omega s_{\max}\right)^2 \lambda \tag{2.52}$$

The total potential energy, V, is identical to the total kinetic energy, so that the total energy of the wave per wavelength ($E = K + V$) is given by:

$$E_\lambda = \frac{\rho}{2} A(\omega s_{\max})^2 \lambda \quad (\text{in J}) \tag{2.53}$$

The acoustic *power*, P, is the energy that propagates per period, T, through the medium, and thus equals (using $\lambda = v_{\text{sound}}T$):

$$P = \frac{E_\lambda}{T} = \frac{\rho}{2} \frac{A(\omega s_{\max})^2 \lambda}{T} = \frac{1}{2} ZA(\omega s_{\max})^2 \quad (\text{in J/s} = \text{W}) \tag{2.54}$$

From this expression [combining with Eq. (2.24) for the maximal pressure change] we find the physical expression for *sound intensity* (the acoustic power per surface area, A, perpendicular to the wave's propagation direction), which is an important acoustic variable in psychoacoustic measurements:

$$I \equiv \frac{P}{A} = \frac{1}{2} Z(\omega s_{\max})^2 = \frac{(\Delta p_{\max})^2}{2Z} \quad (\text{in W/m}^2) \tag{2.55}$$

Because of the enormous range of sound intensities that are relevant for human hearing (ranging from 10^{-12} to over 1 W/m^2), it is customary to quantify the sound intensity on a logarithmic scale, whereby the sound level, β, in decibel (dB), is taken as:

$$\beta \equiv 10\log_{10} \frac{I}{I_0} \text{dB, where } I_0 = 1.0 \times 10^{-12} \text{ W/m}^2 \tag{2.56}$$

The reference level, I_0, is defined as the threshold of hearing of a 1000 Hz pure tone for the average healthy human listener. From this definition it follows that when $I = I_0$, the sound level is 0 dB, and for $I = 1$ W/m^2 it amounts to 120 dB (this is about the average onset of pain for hearing). Also note that doubling the sound intensity leads to an increase in sound level of 3 dB.

The sound level can also be referenced to the maximum of the pressure changes in the wave, through Eq. (2.43). In that case, the sound level is expressed as decibel of sound pressure level (or dB SPL), and it is defined by:

$$\beta_{\text{SPL}} \equiv 20\log_{10} \frac{\Delta p_{\max}}{\Delta p_0} \text{dB SPL} \tag{2.57}$$

In this case, doubling the sound level leads to an increase of 6 dB SPL.

2.5 REFLECTION, TRANSMISSION, AND IMPEDANCE MATCHING

When a traveling acoustic wave, Eq. (2.38), encounters a boundary where the air ends and changes into another medium (eg, water), the acoustic energy may be transmitted into the latter medium, and partly reflected back into the former. The parameter that determines the amount of transmission or reflection is the acoustic impedance of either medium. Suppose the transition is at $x = 0$, and medium 1 ($x < 0$) has acoustic impedance Z_1, whereas medium 2 ($x > 0$) has an acoustic impedance Z_2. We further suppose that a traveling wave is entering from medium 1 with amplitude = 1, wave number = k_1, and frequency = ω_1. The dispersion relation for medium 1 reads $k_1 = \omega_1/v_1$, and the sound velocity in medium 1 is determined by $v_1 = \sqrt{B_1/\rho_1}$. In $x = 0$ this wave will partially reflect (becomes a leftward traveling wave through medium 1, with an amplitude R), and partially transmit (a rightward traveling wave, with amplitude T). In medium 2, the dispersion relation is $k_2 = \omega_2/v_2$, the sound velocity in medium 2 is $v_2 = \sqrt{B_2/\rho_2}$. We initially make no prior assumptions, and therefore allow for potential (unknown and different) phase changes for the reflection and transmission waves.

It's interesting to see that we only need the impedances to fully quantify what's going to happen.

In both media the homogeneous wave equation, Eq. (2.25) holds. In medium 1 we have a superposition of the rightward incoming wave and the reflecting wave, whereas in medium 2 we only have the transmitted wave:

$$\begin{aligned} x < 0 &: s_1(x,t) = \cos(k_1 x - \omega_1 t) + R\cos(k_1 x + \omega_1 t + \varphi_1) \\ x > 0 &: s_2(x,t) = \mathrm{T}\cos(k_2 x - \omega_2 t + \varphi_2) \end{aligned} \tag{2.58}$$

Here, R is called the reflection coefficient, and T the transmission coefficient. Both coefficients are restricted to the [0,1] range. Note that the unknown parameters in Eq. (2.58) are R, T, k_2, ω_2, φ_1, and φ_2. Interestingly, these six unknowns are all determined from only two boundary conditions:

1. The two media are connected, and therefore the wave functions should be continuous. This means that $s_1(0,t) = s_2(0,t)$, for *all times t.*

   From this constraint, we can immediately appreciate that:

$$\cos(\omega_1 t) + R\cos(\omega_1 t + \varphi_1) = T\cos(\omega_2 t + \varphi_2)$$

   and because this condition must hold for all possible times t:

$$\omega_1 = \omega_2; \quad \varphi_1 = \varphi_2 = 0, \quad \text{and} \quad 1 + R = T \tag{2.59}$$

2. Since we may suppose that the transition point has no mass (it's infinitely thin), there is no net force acting on this point, according to Newton's second law ($F = ma$). That means that pressures (force per surface area) immediately left and right of $x = 0$ should be equal [Eq. (2.41)]:

$$-B_1 \frac{\partial s_1}{\partial x}\bigg]_{x=0} = -B_2 \frac{\partial s_2}{\partial x}\bigg]_{x=0}$$

   and therefore:

$$B_1 k_1 \sin(\omega_1 t) - B_1 R k_1 \sin(\omega_1 t) = B_2 T k_2 \sin(\omega_1 t)$$

We thus find:

$$B_1 k_1 (1-R) = B_2 T k_2 \text{ and (from dispersion relations)}: B_1 \frac{(1-R)}{v_1} = B_2 \frac{T}{v_2} \tag{2.60}$$

Combining Eqs. (2.59) and (2.60) then finally yields the reflection and transmission coefficients for the two media:

$$R = \frac{Z_1 - Z_2}{Z_1 + Z_2} \quad \text{and} \quad T = \frac{2Z_1}{Z_1 + Z_2} \tag{2.61}$$

The acoustic impedances of the two media fully determine the reflection and transmission coefficients of this problem!

Note that in case $Z_1 = Z_2$ there is no reflection ($R = 0$), meaning that all acoustic energy is transmitted into medium 2 ($T = 1$). In this case the media are fully compatible (*impedance matching*). However, if $Z_2 \gg Z_1$ (eg, in case of a wall), T will be close to zero, and R approaches the value of -1: there is no transmission of acoustic energy into medium 2, and the entire wave is reflected at $x = 0$, with a phase shift of π radians.

Especially this latter case turns out to be extremely relevant for hearing, as our auditory system (this holds in principle for all land mammals and birds) is confronted with the situation that the inner ear (cochlea) is filled with fluid (water), whereas the acoustic input is received through the air. Therefore, the acoustic wave should somehow cross the transition from air to water. The problem, however, is that the impedance of water is much, much higher than the impedance of air! Let's now see how this works out for hearing, and how evolution has dealt with this problem.

From our earlier analysis, we can take the acoustic impedance of air, $Z_1 = \rho_1 v_1$ and the acoustic impedance of water $Z_2 = \rho_2 v_2$. We also found that the acoustic intensity (power per surface area) of an acoustic wave is determined by $I = p^2/2Z$. The reader can readily verify (Problem 2.10) that the transmitted and reflected sound intensities from medium 1 (air) to medium 2 (water) are written as (Fig. 2.8):

$$I_T = \frac{4Z_2/Z_1}{[1+(Z_2/Z_1)]^2} \quad \text{and} \quad I_R = \frac{[1-(Z_2/Z_1)]^2}{[1+(Z_2/Z_1)]^2} \tag{2.62}$$

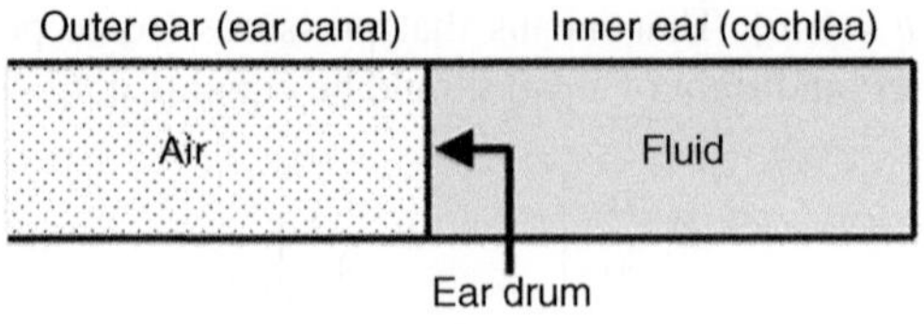

FIGURE 2.8 **Transmission of sound from air to water in case of no adaptive mechanisms to improve the impedance matching problem.**

For air, the impedance at room temperature is about $Z_1 = 420\ \mathrm{Ns/m^3}$, whereas for water the value is $Z_2 = 1.5 \times 10^6\ \mathrm{Ns/m^3}$. The ratio $Z_2/Z_1 \approx 3400$! Substitution of this value into Eq. (2.62) shows that the transmitted acoustic intensity into water would be only 0.001 of the total incoming intensity. The inner ear is not an infinite source of water, but a small bony construction filled with water, and therefore its effective acoustic impedance is somewhat lower (estimated at about $15 \times 10^4\ \mathrm{Ns/m^3}$). Nonetheless, the Z_2/Z_1 ratio is still very high, at about 360. Without any particular measures, the auditory system would be doomed to perceive only the very loudest of sounds! Since this is obviously not the case (as the lowest sound intensities that can be consciously perceived are in the order of 10^{-11}–$10^{-12}\ \mathrm{W/m^2}$!), evolution must have come up with some clever adaptations to deal with the impedance matching problem.

The middle ear (consisting of the tympanic membrane, or eardrum, the three middle-ear bones: malleus, incus, stapes, and the oval window), transfers the acoustic input from the air in the ear canal, to the fluid in the cochlea. This anatomical organization embeds two physical mechanisms to compensate the enormous losses of acoustic energy due to near-total reflection at the air–water boundary: amplification of the pressure through (1) hydraulics and (2) a lever arm (Fig. 2.9). The ratio of the surfaces of the tympanic membrane to that of the oval window is about 20:1. The ratio of the lever arms of the malleus versus incus is about 1.3:1.0. As a result, the transmission of the acoustic power is more efficient. We have just seen, that without the middle ear, the transmitted sound intensity $I_T = 0.001 I_{\mathrm{In}}$ (Fig. 2.8), which is about 30 dB lower than the input sound energy.

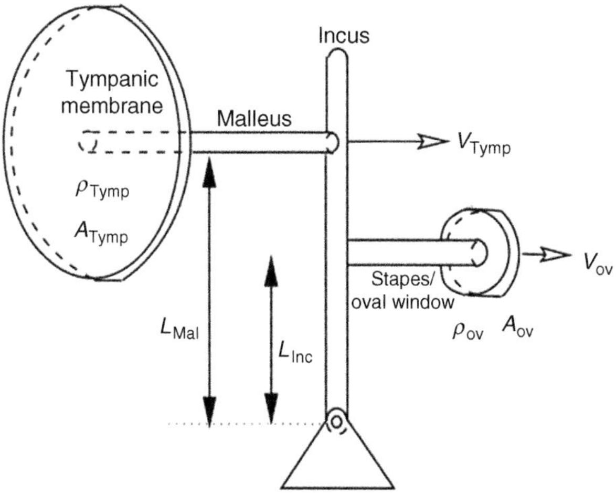

FIGURE 2.9 **Impedance matching in the middle ear: hydraulic pressure amplification through the 20:1 surfaces of tympanic membrane to oval window, in combination with a 1.3:1.0 lever of the hammer and anvil arms.**

For the forces on tympanic membrane and oval window to cochlea the following equilibrium relation should hold:

$$F_{\text{Tymp}} = F_{\text{Oval}} \rightarrow p_{\text{Tymp}} A_{\text{Tymp}} = p_{\text{Oval}} A_{\text{Oval}} \tag{2.63}$$

Due to the lever construction, there is also torque equilibrium:

$$p_{\text{Tymp}} A_{\text{Tymp}} L_{\text{Mal}} = p_{\text{Oval}} A_{\text{Oval}} L_{\text{Inc}} \tag{2.64}$$

Combining the two conditions yields an expression for the ratio between two pressures:

$$\frac{p_{\text{Tymp}}}{p_{\text{Oval}}} = \frac{A_{\text{Oval}} L_{\text{Inc}}}{A_{\text{Tymp}} L_{\text{Mal}}} \tag{2.65}$$

We now replace the pressures by Eq. (2.45): $p = Zv_{\text{Vol}}$, with v_{Vol} the volume velocity in the medium, thereby realizing that the volume velocities at tympanic membrane and oval window are related through the lever ratio:

$$\frac{Z_{\text{Tymp}}}{Z_{\text{Oval}}} = \frac{A_{\text{Oval}}}{A_{\text{Tymp}}} \left(\frac{L_{\text{Inc}}}{L_{\text{Mal}}} \right)^2 \approx (20)(0.78)^2 = 0.04 \tag{2.66}$$

This expression quantifies the transformation of the impedance from tympanic membrane to oval window as a result of the middle-ear mechanics, from the unaided value of 1/3400 to about 1/20. As a result, the ratio between I_T and I_{In} increases from 0.001 to about 0.12, which suggests amplification with a factor of about 100 (ie, a 20 dB improvement!). In reality, the amplification results prove to be even better, 26 dB, for the frequency range between 1.5 and 3.0 kHz, for which the human auditory system is most sensitive.

2.6 SOUND SPECTRUM, FOURIER SERIES, AND BANDWIDTH

Previously (see Section 2.2), we saw that the general solution to the homogeneous wave equation [Eq. (2.25)] is a superposition of spatial–temporal harmonic waves [Eq. (2.32)], for which the coefficients are determined by the boundary and starting conditions, while obeying a linear dispersion relation [Eq. (2.37)]. An example is the case of fixed boundary conditions, which led to Eq. (2.36).

However, to find the full solution for a particular wave problem, that is, specifying the amplitudes and phases of the participating frequency components, also the starting conditions should be incorporated. Here we will analyze such a simple problem, to illustrate the use of discrete Fourier analysis to the homogeneous wave equation with fixed boundary conditions.

The linear superposition of harmonic functions, with discrete frequencies that depend on the boundary conditions, described by Eq. (2.32) [and as an example for fixed boundary conditions, Eq. (2.36)] is called a Fourier series.

To specify this solution, the amplitudes of the sines and cosines (A_k, B_k, C_k, and D_k) in the series need to be determined. These amplitudes are uniquely related to the boundary and starting conditions of the particular wave problem.

Construction of the Fourier series: Suppose that the starting condition for the wave propagation problem (think of the temporal profile of the acoustic wave at $x = 0$, or at any other position) is given by a function, $f(t)$, for which the following requirements hold:

1. $f(t)$ is defined on the interval $t_0 < t < t_0 + T$.
2. $f(t)$ is periodic, with period T, that is, $f(t) = f(t + T)$.

In that case $f(t)$ can be fully represented (approximated to arbitrary precision) by the following Fourier series:

$$f(t) = \frac{a_0}{2} + \sum_{n=1}^{\infty}\left[a_n \cos\left(n\omega_0 t\right) + b_n \sin\left(n\omega_0 t\right)\right] \text{ with } \omega_0 = \frac{2\pi}{T} \tag{2.67}$$

whereby the Fourier coefficients a_n and b_n are determined by:

$$a_0 = \frac{2}{T}\int_{t_0}^{t_0+T} f(t)dt \qquad a_n = \frac{2}{T}\int_{t_0}^{t_0+T} f(t)\cos(n\omega_0 t)dt$$
$$\text{and} \qquad b_n = \frac{2}{T}\int_{t_0}^{t_0+T} f(t)\sin(n\omega_0 t)dt \tag{2.68}$$

In Problems 2.11 and 2.12 the reader may verify (using the orthogonality relations for sines and cosines) the validity of Eqs. (2.67–2.68).

The *discrete spectrum* of signal $f(t)$ is the set of frequencies:

$$f_n = \frac{n\omega_0}{2\pi} \text{ for which } a_n \text{ or } b_n \neq 0 \tag{2.69}$$

For general standing-wave problems, analytical calculation of the coefficients a_n and b_n may be quite tedious. Note, however, that every function $f(t)$ can always be written as the sum of an *even* and an *odd* function, because:

$$f(t) = \frac{1}{2}\left[f(t) + f(-t)\right] + \frac{1}{2}\left[f(t) - f(-t)\right] \equiv g(t) + h(t) \tag{2.70}$$

where $g(t)$ is an even function [ie, $g(t) = g(-t)$], and $h(t)$ = odd [meaning, $h(t) = -h(-t)$]. Whenever a function is even or odd, we need only calculate half of the Fourier coefficients, because:

- for even functions: $\forall b_n = 0$
- for odd functions: $\forall a_n = 0$

It can be quite beneficial in Fourier analysis to check whether the function is even or odd, or whether a simple manipulation can make it even or odd.

Note that exactly the same Fourier analysis can be performed on the *spatial* profile of the wave function, $X(x)$! Instead of using the angular frequency ω_0, and period T, Eqs. (2.67) and (2.68) are then formulated on a spatial interval, L, where the periodicity of X is $2L$ [ie, $X(x) = X(x + 2L)$] and the spatial frequencies are given by the wave numbers, $k_n = n\pi x/L$.

In chapter: Linear Systems Analysis, we will also describe the spectral analysis for nonperiodic, transient functions, for which the period $T\rightarrow\infty$. In that case, the Fourier series (with discrete spectra, at spectral intervals $\Delta\omega = 2\pi/T$) become Fourier integrals (with continuous spectra: $\Delta\omega\downarrow 0$).

Example: As an illustration of the application of Fourier series, we consider the following interesting example: the periodic impulse, with a duration ΔT and period T_0:

$$f(t) = \left\{ \begin{array}{ll} 1 \quad \text{for } 0 \le t < \Delta T & \text{mod}(T_0) \\ 0 \quad \text{for } \Delta T < t \le T_0 & \text{mod}(T_0) \end{array} \right\} \tag{2.71}$$

We first note that $f(t)$ is neither even, nor odd, but that with a simple temporal shift by $t_0 = -T_0/2$, we can make the function even! So, instead of letting the time run between 0 and T_0, we choose our interval on $[-T_0/2, +T_0/2]$. For the Fourier coefficients, Eq. (2.68), we obtain:

$$f(t) = \frac{a_0}{2} + \sum_{n=1}^{\infty} a_n \cos(n\omega_0 t) \quad \text{with } \omega_0 = \frac{2\pi}{T_0} \quad \text{and}$$

$$a_0 = \frac{2}{T_0}\int_{-T_0/2}^{T_0/2} f(t)dt = \frac{2}{T_0}\int_{-\Delta T/2}^{\Delta T/2} 1dt = \frac{2\Delta T}{T_0}$$

$$a_n = \frac{2}{T_0}\int_{-\Delta T/2}^{\Delta T/2} \cos(n\omega_0 t)dt = \frac{2}{n\pi}\sin\left(\frac{n\pi\Delta T}{T_0}\right)$$

Taken together, the Fourier series for Eq. (2.71) is (Fig. 2.10):

$$f(t) = \frac{\Delta T}{T_0} + \sum_{n=1}^{\infty} \frac{2}{n\pi}\sin\left(\frac{n\pi\Delta T}{T_0}\right)\cos(n\omega_0 t) \quad \text{with } \omega_0 = \frac{2\pi}{T_0} \tag{2.72}$$

We can look a bit more closely at the solutions of Eq. (2.72). We first note that the amplitude of the spectral component, a_n, decreases with the spectral number, n: the components at high frequencies have lower amplitudes than those at low frequencies. Second, as long as $n\Delta T \ll T_0$, the sinusoidal modulation can be approximated (first-order Taylor expansion) by $n\pi\Delta T/T_0$. In that case, however, $a_n \approx \frac{2\Delta T}{T_0} = a_0$, that is, constant! For low spectral numbers (low frequencies), the spectrum is therefore almost *flat*.

Third, the sinusoidal modulation of a_n will cross zero for the first time when $n = T_0/\Delta T$. Thus, the longer ΔT is relative to T_0, the sooner the spectral component crosses zero (ie, at smaller n). Suppose we now arbitrarily define that first zero crossing in the spectrum as the *spectral bandwidth* (BW) of signal $f(t)$, then:

$$\omega_{BW} = \frac{T_0}{\Delta T}\omega_0 = \frac{T_0}{\Delta T}\frac{2\pi}{T_0} = \frac{2\pi}{\Delta T} \rightarrow \omega_{BW}\Delta T = 2\pi \tag{2.73}$$

Eq. (2.73) is a very important, even *fundamental*, relation for signal analysis, in particular, for wave functions. It tells us that the bandwidth and signal duration are inextricably intertwined, and therefore cannot be independently manipulated! As a result, short-duration signals inevitably yield a broad spectral bandwidth, and vice versa. Fig. 2.10 illustrates the results of the analysis of this important example for two different pulse durations.

Note that the value 2π in Eq. (2.73) is in fact arbitrary, as it depends on the chosen criterion for the bandwidth (here we took the first zero crossing). The important point is, of course, that bandwidth and signal duration are inversely related. The relation states that it is fundamentally *impossible* to simultaneously have arbitrary good spectral and temporal resolution in a signal.

> A high resolution in the frequency domain comes at the expense of a poor resolution in the time domain, and vice versa!

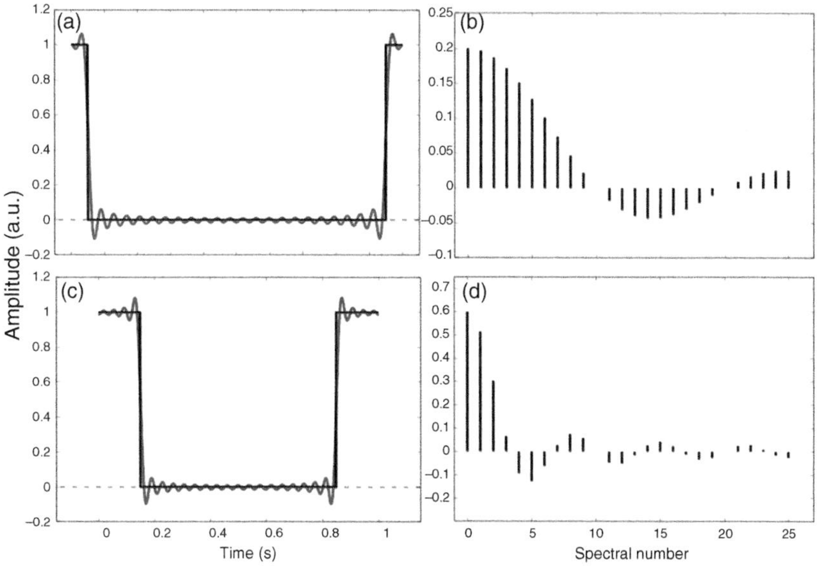

FIGURE 2.10 A short (a, b) and long pulse (c, d) yield quantitatively different spectral amplitude modulations (right), and hence different effective spectral bandwidths. Note the sinusoidal modulation of the spectral numbers, and their decline with increasing n.

In quantum mechanics, similar relations as Eq. (2.73) refer to the wave character of elementary particles like protons and electrons (namely: their energy–life time: $\Delta E \Delta T = \hbar$, and momentum–position trade off: $\Delta p \Delta x = \hbar$, with ħ Planck's constant) and are known as *Heisenberg's uncertainty relations.*

In Problems 2.13–2.15 the reader is encouraged to work out a couple of Fourier problems with and without specified boundary and/or initial conditions.

2.7 PHASE VELOCITY AND GROUP VELOCITY

A harmonic wave traveling in one dimension is described by

$$s(x,t) = A\cos(kx - \omega t) \tag{2.74}$$

The phase of this wave function is

$$\Phi = kx - \omega t \tag{2.75}$$

and the associated *phase velocity* of the wave follows from the dispersion relation:

$$v_\varphi = \frac{\omega}{k} \tag{2.76}$$

The phase velocity indicates how fast a given point with a constant phase (eg, a peak, or a zero crossing) moves through the medium. In a nondispersive medium (constant temperature, bulk modulus, and density) the phase velocity is independent of the wavelength, and therefore constant. In dispersive media the dispersion relation, $\omega(k)$, is nonlinear, and the phase velocity, $v_\varphi = v_\varphi(k)$.

When an (acoustic) signal is meant to transmit *information*, it cannot simply be in the form of a harmonic wave, since such a signal stretches in space and time from $[-\infty, +\infty]$ and cannot transmit anything else than simply its presence or absence. Information, like music, a voice, or a single brief pulse, should be encoded in the wave, for example, as an amplitude modulation, $A(t)$, a frequency modulation, $\omega(t)$, or a phase modulation, $\varphi(t)$, which needs to be decoded by the receiver. The velocity with which such a modulation is transmitted through space and time, the *group velocity,* v_g, is, in general (ie, in dispersive media), not equal to the phase velocity.

Let's illustrate this concept with a simple example. Suppose we emit two harmonic vibrations in $x = 0$, at equal amplitudes, but different frequencies, then, in $x = 0$ the total wave function is:

$$s(0,t) = A\cos\omega_1 t + A\cos\omega_2 t = A_{\text{mod}}(t)\cos\left(\omega_{\text{avg}} t\right) \tag{2.77}$$

where the amplitude modulation and carrier frequency are determined by:

$$A_{mod}(t) = 2A\cos\left(\omega_{\text{mod}} t\right) \text{ with } \omega_{\text{mod}} = \frac{\omega_1 - \omega_2}{2} \text{ and } \omega_{\text{Avg}} = \frac{\omega_1 + \omega_2}{2} \tag{2.78}$$

Similarly, for the traveling waves, Eq. (2.74), the result will consist of a modulation wave multiplied by a carrier wave:

$$s(x,t) = 2A\cos\left(k_{\mathrm{mod}}x - \omega_{\mathrm{mod}}t\right)\cos\left(k_{\mathrm{Avg}}x - \omega_{\mathrm{Avg}}t\right)$$

for which the frequencies and wave numbers are given as:

$$\begin{aligned} \omega_{\mathrm{mod}} &= \frac{\omega_1 - \omega_2}{2} \quad \text{and} \quad \omega_{\mathrm{Avg}} = \frac{\omega_1 + \omega_2}{2} \\ k_{\mathrm{mod}} &= \frac{k_1 - k_2}{2} \quad \text{and} \quad k_{\mathrm{Avg}} = \frac{k_1 + k_2}{2} \end{aligned} \tag{2.79}$$

Suppose that the two frequencies, ω_1 and ω_2 do not differ by much, so that $\omega_{\mathrm{Mod}} \ll \omega_{\mathrm{Avg}}$. We then find the *phase velocity* by setting the phase of the carrier wave constant (eg, zero):

$$v_\varphi = \frac{\mathrm{w}_{\mathrm{Avg}}}{\mathrm{k}_{\mathrm{Avg}}} \tag{2.80}$$

To find the *group velocity*, we set the phase of the modulation wave constant:

$$v_g = \frac{\omega_{\mathrm{Mod}}}{k_{\mathrm{Mod}}} = \frac{\omega_1 - \omega_2}{k_1 - k_2} = \frac{\Delta\omega}{\Delta \mathrm{k}} \tag{2.81}$$

In general, the frequencies obey a dispersion relation, $\omega(k)$, and when the two are close together, we can approximate (Taylor):

$$v_g = \frac{\omega_1(k_1) - \omega_2(k_2)}{k_1 - k_2} = \left.\frac{\mathrm{d}\omega}{\mathrm{dk}}\right]_{k_{\mathrm{Avg}}} + \text{ higher orders} \tag{2.82}$$

Note that for nondispersive media the phase velocity and group velocity are identical. The shape of the total wave function will not change, as all frequency components will travel at the same velocity.

In dispersive media, however, the two velocities will typically differ. Now the shape of the complex wave function (the amplitude modulation) will change, because the different component frequencies travel at different velocities.

2.8 EXERCISES

Problem 2.1: Verify Eq. (2.23) for the speed of gas molecules under adiabatic conditions.

Problem 2.2:

(a) Show that the homogeneous wave equation, Eq. (2.25), indeed holds for *any* function, $s(x,t)$, that can be written as:

$$s(x,t) = s(x \pm vt)$$

This is a very important and fundamental result as it states that any wave shape that travels at a constant velocity through the medium obeys the homogeneous wave equation (and it is a nondispersive medium).

(b) Verify the linearity condition (superposition of solutions) of the homogeneous wave equation.

Problem 2.3: Show Eq. (2.30) by substituting the harmonic traveling wave function after separation of variables.

Problem 2.4: Demonstrate that the standing-waves of the fixed boundary conditions for $s(0,t) = s(L_0,t)$ lead to the solutions given by Eq. (2.36).

Problem 2.5

(a) Follow a similar analysis as you did in Problem 2.4 to find the standing waves under open boundary conditions: $s(0,t)$ and $s(L_0,t)$ = maximum.

(b) Same for mixed boundary conditions: $s(0,t) = 0$ and $s(L_0,t)$ = maximum.

(c) Same for periodic boundary conditions: $s(0,t) = s(L_0,t)$ and $\frac{\partial s}{\partial x}(0,t) = \frac{\partial s}{\partial x}(L_0,t)$.

Problem 2.6: The inhomogeneous wave equation, Eq. (2.50), has harmonic temporal solutions, but nonharmonic spatial solutions. Demonstrate this latter statement.

**Problem 2.7*: In some cases, the inhomogeneous wave equation, Eq. (2.50), can be solved analytically. Consider *fixed* boundary conditions at $x = a$ and $x = 2a$, and substitute for the spatial dependence of the density and bulk modulus:

$$\rho(x) = \frac{m}{x^2}, B(x) = B_0$$

Demonstrate that the standing waves described by Eq. (2.51) (and illustrated in Fig. 2.7) are indeed a solution of the inhomogeneous wave equation, Eq. (2.50).

**Problem 2.8*: A second example of an inhomogeneous standing wave problem is the following: Consider a string of length L with mass density ρ_1 and tension B, that is connected to a second string of the same length and tension, but with mass density ρ_2. Both strings are attached to walls that are $2L$ apart, so that standing waves will arise for fixed boundary conditions (Fig. 2.11).

(a) Show that the frequencies of the eigenmodes of this system obey the following nonlinear relation:

$$v_1 \tan\left(\frac{\omega L}{v_1}\right) = -v_2 \tan\left(\frac{\omega L}{v_2}\right)$$

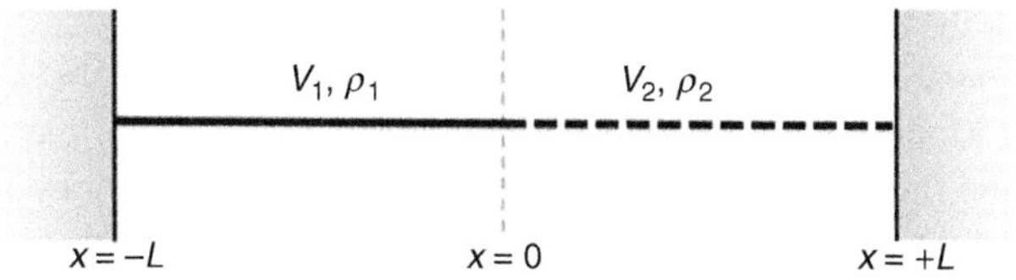

FIGURE 2.11 **Inhomogeneous string, consisting of two equal-length (L) parts with different mass densities, attached in the center at $x = 0$, and fixed boundary conditions at $x = -L$ and $x = +L$.**

(b) Find an expression for the standing wave solutions, $s(x,t)$.
(c) Is it possible that $x = 0$ is a node? If so, when?

Problem 2.9: Show that the total kinetic energy of a harmonic sound wave, described by $s(t) = s_{max}\cos(\omega t)$, confined to a gas cylinder with cross-section A, taken over a full wavelength, λ, is given by Eq. (2.52):

$$K_\lambda = \frac{\rho}{4} A\left(\omega s_{max}\right)^2 \lambda$$

Problem 2.10: Verify the transmitted and reflected intensities, I_T and I_R, respectively, at the boundary of two media with acoustic impedances Z_1 and Z_2, respectively [Eq. (2.52)].

**Problem 2.11*:

Show that sine and cosine obey the orthogonality relations:

$$\int_0^T \cos(n\omega t)\cos(m\omega t) = \int_0^T \sin(n\omega t)\sin(m\omega t) = \frac{T}{2}\delta_{nm}$$

$$\text{and} \int_0^T \sin(n\omega t)\cos(m\omega t) = 0$$

with T the period that is related to the angular frequency by $\omega = \dfrac{2\pi}{T}$

**Problem 2.12*: Using the orthogonality relations of the previous exercise, you can now demonstrate the validity of Eq. (2.68) (the calculation of the discrete Fourier spectrum).

Hint: To calculate a_0, take the time-average of the left- and right-hand sides of Eq. (2.67). To determine the coefficients a_n, b_n multiply both sides of Eq. (2.67) with cos($m\omega t$) and sin($m\omega t$), respectively, and take the time-average.

Problem 2.13: A Fourier series for a periodic function without boundary ~ and initial conditions:

(a) Determine the Fourier series for $f(t) = t^2 - t$, on the interval $0 \le t \le 1$: expand the function such that it becomes an *odd* periodic function.

(b) Idem for the case where the function of (*a*) is made *even*.
(c) Approximate $f(x)$ by the first three Fourier components of the series in (*a*) and (*b*). Compare the predicted values of both series with the actual values of the function in $t = [0.0, 0.2, 0.4, 0.6, 0.8, 1.0]$ and draw the results. Which of the two Fourier series converges fastest to $f(t)$? Why?

Problem 2.14: Consider the following function, $f(t)$ on the interval $0 \le t \le 1$:

$$f(t) = \begin{cases} 1 - ct \text{ for } 0 \le t \le \dfrac{1}{c} \\ 0 \text{ for } \dfrac{1}{c} \le t \le 1 \end{cases} \qquad c \ge 1 \text{ is a constant}$$

(a) Expand $f(t)$ to an even function with period $T = 2$. Draw this function.
(b) Write $f(t)$ as a Fourier series and determine the Fourier coefficients.
(c) Explain what happens to the spectrum of $f(t)$ as $c \to \infty$.

Problem 2.15: A Fourier problem with boundary and initial conditions. Consider a string that has a length of 5π, which is fastened at $x = 0$ and $x = 5\pi$. The string obeys the following wave equation: $\dfrac{\partial^2 y}{\partial t^2} = 25\dfrac{\partial^2 y}{\partial x^2}$ with $y(x,t)$ the transversal deflection of the string from equilibrium over the interval $x \in [0,5\pi]$. The following initial conditions hold for $t = 0$:

$$y(x,0) = \sin(x)[1 + 2\cos(x)] \quad \text{and} \quad \dot{y}(x,0) = 0 \, \forall x$$

Determine $y(x,t)$.

REFERENCES

Crawford, Jr., F., 1968. Waves. Berkeley Physics Course, vol. 3, McGraw-Hill, New York, USA.

Kuwada, S., Yin, T.C.T., Wickesberg, S.E., 1979. Responses of cat inferior colliculus neurons to binaural beat stimuli: possible mechanisms for sound localization. Science 206, 586–588.

Serway, R.A., Jewett, J.W., 2013. Physics for scientists and engineers, ninth ed. Thomson Learning, Belmont, Ca, USA.

Chapter 3

Linear Systems

3.1 MODELING PHYSICAL AND BIOLOGICAL SYSTEMS

Biological systems are complex, yet highly organized. Trying to understand such systems in a quantitative sense, a mere summing up of different elements that comprise the system is inadequate, as the essential aspects of biological function reside in the *interactions* among different subsystems. Not surprisingly, a general mathematical theory for analyzing complex systems, called Systems Theory, arose from the biological sciences.

Before starting to describe the conceptual ideas behind systems theory, let's first have a look at the physicist's method of analyzing Physicists systems. It apply the laws of physics to describe and predict the behavior of a physical system. This approach typically leads to a (set of) differential equation(s) that should be explicitly solved to describe the behavior of the system as a function of time. An example of this approach is the homogeneous and inhomogeneous wave equation described in the previous chapter.

A second illustrative example is the harmonic oscillator, like in Fig. 3.1A, where a mass, m, suspends from a vertical spring with elasticity, K. When we neglect friction, and there is no other external driving force than gravity, only two forces act on this system: the elastic force, described by Hooke's law, and the (constant) gravity acting on the mass (we ignore the mass of the spring). According to Newton's second law, the total force on the system accelerates the mass:

$$F_{\text{Tot}} = m \cdot g - K \cdot y = m \cdot \frac{d^2 y}{dt^2} \tag{3.1}$$

which is a second-order, inhomogeneous (because of gravity), linear differential equation. The equation is *linear*, because the relevant variable, $y(t)$, and its derivative(s) all appear in zero- or first-order (ie, as a constant, or as y^1 and $(\ddot{y})^1$). Its solution is a harmonic function (readily checked), and is described by:

$$y(t) = A\cos(\omega_0 t + \varphi_0) \text{ with } \omega_0 = \sqrt{\frac{K}{m}} \text{ and } \varphi_0 = \text{cos or arccos or } \cos^{-1}\left(\frac{y_0}{A}\right) \tag{3.2}$$

The Auditory System and Human Sound-Localization Behavior. http://dx.doi.org/10.1016/B978-0-12-801529-2.00003-9

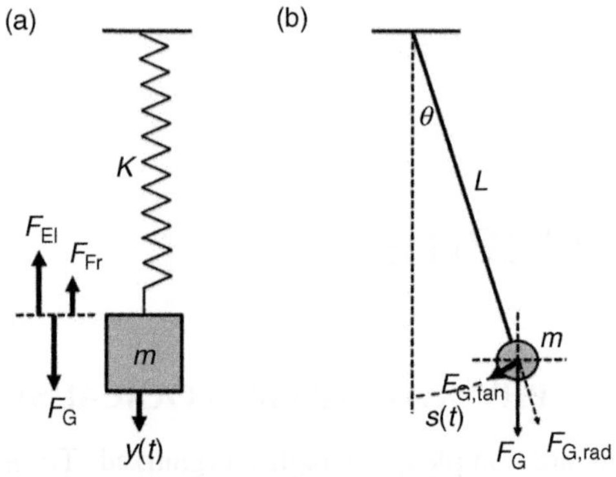

FIGURE 3.1 Two examples of a harmonic oscillator in physics. (A) The mass, m, suspended on a vertical spring, K. The elastic (F_{EL}) and frictional (F_{FR}) forces are in equilibrium with the force of gravity, F_G. (B) The pendulum, mass, m and length, L. The driving force in the system is the tangential component of the force of gravity, $F_{G,tan}$ along the circular path $s(t)$. The variable that describes the motion of the pendulum is the angle with the vertical, $\theta(t) = s(t)/L$ (in radians).

We see that in this solution gravity plays no role, and that the oscillation frequency is solely determined by the mass and the spring's elasticity. Also note that this vibration, once set in motion by an initial amplitude, $A \cos(\varphi_0)$, will last forever: energy is conserved in this system, as there is no dissipation of energy, meaning that no heat is generated or escaping from the system.

If we include friction in this problem, which is given by $\zeta \dot{y}$, with ζ the coefficient of friction, the equation of motion (which remains linear), and its solution reads:

$$F_{Tot} = m \cdot g - K \cdot y - \zeta \cdot \frac{dy}{dt} = m \cdot \frac{d^2 y}{dt^2}$$
$$y(t) = A \cdot \exp\left(-\frac{\zeta t}{2m}\right) \cdot \cos(\omega t + \varphi_0) \text{ and } \omega = \omega_0 \sqrt{(1-\zeta^2)} \tag{3.3}$$

This solution describes a damped oscillation, for which the amplitude of vibration decays exponentially with time. For such a system, kinetic and potential energy are no longer conserved, as the oscillating motion is gradually transformed into heat (by dissipation).

Another example of an oscillator is the physical pendulum shown in Fig. 3.1B. For this system, the equation of motion reads:

$$m \frac{d^2 s}{dt^2} = -m \cdot g \cdot \sin \theta(t) \tag{3.4}$$

where $s(t)$ is the tangent motion along the circular arc defined by the rope, and $\theta(t)$ is the instantaneous angle of the rope with the vertical (Fig. 3.1B). Note that this angle is related to s by $s(t) = \theta(t)L$, so that the equation of motion is *nonlinear*. As a result, it has no simple analytic solutions any more! Indeed, the behavior of the nonlinear pendulum can be quite complex. For example, the oscillation period, $T = 2\pi/\omega$, depends strongly on the initial angle, θ_0, (it can be described by elliptic integrals, but we won't go into this technical issue here). The period is determined by the following integral, which can be approximated for small angles by the first few terms of the Taylor's series:

$$T = 4\sqrt{\frac{L}{g}} \cdot \int_0^{\theta_0} \frac{1}{\sqrt{\cos\theta - \cos\theta_0}} d\theta \approx 2\pi\sqrt{\frac{L}{g}} \cdot \left(1 + \frac{1}{16}\theta_0^2 + \frac{11}{3072}\theta_0^4 + \cdots\right) \tag{3.5}$$

Note that for $\theta_0 = 180°$ the period $T \to \infty$, because then the inverted pendulum attains an (unstable) equilibrium position. Note also that for small θ_0 the solution is a harmonic oscillation that is independent of θ_0. Indeed, this solution is the linear approximation of Eq. (3.4) (where $\sin(\theta) \approx \theta$), which corresponds to the pure harmonic solution, as in Eq. (3.2).

In these examples, the physical model contains a number of known, measurable, physical parameters (eg, mass, friction, gravity, elasticity) that play a role in a particular differential equation for which the solutions depend in a deterministic way on these quantities. Although the solutions for Eqs. (3.1) and (3.3) [and to lesser extent, Eq. (3.4)] are relatively straightforward, these systems are not driven yet by arbitrary external inputs (forces). So suppose that the linear damped oscillator of Eq. (3.3) is driven by an arbitrary, dynamic external force, $F(t)$ (Bechhoefer, 2005):

$$\begin{gathered} m \cdot \frac{d^2y}{dt^2} + \zeta \cdot \frac{dy}{dt} + K \cdot y = F(t) \quad \text{which can be rewritten as:} \\ \frac{d^2y}{dt^2} + 2\zeta \cdot \omega_0 \frac{dy}{dt} + \omega_0^2 \cdot y = \frac{F(t)}{m} \end{gathered} \tag{3.6}$$

Interestingly, Eq. (3.6) can in principle be solved analytically for any force, $F(t)$, when making use of the homogeneous solution, provided by Eq. (3.3) (and see also below, when we discuss the Laplace transform). For example, if the driving force is a *step input*, here described by:

$$\frac{F(t)}{m} = \begin{cases} \omega_0^2 & \text{for} \quad t \geq 0 \\ 0 & \text{for} \quad t < 0 \end{cases}$$

then the step response for the damped oscillator is:

$$y(t) = 1 - \exp(-\zeta\omega_0 t)\frac{\sin\left(\sqrt{1-\zeta^2}\,\omega_0 t + \varphi\right)}{\sin\varphi} \quad \text{with } \varphi = a\cos \text{ or arccos or } \cos^{-1}\zeta \tag{3.7}$$

Note that the time for the step response to adapt to the sudden change in its input signal at $t = 0$, is determined by the exponential factor in the solution of Eq. (3.7), and is of the order of $\tau \approx 1/(\zeta\omega_0)$ s. Although explicit analytic solutions can be obtained for any driving force, $F(t)$, the calculations can become quite tedious.

Unfortunately, in biology, one often cannot resort to the physics method of first principles, because in many situations we know practically nothing, or very little, about the fundamental, mathematical, and physical mechanisms that determine the system's behavior. However, the *black box* (BB) approach provides a very useful mathematical framework that can also be applied to biological systems, and it yields quantitative predictions (and tests) for a system's behavior in response to arbitrary stimuli. Although the BB method lacks (and does not provide) explicit knowledge about the system's internal mechanisms (either physical, or mathematical), and of its (anatomical, functional) structure, it provides nonetheless a general framework that guarantees full predictive power for the system's behavior.

In this chapter we will distinguish the system's theoretical approach for *linear systems* (LS) versus *nonlinear* systems. We will make a hard distinction between these two system types in Section 3.2. For the time being it suffices to state that the LS approach is relatively straightforward: we will show that once we know the response of a linear system to a particular, sufficiently rich, input stimulus, the response to any arbitrary stimulus can be readily predicted as well. This means that as a researcher, one obtains full quantitative power of the system under study by performing a nearly trivial experiment. Unfortunately, such an approach is not possible for nonlinear systems. Hence, one has to resort to certain classes of nonlinear systems: those behaving relatively smoothly over time, and for which the nonlinearity in the input–output relationship can be approached by an analytic function (differential, expandable by a Taylor series around a certain resting point, etc.).

It should be noted that, although stemming from the biosciences, systems theory is by no means confined to biological systems. It has also been applied successfully to a wide variety of other disciplines: technical applications, physics, and also in economic and social sciences.

A key element in systems theory is the use of *BB models*. A BB is an abstraction of the real problem of interest. The basic idea is that the system can only interact with its environment: it communicates with the world by receiving signals (*inputs*), and it reacts to these inputs by creating its own *output* signals, which are sent into the environment. By studying the relationships between inputs and outputs of the system, we gain information about the way in which the system transforms signals, and eventually about its function.

The underlying structure of elements that make up the system is *not* the object of study in pure systems analysis, but is (of course) eventually essential to fully understand the mechanisms of the (biological) system (Fig. 3.2).

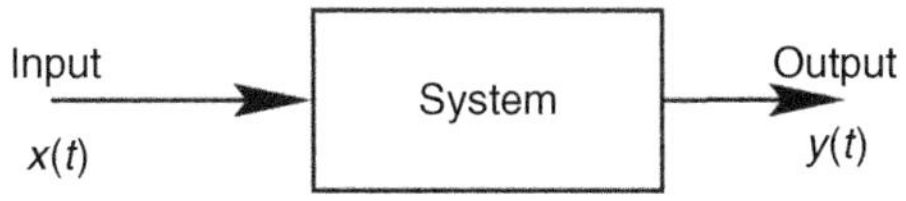

FIGURE 3.2 **The BB concept in system's theory, in which the system communicates with the environment by transforming an input signal, $x(t)$, into an output signal, $y(t)$.** For simplicity, we here only consider one input–one output systems.

The main concept behind systems theory is that when the input–output relations of the BB are *identical* to those of the real system, then the BB model can be considered *equivalent* to the system. In that case, quantitative predictions can be made on the basis of the (mathematically formulated) BB model, to understand the real system's behavior.

Systems identification is the technique that aims to find the functional and mathematical relationship between a system's input and output. In abstract notation:

$$y(t) = \mathcal{F}[x(t,\tau)] \tag{3.8}$$

which states that the system's output (time-dependent) is some function of the time-dependent input signal, where t is the absolute reference time (the present), and time variable τ is measured relative to t (the past and the future). Because of the dependence of relative time, we rewrite Eq. (3.8) preferably as:

$$y(t) = \mathcal{F}[x(t-\tau)] \tag{3.9}$$

F is called a *functional* (a function of a function, resulting in a scalar).

Systems can come in many forms, and can be driven by any number of multiple inputs, while generating any other number of multiple output signals. For ease of description, and mainly with the aim of illustrating the different concepts, rather than striving for completeness, we confine the analysis in this chapter to single input, $x(t)$, single-output, $y(t)$, systems.

We will further restrict the description to *stationary* systems, ie, systems for which the parameters do not vary with time (eg, adaptation phenomena will not be considered). Moreover, we will deal with *causal* systems: such systems only produce an output *after* an input has been provided (so, *predictive* mechanisms are not considered either). For causal systems, relative time, τ, of Eq. (3.9) will only refer to *present* and *past* events in the signal, ie, $\tau \geq 0$.

However, the most important and fundamental distinction to be made in systems analysis is *linear* versus *nonlinear* systems. When a system is linear (defined further), the analytical tools for its identification are fully developed and straightforward. We will see that a single, well-chosen experiment (ie, a single input presentation) will suffice to fully characterize everything there is to know about the system.

In what follows, we discuss the analytical methods used in linear and in nonlinear systems theory using BB analysis tools. Although this exposé involves some mathematics, understanding all technicalities is not essential in order to follow the main arguments. Students with a deeper mathematical background may want to try some of the more technical exercises that are mentioned in the text.

3.2 LINEAR SYSTEMS: THE IMPULSE RESPONSE

A system is *linear* only when it obeys the *superposition principle* (Marmarelis and Marmarelis, 1978; Oppenheim et al., 1983). This important principle entails that the system's response to a weighted sum of an arbitrary number of inputs should exactly equal the *same* weighted sum of the system's responses to each of the individual inputs:

$$\begin{aligned}&\text{If input}: x_k(t) \rightarrow \text{output}: y_k(t)\\ &\text{then if input}: \sum_k a_k \cdot x_k(t) \rightarrow \text{output}: \sum_k a_k \cdot y_k(t)\ \forall t,\ a_k \in \mathbb{R}\end{aligned} \tag{3.10}$$

Note that superposition should hold for all times t! Especially this latter requirement is somewhat counterintuitive, as it means that the duration of the output signal should be independent of the scaling factor, a_k, of the input signal (Problem 3.1). So, for example, supposing that the arm behaves as a linear system, the duration of a 90 degree arm rotation should be exactly the same as for a 1 degree arm rotation! This simple example therefore already suggests that LS may form quite a special class of systems, and that many (if not all) biological systems will often violate the superposition principle. Nevertheless, the linearity requirement of Eq. (3.10) may be approximately met for a certain hopefully relevant range of input signals, in which case the concepts following from the superposition principle may still be applied. We will see in a moment that from Eq. (3.10), we can derive some profound consequences that make the study of LS quite attractive.

As an example, let's demonstrate that the system that transforms an input $x(t)$ according to the simple straight-line relation $y(t) = ax(t) + b$, with a and b real scalars, does *not* obey the superposition principle, and therefore is not a linear system (Fig. 3.3). Instead of delivering a formal mathematical proof, we will show by example that the system is not linear. To that end, we only need to identify one simple case for which Eq. (3.10) is violated. Let's take two arbitrary input signals, $x_1(t)$ and $x_2(t)$, respectively, and define their sum,

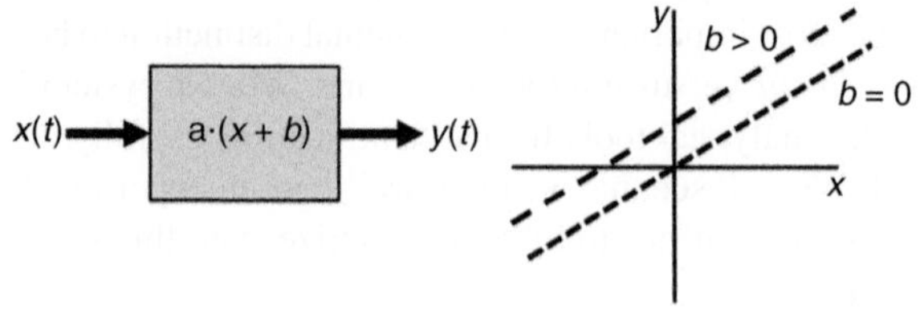

FIGURE 3.3 A straight-line relation between input, $x(t)$, and output, $y(t)$, obeys the superposition principle only when $b = 0$.

$x_1(t) + x_2(t)$, as a third input, for which the system's outputs are $y_1(t)$, $y_2(t)$, and $y_{1+2}(t)$, respectively:

$$y_1(t) = ax_1(t) + b$$
$$y_2(t) = ax_2(t) + b$$
$$y_{1+2}(t) = a(x_1(t) + x_2(t)) + b$$

We then have to verify whether the output to the sum of inputs equals the sum of the individual outputs:

$$y_{1+2}(t) \underset{\text{equal?}}{\Leftrightarrow} y_1(t) + y_2(t)$$

where the right-hand side is computed as

$$y_1(t) + y_2(t) = ax_1(t) + ax_2(t) + 2b$$

Obviously, this is only equal to $y_{1+2}(t)$ for $b = 0$! Hence, the straight line obeys the superposition principle only when it has no offset and passes through the origin!

Impulse response: A very important consequence of Eq. (3.10), central to LS analysis, is that all LS can be fully characterized by their response to the Dirac impulse function, $x(t) = \delta(t)$, the *impulse response*, $y(t) = h(t)$. Let's demonstrate why that is the case, by first introducing the Dirac impulse function and its properties. It is defined in the following way:

$$\delta(t) = \left\{ \begin{array}{cc} \infty & \text{for } t = 0 \\ 0 & \text{for } t \neq 0 \end{array} \right\} \text{ and } \int_{-\infty}^{+\infty} \delta(t)dt = 1 \tag{3.11}$$

Note that from the right-hand expression (which states that the total area of the Dirac pulse is one) it also follows that

$$\int_{-\infty}^{+\infty} f_k \delta(t)dt = f_k \text{ and therefore}: \int_{-\infty}^{+\infty} f(t)\delta(t-\tau)dt = f(\tau) \tag{3.12}$$

where we note that

$$\delta(t-\tau) = \left\{ \begin{array}{cc} \infty & \text{for } t = \tau \\ 0 & \text{for } t \neq \tau \end{array} \right\} \tag{3.13}$$

Despite its weird appearance (a function that is infinitely high for an infinitely short period of time), the Dirac pulse may be readily imagined to be the limit case of a pulse, $B(t)$, with width ΔT, and height, $1/\Delta T$ (Fig. 3.4):

$$B(t) = \left\{ \begin{array}{cc} \dfrac{1}{\Delta T} & \text{for } 0 \leq t \leq \Delta T \\ 0 & \text{elsewhere} \end{array} \right\}$$
$$\delta(t) = \lim_{\Delta T \downarrow 0} B(t) \tag{3.14}$$

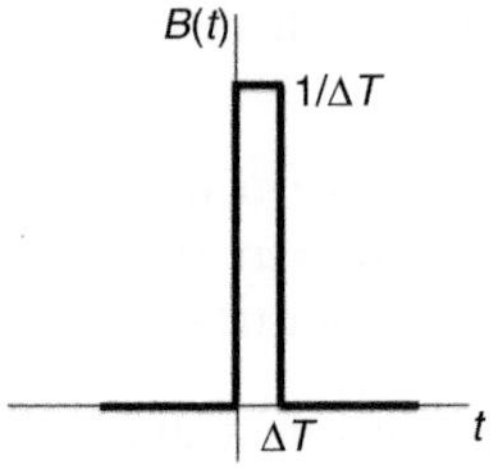

FIGURE 3.4 **Approximation of the Dirac delta function, Eq. (3.12).**

So why is the Dirac delta function so important for LS analysis? The answer lies in the property of Eq. (3.12). Let's play a little bit with that equation, using the fact that the variable t in Eq. (3.12) just serves as a dummy variable for the integration. So here's the trick: rewrite Eq. (3.12) to

$$f(t)=\int_{-\infty}^{+\infty} f(\tau)\delta(t-\tau)\,d\tau \tag{3.15}$$

Although Eq. (3.15) is identical to Eq. (3.12), we may now interpret it in a different way. Eq. (3.15) tells us that we can use the Dirac pulse to "measure" any function $f(t)$ with arbitrary precision. Conceptually it is similar to providing a series of stroboscopic "light pulses" in an otherwise dark disco (or the TL lights in your room, if you happen to never visit disco's...), to capture the movements of the audience at discrete time intervals. In Eq. (3.15) the discrete time intervals are infinitesimally finely spaced, at $d\tau$ intervals (Fig. 3.5).

Eq. (3.15) also states that function $f(t)$ can be constructed by a linear summation (*superposition*) of weighted Dirac pulses, where the function values serve as their weighting factors. This is indeed an important point! If we can write any function as a linear superposition of weighted Dirac pulses, we can use this property to study LS! It's now useful to recall Eq. (3.10), which formulates the superposition requirement for a discrete set of weighted input functions. However, if we replace the discrete summation by Eq. (3.15), we can reformulate the superposition principle as follows:

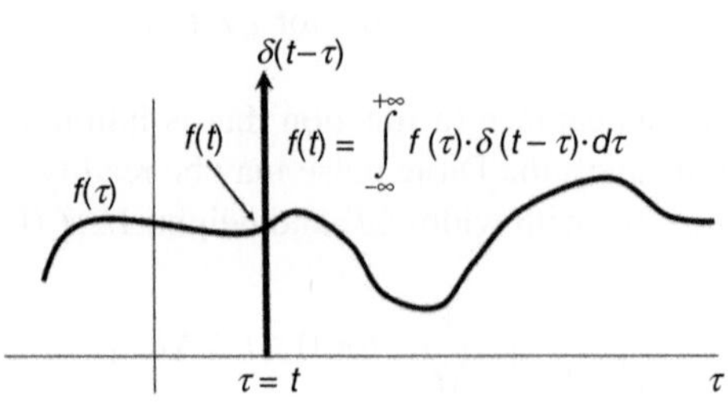

FIGURE 3.5 **The Dirac delta pulse can be used to measure the value of function $f(\tau)$ at a precisely defined time, t.**

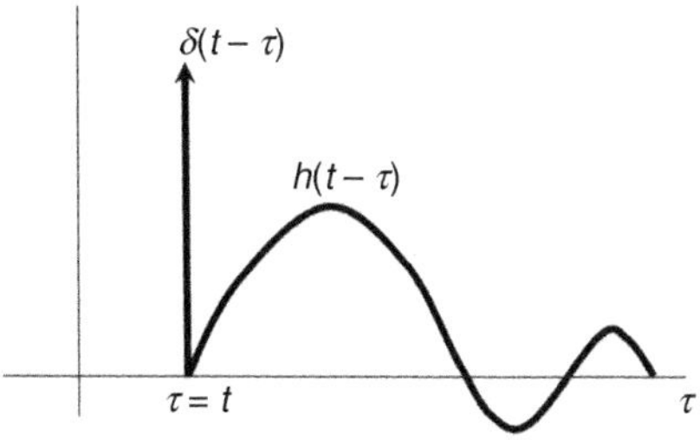

FIGURE 3.6 **Example of the impulse response of a (causal) linear system.** The response follows the impulse (and is zero prior to the pulse).

> *The response of a linear system to a weighted sum of Dirac pulses, is the same weighted sum of Dirac pulse responses (impulse responses, h(t)).*

Let us now formulate this concept mathematically. Suppose that the impulse response of a linear system to a single Dirac pulse is given by:

$$\text{If } x_{\text{Imp}}(t) = \delta(t) \rightarrow y_{\text{Imp}}(t) \equiv h(t) \text{ (impulse response function)} \tag{3.16}$$

Note that, in principle, this function can have any shape, the only practical restrictions being that it has a finite duration, and that it yields a finite integral over time interval $[-\infty,+\infty]$ (Fig. 3.6). We can then write the response of the linear system to an arbitrary input signal, $x(t)$, as follows:

$$\begin{aligned} x(t) &= \int_{-\infty}^{+\infty} x(\tau)\delta(t-\tau)d\tau \rightarrow \\ y(t) &= \int_{-\infty}^{+\infty} x(\tau)h(t-\tau)d\tau \end{aligned} \tag{3.17}$$

where we have simply replaced the Dirac pulses at times $t-\tau$, by the system's impulse responses at those times. This very important, and perhaps formidable looking, result tells us that we only need to know the system's impulse response, $h(t)$, in order to predict the system's output for any *arbitrary* input signal (be it a complex piece of music, or a human voice, birdsong, car noise, really anything)! The lower half of Eq. (3.17) is called a convolution integral, and is fully attributable to the superposition principle, in combination with the property of the Dirac delta function.

As Eq. (3.17) does incorporate the causality constraint, we will introduce this as follows:

$$\text{Causality}: \quad h(s) = 0 \text{ for } s \leq 0 \tag{3.18}$$

In Eq. (3.17) this means that $h(t-\tau) = 0$ for $t-\tau \leq 0$, or, equivalently, whenever $\tau \geq t$. The upper bound of the integral in Eq. (3.17) may therefore be replaced by t. The reader can now readily verify that Eq. (3.17) can then be rewritten into the convolution integral as it is usually presented in the literature:

$$y(t) = \int_0^{+\infty} x(t-\tau)h(\tau)d\tau \tag{3.19}$$

Let's briefly look at the different constituents of this equation, and try to conceptualize their meaning. First, compare Eq. (3.19) with Eq. (3.9) and note that we have now indeed identified the functional F, for all systems that obey the superposition principle. Further, note that two different time variables play a role in this equation: absolute time, t, and relative time, τ.

So, if we consider $y(t)$ as the current value of the system's output, what does Eq. (3.19) tell us? The following table conceptualizes the different symbols of Eq. (3.19):

t:	It is the current, absolute time, which we here call the *present*.
$y(t)$:	It is the current output of the system, measured "now."
τ:	It is measured relative to t, and is therefore the *past* time.
$\tau = 0$:	It is the present, and $\tau = +\infty$ represents the very distant past.
$x(t - \tau)$:	It is the value of the input signal, τ s ago.
$h(\tau)$:	It is the weight of the past input signal, or: the *memory* of the linear system about that event.

Thus: the current output of a linear system, $y(t)$, is determined by the linear sum of weighted memories of past inputs, $x(t - \tau)$. The impulse response function, $h(\tau)$, therefore represents the system's dynamic memory.

In the example of Fig. 3.7, the output may be approximated by the discrete summation:

$$y(t) \approx x(t)h(0) + x(t-\tau_1)h(\tau_1) + x(t-\tau_2)h(\tau_2) + \cdots + x(t-\tau_{\text{end}})h(\tau_{\text{end}}) \tag{3.20}$$

So now that we have seen that *all* LS can be described by the convolution integral of Eq. (3.19), and therefore are uniquely characterized by their impulse response function, $h(\tau)$, how does the simple straight line of Fig. 3.3, $y(t) = ax(t)$, fit into this framework? What is its impulse response function? The reader may verify the following equivalency, which uses the property of Eq. (3.12) of the Dirac delta function:

$$y(t) = ax(t) \Leftrightarrow y(t) = \int_0^{\infty} a\delta(\tau)x(t-\tau)d\tau \Rightarrow h(\tau) = a\delta(\tau) \tag{3.21}$$

Indeed, the impulse response of a scalar gain, a, is the gain times the Dirac impulse function.

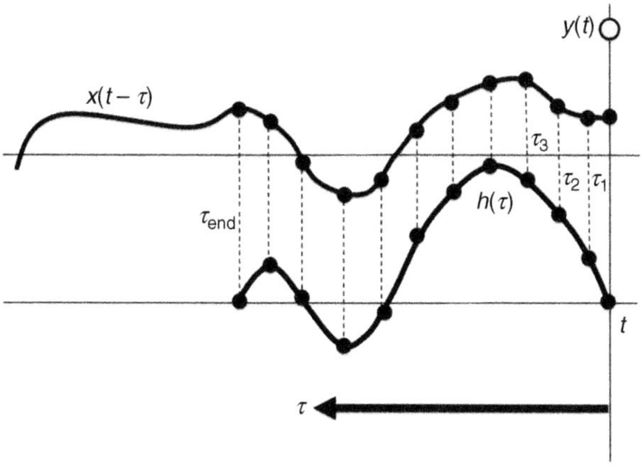

FIGURE 3.7 **Illustration of the convolution principle, where the input signal $x(t)$ is summed and weighted with its past values at $t-\tau$, by the impulse response function, $h(\tau)$.**

The step response: The convolution integral predicts the system's output for any input signal, $x(t)$. To apply it, however, we need to know the system's impulse response function. It is not always advisable to present a Dirac delta pulse [or its approximation, Eq. (3.14)] to the system, as a very high-amplitude input signal could potentially destroy the system! Therefore, other types of input signals should be considered to determine the system's impulse response. An important consideration is the following:

> *Perform a linear transformation, S, on the Dirac delta function to create a new input stimulus, $S(\delta)$. Measure the system's response to this stimulus, $y[S(\delta)]$. The impulse response is then found by applying the inverse transformation on the output: $h(t) = S^{-1}y[S(\delta)]$!*

An example of this idea is the unity step function, also called the Heaviside function, $u(t)$. It is defined as follows:

$$u(t) = \left\{ \begin{array}{ll} 1 & \text{for } t \geq 0 \\ 0 & \text{for } t < 0 \end{array} \right\} \Leftrightarrow u(t) = \int_{-\infty}^{t} \delta(s)ds \tag{3.22}$$

where the right-hand side states that the unity step is the cumulative integral of the Dirac pulse. So let's determine the response of a linear system to the unity step, which is called the system's *step response, $s(t)$*:

$$s(t) = \int_{0}^{+\infty} u(t-\tau)h(\tau)d\tau = \int_{0}^{t} h(\tau)d\tau \tag{3.23}$$

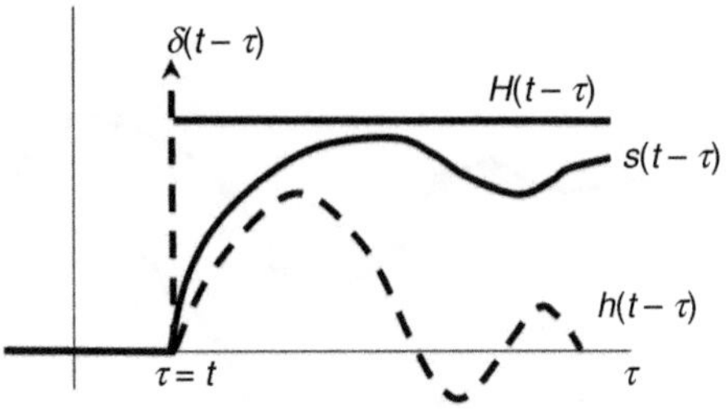

FIGURE 3.8 **Relation between a linear system's impulse response, *h*(*t*), and its step response, *s*(*t*).** The step response is the cumulative integral (surface) of the impulse response; the latter is the instantaneous derivative (slope) of the former.

And indeed, the step response results to be the cumulative integral of the impulse response (Fig. 3.8). To find the impulse response we simply apply the inverse operation of integration, which is differentiation:

$$h(t) \equiv \frac{ds}{dt} \tag{3.24}$$

Note, that if the impulse response is already known (eg, through another method), the step response is simply found by integrating the impulse response (Eq. 3.23). Similar procedures apply for higher-order integration functions of the impulse function, for example, the *ramp*

$$r(t) = \left\{ \begin{array}{cc} t & \text{for } t \geq 0 \\ 0 & \text{for } t < 0 \end{array} \right\} \Leftrightarrow r(t) = \int_{-\infty}^{t} U(\sigma) d\sigma \Rightarrow h(t) = \frac{d^2 r_{RESP}}{dt^2} \tag{3.25}$$

3.3 LINEAR SYSTEMS: THE TRANSFER CHARACTERISTIC

Eq. (3.19) holds for *any* linear system, $h(\tau)$, and for *any* input signal, $x(t)$. So far, we have seen some simple examples (impulse, step), from which it is immediately clear that the response of the system, $y(t)$, looks very different from the input signal (eg, Fig. 3.5, for impulse response). The only exception we have seen so far of a linear system that preserves the shape of the input signal is the scalar gain, Eq. (3.21), which, however, is kind of trivial as it represents a zero-memory system.

One may nevertheless wonder whether there exists a particular type of signal, $x(t)$, the characteristics of which are left unchanged by *all* LS. Such a signal indeed exists, and it is in fact the *harmonic signal*, extensively discussed in the previous chapter. We will see that this interesting and important property is again a direct consequence of the superposition principle, Eq. (3.10).

To see why, let's analyze what happens when presenting a pure harmonic input to an arbitrary linear system, about which we know nothing more than its impulse response. As discussed in chapter: The Nature of Sound, the harmonic input has no particularly interesting time behavior: it started at $t = -\infty$, and will continue until $t = +\infty$, meanwhile repeating its wave form every $T = 2\pi/\omega$ s. Indeed, the only variable of interest when using pure harmonic functions is the (angular) frequency, ω. The amplitude and phase of the harmonic signal are considered its parameters; a simple scaling and shift along the time axis can bring any harmonic function, $x(t) = A\sin(\omega t + \Phi)$ to its standard form. Therefore, without loss of generality, we take the amplitude of the harmonic input $A = 1.0$, and its phase, $\Phi = 0$ rad. We use this signal to compute the output of the linear system, $h(\tau)$, with the convolution integral:

$$x(t) = \sin(\omega t) \Rightarrow y(t) = \int_0^{+\infty} \sin\big(\omega(t-\tau)\big)h(\tau)d\tau \tag{3.26}$$

This result doesn't look very hopeful, until we realize that

$$\sin(a-b) = \sin(a)\cos(b) - \cos(a)\sin(b) \tag{3.27}$$

We plug this into Eq. (3.26), to see that:

$$y(t) = \int_0^{+\infty} [\sin(\omega t)\cos(\omega\tau) - \cos(\omega t)\sin(\omega\tau)]h(\tau)d\tau \tag{3.28}$$

This seems to carry us even further away from any new insights. However, when rewriting Eq. (3.28) as follows:

$$y(t) = \sin(\omega t)\int_0^{+\infty} \cos(\omega\tau)h(\tau)d\tau - \cos(\omega t)\int_0^{+\infty} \sin(\omega\tau)h(\tau)d\tau \tag{3.29}$$

it immediately becomes clear that we're on the right track. The integration variable is the relative time, τ, and not the absolute time, t. Therefore, any function not depending on τ can be taken out of the integration operator! We further note that the two integrals are both *definite* integrals, the results of which will be scalars that depend on the shape of the particular impulse response function (which is a fixed property of the system under study), and on the angular frequency, ω, which in this case is a *variable*, as we calculate the system's output to pure harmonic functions of arbitrary frequencies. Thus, we can succinctly summarize Eq. (3.29) by:

$$y(t) = A(\omega)\sin(\omega t) - B(\omega)\cos(\omega t) \equiv G(\omega)\cos[\omega t + \Phi(\omega)] \tag{3.30}$$

where $A(\omega)$ and $B(\omega)$ are ω-dependent scalar functions resulting from the definitive integrals. The reader may verify the right-hand expression (Exercise); $G(\omega)$ is the system's *amplitude characteristic* (or *gain*), and $\Phi(\omega)$ is its *phase characteristic*, both of which depend on angular frequency. Together, the two functions are the system's transfer characteristic, $H(\omega)$.

In conclusion:

1. The output of any linear system to a pure harmonic input is also a pure harmonic signal with the *same frequency* as the input. Stated differently: harmonic functions are *eigenfunctions* for LS:

Harmonic function in ⇒ same harmonic function out

2. The amplitude, G, and phase, Φ, of the output may vary as a function of angular frequency. The behavior of the amplitude and phase functions is fully determined by the system's impulse response, $h(\tau)$, and is described by the system's transfer characteristic, $H(\omega)$.

A crucial test for linearity of a system is therefore to demonstrate that its output to any harmonic input only yields a harmonic with the *same* frequency, whereas its amplitude and phase may differ from the input. Any deviation from this strict requirement renders a system not linear.

For example, since we already concluded that the system $y(t) = ax(t) + b$ is not linear, we shall now check that indeed it also violates the harmonic requirement. To that end, we provide the harmonic input to this system, and determine the spectrum of its output:

$$y(t) = a\sin(\omega t) + b \quad \text{spectra:} \quad \text{in} = [\omega] \quad \text{out} = [dc(\omega = 0), \omega]$$

We see that the output spectrum is not identical to the input spectrum, because of the extra DC (direct current) from the constant offset, b. So, indeed, the straight line also violates the harmonic requirement, unless $b = 0$.

Another example, which is clearly nonlinear, is $y(t) = ax^2(t)$. Again, apply the harmonic input and check the output spectrum:

$$y(t) = a\sin(\omega t)^2 = \frac{a}{2}[1 - \cos(2\omega t)] \ \text{spectra:in} = [\omega]\text{out} = [dc, 2\omega]$$

In this case, the frequency of the input is even absent in the output harmonic.

Linear systems preserve the input spectrum: frequencies may disappear (as the output amplitude for that particular frequency then happened to be zero), but never can new frequencies appear in the output!

An analysis of LS using harmonic functions and the transfer characteristic, $H(\omega)$, is performed in the *frequency domain*. In contrast, an analysis applying the convolution integral and the impulse response function, $h(\tau)$, takes place in the *time domain*.

Fourier Analysis of LS. Eq. (3.29) provides a direct link to Fourier analysis (introduced for periodic functions in the previous chapter), by noting that the two definitive integrals, in fact, represent the (continuous) Fourier transform of the (transient) impulse response function, $h(t)$:

$$\mathcal{F}\left[h(t)\right] \equiv H(\omega) = \left\{ \int_{0}^{+\infty} \cos(\omega\tau)h(\tau)d\tau, \quad i\int_{0}^{+\infty} \sin(\omega\tau)h(\tau)d\tau \right\} \tag{3.31}$$

where the left-hand side is the real part, $Re\{H(\omega)\}$, of the (complex) Fourier transform, and the right-hand side is its imaginary part, $Im\{H(\omega)\}$.

Formally, the Fourier transform of an (integrable) function, $f(t)$, is defined as:

$$\mathcal{F}\left[f(t)\right] \equiv F(\omega) = \int_{-\infty}^{+\infty} f(t)e^{-i\omega t}dt \tag{3.32}$$

We note that $\exp(-i\omega t) = \cos(\omega t) - i\sin(\omega t)$ (Euler's formula), and $i^2 = -1$ is the imaginary number. Further, any complex number, $z \in \mathbb{C}$, can be written as:

$$z = x + iy = G(\cos\Phi + i\sin\Phi) = Ge^{i\Phi} \tag{3.33}$$

with x and y the real and imaginary parts of z, respectively, and G and Φ are the amplitude and phase of z, respectively.

The astute reader may have noticed that in Eq. (3.31) the integration boundaries run between 0 and $+\infty$, instead of $-\infty$–$+\infty$, as in the actual Fourier transform of Eq. (3.32). Clearly, this difference arises from the causality requirement for the impulse response function, Eq. (3.18). Thus, without changing anything really, we can extend the integration boundaries in Eq. (3.31) to also include the negative time axis, as $h(t) = 0$ for all $t < 0$.

We finally have the necessary tools in our hands to understand why Fourier analysis, rather than the convolution integral, may be a very attractive alternative for LS analysis. Let's therefore determine the Fourier transform of the system's output, given by Eq. (3.19):

$$\begin{aligned}
\mathcal{F}\left[y(t)\right] \equiv Y(\omega) &= \int_{-\infty}^{+\infty} y(t)e^{-i\omega t}dt \\
&= \int_{-\infty}^{+\infty}\left[\int_{0}^{+\infty} x(t-\tau)h(\tau)d\tau\right]e^{-i\omega t}dt \\
&= \left[\int_{-\infty}^{+\infty} x(t-\tau)e^{-i\omega(t-\tau)}d(t-\tau)\right]\left[\int_{-\infty}^{+\infty} h(\tau)e^{-i\omega\tau}d\tau\right]
\end{aligned} \tag{3.34}$$

in which we recognize the following important result:

$$Y(\omega) = H(\omega)X(\omega) \tag{3.35}$$

The Fourier transform of the output of a linear system is determined by the algebraic product of the Fourier transforms of the input signal, and the impulse response function.

From Eq. (3.35) it follows that to find the transfer characteristic of the linear system, one merely has to divide $Y(\omega)$ by $X(\omega)$:

$$\text{Transfer function:} \quad H(\omega) \equiv \frac{Y(\omega)}{X(\omega)} \tag{3.36}$$

The full gain–phase representations of Eqs. (3.35) and (3.36) thus read:

$$\begin{aligned} Y(\omega) = Y_G(\omega)e^{i\Phi_Y(\omega)} &= \left[X_G(\omega)H_G(\omega)\right]e^{i[\Phi_X(w)+\Phi_H(\omega)]} \\ H(\omega) = H_G(\omega)e^{i\Phi_H(\omega)} &= \left[\frac{Y_G(\omega)}{X(\omega)}\right]e^{i[\Phi_Y(\omega)-\Phi_X(\omega)]} \end{aligned} \tag{3.37}$$

3.4 COMBINING LINEAR SYSTEMS IN CONTROL SCHEMES

From Eq. (3.37) it follows that in the frequency domain the amplitude characteristics simply multiply or divide, and the phase characteristics add and subtract. These simple algebraic properties turn out to be very convenient for the analysis of system circuits consisting of interconnected multiple linear subsystems. Let's briefly consider three important examples:

1. subsystems connected in series,
2. parallel subsystems, and
3. feedback systems.

In what follows, we constrain the analysis to systems of two interconnected subsystems only, designated by $h_1(\tau)$ and $h_2(\tau)$ for their impulse responses in the time domain, and by $H_1(\omega)$ and $H_2(\omega)$ for their respective transfer characteristics in the frequency domain (Fig. 3.9).

Of course, multiple systems can be connected in any combination of these three possible schemes. The quantitative analysis of such more elaborate interconnected systems will nonetheless follow the same basic computational strategies as described in the following subsection. For illustrative purposes, we will also include an analysis in the time domain for the series connection, to clarify why an analysis in the frequency domain is often to be preferred.

Series: We apply Eq. (3.36) to the series connection of Fig. 3.8A to determine the transfer characteristic of the total system. First, we see that [Eq. (3.35)] the output of the first system is given by $U(\omega) = H_1(\omega)X(\omega)$, and since this signal serves as input for the second system, $Y(\omega) = H_2(\omega)U(\omega)$. Combining the two, immediately yields the overall transfer function, $H(\omega)$:

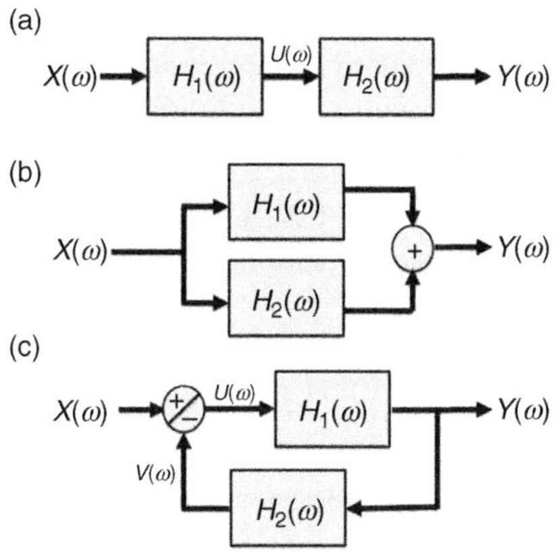

FIGURE 3.9 **There are three essentially different ways to connect two linear subsystems, with transfer characteristics $H_1(\omega)$ and $H_2(\omega)$, respectively.** (A) In series. (B) In parallel. (C) In a (negative) feedback circuit. $U(\omega)$ and $V(\omega)$ are internal, nonobservable signals.

$$H(\omega) \equiv \frac{Y(\omega)}{X(\omega)} = \frac{H_2(\omega)U(\omega)}{X(\omega)} = \frac{H_2(\omega)H_1(\omega)X(\omega)}{X(\omega)} = H_1(\omega)H_2(\omega) \tag{3.38}$$

To obtain the impulse response, $h(\tau)$, for a concatenation of LS, one then merely has to calculate the inverse Fourier transform of Eq. (3.38):

$$h(\tau) = \mathcal{F}^{-1}\left[H(\omega)\right] = \frac{1}{2\pi}\int_{-\infty}^{\infty} H(\omega)e^{i\omega\tau}d\omega \tag{3.39}$$

Alternatively, the impulse response can be found by using the convolution integral (time domain analysis). We then note that:

$$u(t) = \int_0^{+\infty} x(t-\tau)h_1(\tau)d\tau \text{ and } y(t) = \int_0^{+\infty} u(t-\sigma)h_2(\sigma)d\sigma$$

We plug the first convolution integral into the second, to obtain:

$$y(t) = \int_0^{+\infty}\int_0^{+\infty} x(t-\tau-\sigma)h_1(\tau)h_2(\sigma)d\sigma d\tau \tag{3.40}$$

To find the impulse response function, we present the Dirac pulse to the system, yielding:

$$u(t) = h_1(t) \text{ and } \quad y(t) \equiv h(t) = \int_0^{t} h_1(t-\sigma)h_2(\sigma)d\sigma \tag{3.41}$$

In Problem 3.7 the reader can work out a concrete example. Clearly, in the time domain the combined impulse response is not just a simple algebraic multiplication of the individual impulse response functions. Things get ever more tedious when more than two systems are involved in the scheme.

Parallel: The parallel system (Fig. 3.8B) is the simplest case, because the overall characteristic is found by simply adding all transfer characteristics that impinge on the same summing node:

$$H(\omega) \equiv \frac{Y(\omega)}{X(\omega)} = H_1(\omega) + H_2(\omega) \tag{3.42}$$

In a parallel system, the subsystems just add, both in the time domain, and in the frequency domain.

Feedback: In the negative feedback system of Fig. 3.8C, the computations become slightly more tedious, as we have to keep track of multiple dummy variables. Here's how this works in the frequency domain for the two systems (we don't bother to try this in the time domain). Going through the scheme of Fig. 3.8C we can read out the following statements:

1. $U(\omega) = X(\omega) - V(\omega)$
2. $Y(\omega) = H_1(\omega)U(\omega)$
3. $V(\omega) = H_2(\omega)Y(\omega)$

From this set of coupled equations, we eliminate the dummy variables, $U(\omega)$ and $V(\omega)$, respectively, after which we obtain (but check the solution):

$$H(\omega) \equiv \frac{Y(\omega)}{X(\omega)} = \frac{H_1(\omega)}{1 + H_1(\omega)H_2(\omega)} \tag{3.43}$$

There are a few points worth noting about the negative feedback system:

1. $H_1(\omega)\ H_2(\omega)$ is the *open-loop gain* of the feedback system; $\mathrm{H}_1(\omega)$ is the forward gain, and $\mathrm{H}_2(\omega)$ is the feedback gain.
2. For frequencies, ω, where the open-loop gain $\gg 1$, the transfer function becomes $\mathrm{H}(\omega) \approx 1/\mathrm{H}_2(\omega)$, which is independent of the properties of the forward subsystem, and has the inverse characteristic of the feedback gain.
3. For frequencies, ω, where the open-loop gain $= -1$, the system becomes unstable (ie, the transfer $\rightarrow \infty$).

3.5 EXAMPLES

In models of biological systems one typically encounters only a handful of basic transfer functions that often serve as building blocks for elementary subsystems within larger control schemes. It is therefore very useful to familiarize

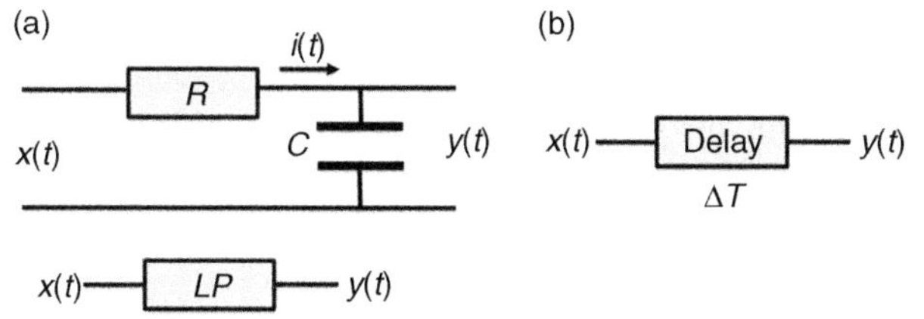

FIGURE 3.10 (A) The RC circuit acts as a linear LP filter. (B) The pure time delay, for which $y(t) = x(t - \Delta T)$.

ourselves with the essential properties of these basic building blocks, such as their impulse and step responses in the time domain, and their gain and phase characteristics in the frequency domain. Here they are:

1. The first-order low-pass (LP) filter
2. The first-order high-pass (HP) filter
3. The second-order band-pass (BP), or band-stop (BS) filter
4. The scalar gain, and the pure time delay
5. The integrator
6. The differentiator

Here, we will analyze the first-order LP filter (Fig. 3.10A), and the pure time delay (Fig. 3.10B), as illustrative examples. The reader is, however, strongly encouraged to work out the other systems in the Exercises.

LP Filter: The system of Fig. 3.10A is called a first-order LP filter because it lets harmonic inputs with a low frequency pass with a high gain (ie, amplitude transfer close to one) and only a small or negligible phase shift (ie, $y(t) \approx x(t)$), but it strongly attenuates harmonics at high frequencies (gain close to zero). Let's see why that is, by resorting to the physics of this system, where the output is the voltage over the capacitor, C, and the input is an arbitrary applied voltage. According to Kirchhoff's law, the three voltages in this circuit (input voltage, voltage across the resistor and across the capacitor) should sum to zero, or:

$$x(t) = V_R + V_C = i(t)R + \frac{q(t)}{C} \tag{3.44}$$

where $q(t)$ is the charge building up on the plates of the capacitor. Note that this is an inhomogeneous, first-order differential equation of the charge, $q(t)$:

$$x(t) = R\frac{dq}{dt} + \frac{q(t)}{C} \tag{3.45}$$

Therefore, the RC circuit indeed represents a linear system, which obeys the superposition principle and can be fully characterized by its impulse response.

To find the impulse response of this system, we should replace $x(t)$ by the Dirac impulse, but instead it's more straightforward in this case to solve Eq. (3.45) for the step response. We thus take $x(t) = V_0$ for $t \geq 0$ and 0 elsewhere. The impulse response is then obtained by differentiation of the step response [Eq. (3.24)].

To solve this inhomogeneous linear differential equation, we have to find a *particular* solution (a concrete function that fulfils the equation), and the *homogeneous* solution to the equation, and then add the two (superposition). The homogeneous solution of Eq. (3.45) is:

$$R\frac{dq_0}{dt} + \frac{q_0}{C} = 0 \Leftrightarrow q_0(t) = Q_0 e^{-(t/RC)} \tag{3.46}$$

Note that the particular solution of Eq. (3.45) would be a constant, and substitution then yields: $q_{\text{part}} = V_0 C$. The total solution for Eq. (3.45) is thus given by:

$$q(t) = Q_0 e^{-(t/RC)} + V_0 C \tag{3.47}$$

The integration constant Q_0 is to be determined from the initial condition that $q(0) = 0$. This yields $Q_0 = -V_0 C$ for the charge over the capacitor, and for its voltage (the step response of the system):

$$y(t) = \frac{q(t)}{C} = V_0[1 - e^{-(t/RC)}] \equiv V_0[1 - e^{-(t/RC)}] \tag{3.48}$$

Note that for $t \to \infty$ the step response approaches the input step value, V_0. Note also that $RC = T_{RC}$ is expressed in seconds, and is called the system's *time constant*. The time constant is the time that it takes the system to reach $[1 - (1/e)] \approx 63\%$ of the final value of the step response (or to drop by 63% of the impulse response). Thus, the two components that make up the first-order RC-system define a single parametric property: the system's time constant.

Now that we have determined the system's step response, we can readily determine its *impulse response*, by differentiation of Eq. (3.48), for $V_0 = 1$:

$$h(\tau) = \frac{1}{T_{RC}} e^{-(\tau/T_{RC})} \quad \text{for} \quad \tau \geq 0 \tag{3.49}$$

From Eq. (3.49) we can show that the tangent to $h(\tau)$ at $\tau = 0$ intersects the ordinate exactly at exactly $\tau = T_{RC}$ (Exercise).

To appreciate why this system is called LP, we have to determine its frequency behavior, expressed by its transfer characteristic. To that end, we calculate the Fourier transform of Eq. (3.49) (use causality constraint on τ):

$$H(\omega) = \mathcal{F}[h(\tau)] = \frac{1}{T_{RC}} \int_0^{\infty} e^{-\tau/T_{RC}} e^{-i\omega\tau} d\tau = \frac{1}{1 + i\omega T_{RC}} \equiv G(\omega) e^{i\Phi(\omega)} \tag{3.50}$$

where the gain and phase characteristics of the system are found to be (but check the solution):

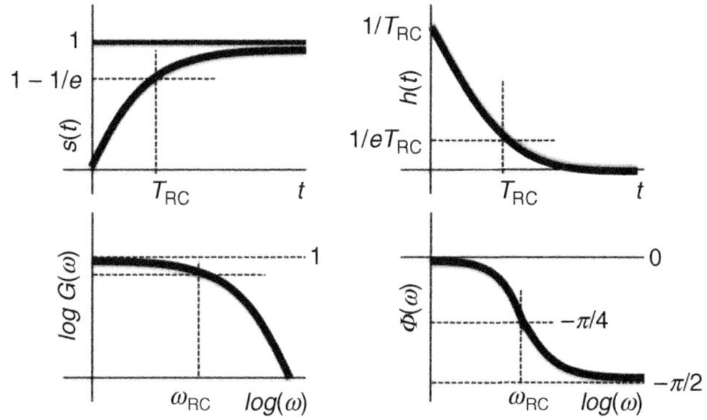

FIGURE 3.11 **Step response and impulse response (top), and gain- and phase characteristics (bottom) for the first-order LP filter of Fig. 3.10A.**

$$G(\omega)=\sqrt{\frac{1}{1+\omega^2T_{RC}^2}}\text{ and }\Phi(\omega)=-\arctan(\omega T_{RC}) \tag{3.51}$$

Indeed, for $\omega\downarrow 0$ the gain of thc system's output approaches $G = 1$, while the phase is 0. When $\omega\rightarrow\infty$ the gain $G\downarrow 0$, and the phase $\Phi\rightarrow-\pi/2$. At the system's *characteristic frequency*, defined by $\omega_{RC} = 1/T_{RC}$, the phase lag is $-\pi/4$ rad, and the gain is 0.707 ($1/\sqrt{2}$) (Fig. 3.11).

Pure time delay: Also the pure time delay, given by $y(t) = x(t-\Delta T)$, is a linear system (the reader should verify this!), so we may ask for its impulse and step response, and for the gain and phase characteristics (Fig. 10B). For the transfer characteristic, we provide the harmonic input to the system (there is no need to explicitly calculate the Fourier transform for this case). Here they are:

$$\begin{aligned}&x(t)=\delta(t)\Rightarrow h(t)=\delta(t-\Delta T) \qquad x(t)=u(t)\Rightarrow s(t)=u(t-\Delta T)\\&x(t)=\sin(\omega t)\Rightarrow y(t)=\sin(\omega t-\omega\Delta T)\quad G(\omega)=1\ \forall\omega\quad\text{and}\quad\Phi(\omega)=-\omega\Delta T\\&\Rightarrow H_{\text{delay}}(\omega)=e^{-i\omega\Delta T}\end{aligned} \tag{3.52}$$

Thus, the pure delay preserves the shape of all signals invariant (just like the scalar gain), the gain of the delay is 1 for all harmonic inputs, but the phase lag is proportional to ω, with a slope given by $-\Delta T$.

3.6 LINEAR SYSTEMS: THE LAPLACE TRANSFORM

The Fourier transform [Eq. (3.32)] requires that the function to be transformed should be integrable. For transient signals, like impulse responses, this is typically the case, but one very important problem with the Fourier transform for systems analysis, is that one of the most common signals, the unity step function, $u(t)$, has a Fourier transform that is not defined (except after using some particular integration tricks . . .):

$$U(\omega)=\int_{-\infty}^{+\infty} u(t)e^{-i\omega t}dt=\int_{0}^{+\infty} e^{-i\omega t}dt=-\frac{1}{i\omega}e^{-i\omega t}\Big]_0^{\infty}=??$$

The Fourier transform is an approximation of arbitrary (integrable) functions with the use of harmonic functions; the complex-exponential weighting function, $e^{i\omega t}$, serves as an orthogonal basis along which the function is expanded. However, in the case of nonintegrable functions, like the unity step, the approximation doesn't work.

It is of course possible to use a different set of orthogonal basis functions to approximate a much larger class of functions. The *Laplace transform* is such an extension of the Fourier transform: instead of relying solely on purely imaginary exponentials (ie, sines and cosines), it uses exponentially decaying harmonics in which the real part of the exponential weighting factor, $\exp(-\sigma t)$, will force many functions to zero for large t:

$$\begin{aligned}\text{Fourier:}\quad & e^{-i\omega t}=\cos(\omega t)-i\sin(\omega t)\\ \text{Laplace:}\quad & e^{-st}=e^{-(\sigma+i\omega t)}=e^{-\sigma t}[\cos(\omega t)-i\sin(\omega t)]\end{aligned} \tag{3.53}$$

The Laplace transform (LT) of a function is defined as follows:

$$\mathcal{L}[f(t)]\equiv F(s)=\int_0^{\infty} f(t)e^{-st}dt \tag{3.54}$$

Note that the LT only uses functions $f(t)$ for $t \geq 0$, which is not a serious constraint for the analysis of causal systems. To visualize the basis functions, $\exp(-st)$, of the LT, Fig. 3.12 shows how they are distributed across the complex s-plane.

Every point in the $s = (\sigma, i\omega)$-plane represents a waveform, $\exp(st)$, as shown. In the right-half plane $\sigma > 0$, and $\exp(\sigma t)$ blows up; in the left-half of the plane ($\sigma < 0$) the waveforms decay exponentially fast to zero. The imaginary ω-axis represents the pure sines and cosines, like in the Fourier transform.

Apart from solving numerical problems for the Fourier transform (integrability), we will see that the LT is also particularly suited to perform systems analysis; the algebraic calculations result to be quite straightforward, and it is relatively easy to use the LT to quickly identify the transfer characteristic, $H(\omega)$, in the frequency domain, and the impulse response function, $h(\tau)$, in the time domain.

For our purpose of LS analysis we don't need to go into the full mathematical background of integrating complex-valued functions. It will be sufficient to get familiar with a couple of basic rules, and some example functions that will be encountered frequently in models of the auditory system and of neurobiological

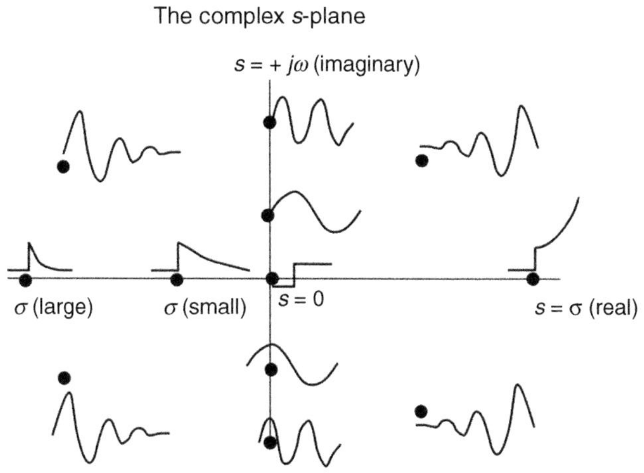

FIGURE 3.12 **Exponentially-weighted sine waves represented in the complex *s*-plane.**

systems in general. Therefore, to get a basic quantitative feeling about LT, we will first calculate a few simple examples, and then proceed with a number of important analytical rules.

Okay, having failed to compute the FT of the unitary step function, then what is its LT? Let's perform the integration:

$$U(s) = \int_0^\infty u(t)e^{-st}\,dt = -\frac{1}{s}e^{-st}\Big]_0^\infty = \frac{1}{s} \quad \text{done!} \tag{3.55}$$

A first important rule is that if we have the transform for one particular waveform (here, the step), we have it for all linearly related waveforms such as its derivative and its integral. For the derivative, use partial integration, and find:

$$\mathcal{L}\left[\frac{df}{dt}\right] = \int_0^\infty \frac{df}{dt}e^{-st}\,dt = \int_0^\infty e^{-st}\,df(t) = sF(s) - f(0_+) \tag{3.56}$$

with $f(0+)$ the function's initial condition.

In short: to differentiate, multiply by s, and to integrate, multiply by $1/s$.

As a result, the LT of the Dirac delta pulse, which is the derivative of the step function, is simply $F(s) = 1$. Conversely, the LT of the ramp, which is the integral of the step, is $F(s) = 1/s^2$.

Like for the FT (Eq. (3.32)), it can now be readily verified that the LT of the output of a linear system [the convolution integral, Eq. (3.19)] is also found by a simple algebraic multiplication of the LT of the input signal with the LT of the system's impulse response:

$$Y(s) = H(s)X(s) \Leftrightarrow H(s) = \frac{Y(s)}{X(s)} \tag{3.57}$$

Another very common signal in system's analysis is the exponential decay, $x(t) = \exp(-at)$, which we already encountered as the impulse response function of the first-order LP RC-system (see Section 3.5). To find its LT:

$$\mathcal{L}\left[e^{-at}\right] = \int_0^\infty e^{-at} e^{-st}\,dt = \frac{1}{s+a} \tag{3.58}$$

To use this new concept, let's get back to the first-order LP filter once more. Its impulse response, and from Eq. (3.58) its associated LT are given by:

$$h(\tau) = \frac{1}{T_{RC}} e^{-(t/T_{RC})} \Rightarrow H_{RC}(s) = \frac{1}{1+sT_{RC}} \tag{3.59}$$

We didn't seem to have gained anything new, but note that we can now readily determine the system's step response in the Laplace domain, by applying Eq. (3.57):

$$S_{RC}(s) = \frac{1}{1+sT_{RC}} \times \frac{1}{s} \tag{3.60}$$

In a moment (see poles and zeros) we will demonstrate how we can obtain the system's transfer characteristic, $H_{RC}(\omega)$, in an elegant and simple way from this LT. To find the step response in the time domain, we only need to find the inverse LT of Eq. (3.60). Here we won't go into the mathematical details of having to compute inverse LTs by integrating in the complex plane. Instead, we use a simple trick, by rewriting the Laplace function into familiar terms for which we already know the time-evolution. In this case, we use basic algebra:

$$\begin{gathered} S_{RC}(s) = \frac{A}{s} + \frac{B}{1+sT_{RC}} \Leftrightarrow A = 1 \text{ and } B = -T_{RC} \\ S_{RC}(s) = \frac{1}{s} - \frac{1}{s+1/T_{RC}} \end{gathered} \tag{3.61}$$

The inverse transformation now results to be almost trivial, since we can immediately use Eqs. (3.55) and (3.58):

$$s_{RC}(t) = u(t) - e^{-(t/T_{RC})} = 1 - e^{-(t/T_{RC})} \quad \text{for} \quad t \geq 0 \tag{3.62}$$

which is indeed the LP filter's step response.

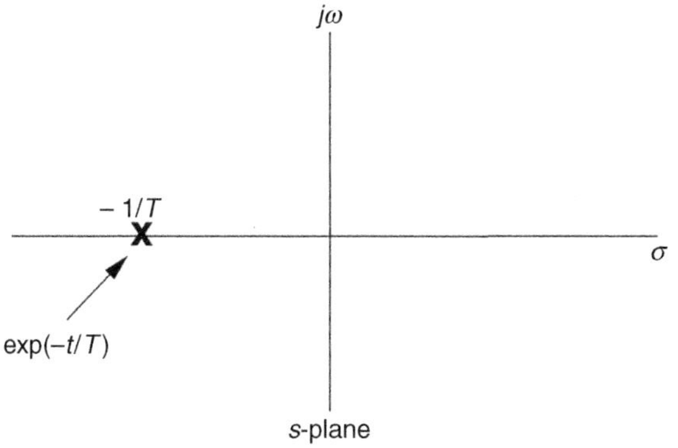

FIGURE 3.13 **The real pole of the first-order LP RC-filter corresponds to the exponential decay (see Fig. 3.11).**

Poles and zeros: We will now introduce a very useful graphic aid for linear system's analysis, which is the use of the *poles and zeros* of the LT, $H(s)$.

We first illustrate the poles-zeros concept for the simple first-order LP filter, Eq. (3.59) (Fig. 3.13). We note that the, $H_{RC}(s) = 1/(sT_{RC} + 1)$, of its impulse response function becomes infinite when $s \to -1/T_{RC}$. We mark this point, called a *pole*, on the real negative σ-axis of the s-plane, with a cross (X).

Interestingly, from Fig. 3.13, this pole can be seen to correspond to the system's *impulse response* signal, $\exp(-t/T_{RC})$ *in the time domain*! As a result, and without having to do any tedious calculations, the location of the pole immediately tells us everything there is to know about this system, as it totally characterizes its dynamics through Eq. (3.19). Hence, it is not surprising that a pole-zero analysis is a popular tool for LS analysis.

The second reason why the pole–zero diagram is so useful for linear system's identification is its direct relation to the system's frequency transfer characteristic, $H(\omega)$. Fourier analysis measures the system's response to harmonics of different frequencies. In the Laplace s-plane the pure harmonics are located on the imaginary axis: $\exp(i\omega)$.

To find the system's response to a harmonic of arbitrary frequency, we choose a test frequency, say ω_0, and determine its distance, $\rho(\omega_0)$, and angle with the positive σ-axis, $\phi(\omega_0)$, to all poles and zeros in the s-plane.

Here's how it works for the first-order LP system: the distance, $\rho(\omega_0)$, between the pole and the test frequency (Fig. 3.14) is:

$$\begin{aligned}\rho_{\text{pole}}(\omega_0) &= \sqrt{\left[\omega_0^2 + \left(\frac{1}{T_{RC}}\right)^2\right]} = \frac{1}{G_{RC}(\omega_0)} \\ \Theta_{\text{pole}}(\omega_0) &= \arctan\left[\frac{\omega}{(1/T_{RC})}\right] = \arctan(\omega T_{RC}) = -\Phi_{RC}(\omega T_{RC})\end{aligned} \tag{3.63}$$

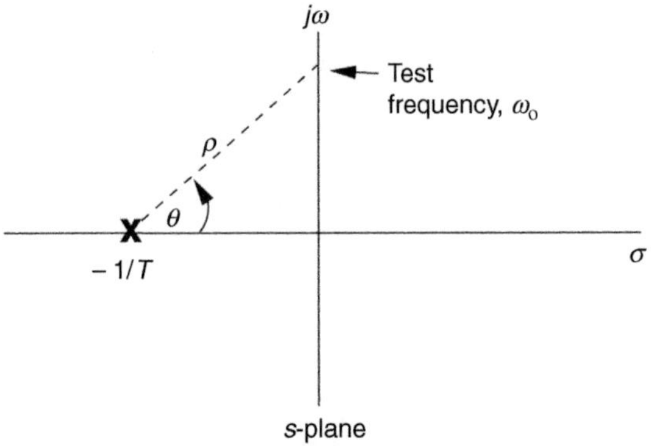

FIGURE 3.14 Finding the gain and phase characteristic of the LP filter with the help of a test frequency on the imaginary axis.

We note [cf. Eq. (3.48)] that the distance of the pole to the test frequency is exactly the *inverse* of the LP filter *gain*, while the angle equals *minus* the system's *phase response*! As the test frequency moves up (higher frequencies), the distance increases, and hence the gain decreases, while the angle approaches $\pi/2$. In the Exercises, you will encounter more examples of basic LS to analyze, such as the HP filter.

For a zero (indicated by an *O* in the *s*-plane) in the Laplace transfer characteristic, the distance $\rho_0(\omega)$ is equal to the gain, whereas the angle, $\Theta_0(\omega)$ equals the system's phase.

Systems with more poles and zeros: From the derivative properties of the LT, it is straightforward to obtain the system's transfer characteristic, $H(s)$, in case a linear differential equation can be specified for the system. Indeed, if functions, $f(t)$ and $g(t)$ have LTs, $F(s)$ and $G(s)$, respectively, the derivative df/dt yields $sF(s)$, the second-order derivative $s^2F(s)$, etc. so that the *n*th order derivative gives $s^nF(s)$.

Consider, for example, the driven, damped harmonic oscillator, Eq. (3.3), for which we can write the corresponding LT:

$$m\frac{d^2y}{dt^2}+c\frac{dy}{dt}+Ky=x(t) \underset{LT}{\Leftrightarrow} ms^2Y(s)+csY(s)+KY(s)=X(s) \quad (3.64)$$

We can thus immediately give the system's Laplace transfer characteristic:

$$H(s)=\frac{Y(s)}{X(s)}=\frac{1}{ms^2+cs+K} \quad (3.65)$$

Suppose that the system's parameters, *m*, *c*, and *K* are such that the quadratic denominator in Eq. (3.65) has two real, negative solutions, which is the case when $c>2\sqrt{mK}$. This is the over-damped situation, defining two exponential time constants in the solution. The two poles (and corresponding time constants) are then found at:

$$c_{1,2} = -\frac{c}{2m} \pm \frac{\sqrt{c^2 - 4mK}}{2m} \equiv \left[-\frac{1}{T_1}, -\frac{1}{T_2} \right], \quad \text{with} \quad T_1 > T_2 \tag{3.66}$$

which directly provides the system's impulse response [it's a second-order system, and is described by the two first-order LP filters in series, eg, Eq. (3.41)].

The corresponding frequency transfer characteristic is then calculated by:

$$G(\omega) = \frac{1}{\rho_1(\omega)\rho_2(\omega)} \quad \text{and} \quad \Phi(\omega) = -\Theta_1(\omega) - \Theta_2(\omega) \tag{3.67}$$

We leave the details of these formulae as an exercise to the reader. But it should be clear by now that the LT provides us with a very convenient method toward full LS identification!

On the other hand, if $c < 2\sqrt{mK}$ the solutions will appear as damped oscillations:

$$c_{1,2} = -\frac{c}{2m} \pm i\frac{\sqrt{4mK - c^2}}{2m} \equiv -q \pm i\omega_1 \tag{3.68}$$

which are two complex-conjugate points in the s-plane.

Without proof, but directly following from the previous examples, we here pose that the Laplace transfer characteristic of arbitrary complex LS will have the following general appearance:

$$H(s) = \frac{a_n s^n + a_{n-1}s^{n-1} + a_{n-2}s^{n-2} + \cdots + a_1 s + a_0}{b_m s^m + b_{m-1}s^{m-1} + b_{m-2}s^{m-2} + \cdots + b_1 s + b_0} \tag{3.69}$$

which is a ratio of two polynomials in s. The numerator will have n zeros; let's call them $-z_1$, $-z_2$, ... $-z_n$. The denominator will have m poles, $-p_1, -p_2, \ldots, -p_m$. So, it is possible to write Eq. (3.69) as follows:

$$H(s) = \frac{(s + z_1)(s + z_2)\cdot\cdots\cdot(s + z_n)}{(s + p_1)(s + p_2)\cdot\cdots\cdot(s + p_m)} \tag{3.70}$$

Graphically, we can represent this by a pole–zero diagram as in Fig. 3.15. Poles and zeros lying off the real axis are complex, and will always occur in conjugate pairs (as in the example of the harmonic oscillator). In this way, the imaginary parts will cancel out (opposite phase), leaving only the real (oscillatory) waveform in the time domain.

From the general Laplace transfer characteristic, Eq. (3.70), we can now readily specify the corresponding frequency transfer characteristic:

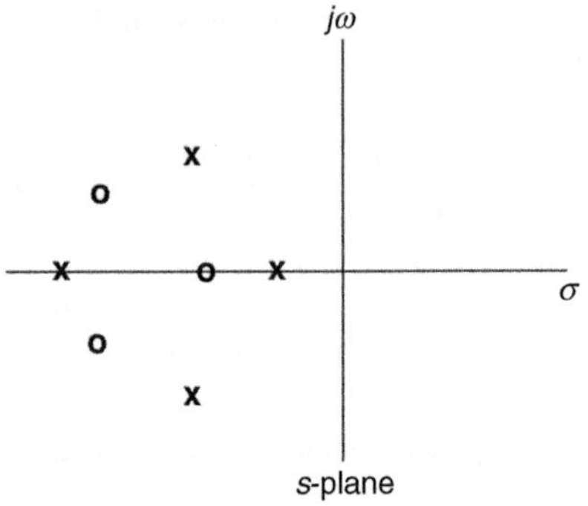

FIGURE 3.15 **Pole–zero diagram of a more general linear system. Poles (*X*) and zeros (*O*) off the real axis always appear as conjugate pairs.**

$$G(\omega)=\frac{\prod_{k=1}^{n}\rho_{\mathrm{Zero},k}(\omega)}{\prod_{k=1}^{m}\rho_{\mathrm{Pole},k}(\omega)} \quad \text{and} \quad \phi(\omega)=\sum_{k=1}^{n}\Theta_{\mathrm{Zero},k}(\omega)-\sum_{k=1}^{m}\Theta_{\mathrm{Pole},k}(\omega) \quad (3.71)$$

That the gains multiply/divide and the phases add/subtract is a direct consequence of the properties of gain and phase as complex numbers:

$$H(\omega)=G(\omega)e^{i\Phi(\omega)}$$

In summary, analysis of the Laplace transfer characteristic with the aid of a poles-zeros plot allows for a rapid, graphical, and complete assessment of a linear system's properties, and provides an efficient route to calculate its response to a variety of stimuli without the need for tedious, technical computations.

Pole-zero analysis of a feedback system: As a final example of an analysis with the help of a zero-pole diagram, we consider the negative feedback system of Fig. 3.16. The feed-forward system is a first-order linear LP filter with time constant T, and the feedback is provided by a simple scalar gain, k, which can be varied.

The transfer function of the total system [response, $R(s)$, divided by the input command] $C(s)$, is:

$$H(s)=\frac{R(s)}{C(s)}=\frac{A/(1+sT)}{1+k[A/(1+sT)]}=\frac{A}{1+kA}\frac{1}{1+s[T/(1+kA)]}=\frac{A'}{1+sT'} \quad (3.72)$$

where we rewrote the transfer function such as to get it into the familiar form of $1/(sT' + 1)$. Now use the pole–zero diagram to see what happens when k increases from zero (= no feedback). The pole of the transfer characteristic is found at $\sigma = -(1 + Ak)/T$, which depends on the gain of the feedback loop!

When $k = 0$, the pole is at $-1/T$ (the regular LP system). As k increases, the pole becomes more negative, and slides out along the negative real axis. In other

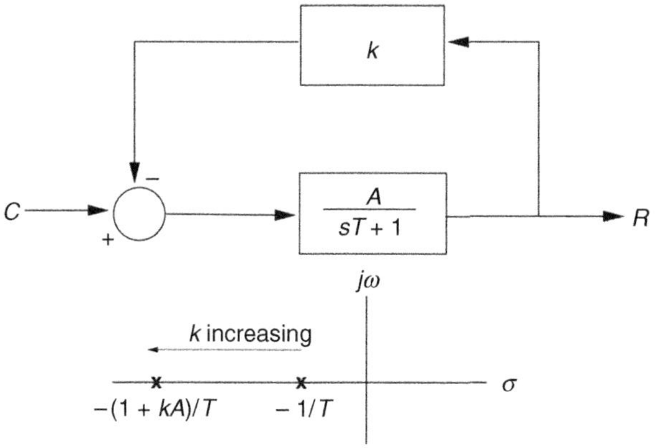

FIGURE 3.16 A sluggish LP filter within a feedback system with variable gain, *k*. As *k* increases, the time constant of the total system (ie, the location of the pole in the system's LT) decreases, leading to a faster system.

words: the system becomes *faster* (a shorter time constant, $T' = T/(1 + Ak)$), but note that the gain, A', also decreases.

3.7 CORRELATION FUNCTIONS AND GAUSSIAN WHITE NOISE

Auto- and cross-correlation: We have discussed several methods to obtain the system's impulse response, $h(\tau)$, by driving the system with different types of stimuli, like the Dirac pulse, the unity step, and the harmonic function [yielding the system's frequency transfer characteristic, $H(\omega)$]. The LT, $H(s)$, provides a convenient analytic tool to obtain both the impulse response function, and the system's frequency response, using relatively simple algebraic and graphical means.

One other useful signal for systems analysis, which we haven't discussed so far, is Gaussian white noise (GWN). Because of its importance also for nonlinear systems identification, we will introduce it here. To that end, we define the notion of auto-correlation and cross-correlation functions for stochastic variables.

Suppose, signals $x(t)$ and $y(t)$ are stochastic variables, which means that their values are drawn from a probability distribution. The expectation value (the average) of a stochastic signal is given by

$$E[x(t)] = \lim_{T\to\infty} \frac{1}{2T} \int_{-T}^{+T} x(t)dt \tag{3.73}$$

We can now define the (second-order) autocorrelation function of a signal as the expectation value of $x(t)$ multiplied with its time-shifted version, $x(t + \tau)$:

$$\Phi_{xx}(\tau) \equiv E[x(t)x(t+\tau)] = \lim_{T\to\infty} \frac{1}{2T} \int_{-T}^{+T} x(t)x(t+\tau)dt \tag{3.74}$$

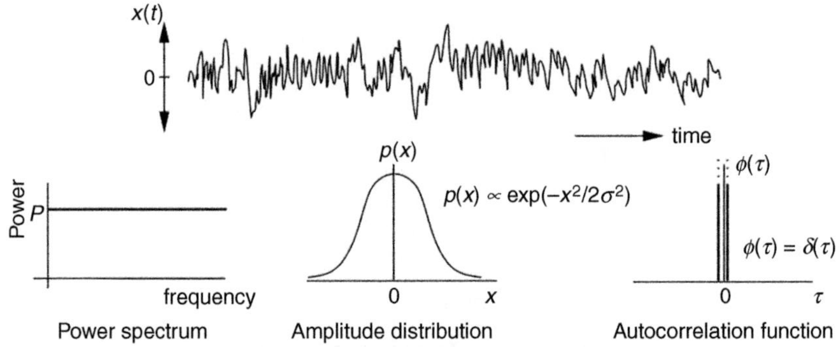

FIGURE 3.17 GWN, and some of its most important properties: a flat power spectrum, at power *P*, a Gaussian amplitude distribution with variance *P*, and the Dirac-delta function as its autocorrelation function.

and, likewise, the (second-order) cross-correlation function between signals $x(t)$ and $y(t)$ is:

$$\Phi_{yx}(\tau) \equiv E\left[y(t)x(t+\tau)\right] = \lim_{T\to\infty} \frac{1}{2T} \int_{-T}^{+T} y(t)x(t+\tau)dt \tag{3.75}$$

The auto- and cross-correlation functions measure how well a signal resembles itself, or some other signal. Note that the type of signal determines the integration limits for the expectation value: we distinguish stochastic, versus periodic, or transient signals (Marmarelis and Marmarelis, 1978; Oppenheim et al., 1983).

For a periodic function, period T, the integration is performed over $t \in [0,T]$, and normalized by the signal's period T; for a transient signal the integration is performed over $t \in [-\infty,+\infty]$, and is not normalized.

The autocorrelation function is always maximal at $\tau = 0$ (and for periodic signals, also for $\pm T$, $\pm 2T$, etc.; see Problem 3.13), and it is symmetrical in its argument, $\Phi_{xx}(\tau) = \Phi_{xx}(-\tau)$.

GWN is a stochastic signal that has a Gaussian amplitude distribution, with average 0, and variance, $\sigma^2 = P$ (Fig. 3.17). The latter is also called the power of the GWN signal. For GWN the auto-correlation function has the very important property (Exercise) that:

$$\text{for} \quad x(t) = \textbf{GWN}: \Phi_{xx}(\tau) = P\delta(\tau) \tag{3.76}$$

We can now use the auto-correlation property of GWN, Eq. (3.76), to determine the impulse response of a linear system! We therefore provide GWN to the system, and calculate the second-order cross-correlation function between the system's output [determined by the convolution integral, Eq. (3.19)], and the input, as follows:

$$\Phi_{yx}(\tau) = \int_{-\infty}^{+\infty} y(t)x(t+\tau)dt = \int_{-\infty}^{+\infty} \left[\int_{0}^{\infty} h(\sigma)x(t-\sigma)d\sigma \right] x(t+\tau)dt \tag{3.77}$$

We notice that the expectation value integrates over time, t, so that the order of the integrals may be reversed:

$$\Phi_{yx}(\tau) = \int_0^{\infty} h(\sigma)\left[\int_{-\infty}^{+\infty} x(t-\sigma)x(t+\tau)dt\right]d\sigma = \int_0^{\infty} h(\sigma)P\delta(\tau-\sigma)d\sigma = Ph(\tau) \quad (3.78)$$

where in the last step we inserted the autocorrelation property of GWN. In other words, to determine the impulse response of a linear system, we simply calculate:

$$h(\tau) = \frac{\Phi_{yx}(\tau)}{P} \quad (3.79)$$

As a final property of auto- and cross-correlation functions, we note their relations to a linear system's transfer characteristic in the frequency domain, by taking the Fourier transform of Eq. (3.77) [note the analogy to Eq. (3.35)]

$$\Phi_{YX}(\omega) = H(\omega)\Phi_{XX}(\omega) \quad \text{and} \quad \Phi_{YY}(\omega) = \left\|H(\omega)^2\right\|\Phi_{XX}(\omega) \quad (3.80)$$

where we see that the auto-spectra (the *power spectral densities*) are real-valued functions (eg, Hartmann, 1998).

The harmonic function (sinusoid) is the only deterministic signal that is preserved by LS. Interestingly, for Gaussian signals, the following observation (given without proof) holds:

> If the input to a linear system is Gaussian, then its output is also Gaussian.

3.8 EXERCISES

Problem 3.1: *linear!* Verify the following important corollary of the superposition principle, Eq. (3.8): suppose a linear system is stimulated with step inputs of different amplitudes, that is;

$$x(t) = AU(t) \quad \text{with} \quad A \in \mathbb{R}$$

and that the unit-step response of the system (to $A = 1$) is given by $s(t)$. After a finite time (defined as the response duration, D) the step response becomes (and remains) constant (ie, $ds/dt = 0, \forall t > D$). Then for any linear system:

(a) D is independent of input amplitude A.
(b) The peak velocity of the step response increases linearly with A.

Problem 3.2: *linear?* A ball, dropped at $t = 0$ from height h_0 to the earth, obeys the following linear differential equation:

$$F = mg = m\frac{d^2y}{dt^2}$$

Verify that the solution for this problem is:

$$y(t) = h_0 + v_0 t - \frac{1}{2}gt^2$$

The duration of the fall toward the earth's surface is then determined by (suppose that the initial velocity $v_0 = 0$):

$$D = \sqrt{\frac{2h_0}{g}}$$

So, although the system is described by a linear differential equation, the response duration depends on the input amplitude, initial height h_0. Thus, in line with what we just have seen in Problem 3.1, we have to consider the system to be *nonlinear*! Do you agree with this statement? Why (not)?

Problem 3.3: HP filter. Analogous to the analysis of the LP filter, described in Section 3.4, we here consider the first-order HP filter. This system can simply be modeled by taking the voltage across the resistor as output of the RC-circuit of Fig. 3.10A.

(a) Show that its impulse response is given by:

$$h_{HP}(\tau) = \delta(\tau) - \frac{1}{T_{HP}}\exp\left(-\frac{\tau}{T_{HP}}\right) \text{with } T_{HP} = RC, \text{ and } \tau \geq 0$$

Give also the HP step response.

(b) Show that its gain and phase characteristics are given by:

$$G_{HP}(\omega) = \frac{\omega T_{HP}}{\sqrt{1+\omega^2 T_{HP}^2}} \quad \text{and} \quad \Phi(\omega) = \arctan\left(\frac{1}{\omega T_{HP}}\right)$$

(c) Also note from Fig. 3.10A that HP = all-pass – LP (from Kirchhoff's law) which provides a shortcut to obtain the results of (1) and (2).

(d) The slow-phase eye-movement response of the vestibular ocular reflex is well approximated by such a filter. The time constant of the vestibular ocular reflex is about $T_V \sim 20$ s. Plot the characteristics (as a Bode plot) and verify its HP properties.

Problem 3.4: Integrator. Without having to use advanced mathematics and Fourier analysis it is sometimes straightforward to compute the transfer characteristic of LS. Try the following example, which is the pure integrator:

$$y(t) = \int_0^t x(\tau)d\tau$$

(a) Show that this is indeed a linear system.
(b) What is its impulse response?
(c) What is the step response?
(d) Determine amplitude and phase characteristic of this system (no need for FT!)
(e) Compare to the first-order LP system. Why is the LP system of Fig. 3.10A also
(f) called a *"leaky integrator?"*

Problem 3.5: *Differentiator.* Consider the differentiator:

$$y(t) = \frac{dx}{dt}$$

(a) Answer the same questions (1–4) as in Problem 3.4.
(b) Compare the differentiator with the HP system of Problem 3.3.

Problem 3.6: *Band pass and band stop.*

(a) Investigate how to create either a BP or a BS second-order filter by combining a first-order LP filter (time constant = T_{LP}) with a first-order HP filter (time constant = T_{HP}). How would you create an allpass filter with the LP and HP filters?
(b) Determine the gain and phase characteristics for the BP, BS, and AP filters.

Problem 3.7: *Series concatenation.*

(a) Calculate the impulse response function by using convolution for the series concatenation of two first-order LP filters, described by:

$$h_1(\tau) = \frac{\exp[-(\tau/T_1)]}{T_1} \quad \text{and} \quad h_2(\tau) = \frac{\exp[-(\tau/T_2)]}{T_2} \quad \text{with} \quad T_1 > T$$

Plot this impulse response function.
(b) Determine (and plot) the transfer characteristic of this system.

Problem 3.8: *Series.* Calculate the impulse response, step response, amplitude and phase characteristics of the series concatenation of a pure integrator, followed by a scalar gain, G, and a pure time delay of T s.

Problem 3.9: *Feedback.* Consider the circuit in Fig. 3.18. Each of the four subsystems is linear, and their transfer characteristics are given by $H_1(\omega)$, $H_2(\omega)$, $H_3(\omega)$, and $H_4(\omega)$, respectively. Suppose that $H_2(\omega)$ is a pure integrator.

(a) Determine the transfer characteristic of the total system. Your answer may only contain the characteristics of the four subsystems. Express also the

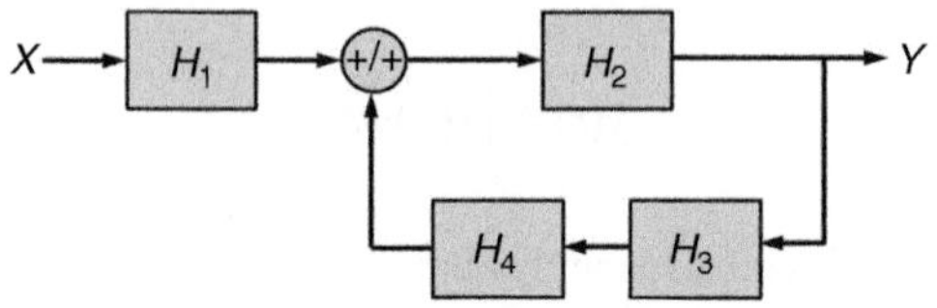

FIGURE 3.18 **A positive feedback scheme consisting of four linear subsystems.**

amplitude and phase characteristics of the total transfer as a function of the amplitude and phase characteristics of the subsystems.

(b) When will the system become unstable?

(c) Suppose that $H_3(\omega)$ is a constant gain, G. Which requirements should hold for system $H_4(\omega)$ to keep the total system unstable?

(d) Suppose that also $H_4(\omega)$ is a constant gain, A. For which frequency will the system be unstable?

Problem 3.10: *Difference differentiation algorithm.* In a computer, you would approximate the ideal differentiator of Problem 3.5 by the difference algorithm:

$$\dot{y}(t) \approx \frac{y(t+\Delta T) - y(t)}{\Delta T}$$

(a) However by doing so, you filter the original signal. Show this by computing the transfer characteristic of this operation.

(b) To reduce noise in an analog signal we add a LP filter (time constant T) in series with a pure differentiator (from Problem 3.5). Determine the transfer characteristic of the total system; give the Bode plot, and analyze how the results depend on T.

Problem 3.11: *Feedback.* Here we analyze the influence of a delay on the transfer characteristic of a linear system with feedback (Fig. 3.19).

(a) Determine the LT of a pure delay: $y(t) = x(t - \Delta T)$, and from that the transfer characteristic (in the frequency domain) of the delay.

(b) Take the system in Fig. 3.19. Determine the total transfer function and the loop gain. The system will spontaneously oscillate, and become unstable, when the loop gain exceeds the value of 1, and at the same time has a phase shift of $-180°$. Make a Bode analysis of the system and estimate the frequency ω_0 where instability kicks in.

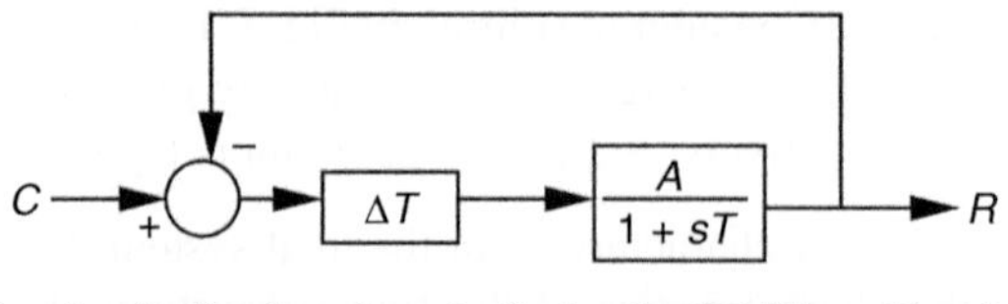

FIGURE 3.19 **Negative feedback system of a first-order LP filter with a delay.**

(c) What happens to the system if A is increased/lowered? What if the time constant T is increased/lowered?

**Problem 3.12*: *Echoes.* Consider the situation of the free field with a sound–source and a microphone. Without reflectors in the environment the microphone will record the sound pressure signal $p(t)$. Suppose the presence of a reflector, which produces an echo of the sound–source. Assume that the echo is a filtered [impulse response $a(\tau)$] and delayed (delay ΔT) version of the original sound, $p(t)$.

(a) Write a time-domain expression for the signal $p^*(t)$ that is picked up by the microphone.
(b) Determine the transfer characteristic that relates the original input sound spectrum, $P(\omega)$, to the measured spectrum with echo $P^*(\omega)$:

$$T(\omega) \equiv P^*(\omega) / P(\omega)$$

(c) The same as in (a) and (b) for the situation of N different reflectors, each with their own impulse response, $a_k(\tau)$ and delay, ΔT_k.
(d) For $N = 1$, how can the distance to the reflector be determined from the power spectrum of the transfer characteristic, $|T(\omega)|^2$?
(e) Draw the power spectrum of white noise with an echo having a 1 ms delay.

Problem 3.13: *Autocorrelation.*
Calculate the auto-correlation functions for the following examples:

(a) $x(t) = P \text{ for } -T \leq t \leq +T \text{ and } x(t) = 0 \text{ elsewhere}$

(b) $x(t) = A\cos\omega t$

(c) $x(t) = A\exp\left(-\frac{t}{T}\right) \text{for } t \geq 0 \text{ and } x(t) = 0 \text{ elsewhere}$

(d) $x(t) = \begin{cases} A\left[\text{sgn(t)}1 - \frac{t}{T}\right] \text{for } -T \leq t \leq +T \\ 0 \quad \text{elsewhere} \end{cases}$

**Problem 3.14*: *Auto- and cross-correlation functions.* For the simple LP filter of Fig. 3.20 we have seen that the relation between the input ($x(t)$ = GWN) and output, $y(t)$, is given by:

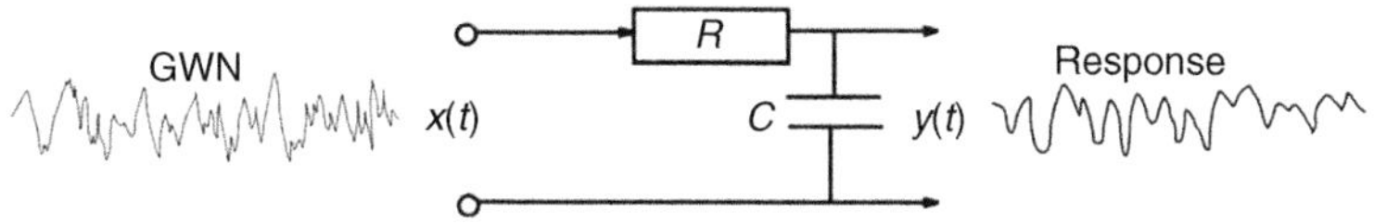

FIGURE 3.20 First-order LP filter with GWN as input.

$$y(t) = \int_0^\infty \frac{1}{RC} \exp\left(-\frac{\tau}{RC}\right) x(t-\tau) d\tau$$

(a) Determine the cross-correlation function $\phi_{yx}(\sigma)$.
(b) Determine the auto-correlation function of the output $\phi_{yy}(\sigma)$.
(c) Determine the response, $y(t)$, if $x(t)$ = the pulse from Problem 3.13a.
(d) Determine the cross-correlation function $\phi_{yx}(\sigma)$ for the pulse input.

REFERENCES

Bechhoefer, J., 2005. Feedback for physicists: a tutorial essay on control. Rev. Mod. Phys. 77, 783–836.

Hartmann, W.M., 1998. Signals, Sound, and Sensation. Springer, New York.

Marmarelis, P.Z., Marmarelis, V.Z., 1978. Analysis of Physiological Signals: the White Noise approach. Plenum Publishing Corp., New York.

Oppenheim, A.V., Willsky, A.S., Young, I.T., 1983. Signals and Systems. Prentice-Hall, Englewood Cliffs, NJ.

Chapter 4

Nonlinear Systems

4.1 NONLINEAR SYSTEMS IDENTIFICATION

We have seen in chapter: Linear Systems Analysis that the analytical advantages of linear systems theory are obvious. We showed that one well-chosen input stimulus, be it the Dirac impulse, Gaussian white noise (GWN), or the harmonic function, can fully characterize the linear system, as the resulting convolution integral (ie, impulse response kernel), the cross-correlation function, and the frequency transfer characteristic all followed directly from the superposition principle of Eq. (3.10) (eg, Bechhoefer, 2005).

Unfortunately, however, the linear systems approach may often prove to be unrealistic in practice, as nonlinearities are abundant in everyday life, and this is certainly true for most biological and physiological systems. For example, nonlinear, logarithmic transformations of sensory inputs, like acoustic power and sound frequency at the cochlea, or light intensity and mean cell density in the retina, allow sensory systems to respond to a much larger range of stimuli than achieved by linear representations. In addition, neurons typically respond in a nonlinear, saturating way to their inputs: no response for inputs below a certain threshold, and a maximum response for strong stimuli. Clearly, for such nonlinear systems, the superposition principle of Eq. (3.10) no longer holds.

In that case the response of the system to a harmonic input will not guarantee to be the same harmonic, with only a frequency-dependent change in its amplitude and phase (see Section 3.3).

For example, consider the simple static (instantaneous, ie, zero-memory) third-order nonlinearity of Fig. 4.1, described by:

$$y(t) = ax(t) + bx^2(t) + cx^3(t) \tag{4.1}$$

and that is stimulated with a superposition of two sinusoids with different frequencies: $x(t) = \sin(\omega_1 t) + \sin(\omega_2 t)$, with $\omega_1 > \omega_2$. It can be readily shown that the output spectrum of this system then consists of the following 13 frequency components (Problem 4.1):

$$\begin{aligned}\{\omega\} = [&0, \omega_1, \omega_2, 2\omega_1, 2\omega_1, (\omega_1 + \omega_2), (\omega_1 - \omega_2)\\ &3\omega_1, 3\omega_2, (2\omega_1 + \omega_2), (2\omega_2 - \omega_1), (2\omega_1 - \omega_2), (2\omega_2 + \omega_1)]\end{aligned} \tag{4.2}$$

The Auditory System and Human Sound-Localization Behavior. http://dx.doi.org/10.1016/B978-0-12-801529-2.00004-0

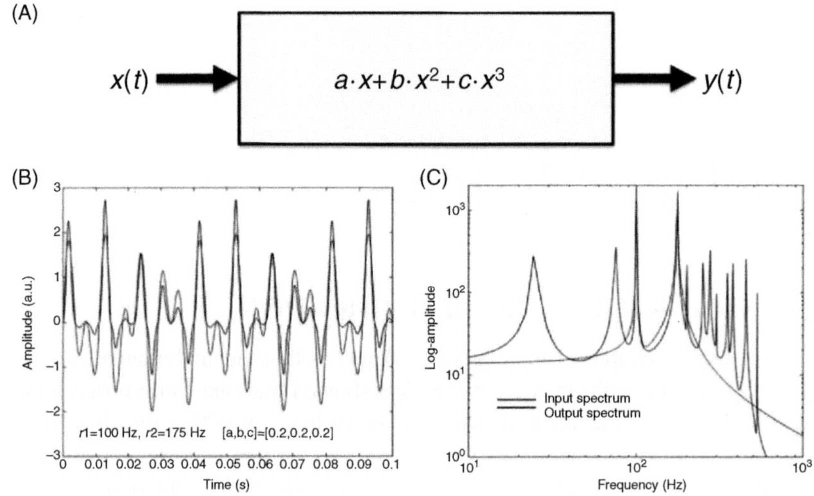

FIGURE 4.1 (A) A third-order nonlinear system is driven by the superposition of two different harmonics (B), each with amplitude 1.0 and frequencies of 100 and 175 Hz, respectively. Red: total input signal; black: system's output, when $a = b = c = 0.2$. (C) The output spectrum of the system (black) contains 13 spectral components (the dc component is not shown).

The six frequencies in the second row of Eq. (4.2) are all due to the third-order nonlinearity, the original frequencies stem from the linear term, while the dc, and the remaining four frequencies in the first row are all caused by the second-order nonlinearity. Fig. 4.1B,C shows a simulation of this nonlinearity, for the case that $a = b = c = 0.2$, and two input frequencies of $f_1 = 100$ Hz and $f_2 = 175$ Hz, respectively.

Because superposition is violated for such systems, so are the convenient consequences that followed from that principle. As a result, the impulse response of a nonlinear system will not allow one to predict the response of the system to arbitrary inputs through the convolution integral of Eq. (3.19). Furthermore, Eq. (4.2) demonstrates that the input and output spectra of nonlinear systems may differ profoundly.

The question therefore arises whether there exists an appropriate test stimulus to efficiently identify the properties of a nonlinear system. This is not a trivial point: systems can be linear in only one way (they should obey the superposition principle), but there are infinitely many ways in which a system may violate linearity.

However, as will be argued in further sections, for many situations GWN could be such a stimulus. GWN has the interesting property that when it is presented for a sufficiently long time, there is a finite probability that any imaginable stimulus is approached to arbitrary accuracy by a particular noise sample (over a finite time interval). As a result, the system is in fact tested with every possible stimulus waveform, as GWN is a *complete* stimulus. Then, two systems will only be identical when they respond to GWN in the same way. The purpose of nonlinear systems identification therefore is to find a mathematical model that responds in exactly the same way to GWN as the real system. Using GWN in systems identification is called *the GWN-method*; the method can be applied to a broad class of systems, provided that the system:

- is time invariant
- has a finite memory (response duration to a transient stimulus should be finite, and integrable).

Systems that change their characteristics over time will not be considered here. However, if a system varies slowly over time, that is, slow with respect to its time constants, the GWN method can still be used quite successfully. Also oscillators, which in principle have an infinitely long memory, can in practice be handled by the GWN method.

Clearly, the example of Fig. 4.1 already shows that for nonlinear systems the superposition principle of Eq. (3.10) is violated. In Fig. 4.2 this principle is further illustrated for the response to two consecutive pulse stimuli. Note that as a result of the causality constraint, the difference between linear and nonlinear behavior can only manifest *after* the onset of the second pulse that is, for time $t \geq t_2 \geq t_1$.

From Fig. 4.2 we also observe that the nonlinear part of the response (for the time-being the two-pulse interaction) can be described as a function of the time *difference* between the observation time and the two pulses. We thus write:

$$y_c(t) = y_a(t) + y_b(t) + \Theta(t_1,t_2) \text{ with } \Theta(t_1,t_2) = 0 \text{ for } t < \max(t_1,t_2) \quad (4.3)$$

and because $\Theta(t_1,t_2) \neq 0$ for t > t_1 and $t > t_2$ we write:

$$\Theta(t_1,t_2) = \hat{k}_2(t-t_1, t-t_2) \text{ with } \Theta(t_1,t_2) = 0 \text{ for } t < \max(t_1,t_2) \quad (4.4)$$

We derived in chapter : Linear Systems Analysis that we could approximate any continuous signal arbitrarily well by an infinite train of weighted pulses: $x(t) = \int_{-\infty}^{+\infty} x(\tau)\delta(t-\tau)d\tau$. In an analogous way, we may now suspect that we can write the response of the nonlinear system at time t as:

- $y_c(t) = y_{\text{Lin}}(t) + y_2(t)$, in which the first term is
- the superposition of responses to each individual (weighted) pulse of the signal, delivered at times, $k\Delta t$:

$$y_{\text{LIN}}(t) = \sum_{k=1}^{\infty} \hat{k}_1(t - k\Delta t)$$

- and to which the nonlinear correction is added that is described by *the interaction between any two pulses*, at all possible inter-pulse time intervals:

$$y_2(t) = \sum_{m=1}^{\infty}\sum_{n=1}^{\infty} \hat{k}_2(t - m\Delta t, t - n\Delta t)$$

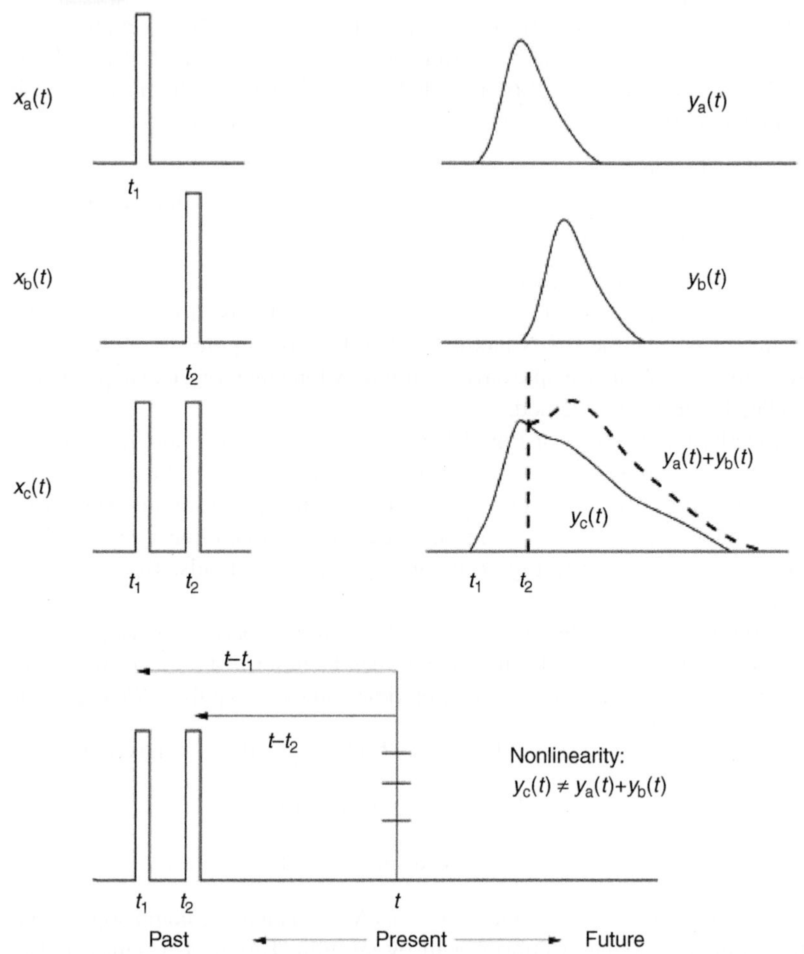

FIGURE 4.2 **The response of the nonlinear system, $y_c(t)$, to the two pulses, $x_a(t)$ and $x_b(t)$, delivered at different times, t_1, and t_2, starts to deviate from the linear sum $y_a(t) + y_b(t)$, of the pulse responses only after $t_2 \geq t_1$.**

Once we accept this idea, it is relatively straightforward to imagine that the response of the nonlinear system could also contain interaction effects from three (and more) pulses within the original input signal! The three-pulse interactions then lead to the following term in the pulses response:

$$y_3(t) = \sum_{m=1}^{\infty}\sum_{n=1}^{\infty}\sum_{p=1}^{\infty}\hat{k}_3(t - m\Delta t, t - n\Delta t, t - p\Delta t)$$

Continuing this argument leads to the presumption that the total response of the system can be represented by the following series:

$$y_c(t) = y_{\text{LIN}}(t) + y_2(t) + y_3(t) + \cdots\cdots$$

In the limit of continuous time ($\Delta t \downarrow 0$) the summations become convolution integrals of weighted Dirac delta pulses. For example,

$$y_2(t) \to \int_0^\infty \int_0^\infty k_2(\tau_1, \tau_2) x(t-\tau_1) x(t-\tau_2) d\tau_1 d\tau_2$$

in which the second-order kernel $k_2(\tau_1, \tau_2)$ measures the strength of the interactions between two Dirac pulses at two past events in the stimulus.

4.2 NONLINEAR SYSTEMS ANALYSIS: THE VOLTERRA SERIES

The French mathematician Vito Volterra demonstrated that for a nonlinear, time-invariant, and analytical system with finite memory the input–output relation for an arbitrary input signal, $x(t)$, can be given by Fig. 4.3 Q1

$$\begin{aligned} y(t) = k_0 + \int_0^\infty k_1(\tau) \cdot x(t-\tau) \cdot d\tau + \\ \int_0^\infty \int_0^\infty k_2(\tau_1, \tau_2) \cdot x(t-\tau_1) \cdot x(t-\tau_2) \cdot d\tau_1\, d\tau_2 + \\ \int_0^\infty \int_0^\infty \int_0^\infty k_3(\tau_1, \tau_2, \tau_3) \cdot x(t-\tau_1) \cdot x(t-\tau_2) \cdot x(t-\tau_3) \cdot d\tau_1\, d\tau_2\, d\tau_3 + \quad \cdots \end{aligned} \tag{4.5}$$

FIGURE 4.3 **Vito Volterra (1860–1940).** *(Source: Wikipedia.)*

where k_0, $k_1(\tau)$, $k_2(\tau_1, \tau_2)$, $k_3(\tau_1, \tau_2, \tau_3)$, etc. are the so-called *Volterra kernels* of the nonlinear system. The Volterra series can also be succinctly written as:

$$y(t) = K_0 + \sum_{n=1}^{\infty} K_n\left[x(t);\, k_n(\tau_1, \cdots, \tau_n)\right] \tag{4.6}$$

The integrals that are represented by K_1, K_2, K_3, etc. are the so-called *Volterra functionals*. The Volterra kernels are symmetric in their arguments: $k_2(\tau_1,\tau_2) = k_2(\tau_2,\tau_1)$, etc. Without loss of generality, the zeroth-order kernel (the offset) is often set to zero: $K_0 = k_0 = 0$. The first-order functional is the convolution integral for a linear system when all the higher-order kernels are zero. Importantly, due to *causality*, a kernel is zero, as soon as one of its arguments becomes negative!

There is an interesting correspondence between the Volterra series of functionals, and the Taylor series approximation of a function:

$$f(x_0 + x) = f(x_0) + \sum_{n=1}^{\infty} \left.\frac{d^n f}{dx^n}\right]_{x_0} \cdot x^n \tag{4.7}$$

As an illustrative example, consider the following nonlinear system without memory (ie, an instantaneous nonlinearity):

$$y(t) = e^{x(t)} \tag{4.8}$$

The Taylor expansion of $y(t)$ around $x_0 = 0$ yields:

$$y(t) = 1 + x(t) + \frac{1}{2}x(t)^2 + \frac{1}{3!}x(t)^3 + \frac{1}{4!}x(t)^4 + \cdots \tag{4.9}$$

Setting Eqs. (4.9) and (4.5) equal for all $x(t)$ and all t then gives the following result for the Volterra kernels of this system (but check for correctness):

- $k_0 = 1$
- $k_1(\tau) = \delta(\tau)$
- $k_2(\tau_1, \tau_2) = \frac{1}{2}\delta(\tau_1)\delta(\tau_2)$
- $k_3(\tau_1, \tau_2, \tau_3) = \frac{1}{6}\delta(\tau_1)\delta(\tau_2)\delta(\tau_3)$
- until we finally obtain:

$$k_n(\tau_1, \tau_2, \cdots, \tau_n) = \frac{1}{n!}\prod_{k=1}^{n}\delta(\tau_k) \tag{4.10}$$

And if the expansion would take place around some other point, $x_0 \neq 0$:

$$k_n(\tau_1, \tau_2, \cdots, \tau_n) = \frac{e^{x_0}}{n!}\prod_{k=1}^{n}\delta(\tau_k) \tag{4.11}$$

Note that the *n*th-order Volterra kernel describes the (causal) interactions between *n* stimulus elements in the past, and thus quantifies the effect of these interactions on the total system's response. However, conversely, the *n*th-order interactions are *not* exclusively determined by the *n*th-order Volterra kernel, but also through all *higher*-order kernels! A simple example illustrates this point, by considering the system's impulse response. From Eq. (4.5), we obtain for the impulse response of a nonlinear system:

$$y(t) = k_0 + k_1(t) + k_2(t,t) + k_3(t,t,t) + \cdots \tag{4.12}$$

The impulse response thus contains the main diagonal elements of *all* Volterra kernels! As a result of this (mixed) contribution of all kernels to the single-pulse response, it is *mathematically impossible* (ill-posed) to determine the Volterra kernels from the impulse stimulus (see Section 4.4). It is therefore desirable to find a method to obtain these kernels *independently*. One way to achieve this is to rearrange the Volterra functionals, Eq. (4.6), for a *specific stimulus*, *x*(*t*), such that each new term in the new series is made *orthogonal* with respect to all the other new terms. This rearrangement of the Volterra series is the so-called *Wiener formulation* of the Volterra series, and the specific stimulus for which the orthogonality constraint holds is *x*(*t*) = GWN.

4.3 NONLINEAR SYSTEMS ANALYSIS: THE WIENER SERIES

We use the concept of functionals to describe the stimulus–response behavior of nonlinear causal systems:

$$y(t) = F_S[x(\tau); -\infty \le \tau \le t] \tag{4.13}$$

with *t* the current, or running, time and *x*(τ) is the *past* input. The system's *memory*, μ, is the time it takes for the effect of a stimulus to disappear, or to become arbitrary small. For systems with a *finite* memory the current response only depends on *recent* past values of the stimulus:

$$y(t) = F_S[x(\tau); t - \mu \le \tau \le t] \tag{4.14}$$

Fig. 4.4 illustrates this idea for the linear RC-circuit, where the memory of the system is measured by its time constant, *T* = *RC*, so that we may take $\mu \sim 5T$. Wiener demonstrated that when the input to a nonlinear (finite memory) causal system is *GWN*, it is possible to rearrange the Volterra functionals in such a way that the new series (the Wiener series) is constructed in which all the terms are mutually *orthogonal*.

Additional properties of GWN: GWN may seem to be a rather peculiar, perhaps not very realistic, signal. However, in practice, many stochastic processes in nature result from the combined contributions of many subprocesses, each of which is stochastic. This is especially true for the brain, where neurons receive stochastic inputs from many other (stochastic) neurons. Interestingly, as a result

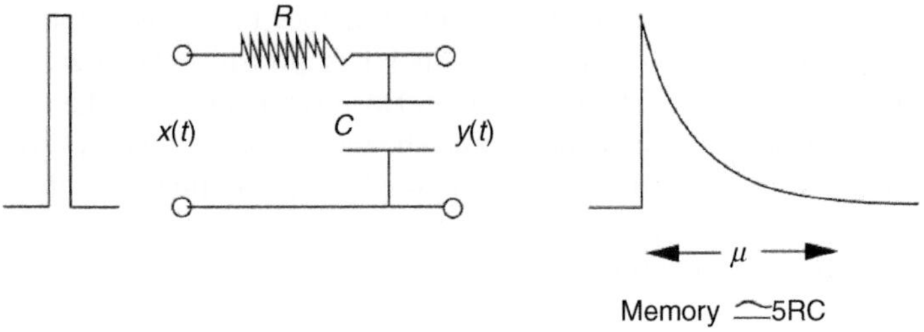

FIGURE 4.4 The system's memory determines how long the current output is still influenced by a stimulus event in the past (here, arbitrarily taken to be about five time constants ago).

of the central limit theorem of statistics, large signal ensembles will then have a Gaussian amplitude distribution with its mean and variance determined by:

$$X=\sum_{n=1}^{N} a_n x_n \mu_X=\sum_{n=1}^{N} a_n \mu_n \text{ and } \sigma_X^2=\sum_{n=1}^{N} a_n^2 \sigma_n^2$$
$$\text{and amplitude distribution: } p(x)=\frac{1}{\sigma_X\sqrt{2\pi}} e^{-\left[(x-\mu_X)^2/2\sigma_X^2\right]} \tag{4.15}$$

In chapter: Linear Systems Analysis, we demonstrated that the second-order autocorrelation function of GWN is the Dirac delta function: $\Phi_{xx}(\tau) = P\delta(\tau)$. In the GWN-method of Wiener's nonlinear systems identification, we will also use higher-order autocorrelation functions of GWN (ie, we need the expectation values of products of more than two GWN signals). For the expectation value of multiple products of Gaussian variables (gn, all taken with zero mean, and power P), the following interesting property holds:

$$\begin{aligned} E\left[g_1 g_2 \cdots g_N\right] &= 0 \quad \text{when} \quad N = \text{odd} \\ E\left[g_1 g_2 \cdots g_N\right] &= \sum \prod E\left[g_i g_j\right] \quad \text{when} \quad N = \text{even} \end{aligned} \tag{4.16}$$

where the pairs $[g_i g_j]$ in the second row of Eq. (4.16) have been drawn in all possible *different* ways from the set of $[g_1, g_2, \cdots, g_N]$ signals. A straightforward combinatorial analysis shows that this can be achieved in:

$$\frac{N!}{(N/2)!} \times 2^{(N/2)} \quad \text{different ways}$$

As an example, for the first five products this boils down to:

$E[g_1] = 0$ and $E[g_1 g_2 g_3] = 0$ and $E[g_1 g_2 g_3 g_4 g_5] = 0$, but
$E[g_1 g_2] = E[g_1 g_3] = \cdots = E[g_4 g_5] = P\delta(\tau)$, and
$E[g_1 g_2 g_3 g_4] = E[g_1 g_2]E[g_3 g_4] + E[g_1 g_3]E[g_2 g_4] + E[g_1 g_4] \cdot E[g_2 g_3] = 3P^2\delta(\tau)$.

Thus, higher-order moments of GWN signals are either zero, or can be simply expressed in terms of *second-order correlation functions*! Because these are all Dirac delta-functions, there is an obvious analytical advantage for using Gaussian stochastic variables in systems identification.

The Wiener expansion: The Wiener functionals are made orthogonal to each other for GWN as the system's input stimulus. In analogy to the Volterra series expansion of the system's output Eq. (4.6), the Wiener series thus reads:

$$y(t) = G_0 + \sum_{n=1}^{\infty} G_n\left[x(t);\, h_n(\tau_1,\cdots,\tau_n),\, P\right] \tag{4.17}$$

where the $G_k[x,h,P]$ are the mutually orthogonal Wiener functionals (specific integral expressions, yet to be determined, see further sections), the $h_k(\tau_1,\ldots,\tau_k)$ are the Wiener kernels, and P is the power of the GWN input signal (P is the signal's variance, σ^2).

The Wiener functionals are constructed from the Volterra series, by an algebraic technique called *Gramm–Schmidt orthogonalization*; two functionals are mutually orthogonal when the expectation value (the time average) of their product is zero.

Briefly, the Gramm–Schmidt orthogonalization procedure starts by setting the zeroth-order term to a constant, that is, $G_0 = k_{00}$. Here, the double subindex nm indicates: the nth-order kernel of the mth-order Wiener functional. The next term in the Wiener series is then written as a first-order *inhomogeneous* functional:

$$G_1 = \int_0^{\infty} k_{11}(\tau)x(t-\tau)d\tau + k_{01}$$

with k_{01} a constant. It's the zeroth-order kernel in the first-order Wiener functional that is to be determined by imposing the (in this case only) orthogonalization constraint:

$$\frac{E[G_0, G_1] = 0 \quad \text{for} \quad x(t)}{k_0\left[\int_0^{\infty} x(t-\tau)k_{11}(\tau)d\tau + k_{01}\right]} \begin{matrix} = \text{GWN} \\ = 0 \end{matrix}$$

In this case, it is clear that $k_{01} = 0$ will be the only solution, as the expectation value of $x(t)$ is zero. Next, the second-order inhomogeneous Wiener functional is set as the sum of zero-, first-, and second-order homogeneous Volterra functionals in the following way:

$$G_2 = k_{02} + \int_0^{\infty} k_{12}(\tau)x(t-\tau)d\tau + \int_0^{\infty}\int_0^{\infty} k_{22}(\tau_1,\tau_2)x(t-\tau_1)x(t-\tau_2)d\tau_1 d\tau_2$$

with the constant, k_{02}, and the first-order kernel, $k_{12}(\tau)$, yet to be determined by orthogonalization of G_2 to *both* G_0 and G_1 (see further sections): $E[G_0, G_1] = E[G_1, G_2] = 0$.

We will not explicitly derive the full procedure for the higher-order Wiener functionals here (in Problem 4.4, the reader is encouraged to finish this for the second-order functional), but instead we here present the first four terms of the Wiener series (up to third order):

$$G_0[h_0; x(t)] = h_0$$

$$G_1[h_1; x(t)] = \int_0^\infty h_1(\tau)x(t-\tau)d\tau$$

$$G_2[h_2; x(t)] = \int_0^\infty\int_0^\infty h_2(\tau_1,\tau_2)x(t-\tau_1)x(t-\tau_2)d\tau_1 d\tau_2 - P\int_0^\infty h_2(\tau_1,\tau_1)d\tau_1$$

$$G_3[h_3; x(t)] = \int_0^\infty\int_0^\infty\int_0^\infty h_3(\tau_1,\tau_2,\tau_3)x(t-\tau_1)x(t-\tau_2)x(t-\tau_3)d\tau_1 d\tau_2 d\tau_3$$

$$-3P\int_0^\infty\int_0^\infty h_3(\tau_1,\tau_2,\tau_2)x(t-\tau_1)d\tau_1 d\tau_2 \tag{4.18}$$

which looks like a formidable set of terms! As an example, let's check orthogonality of the first and second Wiener functionals (you already may have checked this in Problem 4.4!): we thus calculate $E[G_1G_2]$. This is how it goes:

$$E[G_1G_2] = \overline{\int_0^\infty h_1(\tau)x(t-\tau)d\tau \left[\begin{array}{l}\int_0^\infty\int_0^\infty h_2(\tau_1,\tau_2)x(t-\tau_1)x(t-\tau_2)d\tau_1 d\tau_2 + \\ -P\int_0^\infty h_2(\tau_1,\tau_1)d\tau_1\end{array}\right]}$$

where we note that taking the expectation value (ie, the time average) only affects the GWN signals, as the kernels are deterministic functions. Also note that the product between G_1 and the first term of G_2 involves three GWN samples, which yields an expectation value of zero [property of GWN, see Eq. (4.16)]. The same holds for the product of G_1 with the second term of G_2, which only involves one GWN sample. In short, the two functionals are indeed mutually orthogonal. In Problem 4.5 the reader may verify the orthogonality between G_3 and G_1.

Although the Volterra and Wiener kernels describe the same system, and therefore the total (infinite) series of Eqs. (4.4) and (4.15) yield an identical

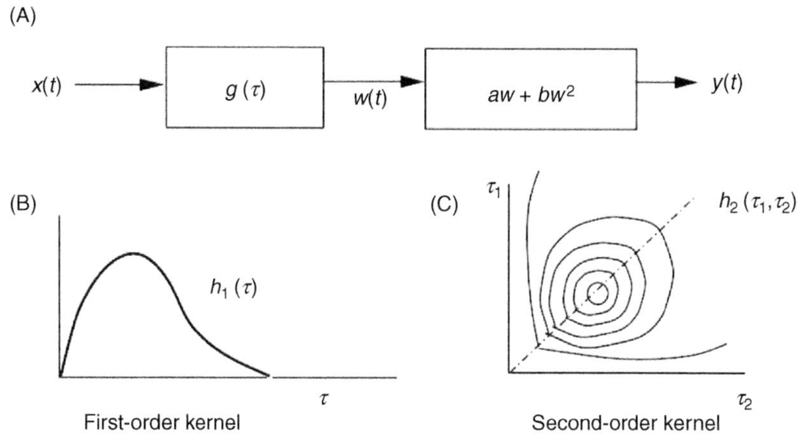

FIGURE 4.5 (A) A series concatenation of a first-order linear filter, followed by a static second-order nonlinearity. (B) The first-order Wiener kernel. (C) The second-order Wiener kernel; note that the latter is symmetric in its arguments.

result, the system *kernels* for either method will differ in general, as they have been constructed in essentially different ways. However, the following holds:

> Systems for which the kernels of order > 2 are zero, have identical Volterra and Wiener kernels.

Fig. 4.5 illustrates a second-order nonlinear system, consisting of a linear filter, with impulse response, $g(\tau)$, followed by a static second-order nonlinearity. The static nonlinearity has no memory, because the system's memory is fully represented by the linear filter. Following the input–output relations of both subsystems, we obtain:

$$\begin{aligned} w(t) &= \int_0^\infty g(\tau)x(t-\tau)d\tau \\ y(t) &= aw(t)+bw(t)^2 \end{aligned} \tag{4.19}$$

from which:

$$y(t) = a\int_0^\infty g(\tau)x(t-\tau)d\tau + b\left[\int_0^\infty g(\tau)x(t-\tau)d\tau\right]^2 \tag{4.20}$$

We now convert the last term into the Volterra-series format:

$$y(t) = a\int_0^\infty g(\tau)x(t-\tau)d\tau + b\int_0^\infty\int_0^\infty g(\tau_1)g(\tau_2)x(t-\tau_1)x(t-\tau_2)d\tau_1 d\tau_2 \tag{4.21}$$

The nonlinear system of Fig. 4.4 thus has the following Volterra (and hence, also Wiener, see Problem 4.6) kernels:

$$\begin{aligned} k_1(\tau) &= h_1(\tau) = ag(\tau) \\ k_2(\tau_1,\tau_2) &= h(\tau_1,\tau_2) = bg(\tau_1)g(\tau_2) \end{aligned} \tag{4.22}$$

Since the Wiener functionals are mutually orthogonal, they provide the *best possible approximation* up to their respective order (in Problem 4.3 you have shown this). In other words, if one decides to truncate the Volterra and Wiener series of a general nonlinear system at the second-order term, the Wiener approximation will be the best in the least-squared error sense. The goal of nonlinear systems identification is therefore to determine the Wiener kernels.

Note that the first-order Wiener functional of a nonlinear system is, in general, *not equal* to the linear response part of the system, represented by the first-order Volterra kernel. Due to the Gramm–Schmidt orthogonalization procedure, the Wiener kernels depend on all higher-order Volterra kernels; as a result, the first-order Wiener kernel already contains a significant contribution of the system's nonlinearity!

Although it is not feasible to perform a precise structural analysis of a system only on the basis of an abstract black box input–output analysis, it is possible to test certain hypotheses about the system when nonlinearities are present. For example, suppose that it is known that the system contains a concatenation of a linear filter and a static quadrator, but that their order is unknown. In that case, a Wiener or Volterra analysis can readily dissociate the two possibilities of Fig. 4.6A versus Fig. 4.6B, as these two alternatives yield different system kernels (ie, the series concatenation is noncommutative). The Volterra kernels for the systems in Fig. 4.6 are (but check):

$$\begin{aligned} k_1^A(\tau) &= k_1^B(\tau) = 0 \\ k_2^A(\tau_1,\tau_2) &= g(\tau_1)\delta(\tau_1-\tau_2) \\ k_2^B(\tau_1,\tau_2) &= g(\tau_1)g(\tau_2) \end{aligned} \tag{4.23}$$

Thus, the first-order kernels are identical, but by studying the second-order kernel, the two different systems can be readily dissociated.

Measuring the Wiener kernels: Orthogonality of the Wiener functionals with respect to GWN means that the Wiener kernels can be measured *independently* from each other. The elegant method of Lee and Schetzen (1965) describes how this is achieved in practice. In brief, the experiment requires (a good approximation of) a GWN input signal, where after the Wiener kernels are calculated offline by applying a (multiple-order) cross-correlation analysis between the system's output and the GWN input. In this analysis, the

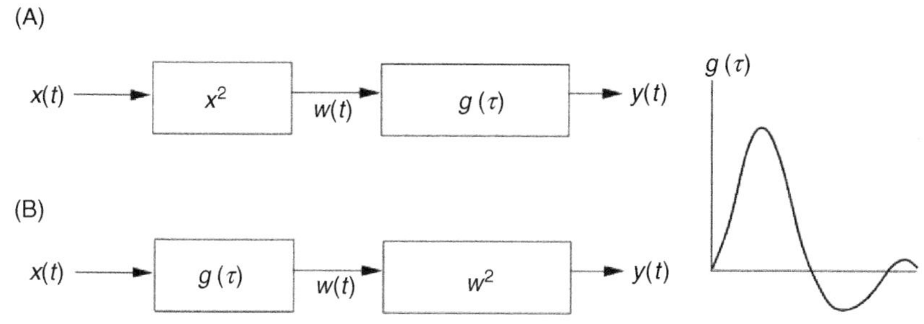

FIGURE 4.6 **In nonlinear systems, the order of the different components in a series concatenation matters: the system kernels for panel (A) differ from those in panel (B).**

autocorrelation properties of GWN [Eq. (4.16)] prove to be particularly convenient. To get the idea, we here describe the cross-correlation method for the first two Wiener kernels.

As a starting point, the Wiener series for the system's output read:

$$y(t) = h_0 + \sum_{n=1}^{\infty} G_n\left[x(t) = \mathrm{GWN}(P); h_n(\tau_1, \cdots, \tau_n)\right]$$
$$\text{and } E\left[G_n G_m\right] = 0 \;\; \text{for } n \neq m \tag{4.24}$$

Measuring h_0: The first observation is that the average of all functionals G_n with $n \geq 1$ is zero, because the noise has zero mean, and all functionals were made orthogonal to the constant, h_0. For example, $E[G_2] = 0$, since

$$\overline{\overline{\int_0^\infty\int_0^\infty h_2(\tau_1,\tau_2)x(t-\tau_2)x(t-\tau_2)d\tau_1\,d\tau_2 - P\int_0^\infty h_2(\tau_1,\tau_1)d\tau_1}} =$$
$$\int_0^\infty\int_0^\infty h_2(\tau_1,\tau_2)\overline{\overline{x(t-\tau_1)x(t-\tau_2)}}d\tau_1 d\tau_2 - P\int_0^\infty h_2(\tau_1,\tau_1)d\tau_1 =$$
$$\int_0^\infty\int_0^\infty h_2(\tau_1,\tau_2)P\delta(\tau_1-\tau_2)d\tau_1\,d\tau_2 - P\int_0^\infty h_2(\tau_1,\tau_1)d\tau_1 = 0$$

Similar calculations can be made for all other Wiener functionals. Thus, if we take the average of the system's output, we immediately measure the zeroth-order Wiener kernel:

$$h_0 = \overline{y(t)} \tag{4.25}$$

Measuring $h_1(\tau)$: The second observation is that each Wiener functional, G_k, is made orthogonal to all homogeneous functionals of $x(t)$ with a degree $< k$. We therefore multiply Eq. (4.22) with the first-order term, $x(t-\sigma)$, and determine the average: in other words, we take the second-order cross-correlation function, $\Phi_{yx}(\sigma)$. Note that all functionals with $k > 1$ will vanish in this average, and $E[x(t-\sigma)h_0] = 0$ too, because the noise has zero mean:

$$\overline{y(t)x(t-\sigma)} = \overline{x(t-\sigma)\int_0^\infty h_1(\tau)x(t-\tau)d\tau} = \int_0^\infty h_1(\tau)\overline{x(t-\sigma)x(t-\tau)}d\tau$$

where we again apply the autocorrelation property of GWN to obtain the first-order Wiener kernel:

$$\Phi_{yx}(\sigma) = Ph_1(\sigma) \Rightarrow h_1(\sigma) = \frac{\Phi_{yx}(\sigma)}{P} \tag{4.26}$$

Measuring $h_2(\tau_1,\tau_2)$: From here on, the method may become clear: we next multiply the output with the second-order term in $x(t)$ and calculate the third-order cross-correlation function:

$$\overline{y(t)x(t-\sigma_1)x(t-\sigma_2)} = \overline{[G_0 + G_1 + G_2]x(t-\sigma_1)x(t-\sigma_2)}$$

because all terms with $k > 2$ will vanish. Since the term with G_1 will involve an odd number (product of three) of GWN signals, that term will vanish too, so only the expectation values for G_0 and G_2 may survive:

$$\begin{aligned}&\overline{y(t)x(t-\sigma_1)x(t-\sigma_2)} = h_0\overline{x(t-\sigma_1)x(t-\sigma_2)} +\\ &\int_0^\infty\int_0^\infty h_2(\tau_1,\tau_2)\overline{x(t-\tau_1)x(t-\tau_2)x(t-\sigma_1)x(t-\sigma_2)}d\tau_1 d\tau_2 -\\ &P\int_0^\infty h_2(\tau_1,\tau_2)x(t-\sigma_1)x(t-\sigma_2)d\tau_1\end{aligned} \tag{4.27}$$

The reader may now readily verify [using the autocorrelation properties of Eq. (4.14)] that

$$\Phi_{yxx}(\sigma_1,\sigma_2) = Ph_0\delta(\sigma_1 - \sigma_2) + 2P^2h_2(\sigma_1,\sigma_2) \tag{4.28}$$

It is possible to get rid of the first term on the right-hand side of Eq. (4.28) by taking the cross-correlation after subtracting the average from the output:

$$y_0(t) \equiv y(t) - h_0 \Rightarrow \Phi_{y_0xx}(\sigma_1,\sigma_2) = 2P^2h_2(\sigma_1,\sigma_2) \tag{4.29}$$

so that the second-order Wiener kernel is obtained by:

$$h_2(\sigma_1,\sigma_2) = \frac{\Phi_{y_0xx}(\sigma_1,\sigma_2)}{2P^2} \tag{4.30}$$

Measuring $h_n(\tau_1, \tau_2, \ldots, \tau_n)$: Lee and Schetzen (1965) described the procedure for the general case, which led to the following recipe for the nth-order Wiener kernel:

- first, subtract from the output, the contribution of all lower-order terms
- then, take the n + first-order cross-correlation for the residual output:

$$h_n(\sigma_1,\sigma_2,\cdots,\sigma_n)=\frac{1}{n!P^n}E\left[\left[y(t)-\sum_{k=1}^{n-1}G_k(h_k;x)\right]x(t-\sigma_1)\cdots x(t-\sigma_n\right] \quad (4.31)$$

Fig. 4.7 provides a schematic illustration of Lee and Schetzen's cross-correlation method up to the second-order Wiener kernel for an arbitrary nonlinear system.

As an exercise, the reader may now verify, by applying the cross-correlation method described previously, that the Wiener kernels for the second-order nonlinear system in Fig. 4.5, are indeed given by Eq. (4.22).

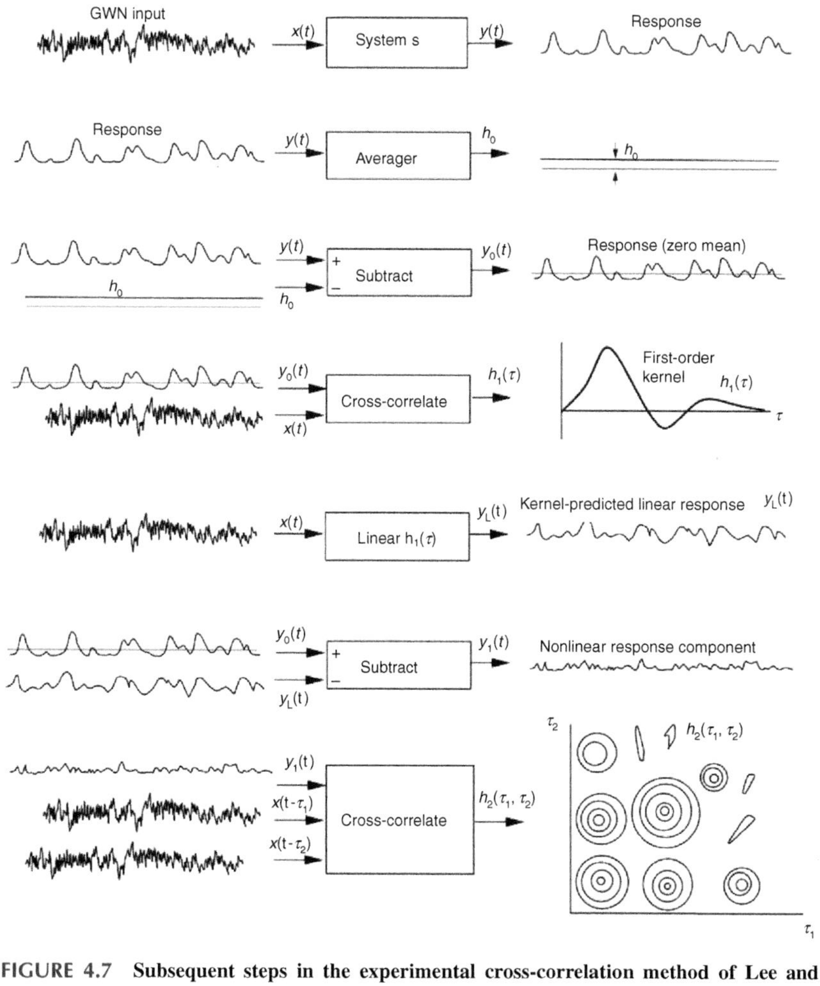

FIGURE 4.7 **Subsequent steps in the experimental cross-correlation method of Lee and Schetzen (1965) to measure the Wiener kernels of a nonlinear system.**

Wiener versus Volterra series: In practical situations, when confronted with an unknown nonlinear system, truncation of the Wiener or Volterra series at a given order (typically, quadratic) is unavoidable. We have seen that in that case the Wiener approximation will provide the best prediction of the output signal, because the extracted low-order Wiener kernels also contain information about potential higher-order nonlinearities. A disadvantage of the Wiener approach is that its applicability is constricted to GWN, for which the functionals were made orthogonal. Thus, in order to make predictions for the system's output to other types of signals, we also need the Volterra kernels. For any system of arbitrary order, there is a unique relationship between the Wiener and Volterra kernels: up to order two, the kernels are identical (see previous section). For order > 2, however, they differ, and the mutual relationships vary, depending on the system's order as well!

The Volterra series consists of homogeneous functionals in the input signal $x(t)$, and is reminiscent to a functional expansion like the Taylor's series [see the example in Eq. (4.9)]:

$$f(x) = k_0 + k_1 x + k_2 x^2 + k_3 x^3 + \cdots \tag{4.32}$$

The Wiener series consists of inhomogeneous, mutually orthogonal, functionals, and is related to the expansion of functions with Hermite polynomials:

$$f(x) = H_0 + H_1(x) + H_2(x) + H_3(x) + \cdots \tag{4.33}$$

where:

$$\begin{aligned} H_0 &= 1 \\ H_1(x) &= x \\ H_2(x) &= x^2 - P \\ H_3(x) &= x^3 - 3Px \\ H_4(x) &= x^4 - 6Px^2 + 3P^2, \text{ etc.} \end{aligned} \tag{4.34}$$

The resemblance of the Hermite polynomials to Eq. (4.18) may be obvious (if so, you may now do Problem 4.7). We also see in Eq. (4.34) that the first-order (linear) term appears in all odd-numbered terms; similarly the second-order nonlinearity figures in all even-numbered polynomials. If we would limit the system's output approximation to a fourth-order expansion with Wiener and Volterra functionals, the Hermite (Wiener) polynomials can be uniquely related to the Taylor (Volterra) coefficients through the following mapping:

$$\begin{aligned} k_0 &= 1 - P + 3P^2 \\ k_1 &= 1 - 3P \\ k_2 &= 1 - 6P \\ k_3 &= k_4 = 1 \end{aligned} \tag{4.35}$$

For a third-order system, however, the Volterra coefficients become:

$$k_0 = 1 - 3P \text{ and } k_2 = k_3 = 1.$$

Example application: The Wiener method has been applied in the auditory system on many occasions (Eggermont, 1993). A recent example is provided by the study of Recio-Spinoso et al. (2005), who recorded the responses to GWN stimuli from auditory nerve fibers in the anesthetized chinchilla with best tunings across the spectral acoustic range. Fig. 4.8 shows an example of the first- and second-order Wiener kernels that were extracted with the cross-correlation method from a mid-frequency fiber (characteristic frequency at $f_C = 2.94$ kHz). The first-order Wiener kernel (Fig. 4.8A) resembles a band–pass filter, ringing at the characteristic frequency (see chapter: The Auditory Nerve).

Across the spectral range, neurons had strong band–pass filter characteristics, but their signal-to-noise ratios decreased rapidly for high frequencies. Note also that the second-order kernel (Fig. 4.8B) shows periodicity that oscillates at the same rate as f_C. This feature was observed across the entire frequency range. As such, chinchilla auditory nerve fibers could be modeled by a simple sandwich configuration (Fig. 4.9) that consists of a linear band–pass filter (*L*), followed by a static nonlinearity (here, a simple rectifier, *NL*), and a low–pass filter (*L*) with a cutoff frequency at about 2.5 kHz. With such a unified model, the responses of chinchilla auditory nerve fibers across the frequency range could be predicted with reasonable accuracy. An example of the prediction of responses with the model is shown in Fig. 4.10 for one of the auditory nerve fibers.

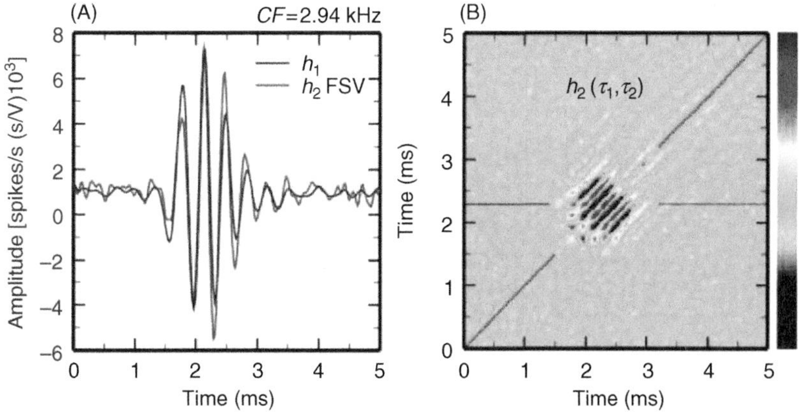

FIGURE 4.8 **Illustrative example of an application of the Wiener method to an auditory nerve fiber recording from anesthetized chinchilla. The cell is tuned to the mid-frequency range, at f_C = 2.94 kHz.** (A) The first-order Wiener kernel has the characteristic of a narrow band pass filter. (B) The second-order Wiener kernel. *(Source: After Recio-Spinoso et al., 2005, with kind permission.)*

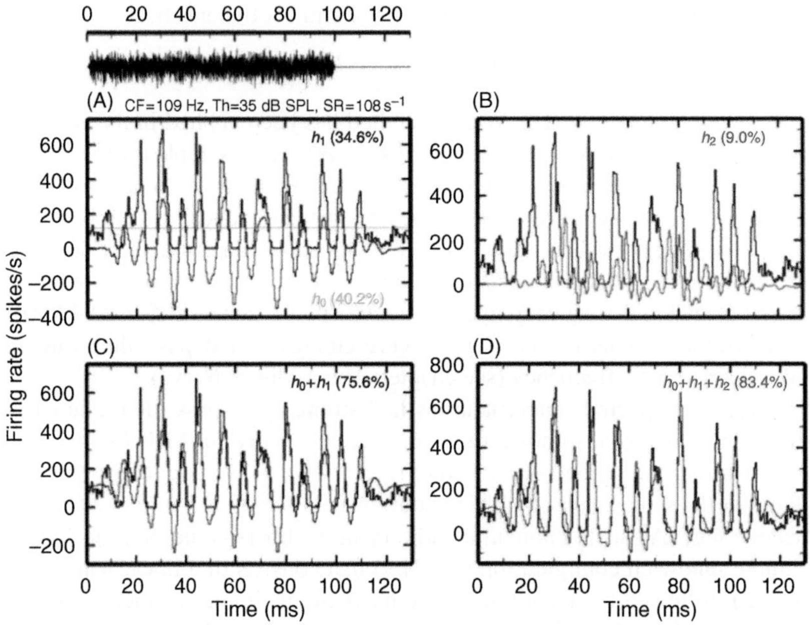

FIGURE 4.9 Comparison of predicted and measured neural responses for the zero-order, first-order and second-order Wiener approximations of a Chinchilla auditory nerve fiber, with a f_C = 109 Hz. The zeroth- and second-order kernels are essential to describe the neural response. The first-order kernel (h_1) explains 35% of the response variability, whereas $h_0 + h_1 + h_2$ can explain 83% of the response. *(Source: After Temchin et al., 2005, with kind permission.)*

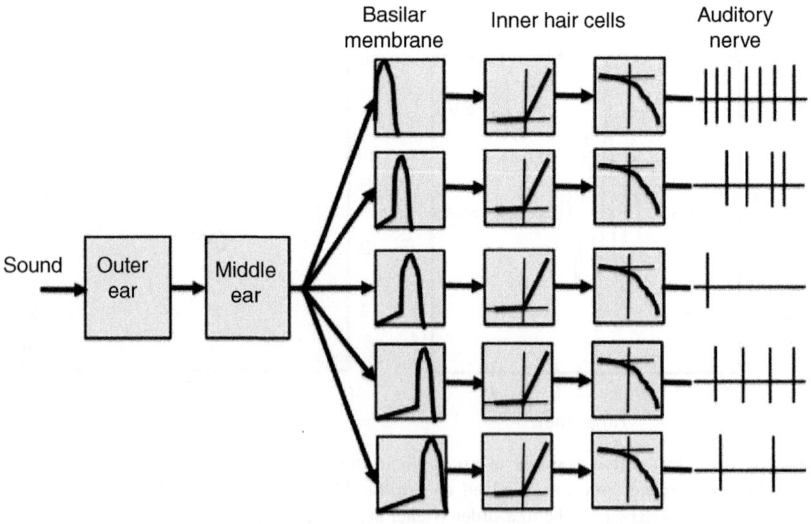

FIGURE 4.10 The cascade of linear and nonlinear subsystems in the auditory periphery, which can be identified through a Wiener analysis with GWN.

Serial cascades: Although it falls beyond the scope of this book, in principle it is possible to disentangle structural models that contain a (known or hypothesized) cascade of linear filters and static nonlinearities. There are plenty of examples of such serial systems in the brain, and the auditory system is no exception. For example, consider the following chain of processes in the auditory periphery (Fig. 4.10).

- The pinna, which acts as a complex and direction-dependent linear filter, with the
- ear canal that effectively acts as a linear filter with multiple resonances around $2n$ kHz, followed by
- the mechanics of the (linear) middle ear (see chapter: The Nature of Sound), and the subsequent
- mechanics on the basilar membrane in the cochlea, which may (in a first approximation) be described by a parallel system of partially overlapping (nonlinear) band–pass filters (see chapter: The Cochlea). Each of these filters is in series
- with an inner hair cell, which acts as a rectifying low–pass filter, and which is finally
- followed by the auditory nerve with its nonlinear, saturating, spike-generation mechanism.

Fig. 4.11 schematically presents part of this cascade, from the band–pass basilar membrane motion to the active (static) nonlinearity that accounts for local outer hair–cell feedback (see chapter: The Cochlea), and the LP filter characteristic of the IHC response. For simplicity, the static nonlinearity is assumed to be of second order (but the analysis remains qualitatively similar for polynomial expansions of arbitrary order, Marmarelis and Marmarelis, 1978).

In this scheme, the output of the IHC is described by:

$$y(t) = \int_0^\infty k(\sigma) w(t-\sigma) d\sigma$$

where the basilar membrane motion in response to cochlear input, $x(t)$, is given as:

$$w(t) = a\int_0^\infty h(\tau)x(t-\tau)d\tau + b\int_0^\infty\int_0^\infty h(\tau_1)h(\tau_2)x(t-\tau_1)x(t-\tau_2)d\tau_1 d\tau_2$$

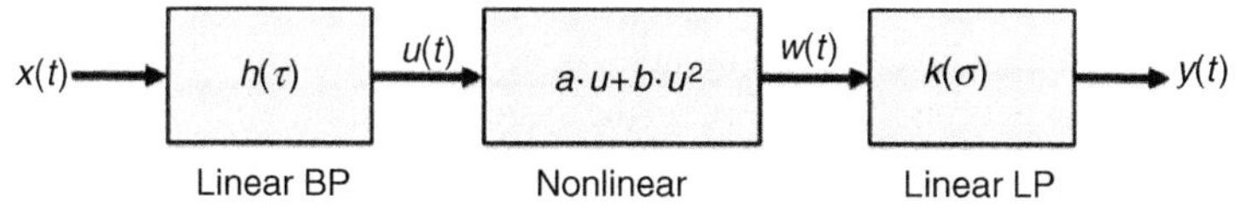

FIGURE 4.11 **The sandwich of a dynamic linear band–pass filter (basilar membrane motion), a second-order static nonlinearity, and a dynamic linear low–pass filter (IHC), as a model for the cochlear partition.**

For the cross-correlation between input and output (which measures the first-order Wiener kernel) we thus obtain:

$$\Phi_{yx}(\lambda)=\overline{y(t)x(t-\lambda)}=\int_0^\infty k(\sigma)\overline{w(t-\sigma)x(t-\lambda)}d\sigma$$

for $x(t) = GWN$ with power P, this reduces to:

$$\Phi_{yx}(\lambda)=aP\int_0^\infty h(\sigma)k(\lambda-\sigma)d\sigma\equiv Ph_1(\lambda)$$
$$\Phi_{yx}(\omega)=aPH(\omega)K(\omega) \tag{4.36}$$

In Problem 4.8 the reader can verify that the second-order Wiener kernel is obtained from the third-order cross-correlation function:

$$\Phi_{yxx}(\lambda_1,\lambda_2)=2bP\left[\int_0^\infty k(\sigma)h(\lambda_1-\sigma)h(\lambda_2-\sigma)d\sigma+A\delta(\lambda_1-\lambda_2)\right] \tag{4.37}$$

where $A=\frac{1}{2}\left[\int_0^\infty k(\sigma)d\sigma\right]\left[\int_0^\infty h^2(\sigma)d\sigma\right]$.

4.4 INDEPENDENT CALCULATION OF VOLTERRA KERNELS

Independent measurement of the Volterra kernels is in general not possible for the Volterra series, for which a given kernel depends on all lower-order kernels. Here we describe an interesting method to independently extract the Volterra kernels from a "meta-model" of the nonlinear system, provided by an artificial neural network (ANN).

A three-layer feed-forward ANN (or multilayer perceptron), containing nonlinear model neurons (eg, with sigmoid input–output characteristics), can be considered a *universal approximator* for arbitrary input–output relations, provided that the network (ie, its number of synaptic connections) is sufficiently complex (Cybenko, 1989). Suppose an ANN model that approximates the input–output relation of a finite-memory nonlinear dynamical system. The ANN thus serves as an abstract model for the dynamical system under study, which can be used to predict the output of the real system for arbitrary input signals. The parameters of the ANN reside in the synaptic connections (weights) of the network neurons between the different layers: from input to hidden layer, and from hidden layer to the single output neuron that represents the system's output, $y(t)$.

If true, a deterministic relation should exist between the Volterra series, and the internal representation of the system's parameters (ie, the Volterra kernels) in the ANN. Wray and Green (1994) demonstrated that the synaptic weights in the ANN, trained by the backpropagation algorithm (Rumelhart et al., 1986), can indeed be directly mapped onto the discrete approximations of the system's

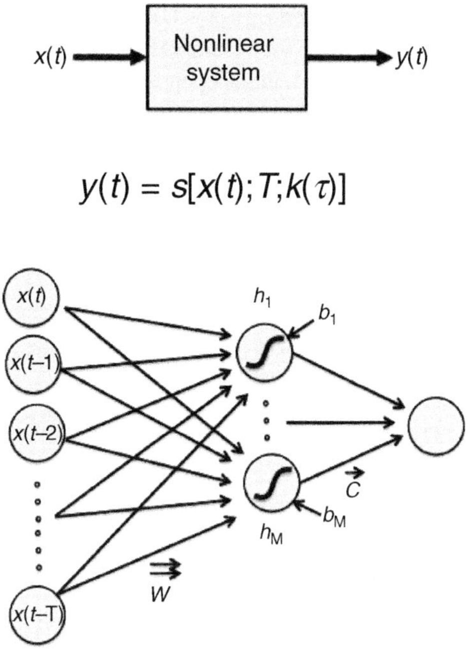

FIGURE 4.12 **An appropriately trained three-layer ANN is equivalent to the real nonlinear system, as it can produce the same input–output relations.** The input to the ANN represents the system's memory of the input in T unit time steps. The synaptic weights in the network to the (nonlinear) hidden units and (linear) output unit embed the system's Volterra kernels.

Volterra kernels. This interesting property of ANN's allows one to independently extract the system's nth order Volterra kernel (and hence, the system's nth order nonlinearity), without first having to calculate the other system kernels, or to resort to the (truncated) Wiener series (Fig. 4.12).

To quantify the equivalence between the Volterra functionals and the synaptic weights in the ANN, the former is described in discrete form, in which we also incorporate the system's finite memory of T time samples (ie, the integration boundaries of the convolution integrals in Eq. (4.5) run from 0 to T):

$$y(t)=k_0+\sum_{\tau=0}^{T}k_1(\tau)x(t-\tau)+\sum_{\tau_1=0}^{T}\sum_{\tau_2=0}^{T}k_2(\tau_1,\tau_2)x(t-\tau_1)x(t-\tau_2)+$$
$$\sum_{\tau_1=0}^{T}\sum_{\tau_2=0}^{T}\sum_{\tau_3=0}^{T}k_2(\tau_1,\tau_2,\tau_3)x(t-\tau_1)x(t-\tau_2)x(t-\tau_3)+\cdots\cdots \tag{4.38}$$

In the three-layer ANN, the neurons in the hidden layer have a nonlinear input–output characteristic. This characteristic is typically described by a sigmoid function, like the hyperbolic tangent:

$$\sigma(x)=\tanh(x)\approx x-\frac{1}{3}x^3+\frac{2}{15}x^5-\frac{17}{315}x^7+\cdots \tag{4.39}$$

The input to each hidden neuron ($n = 1 \cdots M$) is determined by the weighted output of the input layer with its T input samples (represented by a $(T+1) \times M$ synaptic weight matrix, $[W]$), and a unit-dependent offset (bias), b_n:

$$h_n[\vec{x}(t)] = \tanh\left[b_n + \sum_{k=0}^{T} w_{kn} x(t-k) \right] \tag{4.40}$$

This nonlinearity can be expanded by its Taylor series, including the bias:

$$h_n(x) = (x+b_n) - \frac{1}{3}(x+b_n)^3 + \frac{2}{15}(x+b_n)^5 - \frac{17}{315}(x+b_n)^7 + \cdots \tag{4.41}$$

The output of the neural network performs a simple linear weighting (through a $M \times 1$ weight vector, $\vec{c}$) of the hidden layer's output:

$$y[\vec{x}(t)] = \sum_{k=1}^{M} c_k h_k[\vec{x}(t)] \tag{4.42}$$

Combining Eqs. (4.42) and (4.41) leads to a polynomial expansion of the input–output relation of the ANN, which can be directly compared (by identifying each of the terms) to the Volterra kernels of the system [Eq. (4.38)]. In general, we can describe the polynomial expansions for nonlinear hidden-layer neurons as:

$$h_n(x) = \sum_{k=0}^{\infty} a_{kn} x^k \tag{4.43}$$

[Eq. (4.41) holds for the case of the hyperbolic tangent]. The Volterra kernels can then be expressed in the synaptic weights and polynomial coefficients by:

$$k_0 = \sum_{m=1}^{M} c_m a_{0m} \quad k_1(\tau) = \sum_{m=1}^{M} c_m a_{1m} w_{\tau m}$$

$$k_2(\tau_1, \tau_2) = \sum_{m=1}^{M} c_m a_{2m} w_{\tau_1 m} w_{\tau_2 m} \tag{4.44}$$

and for the nth order Volterra kernel:

$$k_n(\tau_1, \tau_2, \cdots, \tau_n) = \sum_{m=1}^{M} c_m a_{nm} w_{\tau_1 m} w_{\tau_2 m} \cdots w_{\tau_n m} \tag{4.45}$$

Fig. 4.13 exemplifies this approach to the cascade of a linear filter and a squarer. You may verify (Problem 4.9) that analytical Wiener–Volterra

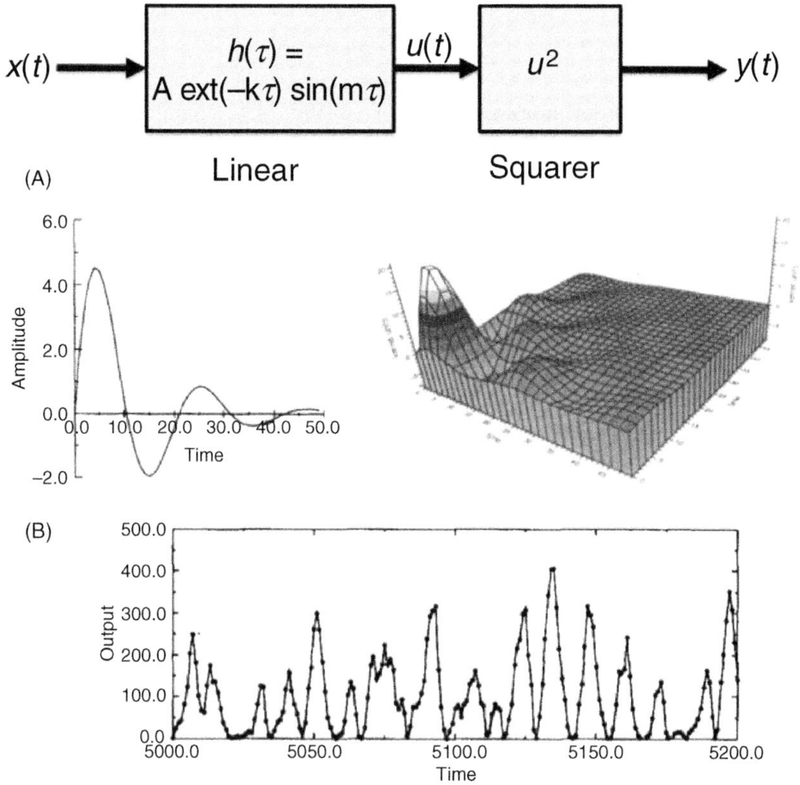

FIGURE 4.13 **Nonlinear systems identification is possible when using a three-layer ANN, consisting of 50 input units (representing 50 time samples of the input signal that lasted 4000 samples), 10 hidden units with hyperbolic tangent characteristics, and one linear output unit.** (A) The linear filter is shown on the left ($A = 6.67$; $k = 0.08$; and $m = 0.3$); the system's only nonzero kernel is the second-order Volterra (or Wiener) kernel, here extracted from the network, and shown on the right. (B) Comparison of the network's output (dots) and real system's output (continuous line). *(Source: After Wray and Green, 1994, with permission.)*

analysis gives for the second-order kernel (the only surviving one in the Volterra series):

$$k_2(\tau_1, \tau_2) = A^2 e^{-k(\tau_1+\tau_2)} \sin(m\tau_1)\sin(m\tau_2) \tag{4.46}$$

The neural network, trained on Gaussian colored noise, on randomly drawn input sequences of 50 samples, is able to precisely follow the predicted output of the system. Fig. 4.13B shows the network's output superimposed on the analytically calculated output, using Eq. (4.46) and the linear filter kernel. Fig. 4.13A shows the extracted second-order Volterra kernel of the ANN; the network's kernel is virtually identical to the analytically calculated kernel of Eq. (4.46).

4.5 EXERCISES

Problem 4.1:

(a) For the third-order nonlinear system of Fig. 4.1, calculate the output spectrum when the input is given by: $x(t) = \sin(2\pi f_1 t) + \sin(2\pi f_2 t)$, where $f_2 > f_1$.

(b) Write a Matlab script (or use your preferred programming environment) to calculate and plot the amplitude spectrum for the output of a finite-order (up to $n = 5$) polynomial nonlinearity in response to the stimulus of (a) for arbitrary frequencies.

Problem 4.2: In analogy to the Taylor expansion for the exponential function, give the Volterra kernels for the following nonlinear functions:

(a) $y(t) = \log[x(t) + C]$ for $C = 1$.
(b) $y(t) = \exp[x(t)]$ preceded by a linear system with impulse response, $g(\tau)$
(c) $y(t) = \cos[x(t)]$

Problem 4.3: The Wiener functionals are orthogonal to each other. As a consequence, if the series is terminated at a certain order, say N, it is the *best possible* approximation of the system up to that order in the least-squared error sense. We are going to show this by the following example. Assume that signal $x(t)$ (defined on the interval $[0, T]$) is represented by an infinite series of orthogonal signals:

$$x(t) = \sum_{k=0}^{\infty} a_k f_k(t)$$

with $\{fk(t)\}$ basis functions that are orthonormal with respect to a given weighting function, $w(t)$, on the interval $[0,T]$. This means that the following inner-product condition holds:

$$\int_0^T f_k(t) f_m(t) w(t)\, dt = \delta_{km} \text{ with } \delta_{km} = \begin{matrix} 1 \text{ for } k = m \\ 0 \text{ for } k \neq m \end{matrix}$$

For simplicity, we take $w(t) = 1$ (think of the Fourier representation, as an example).

(a) Check that the coefficients a_k are determined by:

$$a_k = \int_0^T x(t) f_k(t)\, dt$$

Now suppose that we approximate $x(t)$ by taking the first N terms of the series:

$$\tilde{x}(t) \approx \sum_{k=0}^{N} a_k f_k(t)$$

As a measure for the error in this approximation, we take the mean squared error (MSE), ε:

$$\varepsilon = \int_0^T [x(t) - \tilde{x}(t)]^2 \, dt$$

(b) Show that the a_k for which the error reaches its minimum are given by your answer in (a). In other words, the orthogonal approximation is the most efficient.

Problem 4.4: Use the autocorrelation properties of GWN [Eq. (4.16)] and Gramm–Schmidt orthogonalization to determine the second-order Wiener functional, by making the inhomogeneous functional

$$G_2[x(t), P] = \begin{matrix} k_{02} + \int_0^\infty k_{12}(\tau) x(t-\tau) d\tau + \\ \int_0^\infty \int_0^\infty k_{22}(\tau_1, \tau_2) x(t-\tau_1) x(t-\tau_2) d\tau_1 d\tau_2 \end{matrix}$$

orthogonal to the zeroth- and first-order Wiener functionals of Eq. (4.17) [Eq. (4.18) for the answer].

Problem 4.5: Verify that the third-order Wiener kernel is indeed orthogonal to the first-order Wiener kernel [Eq. (4.18) for their formulations].

Problem 4.6: Apply the cross-correlation method of Lee and Schetzen (1965) to determine the Wiener kernels of the system in Fig. 4.5, and verify that they are indeed identical to the Volterra kernels of the system.

Problem 4.7: Use the Hermite polynomials to write down the full expression for the fifth Wiener functional, $G_4[h_4, x(t); P]$.

Problem 4.8: Determine the second-order Wiener kernel for the model of Fig. 4.9 (BM to IHC response).

Problem 4.9: In the model of Fig. 4.12 the linear filter is described by the following function:

$$g(t) = \frac{a}{m} \exp(-kt) \sin(mt)$$

with $a = 2$, $m = 0.3$, and $k = 0.08$.

Calculate the Volterra kernels for this system.

**Problem 4.10*: In this Matlab computer exercise you perform the Lee and Schetzen (1965) cross-correlation technique to identify a nonlinear system on the basis of a single pair of input–output data. All you know is that the system either consists of a L—NL, or a NL—L cascade, in which NL is a static second-order nonlinearity:

$$y(t) = ax(t) + bx(t)^2$$

and the linear filter has the following appearance:

$$y(t) = m\exp(-kt)\sin(nt)$$

You may obtain the input and output noise signals from the model system (sampled at 1 ms temporal resolution, with a total signal duration is 4.0 s) from the book's website, where you will also find some relevant tips, tricks, and scripts to implement cross-correlation functions and linear convolutions.

Estimate the parameter values of the system, [a, b, m, k, and n], as well as the order in which the subsystems were concatenated.

Problem 4.11: When GWN is filtered by a first-order low–pass *linear* system with finite bandwidth and impulse response $k(\tau)$, the output is also band limited. Yet, it still has Gaussian statistics, and hence is a Gaussian process. Show this, by determining the expectation values $<y>$, $<y^2>$, $<y^3>$, and $<y^n>$ of the system's output, $y(t)$, for independent output samples, and use the properties of GWN autocorrelations as expressed in Eq. (4.16). Distinguish n = odd and n = even.

If the moments of independent output samples are the same as for GWN, then they represent the same stochastic process.

REFERENCES

Bechhoefer, J., 2005. Feedback for physicists: a tutorial essay on control. Revs. Mod. Phys. 77, 783–836.

Cybenko, G., 1989. Approximation by superpositions of a sigmoidal function. Maths. Control Sigs. Sys. 2, 303–314.

Eggermont, J.J., 1993. Wiener and Volterra analyses applied to the auditory system. Hear. Res. 66, 177–201.

Lee, Y.W., Schetzen, M., 1965. Measurement of the Wiener kernels of a nonlinear system by cross-correlation. Int. J. Control 2, 237–254.

Marmarelis, P.Z., Marmarelis, V.Z., 1978. Analysis of Physiological Signals: the White Noise approach. Plenum Publishing Corporation, New York.

Recio-Spinoso, A., Temchin, A.N., Van Dijk, P., Fan, Y.-H., Ruggero, M.A., 2005. Wiener-Kernel analysis of responses to noise of Chincilla auditory nerve fibers. J. Neurophysiol. 93, 3615–3634.

Rumelhart, D., McLelland, J.L., PDP Research Group (Eds.), 1986. Parallel Distributed Processing: Explorations in the Microstructure of Cognition, MIT Press, Cambridge, MA, USA.

Temchin, A.N., Recio-Spinoso, A., Van Dijk, P., Ruggero, M.A., 2005. Wiener kernels of Chinchilla auditory-nerve fibers: verification using responses to tones, clicks, and noise and comparison with basilar-membrane vibrations. J. Neurophysiol. 93, 3635–3648.

Wray, J., Green, G.G.R., 1994. Calculation of the Volterra kernels of non-linear dynamic systems using an artificial neural network. Biol. Cybern. 71, 187–195.

Chapter 5

The Cochlea

5.1 INTRODUCTION: FROM ACOUSTIC INPUT TO TRAVELING WAVE

The acoustic pressure signal that drives the motion of the tympanic membrane and middle-ear bones (see chapter: The Nature of Sound), results in an amplified vibratory motion of the stapes, which is attached to the oval window at the cochlear *base*. The cochlea is filled with an ionic fluid that can flow between the interconnected bony structures of the cochlea and vestibular canals (Fig. 5.1). The cochlea is divided into three chambers, separated from each other by bony walls, and called *Scala Vestibuli, Scala Media*, and *Scala Tympani*, respectively (Fig. 5.4). The partitions are connected at the far end of the cochlea (the *apex*), by an opening, known as *helicotrema*.

The movements of the stapes exert a time varying–inward outward directed force at the basal end of the Scala Vestibuli (Fig. 5.2). Because fluid is incompressible, the resulting pressure is released with an opposite sign at the elastic *round window* at the base of the Scala Tympani. This means that across the *Scala Media*, the absolute *pressure difference* between Scala Vestibuli and Scala Tympani varies from a maximum value at the cochlear base (at oval and round windows), to zero at the helicotrema, which acts as a pressure short-circuit. It is this time-varying pressure difference across the length of the central cochlear partition, $\Delta p(x,t)$, which acts as the *driving force* for hearing. For a pure tone, that is, a harmonic sound with frequency ω rad/s, the pressure from the stapes on the fluid at the base of the Scala Vestibuli ($x = 0$) is described by

$$p_{\text{base}}(t) = A_{\text{base}} \sin(\omega t) \tag{5.1}$$

with A_{Base} the pressure amplitude at the base. The pressure *difference* at the cochlear base is then given as (Fig. 5.3):

$$\Delta p_{\text{base}}(t) = 2A_{\text{base}} \sin(\omega t) \tag{5.2}$$

Along the cochlear partition (with coordinate, x), this pressure difference gradually falls to zero when it reaches the apex (in the human cochlea, at about

The Auditory System and Human Sound-Localization Behavior. http://dx.doi.org/10.1016/B978-0-12-801529-2.00005-2

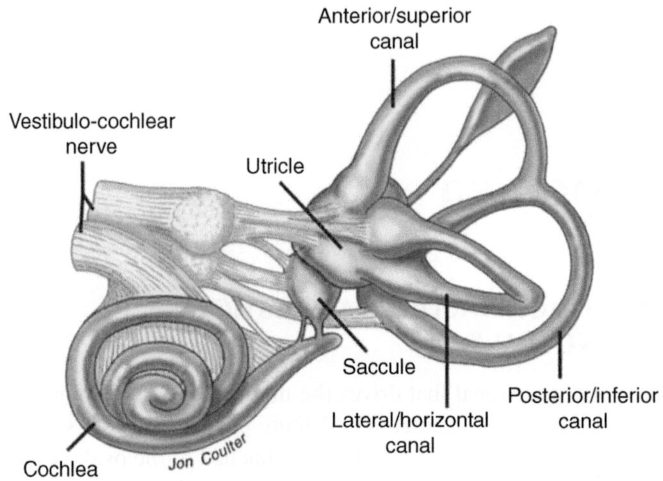

FIGURE 5.1 The bony structures of the inner ear. It shows the vestibular organs (the three semicircular canals, which measure three-dimensional head rotations, and the otoliths: saccule and utricle, which respond to linear head accelerations and gravity), and the cochlea, the sensory organ for audition. Output is transmitted to the CNS by auditory and vestibular nerves. *(Source: Jon Coulter, with kind permission.)*

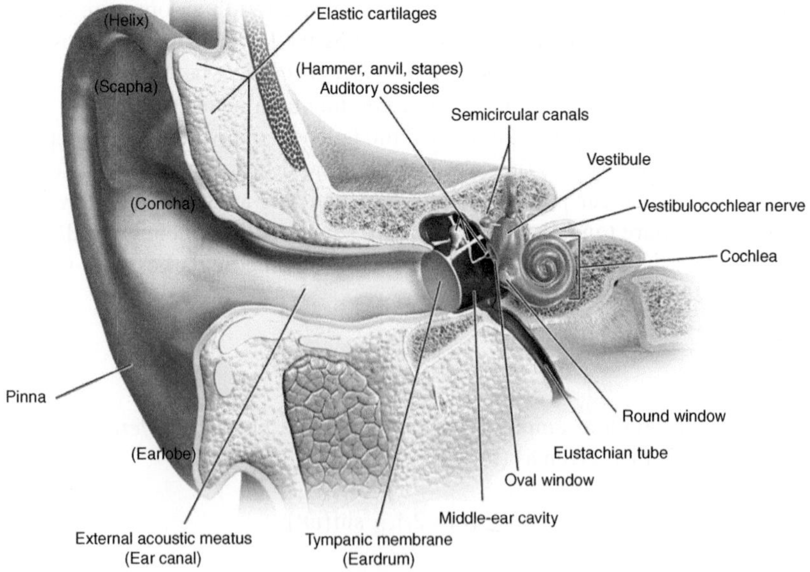

FIGURE 5.2 Anatomy of the ear: outer ear, middle ear, and cochlea. The cochlea is acoustically stimulated by the vibratory movements of the stapes at the oval window. The round window moves in antiphase with respect to the oval window. *(Source: Wikipedia.)*

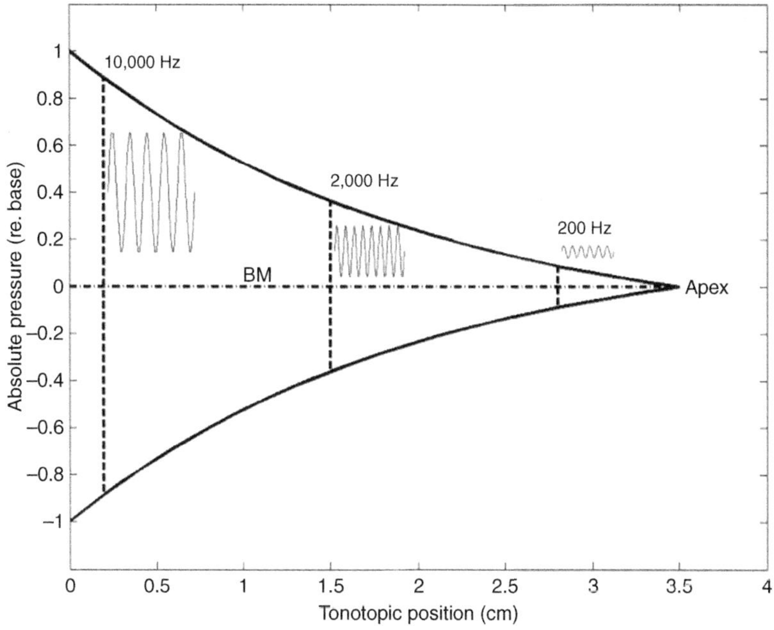

FIGURE 5.3 **The amplitude of the pressure across the BM decays with distance from the base at $x = 0$** [Eq. (5.4)]. The oscillating driving forces at different locations therefore have very different amplitudes too (here illustrated to scale for three frequencies at $x = 0.2$, 1.5, and 2.8 cm from the base; here: $\beta = 0.5\ \text{cm}^{-1}$, and $x_{\text{apex}} = 3.5$ cm).

$x_{\text{apex}} = 3.5$ cm from the base), making the pressure difference strongly location-dependent. To give a first, very simple idea, it may appear like:

$$\Delta p(x,t) = 2A_{\text{base}}a(x)\sin(\omega t) \tag{5.3}$$

with $a(x)$ a function that decays monotonically from one to zero between base and apex (see Section 5.3, where we describe a full linear model for the cochlear hydro-dynamics). For example, if the pressure difference were to fall exponentially, with a spatial constant of $\beta\ \text{cm}^{-1}$, it would read (Fig. 5.3):

$$\Delta p(x,t) = 2A_{\text{base}}\frac{e^{-\beta x} - e^{-\beta x_{\text{apex}}}}{1 - e^{-\beta x_{\text{apex}}}}\sin(\omega t) \tag{5.4}$$

As the speed of sound in water is 1,500 m/s, the pressure along Scala Vestibuli and Tympani develops virtually instantaneously. However, due to the hydrodynamic interactions between the fluid and the acoustic impedance of the basilar membrane (BM), it takes about 5–6 ms for the time-varying pressure difference to develop along the cochlear partition from base to apex (Section 5.3).

The BM (Fig. 5.4) responds to the dynamic pressure difference with a vibratory motion that has the same frequency as the driving acoustic (pure-tone) input. Along the length of the cochlea, the envelope of the BM vibrations appears as a

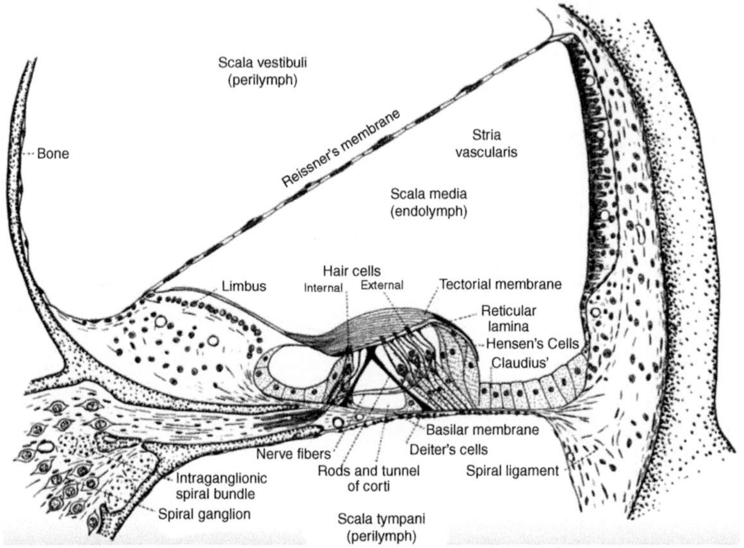

FIGURE 5.4 Cross section of the cochlear partition, showing the organ of Corti within Scala Media. Important structures are the BM, the inner hair cell, the three outer hair cells (OHCs) and the tectorial membrane to which the OHCs' stereocilia attach. Reissner's membrane is considered acoustically transparent; Scala Media and Scala Vestibuli form a single acoustic compartment. *(Source: Davis and Associates, 1953, with kind permission).*

traveling wave that propagates from base to apex (Fig. 5.5) with a group velocity that depends on the location along the BM, $v_g(x)$, and an envelope amplitude that gradually grows to its peak at a frequency-dependent location, $A_{\text{peak}}(x)$: high-frequency harmonics peak near the BM base, while low-frequency tones reach

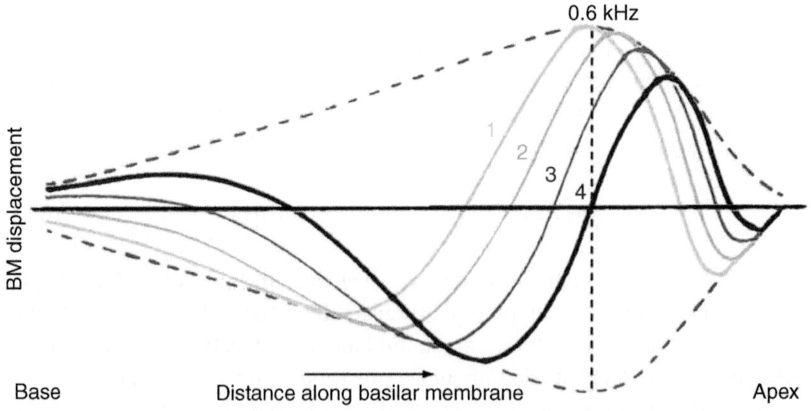

FIGURE 5.5 A pure-tone sound causes a traveling wave that propagates along the BM from base to apex. The wave reaches its peak amplitude at a frequency-dependent location, after which it quickly drops to zero. Here the wave is shown at four time points. The *red dashed line* delineates the wave's envelope. Note the change of the wavelength along the length of the BM, which cooccurs with a systematic decrease in the wave's group velocity, until it comes to a full stop beyond the peak.

their peak amplitude near BM apex. The wave amplitude quickly drops to zero beyond the peak.

The time it takes for the traveling wave to reach its peak amplitude varies with location too. Relative to the cochlear base, the lowest frequencies near the BM apex reach their peak amplitude about 5–6 ms later, which leads to substantial *dispersion* of the acoustic signal when it contains multiple frequencies!

These interesting properties of the traveling wave all result from location-dependent properties of the BM impedance. In first (ie, *linear*) approximation, the BM responds to the local pressure variations as a second-order, damped oscillator (chapter:Linear Systems; see also further, Section 5.5), with a location-dependent elasticity (or: compliance, C): $Z(\omega) = Z[\omega, C(x,\omega)]$.

As a result of these local micromechanics, sound frequencies are mapped *tonotopically* along the BM, from high (base) to low (apex), running (in the human inner ear) from $f_{max} \approx 20$ kHz, down to $f_{min} \approx 50$ Hz. The tonotopic mapping is not linear, but may be approximated by a logarithmic function, in which each doubling (or halving) of the frequency (ie, at an octave interval) occupies a roughly constant spatial extent on the BM of about $\lambda \approx$ 4–5 mm. An approximate description for the tonotopy may therefore be (Fig. 5.6):

$$x_{BM} = -\lambda \log_2 \left(\frac{f}{f_{max}} \right) \text{ cm} \tag{5.5}$$

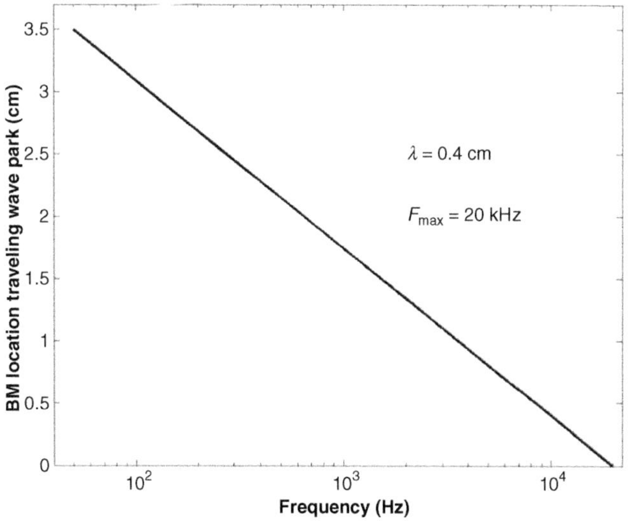

FIGURE 5.6 The tonotopy along the BM may be approximated by Eq. (5.5). Here, $\lambda = 0.4$ cm, and $f_{max} = 20$ kHz.

The BM wave dynamics are quite different from the simple (nondispersive) constant-amplitude and constant velocity traveling waves that arise, for example, in an elastic spring or a rope under constant tension, or in air with a constant bulk modulus (as outlined in chapter: The Nature of Sound), as such mechanical systems obey the simple homogeneous and nondispersive wave equation, Eq. (2.25).

Perhaps a better way to describe the BM traveling wave would be that of a shallow-water wave that breaks on the beach: in such water waves, the wave velocity strongly depends on depth, so that these waves obey a nonlinear dispersion relation. As the wave velocity decreases with decreasing depth (impedance), the wave amplitude increases (because of the incompressibility of water), until the wave crest topples over and breaks. The properties of water waves result from a few basic physical principles that apply to fluids (ie, in easily deformable, but incompressible media). The next section summarizes some of the interesting physics underlying traveling shallow-water waves, as a nice metaphor for the BM traveling wave.

5.2 BASIC PHYSICS UNDERLYING WATER WAVES

The tsunamis that devastated South-East Asia on Christmas eve 2004 (Fig. 5.7), and more recently on Mar. 11, 2011 near Sendai, in Japan, gave new meaning to the concept of "water waves." In the beginning, the story went that waves of many meters high had traveled over the ocean at a tremendous speed. However, the actual tsunami wave in the middle of the ocean is usually not high at all (typically, only several tens of cm, and hardly noticeable if you happen to be on a boat), but the wave crest extends over a vast distance (easily extending over a few hundred kilometers!). Moreover, the speed of this traveling, pulse-shaped, wave is about

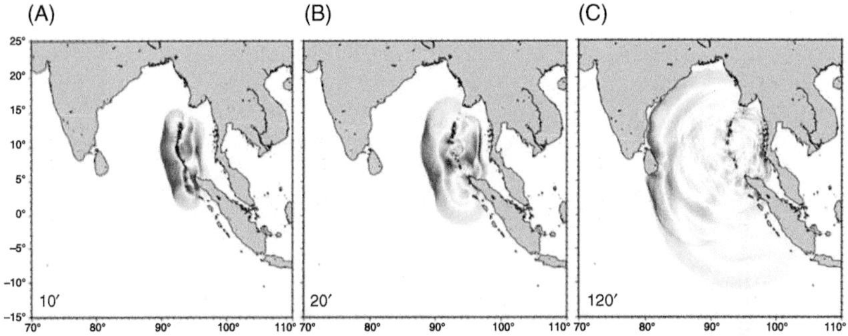

FIGURE 5.7 The tsunami in the Indian Ocean caused by an extremely strong earthquake on the ocean floor at a depth of about 5 km near Sumatra on Christmas eve, 2004. The three consecutive shots (taken from NASA satellite observations) show the amplitude of the traveling wave (*red*, positive; *blue*, negative) at respectively (A) 10 min, (B) 20 min, and (C) 120 min after the onset of the earthquake. The speed of this nondispersive "shallow"-water wave is about 1000 km/h, and its wavelength is several hundred kilometers long. Note the reflective waves (*blue*) that bounce off the small islands as given in part B. *(Source: dr Kenji Satake, Tokyo, with kind permission).*

800–1000 km/h (ie, over 250 m/s!). When such a tsunami approaches the coast, the wave crest becomes higher and higher, while its speed decreases strongly. Finally, the wave breaks on the coast, thereby transferring an enormous amount of mechanical energy. What are the physical mechanisms behind this type of waves?

Here we spend a brief time on the physics underlying water waves. Although the model will be strongly simplified, it will nevertheless be quite useful to understand the most important properties of water waves. Some of these properties are also relevant for understanding the basic hydrodynamics within the cochlea that are described in the next section, and to appreciate the nondispersive traveling wave described earlier.

Dry water: The first simplification in the physical model of water waves is that of so-called "dry," nonviscous, water, which means that we ignore internal friction between the different water layers. We will further limit the description to relatively small wave amplitudes, evoked by a harmonic, sinusoidal, perturbation. Third, we assume that the water wave propagates essentially in one dimension (the x-direction): this allows for a description of a straight wave front, parameterized by a single wavelength, in which the peaks and valleys follow straight, parallel lines, perpendicular to the propagation direction in the (x,z)-plane. We will see that despite these simplifications the dispersion relation of the water waves, $\omega(k)$, with $k = 2\pi/\lambda$ the wave number, can be nonlinear (Crawford, 1968).

At equilibrium the water surface is flat and horizontal. When a perturbation causes a wave, two restoring forces will cause the peaks of the wave to return to equilibrium: gravity, g, and the surface tension, T, of the water. Moreover, because water is incompressible, the excess of water in the wave's peak has to come from neighboring troughs. As a result, the water particles (here described as tiny water volumes, think of infinitesimal "droplets") will undergo combined *longitudinal* and *transversal* movements. We will derive the following properties of water waves:

- when the water depth $h \ll \lambda$ we speak of *shallow-water waves*, or tidal waves. The particles will move along straight lines, and the wave speed is independent of λ, only depending on h (nondispersive waves).
- when $h \gg \lambda$ we have *deep-water waves*. Now, the particles move in circles with a radius that depends on their equilibrium position in depth. Furthermore, the wave speed depends on λ (dispersive waves).

Consider an infinitely large reservoir at a uniform depth, h. In equilibrium the water surface is horizontal, and lies in the plane $y = 0$; the bottom is at $y = -h$. Position (x,y) refers to the *equilibrium* position of a water particle, where $x \in [-\infty,+\infty]$ and $y \in [-h,0]$. A water drop is thus conveniently labeled by its equilibrium location, irrespective of where it actually is during the wave motion. The water wave is a 2D movement of particles within in the (x,y)-plane (we may ignore the z-direction), which is described by a wave-vector function:

$$\vec{\psi}(x,y,t) = \psi_x(x,y,t)\hat{x} + \psi_y(x,y,t)\hat{y} \tag{5.6}$$

with $\hat{x}$ the longitudinal movement direction of the particle, and $\hat{y}$ its transversal movement direction. The speed of the water droplet at (x,y) is therefore:

$$\vec{u}(x,y,t)=\frac{\partial\vec{\psi}(x,y,t)}{\partial t}=\frac{\partial\psi_x}{\partial t}\hat{x}+\frac{\partial\psi_y}{\partial t}\hat{y}\equiv v\hat{x}+w\hat{y} \tag{5.7}$$

We now impose two important physical boundary conditions on the dry water:

1. Conservation of mass, and
2. Absence of turbulent motion (rotation-free water).

The first condition uses the incompressibility of water, and states that the total amount of water that enters and leaves a certain water-filled volume per unit of time should be zero. This constraint leads to the *continuity equation*, which for motion in 2D (the x–y plane) reads:

$$\rho\left[\frac{\partial v}{\partial x}+\frac{\partial w}{\partial y}\right]=\rho\left(\vec{\nabla}\vec{u}\right)=0 \tag{5.8}$$

with ρ the density of water, and $\vec{u}=(v,w)$ is the velocity of the water particles in the x and y directions, respectively. Using Eq. (5.7) yields

$$0=\rho\left(\vec{\nabla}\frac{\partial\vec{\psi}}{\partial t}\right)\Rightarrow\left(\vec{\nabla}\vec{\psi}\right)=\text{constant}=0. \tag{5.9}$$

The integration constant is zero, because the water volume is assumed to be homogeneous (ie, there are no "air bubbles").

The second condition means that at infinitesimal scale there is no net rotation in the flow field of the liquid, which is mathematically formulated as:

$$\vec{\nabla}\times\vec{u}=0\Rightarrow\vec{\nabla}\times\vec{\psi}=0\quad\text{and in}\quad 2\text{D:}\left(\frac{\partial\psi y}{\partial x}-\frac{\partial\psi x}{\partial y}\right)\hat{z}=0 \tag{5.10}$$

The two constraints of Eqs. (5.9) and (5.10) provide the necessary boundary conditions to determine solutions for harmonic traveling waves. We therefore propose as tryout solutions for the horizontal and vertical components of the wave vector:

$$\begin{aligned}\psi_y(x,y,t)&=\text{A}\cos(\omega t-kx)f(y)\\ \psi_x(x,y,t)&=\text{A}\sin(\omega t-kx)g(y)\end{aligned} \tag{5.11}$$

where the unknown functions $f(y)$ and $g(y)$ should be determined by imposing the constraints of Eqs. (5.9) and (5.10). It is left as an Exercise to the reader to show that the full solution then is given by:

$$\begin{aligned}\psi_y(x,y,t)&=\text{A}\cos(\omega t-kx)\left[e^{ky}-e^{-2kh}e^{-ky}\right]\\ \psi_x(x,y,t)&=\text{A}\sin(\omega t-kx)\left[e^{ky}+e^{-2kh}e^{-ky}\right]\end{aligned} \tag{5.12}$$

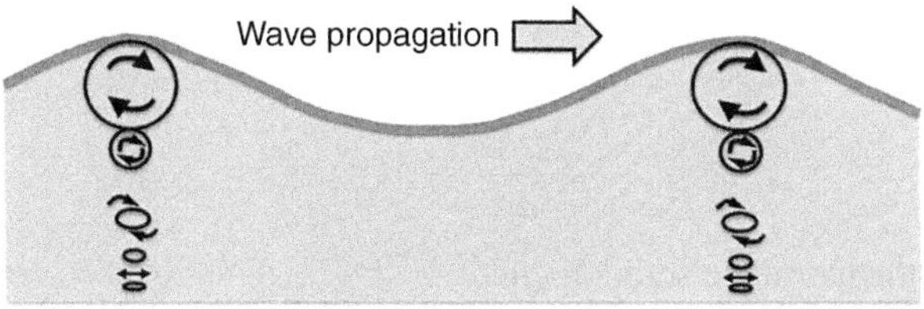

FIGURE 5.8 **Movement of individual water particles in a deep-water harmonic traveling wave at different depths.**

This wave function predicts *elliptic trajectories* for the water particles, where the longitudinal movement along the wave-propagation direction (x) is harmonic, and the transversal movement (y) depends on the location of the particle, on the wavelength, and on the basin depth (Fig. 5.8). To get more insight into the behavior of the water particles, we highlight two extreme cases: very short versus very long wavelengths with respect to the depth, h. The former case refers to *deep-water* waves, the latter to *shallow-water* waves. The general solution of Eq. (5.12) incorporates these two extremes.

Deep-water waves: When $h \ll k$, the factor containing $\exp(-2kh) \ll 1$ and may be neglected. In that case

$$\begin{aligned}\psi_y(x,y,t) &= \mathrm{A}\cos(\omega t - kx)e^{ky}\\ \psi_x(x,y,t) &= \mathrm{A}\sin(\omega t - kx)e^{ky}\end{aligned} \tag{5.13}$$

which describes harmonic *circular* motion of the water particles, with a radius that decreases exponentially with y (note that $y < 0$!). At the top op the wave crest the particles move forward, while in a trough they move backward. Note that $\exp(ky) = \exp(-|y|/\bar{\lambda})$ where $\bar{\lambda} \equiv \lambda/2\pi$ is the *reduced* wavelength (Fig. 5.8).

Shallow-water waves: In this case $y \ll \bar{\lambda}$ and $h \ll \bar{\lambda}$, which leads to the following solution (see Exercises):

$$\begin{aligned}\psi_y(x,y,t) &= 2\mathrm{A}\cos(\omega t - kx)\cdot k(y+h)\\ \psi_x(x,y,t) &= 2\mathrm{A}\sin(\omega t - kx)\end{aligned} \tag{5.14}$$

This reduces to purely horizontal particle motion on the bottom of the reservoir (where $y = -h$), and to elliptical motion on the surface (where $y = 0$).

Dispersion relation: To get at the dispersion relation for water waves we have to incorporate the forces that act on the water particles. Two forces play a role in dry water waves: gravity, g, and surface tension, T (where we neglect viscosity as a third force). We use the fact that for general harmonic motion (be it a pendulum, an electric circuit, water, etc.) the following statement holds (Crawford, 1968):

> The total restoring force per unit displacement per unit mass = ω^2.

For example, Newton's second law describes the (linearized) pendulum (Eq. 3.4) by:

$$ML\frac{d^2\psi}{dt^2} = -Mg\psi$$

from which the harmonic solution yields $\omega^2 = g/L$. This can also be written as:

$$\omega^2 = \frac{Mg\psi}{(L\psi)M} = \frac{\text{restoring force}}{\text{unit of displacement} \times \text{mass}}$$

For a given vibrational mode of the water waves, all particles have the same frequency, and hence the same ω^2. Thus, we can obtain the dispersion relation between the spatial wave mode (k) and the driving frequency (ω), by analyzing the movements of a single water particle.

Gravitational waves: In case the gravitational forces dominate the surface tension, we have so called *gravitational* water waves.

The highlighted gray volume of water in Fig. 5.9 undergoes a force that is proportional to the pressure difference along x: $\Delta p(x) = p(x + \Delta x) - p(x)$. This pressure gradient is caused by the difference in height (and hence by the volume) of the water, determined by the wave shape: $\psi_y(x + \Delta x) - \psi_y(x)$.

In other words:

$$\begin{aligned} F_x &= -L\Delta y p(x) \\ &= -L\Delta y \rho g\left[\psi_y(x+\Delta x) - \psi_y(x)\right] \\ &\approx -\Delta M g \left.\frac{\partial \psi_y}{\partial x}\right]_{y=0} \end{aligned} \tag{5.15}$$

The partial derivative is obtained from the general spatial solution of the traveling water waves, Eq. (5.12). The restoring force equals the

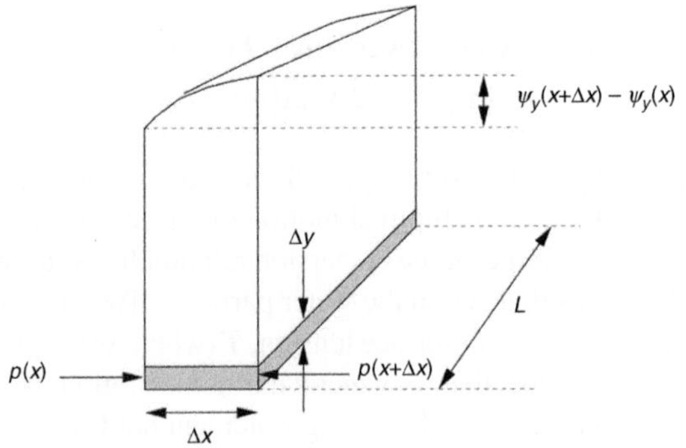

FIGURE 5.9 A small volume of water (Δx,Δy,L) at depth y, subjected to the pressure from the force of gravity by the water volume above it. The pressure difference is due to the mass of the wave crest, and is given by $\rho g \Delta \psi_y$.

acceleration on the mass, ΔM, and for harmonic motion at frequency ω, this yields

$$F_x = \Delta M \left.\frac{\partial^2 \psi_x}{\partial t^2}\right]_{y=0} = -\Delta M \omega^2 \psi_x\big]_{y=0} \tag{5.16}$$

Combining the results gives the dispersion relation for gravitational waves:

$$\omega^2(k) = gk\frac{1-e^{-2kh}}{1+e^{-2kh}} = gk\tanh(kh) \tag{5.17}$$

In the Exercises the reader can show that the phase velocity, $v_\varphi = \omega/k$, is then given by:

- for deep-water waves:

$$v_\varphi = \sqrt{g\lambda}\ \text{(dispersive waves)} \tag{5.18}$$

- for shallow-water waves:

$$v_\varphi = \sqrt{gh}\ \text{(nondispersive waves)} \tag{5.19}$$

Deep-water waves are dispersive: when the wave is determined by a superposition of multiple frequencies, its shape will change as the wave progresses. For shallow-water waves the shape is preserved, as the different frequencies all travel at the same phase velocity that only depends on (assumed constant) depth. The tsunami should therefore be considered a *shallow-water wave*, despite the fact that the depth at which it originates may be 5 km or more! This explains why tsunamis can propagate unchanged over vast distances across the ocean (Fig. 5.7), and why they are so rare….

Surface tension: At the transition between water and air, where the water molecules lose their tight bonds to each other, and bond to air molecules, the water surface behaves somewhat as a stretched elastic membrane. This membrane resists deformation, and hence the restoring force is proportional (by tension constant, T) to the amount of deformation from a flat surface, which is quantified by the surface *curvature*. For a sinusoidal wave shape, the curvature is k^2 (to be checked in the Exercises), so that the restoring force due to the surface tension of a sinusoidal shape becomes:

$$F_x = -\Delta M \frac{Tk^2}{\rho}\left.\frac{\partial \psi_y}{\partial x}\right]_{y=0} \tag{5.20}$$

Combining both the influence of gravity and surface tension eventually leads to a more complete dispersion relation for water waves (see Exercises):

$$\omega^2(k) = \left(gk + \frac{Tk^3}{\rho}\right)\tanh(kh) \tag{5.21}$$

Depending on the strength of the tension, T, the shallow-water wave can become dispersive, as with the approximation $hk \ll 1$ and tanh $(kh) \approx kh$:

$$\omega^2(k) \approx \left(ghk^2 + \frac{Thk^4}{\rho} \right) \Rightarrow \omega(k) = k\sqrt{gh + \frac{Th}{\rho}k^2} \tag{5.22}$$

The phase- and group velocities thus become:

$$\begin{aligned} v_\varphi &\equiv \frac{\omega(k)}{k} = \sqrt{gh + \frac{Th}{\rho}k^2} = v_\varphi(k) \\ v_g &\equiv \frac{d\omega}{dk} = v_\varphi + \frac{2Th\frac{k^2}{\rho}}{v_\varphi} = v_g(k) \end{aligned} \tag{5.23}$$

In the typical situation, however, the terms containing Thk^2 can be neglected, so that the phase velocity is independent of the wavelength, only depending on the depth, h, and phase- and group velocities are identical. This describes non-dispersive waves, as long as h = constant. However, if h varies, the propagation velocity of the wave will change too.

5.3 THE LINEAR COCHLEAR MODEL (VON BÉKESY AND ZWISLOCKI)

To develop a hydrodynamic model of the cochlea, we follow the approach of Josef Zwislocki, who incorporated the Nobel Prize-winning experimental work of Georg Von Békésy (1961, Nobel Prize in *Physiology or Medicine*; Fig. 5.10).

FIGURE 5.10 Géorg Von Békésy 1899–1972, (A) and Josef J. Zwislocki (B).

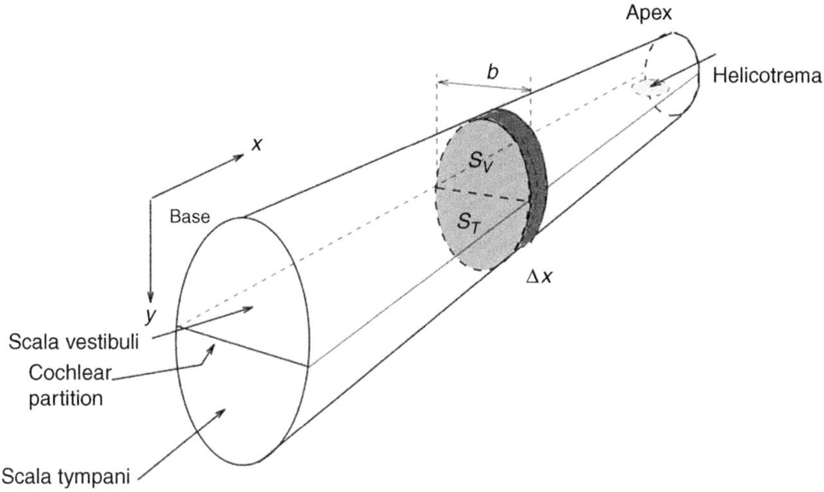

FIGURE 5.11 Schematic representation of the physical model of the cochlea. S_V, Scala vestibuli; S_T, scala tympani. The scalas are separated by the cochlear partition (BM, width b). At the apex the scalas are connected by the helicotrema. The highlighted part at location x, has length Δx, width $b(x)$, and scala cross sections $S_V(x)$ and $S_T(x)$.

In this model we imagine the cochlea as a linearly stretched compartment (Fig. 5.11). The input signal is the in- and outward movement of the stapes plate at the oval window. The cochlea is filled with fluid, and separated (for cross section, see Fig. 5.3) by a triangular structure, called the cochlear partition, or *organ of Corti*. In cochlear models this partition is flattened to represent an elastic, centrally running partition (the BM with supporting structures) between Scala Vestibuli (S_V) and Scala Tympani (S_T). In other words, Reissner's membrane (RM) does not interact with the acoustic hydrodynamics within the cochlea (it is acoustically transparent); as a result, the fluid within Scala Media can be incorporated within the compartment of Scala Vestibuli [see, however, recent evidence from Reichenbach et al. (2012), implicating RM in oto-acoustic emissions]. At the apex, the two compartments are connected by the helicotrema. When the stapes moves inward, the round window at the base of S_T moves outward. All elastic properties of the central partition are assigned to the BM: a high stiffness near the stapes (the base), decreasing gradually toward the helicotrema (apex).

The movements of the BM can be measured with the Mössbauer technique, or with laser interferometry. Measured speeds at the threshold of hearing are of the order of 0.05 mm/s at a frequency of 20 kHz; this corresponds to BM displacements of only 4 Å (check!).

We consider one-dimensional fluid motion, in which the compression wave propagates from the oval window in the x-direction toward the helicotrema. Locally, the fluid particles have velocity $v(x)$. In what follows, we apply continuity Eq. (5.9), which describes mass conservation.

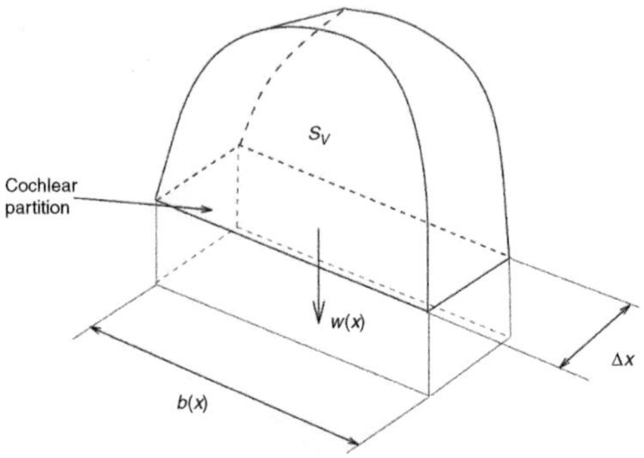

FIGURE 5.12 Scheme of the local volume change at x in the scala vestibuli due the up–down movement $w(x)$ of the BM. As an approximation, we consider the BM movement as a linear displacement, that is, without a change in shape of the partition.

Fig. 5.12 shows a schematic cross section of the Scala Vestibuli at location x along the BM. All variables are location-dependent: the width of the BM, $b = b(x)$, the cross section $S_V = S_V(x)$, and $S_T = S_T(x)$, the (horizontal) velocity of the fluid, $v(x)$, and the BM is supposed to move linearly upward and downward at velocity, $w(x)$. Note that what happens to the BM in S_T is a mirror image of what happens in S_V.

The mass-inflow per time unit to the volume element in S_V at location x is equal to the change in volume (velocity times cross section) times the density (which is constant because of incompressibility):

$$\Delta m(x) = v_V(x) S_V(x)(x)\rho \tag{5.24}$$

and at location $x + \Delta x$, to which we apply a first-order Taylor expansion:

$$\begin{aligned} \Delta m(x+\Delta x) &= v_V(x+\Delta x) S_V(x+\Delta x)\rho \\ &= \left[v_V(x) + \frac{\partial v_V}{\partial x}\Delta x \right]\left[S_V(x) + \frac{\partial S_V}{\partial x}\Delta x \right]\rho \\ &= \Delta m(x) + \rho \frac{\partial}{\partial x}(v_V S_V)\Delta x \end{aligned} \tag{5.25}$$

The net change in mass resulting from the pressure wave is therefore:

$$\Delta M^x(x) = m(x) - m(x+\Delta x) = -\rho \frac{\partial}{\partial x}(v_V S_V)\Delta x \tag{5.26}$$

The volume change as a result of the BM movement equals $w(x)b(x)\Delta x$, so that the change in mass per unit time for the vertical movement (in Fig. 5.11 the membrane moves downward, so a mass increase for S_V):

$$\Delta M^y(x) = \rho w(x) b(x)\Delta x \tag{5.27}$$

According to the continuity equation, the total mass change is zero:

$$\rho\Delta x\left[w(x)b(x)+\frac{\partial}{\partial x}(v_V S_V)\right]=0 \tag{5.28}$$

For Scala Tympani, we obtain a similar condition, with only the BM velocity inverted:

$$\rho\Delta x\left[-w(x)b(x)+\frac{\partial}{\partial x}(v_T S_T)\right]=0 \tag{5.29}$$

To determine the relation between BM movement, $w(x)$, and the pressure wave in the cochlea, we have to analyze the force balance on the volume elements in S_V and S_T. The three forces that play a role in this problem are the *net pressure* on the volume element, the *inertia* (Newton's second law), and the *viscous forces* on the fluid. Since the volume element does not accelerate through the cochlea, the sum of these forces adds to zero. First, we consider S_V:

1. Suppose that the pressure in x is $p_V(x)$, then the force at the cross-sections in x and $x + \Delta x$ is, in first-order Taylor approximation:

$$\begin{aligned} F_V^p(x) &= p_V(x)S_V(x) \\ F_V^p(x+\Delta x) &= p_V(x)S_V(x)+\frac{\partial(p_V S_V)}{\partial x}\Delta x \\ \Rightarrow F_V^{\Delta p}(x) &= -\frac{\partial(p_V S_V)}{\partial x}\Delta x \approx -S_V\frac{\partial p_V}{\partial x}\Delta x \end{aligned} \tag{5.30}$$

   where the approximation holds when the cross-section varies slowly with x.
2. The impulse of the liquid in the volume element (mass × velocity) is $I(x)=\left[\rho S_V(x)\Delta x\right]v_V(x)$, so that according to Newton's law:

$$F_V^N(x)\equiv\frac{\partial I(x)}{\partial t}=\rho\frac{\partial(S_V v_V)}{\partial t}\Delta x\approx\rho S_V\frac{\partial(v_V)}{\partial t}\Delta x \tag{5.31}$$

   where the cross section is assumed to vary little over time.
3. In a viscous liquid the frictional force is proportional to the speed of the liquid particles:

$$F_V^{\text{visc}}(x)=-\left[R_V(x)S_V(x)\right]v_V(x)\Delta x \tag{5.32}$$

   with $R_V S_V$ is the total resistance per unit length.

   The force balance thus reads:

$$F_V^N(x)-F_V^{\Delta p}(x)-F_V^{\text{visc}}(x)=0 \tag{5.33}$$

and reduces to:

$$\rho\frac{\partial v_V}{\partial t}+R_V v_V=-\frac{\partial p_V}{\partial x} \quad \text{together with}$$
$$wb+S_V\frac{\partial v_V}{\partial x}=0 \tag{5.34}$$

The force balance and continuity equation for S_T gives a very similar result, as the only difference is the sign of the movement, w, of the BM:

$$\rho\frac{\partial v_T}{\partial t}+R_T v_T=-\frac{\partial p_T}{\partial x}$$
$$-wb+S_T\frac{\partial v_T}{\partial x}=0 \tag{5.35}$$

These coupled differential equations can be readily uncoupled by applying the following trick: differentiate the force equation with respect to x, and the continuity equation with respect to t. We further assume that the resistance varies slowly with x, so that its spatial derivative may be neglected. Adding the two equations then gives for each of the two compartments (but check):

$$-\frac{b}{S_V}\left(\rho\frac{\partial w}{\partial t}+R_V w\right)=-\frac{\partial^2 p_V}{\partial x^2}$$
$$+\frac{b}{S_T}\left(\rho\frac{\partial w}{\partial t}+R_T w\right)=-\frac{\partial^2 p_T}{\partial x^2} \tag{5.36}$$

If we now take the difference between S_V and S_T, approximate that $R_V=R_T\approx R$, and define the effective cross section $(1/S)=(1/S_V)+(1/S_T)$, we get:

$$\frac{b}{S}\left(\rho\frac{\partial w}{\partial t}+Rw\right)=-\frac{\partial^2 \Delta p}{\partial x^2} \tag{5.37}$$

where $\Delta p(x,t)$ is the instantaneous pressure difference across the BM, and $w(x,t)$ is the BM velocity. These two quantities are related through Eq. 2.46:

$$p(x,t)=b(x)w(x,t)Z(x,\omega) \tag{5.38}$$

This relation allows us to eliminate the BM velocity, leading to an equation that only contains the instantaneous pressure difference:

$$\frac{1}{SZ(x,\omega)}\left[\rho\frac{\partial \Delta p(x,t)}{\partial t}+R\Delta p(x,t)\right]=\frac{\partial^2 \Delta p}{\partial x^2} \tag{5.39}$$

This is the central equation of cochlear hydrodynamics, with a minimum of approximations and assumptions. Note that Eq. (5.39) a *linear* differential equation in the pressure difference, so that it describes essentially a *linear* cochlear model! Despite this apparent oversimplification, the equation is still quite complex, as S, Z, and R are all functions of x. This makes an analytical treatment of Eq. (5.39) far from trivial.

Zwislocki's approximations: To analyze Eq. (5.39), Zwislocki made a number of simplifying assumptions and substitutions, taken from the experimental findings of Von Békésy.

As discussed in chapter: Linear Systems, linear systems can often best be dealt with in the frequency (or Laplace) domain. In that case, we assume that the sound input is a harmonic signal. In complex notation this is conveniently written as:

$$\Delta p(x,t) = P(x)e^{i\omega t} \tag{5.40}$$

Substitution into Eq. (5.39), and collecting the terms then yields

$$\frac{(R+i\rho\omega)}{SZ(x,\omega)} = \frac{1}{P}\frac{d^2P}{dx^2} \tag{5.41}$$

and using the impedance relation, Eq. (5.38), gives an equation for the BM velocity:

$$\frac{b(R+i\rho\omega)}{S} = \frac{1}{W}\frac{d^2W}{dx^2} \tag{5.42}$$

To arrive at the total displacement of the BM as function of distance to the stapes, $y(x)$, we integrate $w(x,t)$ after taking the inverse FT:

$$y(x) = \int_0^t w(x,\tau)d\tau \text{ and in the frequency domain: } Y(\omega) = \frac{W(x,\omega)}{i\omega} \tag{5.43}$$

Zwislocki proposed a number of physical-realistic descriptions for $R(x)$, $S(x)$, and $Z(x,\omega)$. For example, Fig. 5.13A suggests that the cochlear cross-section as function of the distance to the oval window can be reasonably well approximated by an exponential function:

$$S(x) = S_0 e^{-ax} \quad \text{with } S_0 \approx 0.0125 \text{ cm}^2 \quad \text{and } a \approx 0.5 \text{ cm}^{-1} \tag{5.44}$$

Analysis of the experimental results of the viscous component led to:

$$R(x) = R_0\sqrt{\omega}e^{-ax^2} \text{ with } R_0 \approx 2.24 \text{ (g/cm)}^3\text{s}^{-0.5} \tag{5.45}$$

The most general model for the BM impedance proposes that a local membrane element has an equivalent *mass* $M(x)$, an *elasticity* (or

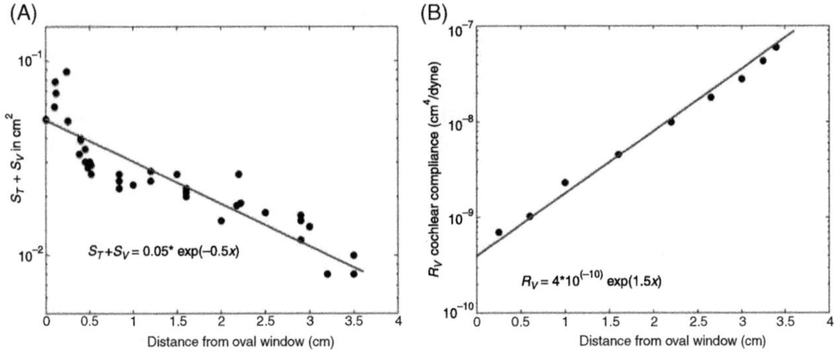

FIGURE 5.13 (A) Cross-section of scala vestibuli and scala tympani as a function of x for the human cochlea. (B) Static compliance of the BM as function of x.

compliance), $C(x)$, and a *damping* $R_m(x)$. The membrane impedance is then described by:

$$Z(x,\omega)=R_m(x)+i\left[\omega M(x)-\frac{1}{\omega C(x)}\right] \tag{5.46}$$

Following the measurements of Von Békésy, Zwislocki proposed the following approximations:

$$\begin{gathered} M(x)\approx 0 \\ R_m(x)\approx R_m = 468 \\ C(x)=C_0 e^{-hx} \quad \text{with } C_0 \approx 4\times 10^{-10}\ \text{cm}^4/\text{dyne},\ h\approx 1.5\ \text{cm}^{-1} \end{gathered} \tag{5.47}$$

[for data on $C(x)$, see Fig. 5.13B]. Substitution of these experimentally inspired relations in Eq. (5.39) then yields the following, formidable equation:

$$\frac{1}{P}\frac{d^2P}{dx^2}=\frac{R_0\sqrt{\omega}\exp\left(\frac{ax}{2}\right)+i\omega\rho}{S_0\exp(-ax)\left[R_m-i\exp(-hx)/(\omega C_0)\right]} \tag{5.48}$$

Finally, after applying the following approximations (which were supported by numerical substitution of the measured values):

- $\omega R_m C_0 \ll 1$
- $R_0 R_m C_0 \ll \rho$
- $\omega R_m C_0 + R/(\omega\rho) \ll 1$

Eq. (5.48) then finally reduces to a relatively standard differential equation (Exercise):

$$-\frac{1}{P}\frac{d^2P}{dx^2}=\frac{\omega^2 C_0\rho}{S_0}e^{(h+a)x} \tag{5.49}$$

This equation can be transformed into a standard Bessel equation, for which analytical solutions are available.

The solution following from Eq. (5.49) (normalized to the pressure at the stapes, $x = 0$), after substituting the measured values, is eventually given by (Dallos, 1973):

$$\text{Amplitude: } |P(x,f)| = P_0 \exp\left[-\frac{1}{2}x - 6.7\times 10^{-10} f^2\left(e^{2.5x}-1\right) - 1.84\times 10^{-3}\sqrt{f}\left(e^{1.25x}-1\right)\right]$$

$$\text{Phase: } \Phi(x) = \Phi_0 - 1.12\times 10^{-3} f\left(e^{x}-1\right) \qquad (5.50)$$

where x is in centimeters, and f in Hertz. The pressure amplitude decreases monotonically when both f and x increase. At a constant driving frequency, the amplitude of the pressure wave decreases in the direction of the helicotrema [where it becomes vanishingly small; cf. the highly simplified Eq. (5.3)].

At a fixed position, the pressure decreases with increasing frequency (Fig. 5.14). In other words:

- High frequencies give rise to a significant pressure amplitude over only a restricted range of the BM.
- Low frequencies cover a wider spatial range (Fig. 5.14A).
- Locations away from the stapes are progressively delayed. This phase-lag is strongly frequency-dependent (Fig. 5.14B).

The main quantity of interest, however, is the vertical displacement of the BM $y(f, x)$ (see previous sections), which can be obtained from the pressure wave by the impedance equation, Eq. (5.38), and subsequent integration, Eq. (5.43). Using the same approximations as above, the amplitude of the impedance yields:

$$\frac{1}{|Z|} \approx \omega C\sqrt{1+\omega^2 R_m C^2} \approx \omega C = 2\pi f \times 4\times 10^{-10} e^{x} \qquad (5.51)$$

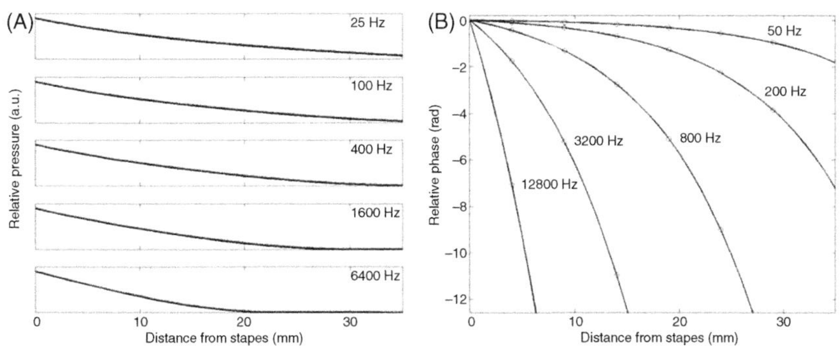

FIGURE 5.14 Properties of the pressure wave. Calculated according to Eq. (5.50) for five different frequencies: relative amplitude (A) and relative phase (B). Note the strong frequency- and location-dependence of both quantities.

The displacement of the BM is then [from Eq. (5.43)]:

$$|y(x,f)| = \frac{\omega C|P(x,f)|}{\omega}$$

This determines the relative amplitude of the BM traveling wave with respect to the stapes.

$$|y(x,f)| = |y(0,f)| \exp\left[x - 6.7\times 10^{-10} f^2 \left(e^{2.5x}-1\right) - 1.84\times 10^{-3}\sqrt{f}\left(e^{1.25x}-1\right)\right] \quad (5.52)$$

The phase of the BM displacement (with the used approximations) is equal to that of the pressure wave, Eq. (5.50). Fig. 5.15 shows that the linear model of Eq. (5.52) predicts a nice monotonic relationship between the location of the maximum of the traveling wave, and the driving frequency of the stapes. It demonstrates the asymmetry of the wave shape, with its steep roll-off, which has a similar appearance as a shallow water wave breaking on the beach (Section 5.2). It also accounts for the logarithmic tonotopy of the BM (right).

The cochlea thus functions as a mechanical frequency analyzer that acts as a tonotopically organized, parallel set of frequency specific–band pass filters. The temporal behavior of the BM vibration along the partition is best described as a *traveling wave*.

Note, however, that the wavelength is not constant along the partition, but decreases as a function of x. This behavior requires an analysis of the wave's

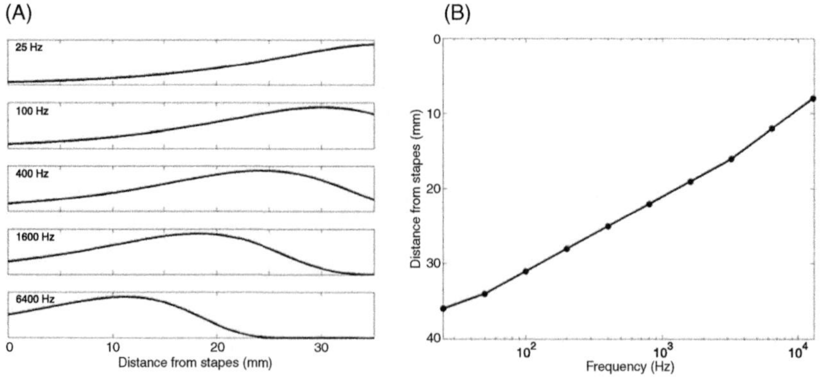

FIGURE 5.15 Envelope of the traveling wave along the length of the BM for five different frequencies, according to Eq. (5.52). Notice that the width of the wave, as well as the location of the peak systematically shifts with sound frequency (A). (B) Tonotopy along the BM for the peaks of the traveling wave (cf. with Fig. 5.6).

phase velocity (chapter: The Nature of Sound). The BM oscillates with $y(x,t)$, so that the phase velocity is defined by:

$$v_\varphi = \frac{dx}{dt} \tag{5.53}$$

The instantaneous phase of the traveling wave is determined by [making use of Eq. (5.50)]

$$\Phi(t) = \omega t + \Phi(x,\omega) = \omega t + \Phi_0 - 1.8\times 10^{-4}\,\omega\left(e^x - 1\right) \tag{5.54}$$

from which the phase velocity is found by differentiation:

$$v_\varphi \equiv \frac{\partial \Phi(t)}{\partial t} = \omega - 1.8\times 10^{-4}\,\omega e^x \frac{dx}{dt} \tag{5.55}$$

For locations that are precisely one wavelength apart, the phase is constant, so that $\partial\Phi/\partial t = 0$ yields the following important relation for the traveling wave:

$$v_\varphi = 5.6\times 10^3 e^{-x} \quad \text{cm/s} \tag{5.56}$$

This interesting result demonstrates that:

- The propagation speed of the traveling wave toward the helicotrema decreases exponentially with distance from the base.
- The speed of the wave is independent of sound frequency.
- The time it takes for the wave to reach a given point, x_0, on the cochlea is $T_0 = 1.8\times 10^{-4}(e^{x_0} - 1)$ s, taking about 5.7 ms to reach the helicotrema. This is in good agreement with the experimental observations of Von Békésy.

Note that according to the linear model, the traveling wave along the BM is entirely due to the interaction of *independent* BM elements with the local pressure difference in the fluid. The wave does not arise because BM elements transfer elastic energy in the longitudinal direction from one location to the next (like in a spring, in coupled oscillators, or in a rope); instead, the exchange of mechanical energy is with the local fluid only. Clearly, this description is an approximation of the real cochlear mechanics, as the membrane elements are connected to each other, and therefore must transfer mechanical energy. Yet, the simple linear model already yields realistic cochlear behavior, which is quite remarkable, and far from trivial. Moreover, it has important consequences for hearing. For example, if a certain portion of the BM cannot be set into motion, it will *not* affect the traveling wave!

Also a local rupture of the membrane, or ossification, leaves the traveling wave untouched. As a result, hearing of frequencies beyond the damage is unaffected! This has also been confirmed experimentally (eg, the "disco dip" at around 4 kHz).

5.4 THE ACTIVE, NONLINEAR COCHLEA: ROLE OF OUTER HAIR CELLS

Despite the apparent success of the linear cochlear model it cannot deal with some real fundamental problems. For example, Fig. 5.15 indicates that the extent of the traveling wave around the maximum of the BM vibration occupies about 1/3 of the total cochlear partition, which suggests that

1. fibers in the auditory nerve should be very broadly tuned, and
2. frequency selectivity of the auditory system is quite poor.

However, the frequency resolution of the human auditory system is actually very high (better than 0.5% over the full frequency range from 50 to 20 000 Hz). To measure the frequency selectivity of a given BM location one can determine its *tuning curves*, which quantify how the local amplitude of the vibration depends on the (relative) frequency.

Fig. 5.16 shows the normalized mechanical tuning curves, measured at five different BM locations by Von Békésy (apical data), and with the Mössbauer technique by Johnstone and Boyle (1967; at 18 kHz). The tuning curves clearly appear as band-pass filters, whereby their widths decrease with increasing frequency.

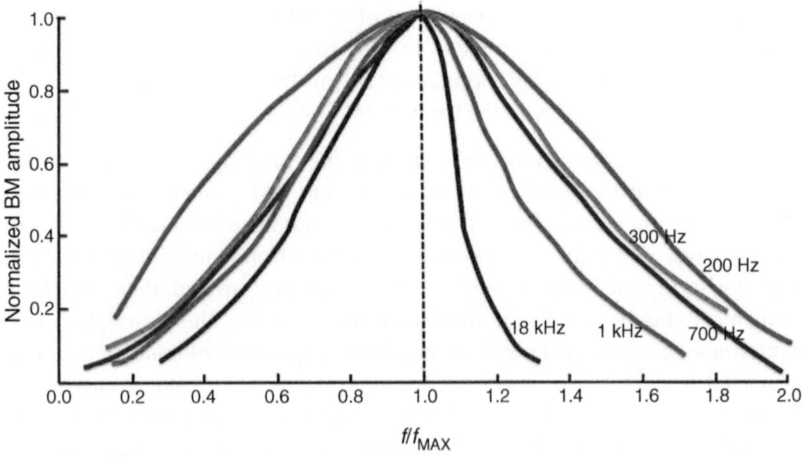

FIGURE 5.16 Normalized mechanical tuning curves for different cochlear positions as reported by Johnstone and Boyle (1967).

The effective bandwidth of a filter is quantified by its Q-factor, which is defined by:

$$Q_{10\,\mathrm{dB}} \equiv \frac{f_{max}}{BW_{10\,\mathrm{dB}}} \tag{5.57}$$

with $BW_{10\ \mathrm{dB}}$ the bandwidth at 10 dB below the maximum response. The linear model predicts Q-factors between 1.0 and 2.5; the measured BM data in Fig. 5.16 suggest Q-factors up to 3.5.

However, when comparing the (passive) BM filters to the tuning curves of auditory nerve fibers ($Q_{10\ \mathrm{dB}}$~8.5), a dramatic difference is evident (Fig. 5.17A). In particular, the low-frequency slope of the BM filter is far too low, and nerve fiber tuning curves are much sharper than predicted from the BM tuning characteristics.

A second feature that cannot be explained by the linear model concerns the extreme sensitivity of the auditory system, especially for stimuli at very low intensities. Moreover, responses saturate when the stimulus levels become too high.

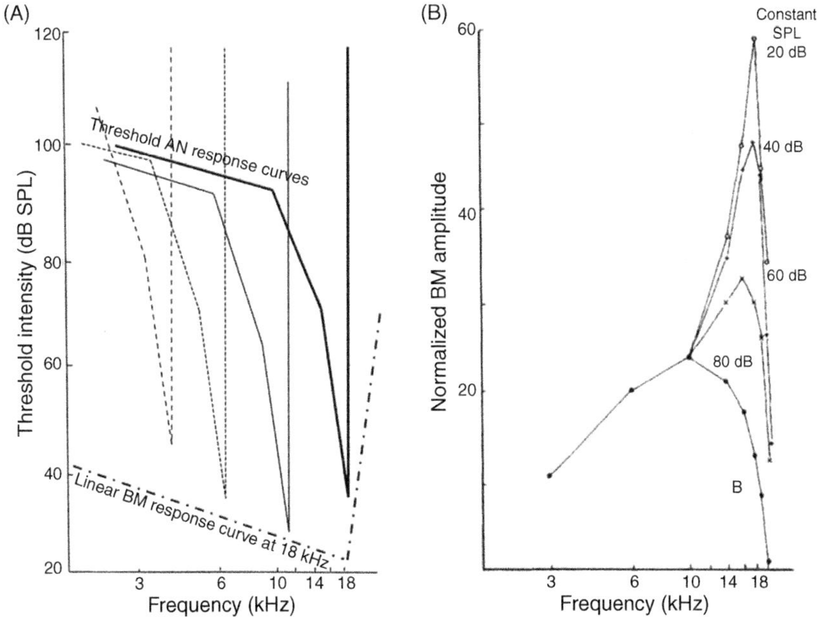

FIGURE 5.17 (A) Stylized tuning curves of four primary auditory nerve fibers, compared to the BM tuning curves (dotted line at the bottom). (B) Normalized gain functions of the BM for a 18.5 kHz tone presented at four different sound levels, between 20 and 80 dB SPL. The gain is defined as the BM velocity divided by the stapes input velocity. Note strong sharpening of the tuning curves at the lowest sound levels, indicative of nonlinear behavior. Note also the systematic shift of the peak to a higher best frequency, that is, to a more basal BM location, suggesting that the nonlinearity boosts the BM vibration at about 1/3–1/2 octave closer to the cochlear base than the passive tuning curve. *(Source: After Johnstone et al., 1986, with kind permission).*

These properties suggest the presence of a profound nonlinearity, as linear models do not produce different sensitivities for different sound levels, as the output scales linearly with the input strength (chapter: Linear Systems). Moreover, the nonlinearity appears to betray an *active* mechanism, as the highly increased sensitivity (high gain, in combination with sharpening of the BM tuning curve) may not be achievable by passive, dissipative, mechanisms without adding extra energy.

Indeed, measurements of the BM movements in the living cochlea with the Mössbauer technique demonstrated the existence of a spectacularly increased sensitivity for low sound levels, and saturation (stimulus-independent sensitivity, hence a linear response) for high levels (Fig. 5.17B). In the dead cochlea, however, these different response regimes disappear: in that case the BM again behaves as a linear system.

These important observations call for an essential modification of the linear Von Békésy–Zwislocki model. A crucial element in the nonlinear response behavior of the living cochlea is the involvement of outer hair cells (OHCs). Fig. 5.18B shows a magnified cross section of the organ of Corti with the inner (IHC) and OHCs.

The human cochlea contains about 12,000 OHCs and 3,500 IHCs. Against each IHC there are three OHCs. The flat top of the hair cells is covered with hair

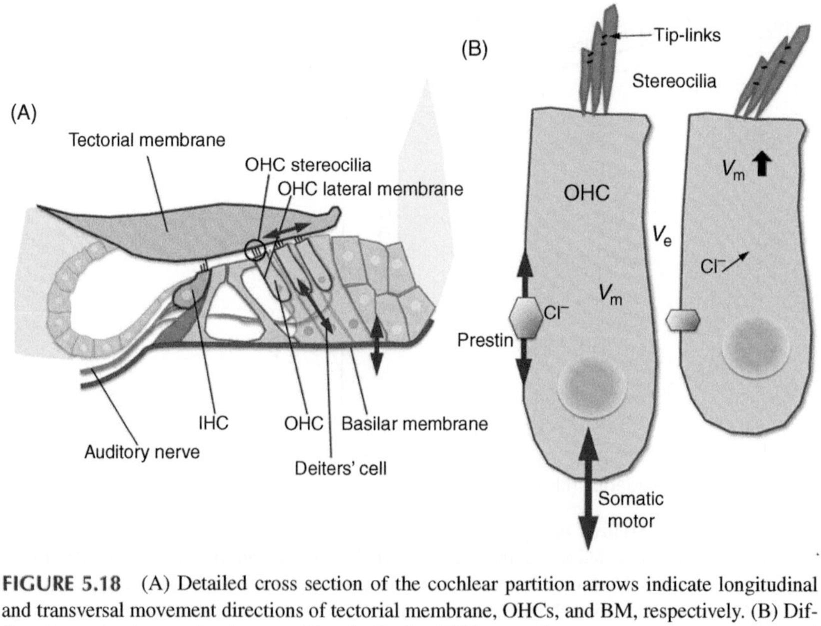

FIGURE 5.18 (A) Detailed cross section of the cochlear partition arrows indicate longitudinal and transversal movement directions of tectorial membrane, OHCs, and BM, respectively. (B) Different transduction mechanisms could underlie the OHC's somatic motor response: movement of the stereocilia, in combination with a rapid configuration change of prestin molecules. The resulting potential change across the cell length induces a motor response (length change) of the cell which in turn provides mechanical feedback (force) to the scala vestibuli/tympani.

bundles (*cilia*), which are made of sturdy filaments. On the OHC the cilia are arranged according to a wedge-shaped V or W form. Due to the fluid motion, the hair cell bundles can pivot around their pedestal. The resulting mechanical change of angle of the stereocilia causes a change of the internal potential of the hair cell. In the IHC this potential change can lead to the production of action potentials in the primary auditory nerve, according to the Hodgkin–Huxley spike-generation mechanism.

However, in the OHC the bending of the cilia induces a *length change* of the cell (Fig. 5.18B). This so-called *electro-motility* of the OHC has the *same* frequency as the acoustic signal that caused the cilia to bend, and it has in *in vitro* recordings that been established the motility responses can exceed well over 20 kHz (Scherer and Gummer, 2004)! As a result, the fast motion of the OHC can have an immediate effect on the vibratory motion of the BM, which was primarily induced by the traveling wave. In other words, the OHCs could provide a selective, local mechano-acoustic feedback to the BM, which is strongly frequency-selective (and could thus embody the long-sought active filter).

A further remarkable observation is that the *rest length* of the OHC is uniquely related to the resonance frequency of its mechanical response. Long OHCs are found near the apex, they have long cilia, and are mechanically tuned to low frequencies. At the base, the OHCs are short, are tuned to high-frequencies, and have short cilia (Brendin et al., 1989; Fig. 5.19A). Interestingly, these relationships, which do not apply to the IHCs, extend across many different species (Dannhof et al., 1991; Fig. 5.19C). Whether this relationship has a functional role in the OHC feedback mechanism, however, is not known.

It is thought that OHC mechanical feedback to the BM could act as a *positive* feedback, by locally amplifying the BM motion (ie, by providing *negative damping*). This mechanism causes increased sensitivity for low-intensity sounds, given that the OHC motility also depends on the absolute sound level through an additional efferent *neural* feedback pathway from the superior olive in the brainstem. It is therefore generally assumed, albeit still heavily debated (Ashmore et al., 2010), that the motor-mechanism of the OHCs forms the basis for the extreme sensitivity and selectivity of the cochlea. Fig. 5.20 provides a physical representation of the cochlear amplifier, as proposed by Nobili et al. (1998).

To model the role of the OHCs in the living cochlea, Nobili et al. (1998) proposed the following formalism. The idea is that the organ of Corti (BM and supporting structures) forms an uncoupled chain of position-dependent oscillators that interact with the surrounded fluid, and undergo a nonlinear local feedback from the electro-motility response of OHCs.

We saw in chapter: Linear Systems that the basic equation of motion for a second-order damped oscillator (here taken at location x_n, and not coupled through a medium) is:

$$m_n \frac{d^2 y_n}{dt^2} + \gamma_n \frac{dy_n}{dt} + k_n y_n = f_n(t) = -G_n a_s(t). \qquad (5.58)$$

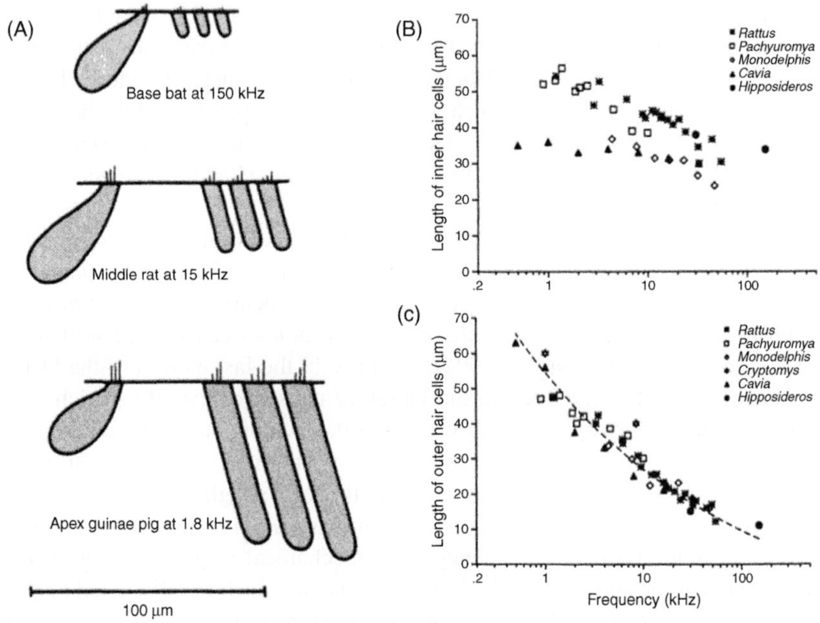

FIGURE 5.19 (A) The length of OHCs relates to their best frequency. Three examples from three different species (bat, rat, and guinea pig), taken from different locations along the BM. (B) The best frequencies of IHCs do not relate in a unique way to their size. (C) For OHCs, however, the same quantitative relation is found across species. *(Source: After Dannhof et al., 1991, with kind permission.)*

with m_n, γ_n, and k_n the local mass, viscosity, and elasticity of the oscillator at (discretized) location n. The external driving force, $f_n(t)$, is proportional to the (negative) acceleration of the stapes, $a_s(t)$, which is transmitted through the fluid as a local force to the oscillator (it is amplified by a local constant G_n, which is reminiscent to the pressure drop shown in Fig. 5.3).

In a more realistic cochlear model, however, the fluid at nearby locations does provide additional (*passive*) forces to the oscillator at n: (1) the resulting lateral hydrodynamic forces lead to motion of N oscillators at nearby sites, j, which in turn influence the oscillator at site n through (negative) fluid coupling (Fig. 5.21) and (2) shear forces that are due to differences in velocity of the fluid particles around the BM site. Together, these hydrodamic couplings provide the mechanism for Von Békésy's traveling wave (developed in Section 5.3), as illustrated in Fig. 5.21B.

In addition, the OHCs provide a feedback force, OHC_n, which opposes the dissipative damping through the electro-motility of the OHC cell body. This motility follows the rapid synchronized motion of the cell's stereocilia that undergo a displacement $z_n(t)$. The positive feedback (ie, effective negative damping) to the BM is mediated via the supporting Deiters' cells (Fig. 5.20A). Taken

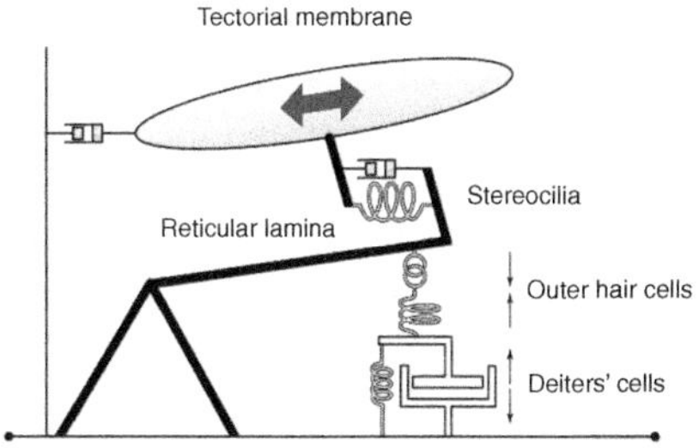

FIGURE 5.20 Physical model of the Organ of Corti (cf. with Figs. 5.4 and 5.18A), describing the coupling of OHCs to the BM. *(Source: Nobili et al., 1998, with kind permission.)*

together, the equation of motion of the *active* cochlear partition at site n thus reads:

$$\sum_{j=1}^{N}(G_n^j + m_n\delta_n^j)\frac{d^2y_j}{dt^2} + \gamma_n\frac{dy_n}{dt} + \mathrm{OHC}_n(z_n) + k_ny_n + s_n\left(2\frac{dy_n}{dt} - \frac{dy_{n+1}}{dt} - \frac{dy_{n-1}}{dt}\right) = -G_na_s(t) \quad (5.59)$$

where δ_n^j is the Kronecker delta (it is 1 for $j = n$, and 0 otherwise) (Fig. 5.21).

Fig. 5.17B shows how the location of the excitatory peak of an auditory nerve fiber systematically shifts in the basal direction (to higher frequencies) for lower sound levels. The total shift over an 80 dB intensity range amounts to 1/3–1/2 octave, and suggests that the feedback mechanism of the cochlear

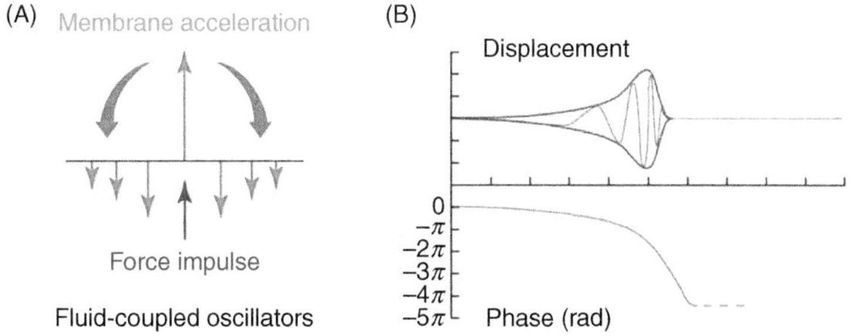

FIGURE 5.21 (A) A force impulse applied locally at BM site n (*red arrow*) leads to an instantaneous acceleration of the BM element, but also to an oppositely directed hydrodynamically mediated force (*blue arrows*) to nearby sites (*downward green arrows*). (B) Because of location-dependent mechanical properties of the BM (modeled as an exponential decay of its stiffness), the resulting BM movement is a traveling wave. The wavelength of the oscillations decreases from base to apex, reaching a critical point at a location that depends on the driving frequency (right). The phase delay (*green curve*) increases with distance from the stapes. *(Source: Nobili et al., 1998, with kind permission.)*

amplifier is in fact generated by a *second* set of tuned filters. Clearly, these filters involve a role for the OHCs, but it has been suggested that the coupling occurs between OHCs in combination with the mechanical properties of the tectorial membrane (TM). The TM attaches firmly to the long stereocilia of OHCs, and is in turn influenced by BM motion through hydrodynamic coupling (Fig. 5.22). Further, also the elastic TM can behave as an oscillator (eg, Zwislocki and Kletsky, 1979; Russell et al., 2007; see also Fig. 5.18A). Following Eq. (5.59), this behavior can thus be described as:

$$m_n^{\mathrm{TM}} \frac{d^2 z_n}{dt^2} + \gamma_n^{\mathrm{TM}} \frac{dz_n}{dt} + k_n^{\mathrm{TM}} z_n = -C_n \frac{d^2 y_n}{dt^2} \tag{5.60}$$

with C_n a constant local hydrodynamic coupling. Note that at resonance $(m_n^{\mathrm{TM}})[(d^2 z_n)/(dt^2)] = -k_n^{\mathrm{TM}} z_n$ (Newton's second law), so that Eq. (5.60) reduces to:

$$z_n = -\frac{C_n}{\gamma_n^{\mathrm{TM}}} \frac{dy_n}{dt} \tag{5.61}$$

As a result, at resonance (ie, locally, and driven by the pure tone acoustic input) the displacement of the stereocilia can indeed lead to a *negative damping* (ie, amplification) of the BM motion (Fig. 5.23).

Two mechanisms could underlie the remarkable high-frequency mechanical response of the OHC, which could presumably work together: (1) the cell-body motility induced by prestine proteins, which provides the feedback force to the BM, and (2) the stereocilia themselves that could drive the cell contractions at the correct phase (Ashmore et al., 2010; Fettiplace and Hackney, 2006; Hudspeth, 1985; Hudspeth 2014).

Yet, despite the in vitro evidence of high-frequency OHC responses, it still remains to be established whether the cochlear amplifier results from an active mechanism [that adds energy (negative damping) to the BM; Dallos, 1992;

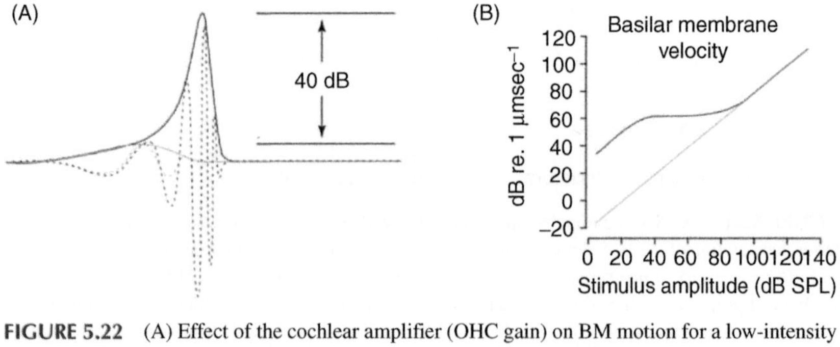

FIGURE 5.22 (A) Effect of the cochlear amplifier (OHC gain) on BM motion for a low-intensity sound, at high OHC gain (*red curve*), and at low gain (*blue curve*). (B) The *blue curve* shows the linear BM response (independent of sound level); the *red curve* shows the result of a level-dependent (saturating) gain of OHC feedback. *(Source: Nobili et al., 1998, with kind permission.)*

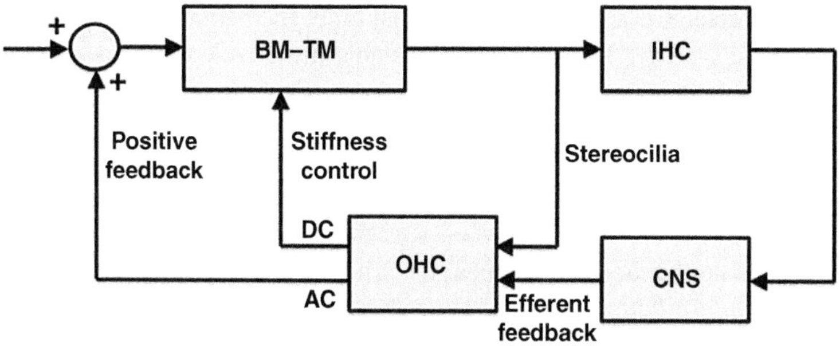

FIGURE 5.23 Black-box control scheme to summarize the relevant functional elements of the active cochlear partition. It includes a DC parametric stiffness regulation, as well as cycle-to-cycle AC positive feedback from the OHCs to BM at the acoustic frequency. The OHCs in turn are controlled by an efferent feedback signal from the CNS that saturates their electro-motility response at high sound levels. The passive properties of the cochlea are provided by the hydrodynamic interactions with BM–TM.

Dallos and Evans, 1995; Ashmore 2008], or, instead, is due to active compression of the BM response (adding damping, eg, Van der Heijden, 2014). In other words: do OHCs provide precise in-phase high-frequency positive feedback to the BM, or do they stiffen to attenuate the BM response at high intensities? In vivo recordings of OHC motion at high frequencies are needed to settle this important issue (Ashmore et al., 2010).

A critical nonlinear oscillator? Taken together, four nonlinear cochlear phenomena underlie the function of the inner ear: (1) a high amplification increases the sensitivity of the system at low sound levels; (2) frequency selectivity increases the spectral resolution of the system; (3) a compressive nonlinearity strongly increases the dynamic range for sound input pressure levels. According to Hudspeth and coworkers, all three phenomena can be captured by a single nonlinear mechanism: a critical oscillator that is tuned to operate closely to a so-called Hopf bifurcation; (4) As an epiphenomenon of this critical system, spontaneous oscillations may occur, which are indeed recorded as spontaneous oto-acoustic emissions (OAEs) by the mammalian (and nonmammalian) ear. Interestingly, the dependence of the nonlinear distortion products of so-called evoked OAEs on the intensities of pairs of input tones is explained by the same Hopf mechanism (Camalet et al., 2000; see also Reichenbach et al., 2012 for an additional hydrodynamic mechanism that mediates the wave propagation of nonlinear distortion products like $2f_1 - f_2$ frequency components towards the middle ear along Reissner's membrane).

A Hopf bifurcation can occur in a dynamical system that has a cubic nonlinearity and is described by the following generic form:

$$\frac{dz}{dt} = z\left(\lambda + i + b|z|^2\right) \quad \text{with} \quad [z = x + iy,\ b = \alpha + i\beta] \in \mathbb{C}, \quad \text{and} \quad \lambda \in \mathcal{R} \tag{5.62}$$

The parameter λ is a control variable that tunes the behavior of $z(t)$. If $\lambda > 0$, and $\alpha < 0$ the system's solution attains a stable limit cycle (called supercritical oscillation), which is described by:

$$z_{LIM}(t) = \sqrt{\frac{\lambda}{|\alpha|}} e^{i(1+\beta r^2)t} = re^{i\omega_0 t} \tag{5.63}$$

When this nonlinear system is near the critical bifurcation point at $\lambda = 0$, and is driven by an oscillatory stimulus at its characteristic frequency:

$$f(t) = F\exp(i\omega_0 t) \tag{5.64}$$

so that

$$\frac{dz}{dt} = z\left(\lambda + i + b|z|^2\right) + f(t) \tag{5.65}$$

the response amplitude varies with stimulus amplitude, F, through a compressive power law relation:

$$|z| \propto F^{1/3} \tag{5.66}$$

This compressive nonlinearity nicely corresponds to the amplitude-dependent gain of the cochlear amplifier seen in the recordings of Fig. 5.17: a high gain for low-intensity sounds, and a low gain for high-intensity sounds, as the gain is defined by (Fig. 5.24):

$$G = \frac{d|z|}{dF} \propto \frac{1}{F^{2/3}} \tag{5.67}$$

Note that the discrete nonlinear feedback model of Nobili et al. (1998), described by Eq. (5.59), can be rewritten such that the OHC nonlinearity, OHC(z_n), in their model contains the cubic nonlinear term that is required for the Hopf bifurcation in Eq. (5.62) (Hudspeth et al., 2010).

5.5 EXERCISES

Problem 5.1:

(a) Show that Eq. (5.12) is indeed a solution that obeys the dry-water constraints Eqs. (5.9, 5.10).
(b) Make the deep-water wave approximation to show Eq. (5.13).
(c) Same for the shallow-water wave approximation of Eq. (5.14).

Problem 5.2:

(a) Derive the dispersion relation for gravitational water waves, Eq.(5.17).

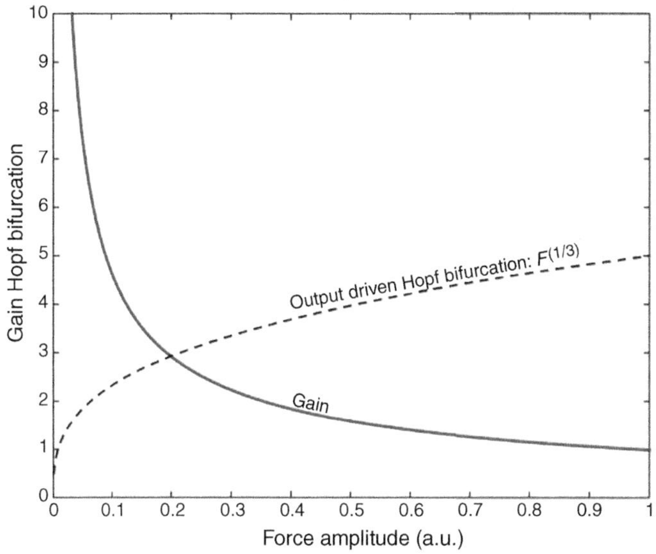

FIGURE 5.24 Response gain (*red line*) and output (*dashed line*) of the driven Hopf bifurcation, as described by Eqs. (5.66 and 5.67).

(b) Compute the phase velocity for deep-water waves. What is the group velocity?
(c) The same for shallow-water waves.

Problem 5.3:

(a) Apply the combined effects of gravity and tension to determine the total dispersion relation of dry water, Eq. (5.21).
(b) Determine the phase- and group velocity.
(c) Show that there is one particular wave length, for which phase- and group velocities are identical (which wave length? How high is this velocity?).

Problem 5.4: Consider an infinite basin of dry water, stretching from $x \in [-\infty, +\infty]$. Ignore the z-dimension. For $x < 0$ the depth of the basin is $h = h_1$, so that $y_{\text{bottom}} = -h_1$. At $x = 0$, the bottom profile suddenly jumps upward to $h = h_2 < h_1$. A traveling water wave comes from the left ($x < 0$) at amplitude A, and moves rightward. We consider the deep-water case, that is, $\lambda \ll h_{1,2}$.

(a) Compute the reflection, R, and transmission, T, at $x = 0$ (use: *impedance*).
(b) Same for the shallow-water case, $\lambda \gg h_{1,2}$.

Problem 5.5: Negative dynamic feedback with a delay can cause instability problems. In this Exercise we analyze the influence of a delay on the transfer characteristic of a linear system with feedback (Figs. 5.25 and 5.26).

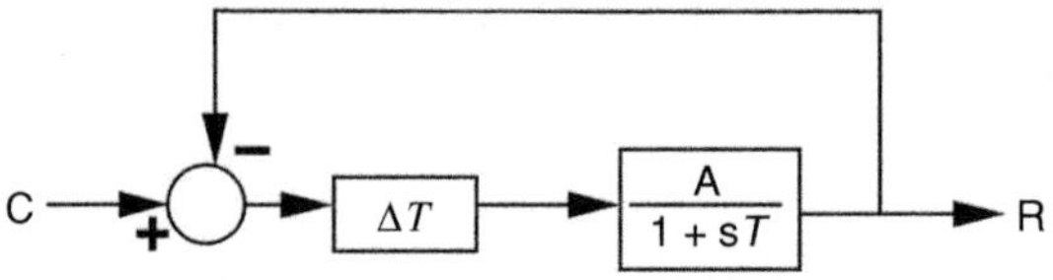

FIGURE 5.25 **Dynamic feedback model of a low-pass filter with a delay.**

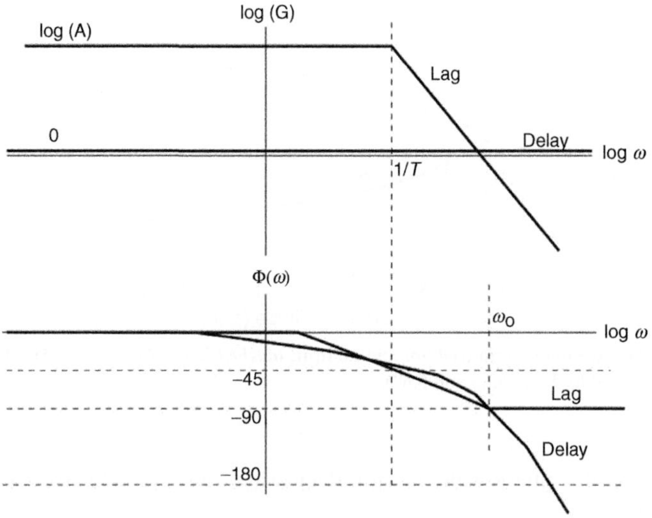

FIGURE 5.26 **Bode plot for the two subsystems.**

(a) Determine the Laplace transform of a pure delay: $y(t) = x(t - \Delta T)$, and from that calculate its transfer characteristic in the frequency domain.

(b) Consider the system shown in Fig. 5.25. Determine the total transfer function and the loop gain. The system will spontaneously oscillate, and thus become unstable, when the loop-gain exceeds the value of 1, and at the same time has a phase shift of −180 degrees. Perform a Bode analysis on this system and estimate the frequency $\omega 0$ where instability kicks in (Fig. 5.26).

(c) What happens to the system if A is increased/lowered? What if the time constant T is increased/lowered?

Problem 5.6: The wave equation of the BM as deduced by Von Békésy and Zwislocki is linear, which means that the superposition principle should hold, and that the output amplitude of the BM is independent of stimulus amplitude. However, when you listen (carefully) to a superposition of two frequencies, say $f_1 = 440$ Hz (musical "A") and $f_2 = 523$ Hz (musical "C"), you can hear the presence of a third tone with a frequency that is close to the musical "F" (349 Hz)! This additional tone is a *combination tone*, and tends to have a frequency of $2f_1 - f_2 = 357$ Hz. This combination tone is a manifestation of a nonlinearity in the system, and is due to the nonlinear cochlea (note that the effect disappears when the two tones are presented to different ears!).

Suppose, for simplicity, that the output of the BM, $q(t)$, depends on the instantaneous pressure, $p(t)$, through the following nonlinear third-order relation (a third-order Taylor approximation on the nonlinear transfer function):

$$q(t) = ap(t) + bp^2(t) + cp^3(t) \tag{5.68}$$

(a) By presenting a superposition of two harmonic waves at the input, say

$$p(t) = \cos(\omega_1 t) + \cos(\omega_2 t)$$

show that the output of the system can described by a spectrum that contains 13 frequencies! Determine also their relative amplitudes. Which nonlinear term causes the observed "F" percept?

(b) What happens to the amplitude(s) of the distortion products if the input is given by $p(t) = A\cos(\omega_1 t)$

Problem 5.7: Analysis of a Hopf bifurcation. Consider the Hopf bifurcation in polar coordinates (with a and μ nonzero real-valued parameters):

$$\begin{aligned} \frac{dr}{dt} &= r\left(a - \mu r^2\right) \\ \frac{d\phi}{dt} &= \omega_0 \end{aligned} \tag{5.69}$$

Investigate the stability of the fixed points as function of the parameters. When do we see a limit cycle? What determines its amplitude and frequency?

REFERENCES

Ashmore, J., 2008. Cochlear outer hair cell motility. Physiol. Rev. 88, 173–210.

Ashmore, J., Avan, P., Brownell, W.E., Dallos, P., et al., 2010. The remarkable cochlear amplifier. Hear. Res. 266, 1–17.

Brendin, L., Flock, A., Canlon, B., 1989. Sound-induced motility of isolated cochlear outer hair cells is frequency-specific. Nature 342, 814–816.

Camalet, S., Duke, T., Jülicher, F., Prost, J., 2000. Auditory sensitivity provided by self-tuned critical oscillations of hair cells. Proc. Natl. Acad. Sci. USA 97, 3183–3188.

Crawford, Jr., F., 1968. Berkeley physics course. Waves, vol. 3, McGraw-Hill, New York.

Dallos, P., 1973. The Auditory Periphery. Biophysics and Physiology. Academic Press, New York.

Dallos, P., 1992. The active cochlea. J.Neurosci. 12, 4575–4585.

Dallos, P., Evans, B.N., 1995. High-frequency motility of outer hair cells and the cochlear amplifier. Science 267, 2006–2009.

Dannhof, B.J., Roth, B., Bruns, V., 1991. Length of hair cells as a measure of frequency representation in the mammalian inner ear? Naturwissenschaften 78, 570–573.

Fettiplace, R., Hackney, C., 2006. The sensory and motor roles of auditory hair cells. Nat. Rev. Neurosci. 7, 19–29.

Hudspeth, A.J., 1985. The cellular basis of hearing: the biophysics of hair cells. Science 230, 745–752.

Hudspeth, A.J., 2014. Integrating the active process of hair cells with cochlear function. Nat. Rev. Neurosci. 15, 600–614.

Hudspeth, A.J., Jülicher, F., Martin, P., 2010. A critique of the critical cochlea: Hopf – a bifurcation – is better than none. J. Neurophysiol. 104, 1219–1229.

Johnstone, B.M., Boyle, A.J.T., 1967. Basilar membrane vibration examined with the Mössbauer technique. Science 158, 389–390.

Johnstone, B.M., Patuzzi, R., Yates, G.K., 1986. Basilar membrane measurements and the travelling wave. Hearing Res. 22, 147–153.

Nobili, R., Mammano, F., Ashmore, J., 1998. How well do we understand the cochlea? Trends Neurosci. 21, 159–167.

Reichenbach, T., Stefanovic, A., Nin, F., Hudspeth, A.J., 2012. Waves on Reissner's membrane: a mechanism for the propagation of otoacoustic emissions from the cochlea. Cell Reports 1, 374–384.

Russell, I.J., Legan, P.K., Lukashkina, V.A., Lukashkin, A.N., Goodyear, R.J., Richardson, G.P., 2007. Sharpened cochlear tuning in a mouse with a genetically modified tectorial membrane. Nat. Neurosci. 10, 215–223.

Scherer, M.P., Gummer, A.W., 2004. Vibration pattern of the organ of Corti up to 50 kHz: evidence for resonant electromechanical force. Proc. Natl. Acad. Sci. USA 101, 17652–17657.

Van der Heijden, M., 2014. Frequency selectivity without resonance in a fluid waveguide. Proc. Natl. Acad. Sci. USA 111, 14548–14552.

Zwislocki, J.J., Kletsky, E.J., 1979. Tectorial membrane: a possible effect on frequency analysis in the cochlea. Science 204, 639–641.

Chapter 6

The Auditory Nerve

6.1 INTRODUCTION: TUNING OF AUDITORY NERVE FIBERS

The first stage of neural processing on the cochlear output is encountered at the auditory nerve (AN) fibers that project their axons to the ipsilateral cochlear nucleus (CN). The frequency tuning curve of an AN fiber is often described by an excitation threshold curve, which corresponds to the minimum sound level that is needed for a particular pure tone (frequency, f_T) to excite the fiber above its spontaneous firing rate. Such threshold tuning curve has the appearance of an asymmetrical "V"-shape (Fig. 6.1A), which reaches its minimal threshold at the fiber's "characteristic frequency," f_{cf}. Plotted on log–log scale, the threshold rapidly increases for neighboring frequencies: the steepest increase is obtained for frequencies $f_T > f_{cf}$, (corresponding to BM locations basal to the f_{cf} site) and for $f_{cf}/2 < f_T < f_{cf}$, (more apical BM sites), while a much shallower threshold increase is obtained for frequencies more than an octave below f_{cf} (for $f_T < f_{cf}/2$).

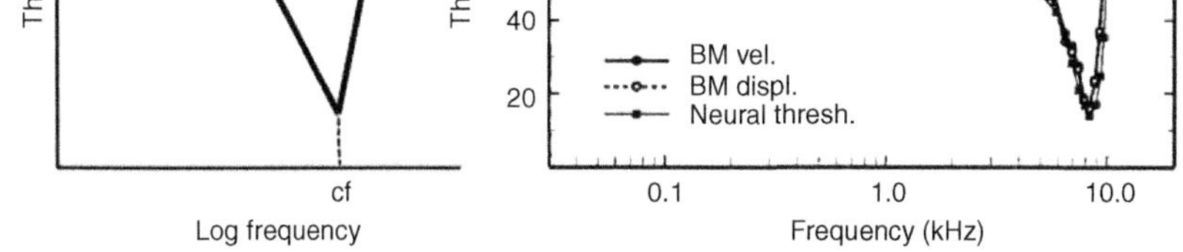

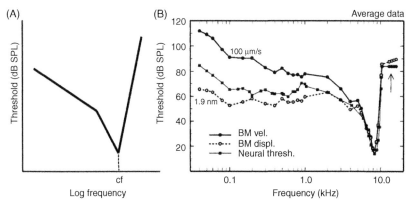

FIGURE 6.1 (A) Threshold tuning curve of a hypothetical AN fiber, highlighting the sharp tuning of the neuron to its characteristic frequency (f_{cf}), at which the threshold for neural excitation is minimal. (B) Example of a recorded amplitude response from a chinchilla AN fiber with f_{cf} = 9.2 kHz (*filled squares*), compared to the movement of the basilar membrane (*open and closed circles*, for BM displacement and peak velocity, respectively). Note close correspondence between the different measurements *(Source: From Ruggero et al., 2000, with kind permission.)*

The Auditory System and Human Sound-Localization Behavior. http://dx.doi.org/10.1016/B978-0-12-801529-2.00006-4

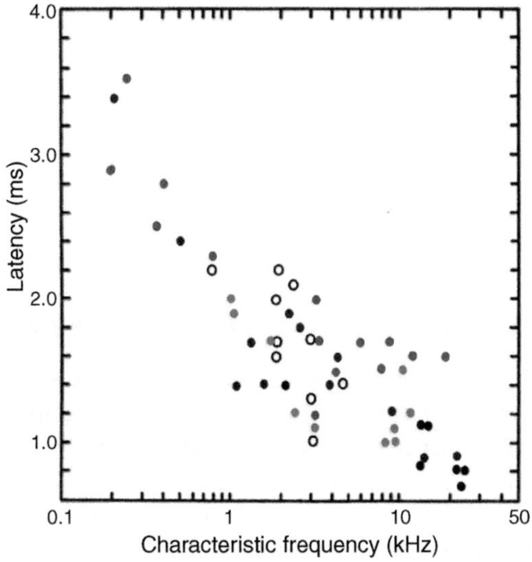

FIGURE 6.2 The response latency of AN fibers is inversely related to *cf.* Data from guinea pig, as reported by Evans (1972).

The response latency of AN fibers is inversely related to their log(f_{cf}), which is due to the delay of the traveling wave along the BM. In the guinea pig, these delays run between 0.4 (high f_{cf}) and 3.6 (low f_{cf}) ms (Fig. 6.2).

The threshold tuning curves of (basal) AN fibers closely resemble the BM displacement and peak-velocity tuning curves, as measured with the Mössbauer technique in the alive cochlea of the chinchilla (Ruggero et al., 2000; Fig. 6.1B). As a result, the detailed BM motion characteristics (both amplitude and phase) might in principle be reconstructed from the neural responses of AN fibers, as relatively minor signal transformations seem to underlie the BM to AN response (see Section 6.7, for a nice example of such a procedure).

However, at high sound levels additional micromechanical interactions, which are not directly due to BM motion (eg, vibrations within the organ of Corti, vibrations of the tectorial membrane, or IHC physiology), appear to come into play as well, as the AN phase curves sharply deviate for levels >80 dB SPL, leading to a full phase reversal of 180 degree (Ruggero et al., 2000).

6.2 REVERSE CORRELATION AND THE GAMMATONE FILTER

The reverse-correlation (*revcor*) function can be used to measure the effective linear filter of an (auditory) neuron from its stimulus–response behavior (Ringach and Shapley, 2004). Interestingly, this even holds when the response is generated through a static nonlinearity (Price, 1958). Suppose that the neuron emits a train of spikes (symbolized by Dirac delta functions, thus making it a *point process*) in response to stimulus $x(t)$:

$$s(t) = \sum_{n=1}^{N} \delta(t - t_n) \tag{6.1}$$

The cross-correlation (see chapter: Linear Systems) between stimulus and neural response is then determined by

$$\Phi_{sx}(\tau) = \int_{-\infty}^{+\infty} x(t)s(t+\tau)dt = \sum_{n=1}^{N} x(t_n - \tau) \tag{6.2}$$

Given that a neuron's spiking results from a stochastic process, described by a probability density function, $p(t)$, the expectation value of the revcor function becomes:

$$< \Phi_{sx}(\tau) >= \int_{-\infty}^{+\infty} x(t-\tau)p(t)dt \tag{6.3}$$

De Boer and De Jongh (1978) applied the revcor analysis to estimate the linear filter underlying AN responses. Fig. 6.3 illustrates the underlying conceptual model of their analysis.

This simple three-stage cascade model consists of a linear filter stage (containing the characteristics of the outer ear, ear canal, middle ear, and a linear band-pass filter in the cochlea) that is described by the joint impulse response function, $h(\tau)$. The linear stage yields an output, $y(t)$, which feeds into a static nonlinearity (representing cochlear compression and the active amplifier).

The output of the nonlinearity is a positive-only signal, $z(t)$, that provides the input to a stochastic spike generator, for which the probability of firing a spike is assumed to be proportional to $z(t)$:

$$p(t) = f_{cf}(x(t) * h(t)) \tag{6.4}$$

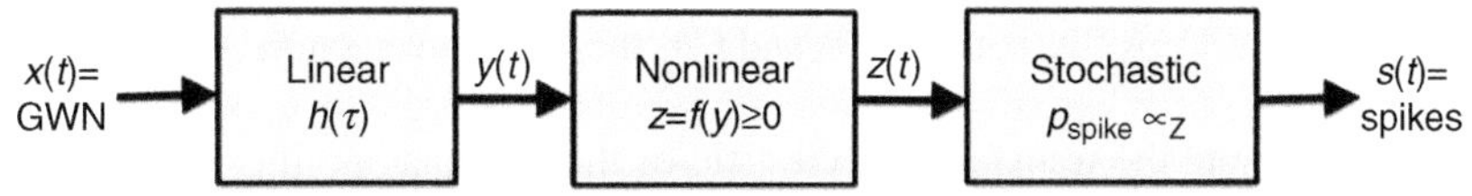

FIGURE 6.3 Cascade model of the peripheral auditory system, from sound input to spiking output of the AN. A linear filter with overall impulse response $h(\tau)$, is followed by a static nonnegative nonlinearity (jointly accounting for all cochlear and neural nonlinearities), and a stochastic spike generator. If the input is GWN, the linear impulse response follows from cross-correlating $s(t)$ and $x(t)$. *(Source: After De Boer and De Jongh, 1978.)*

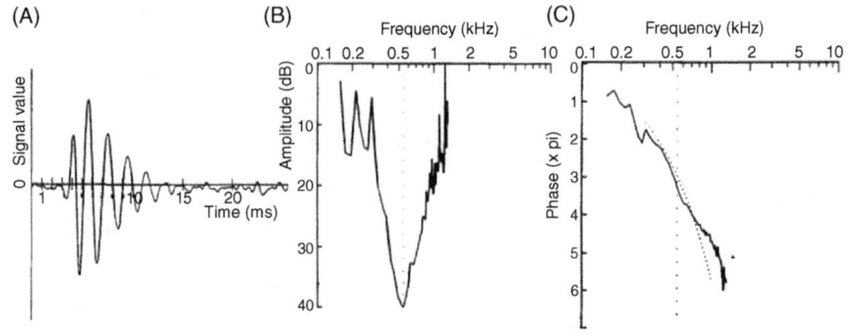

FIGURE 6.4 (A) Reconstructed revcor impulse response of an AN response. The response starts 3 ms after stimulus onset. Note the substantial ringing of the response, indicative of a band-pass filter characteristic. The characteristic frequency of the ringing is 0.5 kHz. (B) Amplitude response curve (*plotted in inverted format*). (C) Corrected phase–response curve (the pure time delay of $\Delta T = 3$ ms, yielding an extra phase shift of $-\omega\Delta T$ (*dotted line*), has been subtracted from the response phase). *(Adapted from De Boer and De Jongh, 1978, with kind permission.)*

with * indicating time-convolution (see chapter: Linear Systems). Interestingly, when $x(t)$ = GWN, with its autocorrelation given by Eq. 3.76, the combination of Eqs. (6.3) and (6.4) simply yields (Price, 1958):

$$<\Phi_{sx}(\tau)>= ch(\tau) \quad (6.5)$$

In other words, despite the presence of a static nonlinearity, and a stochastic pulse generation stage, the revcor function yields a direct measure for the linear filter of the model. An example of this analysis, taken from recordings in cat AN by De Boer and De Jongh (1978), is illustrated in Fig. 6.4.

The impulse-response functions of AN fibers indeed have a band-pass appearance, and resemble amplitude-modulated harmonic waves. They have a striking resemblance to so-called *gammatones* (at least for cells tuned to lower frequencies, that is, < 3–4 kHz), for which the modulation envelope is described by a gamma function: $\Gamma(t) \sim t^{n-1}\exp(-\alpha t)$. The gammatone filter (order n) has an impulse response function that is described by (Hartmann, 1998):

$$h(\tau) = a\tau^{n-1}\exp(-2\pi b\tau)\cos\left(2\pi f_{cf}\tau + \varphi\right) \quad (\tau \geq 0) \quad (6.6)$$

where $n \geq 1$ is the order of the filter, a (amplitude scaling) and b (the filter's band width) are positive parameters, and f_{cf} is the filter's characteristic frequency (Fig. 6.5A).

To calculate the transfer characteristic of the gammatone filter we use the following property of the Fourier transform, which applies to all modulated harmonic functions:

$$h(t) = g(t)\cos\left(2\pi f_{cf}t\right) \Rightarrow H(f) = \frac{1}{2}\left[G\left(f + f_{cf}\right) + G\left(f - f_{cf}\right)\right] \quad (6.7)$$

with $G(f)$ the Fourier transform of amplitude modulation function $g(t)$.

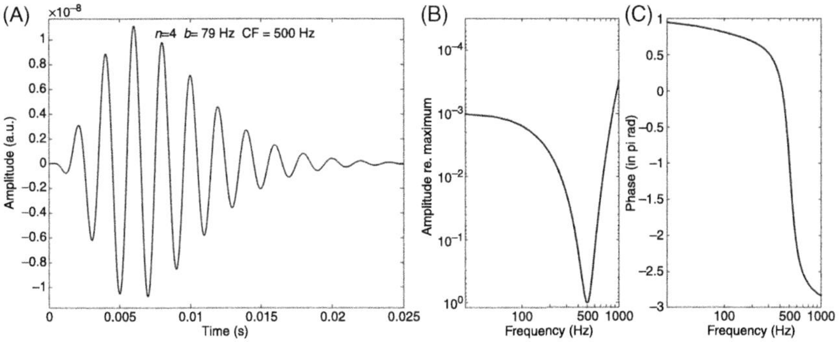

FIGURE 6.5 (A) Example of a gammatone impulse response [Eq. (6.6)] with $n = 4$, $a = 1$, $b = 79$ Hz, $\varphi = 0$, and $f_{cf} = 500$ Hz. (B) Amplitude spectrum (y-scale inverted) and (C) phase spectrum (shifted by π radians, so that $\varphi = 0$ at resonance) of the filter. Compare with Fig. 6.4, which is presented at approximately the same scale.

Using the property of Eq. (6.7), the frequency response of the gammatone filter (Fig. 6.5B) can be approximated (Exercise) by a cascade of n first-order identical low-pass filters, and is thus described by:

$$H(f) \approx \left[\frac{1}{1 + j\dfrac{f - f_{cf}}{b}} \right]^n \quad (f > 0) \tag{6.8}$$

Fig. 6.5 shows an example of the impulse response and frequency characteristics (amplitude/phase) of a fourth-order gammatone, with $f_{cf} = 500$ Hz. In Fig. 6.6 we present a collection of six gammatones over a range of characteristic frequencies. The selected bandwidths correspond to the psychophysical

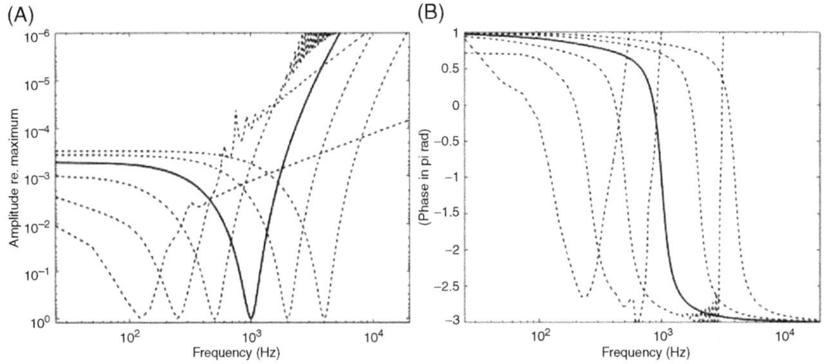

FIGURE 6.6 **Transfer characteristics for six gammatone filters ($n = 4$), with $f_{cf} = 0.125$, 0.25, 0.5, 1, 2, and 4 kHz, respectively.** The filter bandwidths were computed from Eq. (6.9). The solid curves correspond to the filter at $f_{cf} = 1$ kHz. (A) Amplitude. (B) Phase.

Cambridge relation (and associated AN-fiber bandwidths), and are determined by the following equation:

$$b = 24.7\left(1 + 4.37 f_{cf}\right) \text{ with } f_{cf} \text{ in kHz} \tag{6.9}$$

6.3 PHASE LOCKING

An AN fiber receives direct synaptic input from its IHC, which only delivers an excitatory signal when the BM moves upward (ie, at pressure rarefactions, when the stapes moves outward), because the stereocilia effectively act as nonlinear rectifiers. For a downward BM movement the fiber is slightly suppressed by the IHC. As a result, the fiber may fire an action potential within the positive part of the waveform of the stimulus only. For a pure tone, this will always be roughly at the same phase within its periodic cycle. This remarkable feature is termed *phase locking*, and is observed for low-frequency (below 3–4 kHz) AN fibers only.

An IHC is connected to about 10 AN fibers, each having a different excitatory threshold. Because spiking of a neuron is a stochastic process, and AN fibers have a refractory period of about 1 ms, a given fiber will typically not fire an action potential for every stimulus period, but may often skip one or more periods before it fires the next action potential (Fig. 6.7). Due to the stochasticity of spike-generation, the inter-spike intervals obtained from phase-locked AN

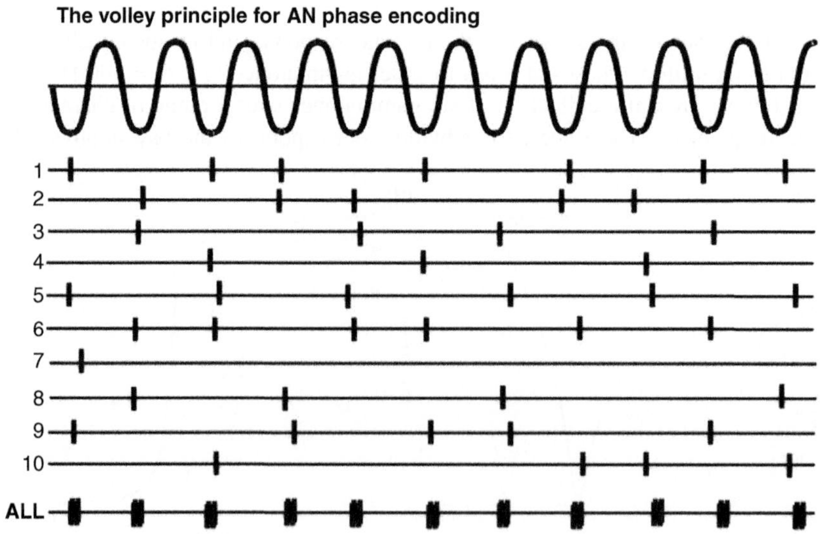

FIGURE 6.7 **The *volley principle* for equally innervated AN fibers allows for the transmission of accurate timing information about the acoustic stimulus, even though individual fibers may skip one or more cycles of the stimulus, and show some temporal jitter.**

fibers will not resemble a delta-peak at exactly one phase of the stimulus period (perfect phase locking), but spikes will occur at multiples of the stimulus period with a certain amount of jitter.

However, the 10 fibers together may encode each stimulus period quite faithfully, even for frequencies that far exceed 1 kHz (1/refractory-period). Through this so-called "*volley principle*," the AN can transmit accurate timing information about the acoustic stimulus to the subsequent stages in the auditory pathway: the CN.

At high characteristic frequencies (>3 kHz), the low-pass filter properties of the IHC's membrane no longer allows for precise locking to a particular phase of the acoustic stimulus. Instead, the output of IHCs that drive high-frequency fibers is dominated by a prominent dc (resulting from the rectification provided by the stereocilia; Fig. 6.8). The corresponding AN fibers respond with a stochastic spike train (Poisson-like) to this input, where the firing rate relates to the absolute value of the dc (which, in turn, is related to the sound level; see also the next section).

Clearly, phase locking to pure tones in high-frequency fibers is lost, although they may still respond quite precisely to the phase of the *stimulus envelope* of complex sounds that contain low-frequency amplitude modulations of its preferred high f_{cf} (see Section 6.7).

This sensitivity of high-frequency AN fibers to low-frequency envelope modulations is due to nonlinearities in the transduction mechanisms of the neuron itself (see Section 6.7).

The quality of a cell's phase locking to a periodic stimulus, period *T*, is usually quantified by the *vector strength*, *VS*, which is defined as:

$$VS(T) = \frac{\sqrt{\left(\sum_{n=1}^{N} \cos\varphi_n\right)^2 + \left(\sum_{n=1}^{N} \sin\varphi_n\right)^2}}{N} \tag{6.10}$$

where each spike $n \in [1,N]$ is treated as a unit vector with phase $\varphi_n \in [0,2\pi]$ relative to the stimulus period, *T* (see Fig. 6.9B for an illustration of this concept). When $VS(T) = 0$ there is no phase locking, and the spikes are distributed homogeneously over the stimulus period. When $VS(T) = 1$ there is perfect phase locking: all spikes fall exactly on the same phase in the stimulus (no jitter). For AN fibers, the spikes will jitter around the maximum sound pressure of the tone, so that typically the vector strength will be <1 (the fiber in Fig. 6.9 has a $VS = 0.64$).

An equivalent way to characterize the amount of phase locking is by calculating the Fourier Transform of the histogram of the inter-spike intervals, all wrapped onto a single stimulus period (see bottom panel Fig. 6.9B). The alternative definition of the vector strength is then:

$$VS(T) = \frac{A_1(T)}{A_0(T)} \tag{6.11}$$

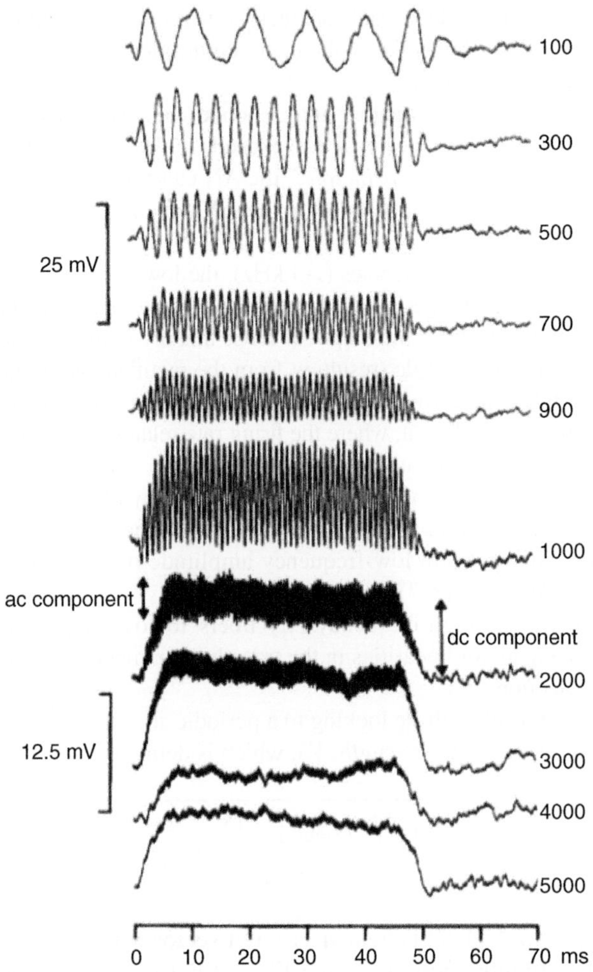

FIGURE 6.8 Membrane response of an IHC to tones of different frequencies consists of an ac-component at the driving frequency, and a dc, due to rectification by the stereocilia. The low-frequency response is dominated by a clear ac component, which is in phase with the tone. At frequencies above 3 kHz, the ac component is lost (due to membrane low-pass filtering), and the dc component remains. *(Source: After Palmer and Russell, 1986, with kind permission.)*

with A_1 the amplitude of the first Fourier component of the histogram (the sinusoid with frequency $f_1 = 1/T$ that best fits the histogram's shape), and A_0 the dc of the FT, which corresponds to the mean firing rate during the period. Phase locking of individual AN fibers deviates significantly from the maximum value of 1.0, and scatters, on average, around 0.65.

Interestingly, the spherical bushy cells (SBCs) in the CN, which receive their inputs exclusively from multiple AN fibers, result to have much sharper phase

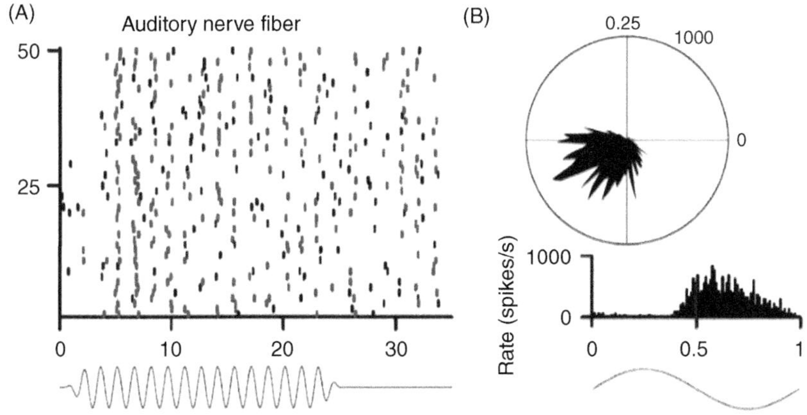

FIGURE 6.9 Phase locking in a cat AN fiber with *cf* = 0.67 kHz. (A) Repetition of 50 trials to a 25 ms tone pip shows the phase-locked spikes (*red*). (B) Vector strength of the fiber, shown as a PSTH within the sound's normalized period (bottom), and as a polar plot [cf. Eq. (6.8)]. *(Source: After Joris and Smith, 2008, with kind permission.)*

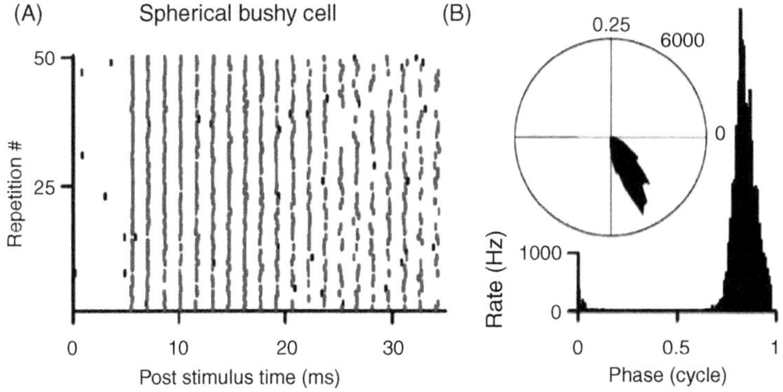

FIGURE 6.10 Enhanced ("picket-fence") phase locking at the level of spherical bushy cells (CN). Same format as Fig. 6.9. For this cell (same f_{cf}), $VS = 0.93$. *(Source: Joris and Smith, 2008, with kind permission.)*

locking, allowing them to respond for virtually *every cycle* of a low-frequency sound, but also at a sharper defined phase (higher precision), as $VS \sim 0.9$ for these neurons (Fig. 6.10). Presumably, two factors could contribute to this improved phase locking: (1) the volley principle of Fig. 6.7 leads to a more faithful cycle-by-cycle input to spherical bushy cells and (2) the unique synaptic specialization of these cells (embodied by the giant Held synapse) allows them to have a less variable response to their input, thereby strongly reducing spike jitter.

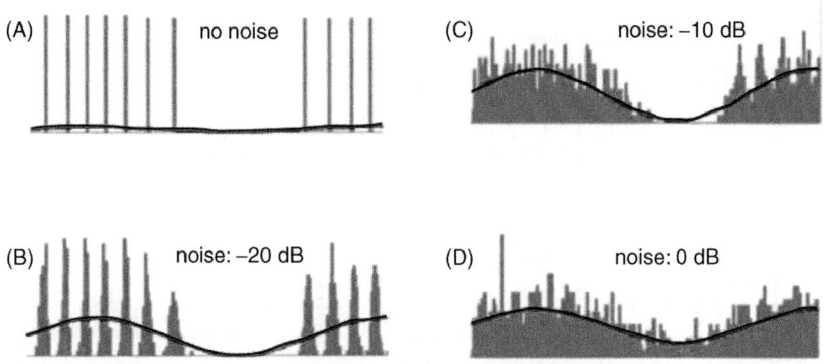

FIGURE 6.11 Influence of noise in the spike timings on the transmission of an acoustic signal (here: a sinusoid). In (A) the inter-spike intervals are constant and without noise (high precision); the sinusoidal input signal is not apparent from the spike-density function of the highly peaked histogram (*black line*). (B–D) With increasing amounts of noise in the spike timings (less precision) the spike-density function provides a better approximation of the original input.

Note that the fact that phase locking at the SBCs is substantially enhanced with respect to the AN fibers does not necessarily mean that the former provide a better representation of the acoustic input than the latter. Clearly, their tuning to specific timing events of the stimulus is sharper, but in order to reconstruct the original acoustic information from the spiking patterns of auditory neurons, the more stochastic phase-locked representation of AN fibers does a much better job than the picket-fence phase-locked responses of spherical bushy cells. The usefulness of noise in obtaining an improved representation of the temporal evolution of the stimulus is illustrated in Fig. 6.11, in which a computer simulation demonstrates that with increasing noise in phase-locking (Fig. 6.11B–D), the spike-density function of the neural activity provides a much better approximation for the sinusoidal input signal than the neuron without noise (Fig. 6.11A).

As the SBCs project to the medial superior olivary (MSO) complex, where neurons are sharply tuned to the *binaural* coincidence detection of spikes, a precise timing of the spikes with respect to acoustic events (like an interaural timing difference, see chapter: Acoustic Localization Cues) is clearly beneficial. It is therefore commonly thought that both mechanisms of phase locking in the auditory system (jittery tuning vs. picket-fence tuning) may in fact underlie different functionalities within the auditory processing streams.

6.4 RATE-LEVEL TUNING OF THE AUDITORY NERVE

AN fibers can be categorized according to their spontaneous firing rates (SR) into low SR, medium SR, and high SR fibers. The dynamic range of an individual high SR fiber is in the order of 20 dB, and is therefore insufficient to encode the full dynamic range of audible acoustic inputs (about 100 dB). As these fibers, which are in the majority, have low recruitment thresholds, they can only

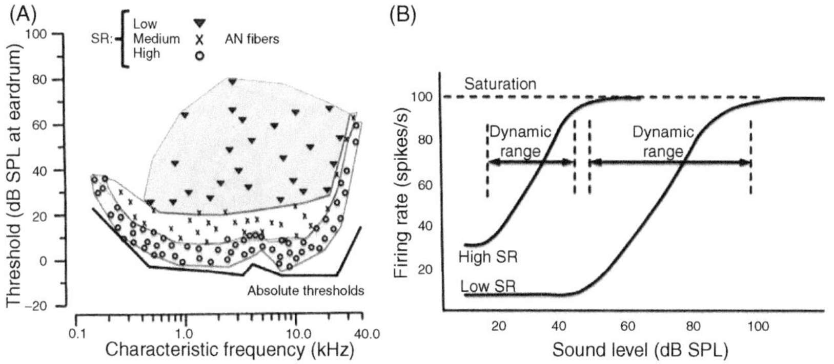

FIGURE 6.12 (A) Schematized distribution of AN fibers according to their recruitment thresholds and characteristic frequencies, as reported by Liberman and Kiang (1978) for cat. High: SR > 18 spikes/s; low: SR < 0.5 spikes/s. Solid line represents the cat's absolute hearing threshold. (B) Schematic rate-level curves: high SR fibers respond differentially over a small dynamic range, but have low recruitment thresholds. Low SR fibers respond over a large dynamic range, and have high thresholds.

encode the loudness of low-intensity sounds. On the other hand, the fibers with a low SR are recruited at higher sound levels (higher thresholds), and they can differentiate sound levels over a much larger dynamic range (Fig. 6.12).

As a result of the differences in thresholds and dynamic range of AN fibers, the full range of audible sound levels is encoded by subsequent recruitment of the different subpopulations. Fig. 6.13 gives a schematic account (after Sachs and Abbas, 1974) on how the compressive nonlinearity in the cochlear transduction influences the rate-intensity encoding for the three different classes of SR-fibers. For example, for a sound at 50 dB SPL virtually all high-SR fibers will be saturated, the medium SR-fibers fire at about half their maximum rates, while only few low-SR fibers will start to become recruited. This distribution in the population activity can thus relate to the perceived loudness associated with an intensity of 50 dB SPL. At about 70 dB SPL also the medium SR fibers become saturated, but the low SR fibers still respond differentially (nearly linearly) to the changes in absolute sound level. The stretching in dynamic range across fiber types could thus result from nonlinear compression at the level of the cochlea.

The rate-level function of an AN fiber according to the model of Sachs and Abbas (1974) is described by:

$$R(p,\Theta) = SR + R_{\text{MAX}} \frac{(p-\Theta)^{\alpha}}{S+(p-\Theta)^{\alpha}} \quad \text{for } p \geq \Theta \tag{6.12}$$

where p is the sound level (in dB SPL), α the exponent of the compressive nonlinearity (eg, $\alpha = 0.3$), Θ is the fiber's recruitment threshold, and SR the spontaneous firing rate. Further, R_{MAX} is the maximum firing rate above SR

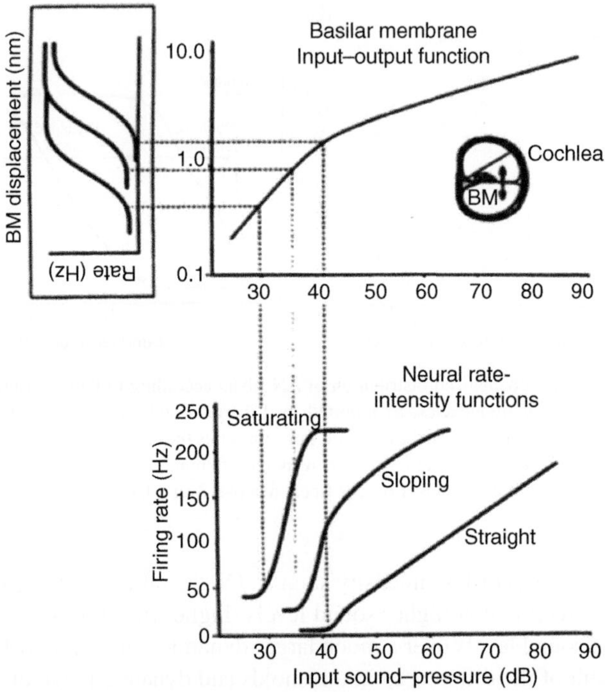

FIGURE 6.13 **A simple mechanistic model proposed by Sachs and Abbas (1974) for the observed stretching in dynamic range for low-SR fibers, when compared to the medium- and high-SR fibers.** As the BM displacement has a nonlinear relation with sound level, the shape of the rate-level functions of AN fibers depend strongly on their recruitment threshold.

(the fiber's saturation level is therefore $R_{SAT} = R_{MAX} + SR$), and S is a measure for the fiber's sensitivity (it relates to its rate-level slope): when $(p-\Theta) = S^{(1/\alpha)}$, the firing rate is at half its maximum level. In normalized form, the rate-level function can be reformulated as:

$$R_{NORM}(p,\Theta) = \frac{R(p,\Theta) - SR}{R_{MAX}} = \frac{1}{1 + S(p-\Theta)^{-\alpha}} \quad \text{for } p \geq \Theta \qquad (6.13)$$

For a recent account on the limitations of this simple model, and of possible extensions to overcome these limitations, see Heil et al. (2011).

6.5 TWO-TONE SUPPRESSION

Two-tone suppression is a nonlinear phenomenon that can be observed in the firing rates of AN fibers (Arthur et al., 1971). The open symbols in Fig. 6.14 show the threshold curve of an AN fiber that has its characteristic frequency at 8 kHz.

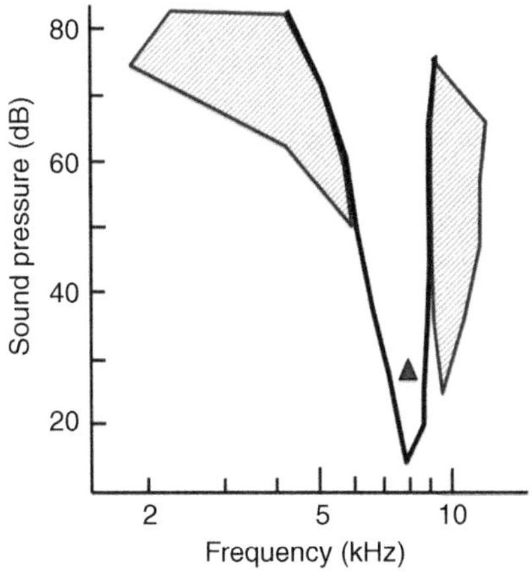

FIGURE 6.14 **Schematized example of two-tone suppression in an AN fiber, as first reported by Arthur et al. (1971).** *Black line*: Tuning curve. *Blue triangle*: Supra-threshold tone presented at the fiber's best frequency (8 kHz); when a second tone is presented within the inhibitory side bands (*red hatched areas*) activity decreases by more than 20%.

When the f_{cf} tone is presented slightly above threshold (triangle) the fiber will fire at a certain rate. If a second tone is presented simultaneously with the f_{cf} tone, say at 5 kHz, the firing rate of the fiber will decrease with respect to the f_{cf}-alone case. This interaction reflects the presence of a nonlinear mechanism, because linear systems will never produce a decrease in the firing rate when *adding* another tone. For linear systems, the rate should at least stay the same, or increase.

For the inhibition to become expressed, the second tone has to be presented at an intensity that falls within the hatched areas. The level dependence of two-tone suppression is another manifestation of its nonlinear character.

Ruggero et al. (1992) demonstrated with the Mössbauer technique that two-tone suppression is caused by nonlinear interactions at the level of the BM. Their experiments showed that the phenomenon requires an active cochlea, as it disappears after cochlear damage, and after the animal's death.

6.6 MODELING THE AUDITORY NERVE

The response patterns of the population of AN fibers represent the full content of acoustic information transmitted to the auditory system. As far as audition goes: it's all there is! To interpret and model neural activity at subsequent stages along the auditory pathway, it is therefore crucial to have an adequate model that accounts for AN population responses. An efficient AN model should capture the different properties that underlie the tuning characteristics of AN fibers, as well as the temporal spike trains across the audible frequency domain for a wide range of acoustic inputs (simple as well as complex sounds over a large range of intensities), but at a minimal computational burden.

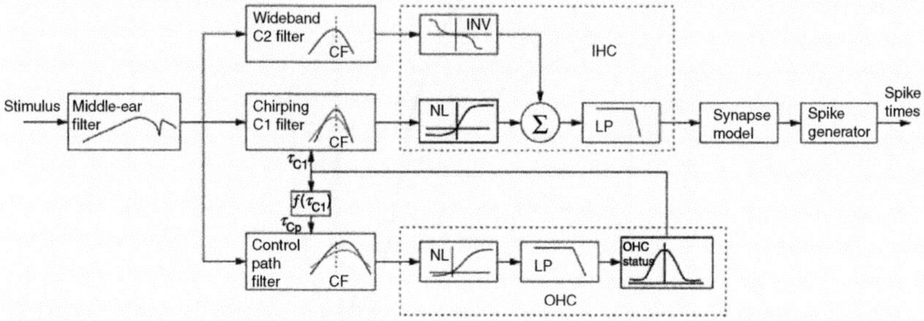

FIGURE 6.15 Neurocomputational model of Zhang et al. (2001) that explains the detailed spiking patterns of AN fibers for complex sounds over a wide range of sound levels. *(After Zhang et al., 2001, with kind permission.)*

The various nonlinear response behaviors of AN fibers to changes in sound level, like the compressive gain, dynamic bandwidths, the phase changes of responses in their phase locking, but also two-tone suppression, should all be faithfully captured by such a model, yet without having to resort to the detailed and tedious hydromechanics and micromechanical properties of the cochlea and cochlear amplifier, as described in chapter: The Cochlea.

Zhang et al. introduced, and in later studies further developed, a phenomenological neurocomputational model that explains the spike trains of AN fibers in response to a wide variety of acoustic inputs, and captures the essential physiological characteristics of the AN response with a minimum of neurobiological detail (Zhang et al., 2001; Zilany et al., 2009).

Fig. 6.15 shows the basic outline of their model, in which the acoustic stimulus (movement of the stapes) drives three parallel mechanisms:

1. A direct path that embodies the band-pass filters of the cochlea (the gammatone filter bank of the BM, like described in Section 6.2), followed by a nonlinear transduction stage of the IHC. This nonlinearity accounts for the rectifying nature of the IHC membrane potentials (the dc component), whereas the subsequent low-pass filter accounts for the high-frequency loss of phase locking in the ac-output of the IHC (Fig. 6.8).
2. A wider band-pass filter extending to neighboring frequencies beyond the gammatone band to account for the nonlinear effects of two-tone suppression (Fig. 6.14; the nonlinear inverter block in the model).
3. A nonlinear feedback control path that incorporates a model for the active amplifier, the OHCs, and which in turn influences the tuning characteristics of the BM (and its own) gammatone filters. Depending on the status of the OHC the time constant of the local gammatone filter is changed, leading to a broadening of its bandwidth, and a decrease of the gain.

Finally, the IHC drives the spike generator of the AN fiber through an adaptive synaptic connection (implemented by the box: "Synapse model").

Briefly, the gammatone filters of the BM partition are described by a concatenation of a dynamic third-order gammatone, and a static, linear (first-order) gamma filter. The transfer characteristics of these filters are adapted from Eq. (6.8), and are described as follows:

$$G_{\mathrm{DYN}} \cong \frac{\tau(t)^3 e^{-j\omega\Delta T}}{\left[1+j\tau(t)\left(\omega-\omega_{cf}\right)\right]^3} \quad \text{and} \quad G_{\mathrm{LIN}} \cong \frac{\tau e^{-j\omega\Delta T}}{\left[1+j\tau\left(\omega-\omega_{cf}\right)\right]} \tag{6.14}$$

where $\tau(t)$ is the time-varying time constant of each dynamic gammatone filter, ΔT is the transmission delay of the peak of the BM traveling wave (see Fig. 6.2), and ω_{cf} is the characteristic frequency of the AN fiber to be modeled.

The gain and bandwidth of the dynamic BM filters are controlled by the time-varying time constant, $\tau(t)$, which is determined by the output of the OHC control path. It is assumed to vary between two extreme values: a narrow bandwidth (high gain) value for low sound levels, and a wide bandwidth (low gain) value at high sound levels:

$$\tau_{\mathrm{Narrow}} = \frac{2Q_{10}}{\omega_{cf}} \le \tau(t) \le \tau_{\mathrm{Narrow}} 10^{-\frac{GAIN(cf)}{60}} = \tau_{\mathrm{Wide}} \tag{6.15}$$

where the quality factor, Q_{10}, is specified by the psychoacoustic Cambridge relation [Eq. (6.9)]. The compressive gain of the cochlear amplifier at f_{cf}, $GAIN(f_{cf})$, varies monotonically between 15 dB for low frequencies (<1000 Hz), to 70 dB for the high frequencies, a behavior that is based on available cat data. The time constant for the static linear filter of Eq. (6.14) is taken as $\tau = \tau_{\mathrm{Narrow}}$.

The wide-band filter in the suppression pathway is described by a third-order gammatone with its f_{cf} shifted by 1.2 mm basal to the fiber's f_{cf} (ie, toward higher frequencies). Its bandwidth is controlled by Eq. (6.15). The low-pass filter (with a cut-off frequency at 3 kHz; see Fig. 6.8) of the IHC is driven by the difference between the C1 and C2 pathways, which implements two-tone suppression.

The output of the OHC control path is the time-varying time constant, which is generated by a nonlinearity, driven by the output voltage of a LP filter (cut-off at ~ 800 Hz), $V_{LP}(t)$:

$$\tau(t) = \tau_{\mathrm{Narrow}} \left\{ R_0 + \left(1-R_0\right)\left[\frac{(\tau_{\mathrm{Wide}}/\tau_{\mathrm{Narrow}}) - R_0}{1-R_0}\right]^{\beta\left|V_{LP}(t)\right|} \right\} \tag{6.16}$$

in which $R_0 = 0.05$, and $\beta = 1/DC$, where $DC = 0.37$ is the estimated dc output of the OHC's low-pass filter at high sound levels (Exercise).

The IHC-AN synaptic stage in the model accounts for the different adaptation effects (occurring at different time scales, from milliseconds to seconds) observed in AN fiber responses to sustained sounds. The model captures an exponential adaptation mechanism (corresponding to AN onset effects at two timescales: 2 and 60 ms, respectively), and a power–law adaptation mechanism (accounting for sustained effects that act on a longer time scale, of seconds). This latter slow mechanism also accounts for the variation in spontaneous firing rates, by implementing a fractional Gaussian ($\sim 1/f$) noise generator. For a detailed account of this part of the model, see Zilany et al., (2009).

Finally, the spike generator of the AN fiber is modeled by a renewal process, resulting in a nonhomogeneous Poisson process with refractoriness. With $s(t)$ as the output of the synapse, the firing rate of the AN is then determined by:

$$R(t) = s(t)\left[1 - H(t)\right] \tag{6.17}$$

where $H(t)$ (implementing history effects) is described as the sum of two exponentials:

$$H(t) = \begin{cases} c_0 e^{-\gamma_1(t-t_1-R_A)} + c_1 e^{-\gamma_2(t-t_1-R_A)} & \text{for } t - t_1 \geq R_A \\ 1 & \text{elsewhere} \end{cases} \tag{6.18}$$

(t_1 is the timing of the previous spike, and R_A is the refractory period; $c_{1,2}$ and $\gamma_{1,2}$ are parameters).

The complete model of Fig. 6.15 accounts for a large variety of (nonlinear, nonstationary) response patterns of AN fibers, and can serve as an efficient and adequate descriptor for the neural representation of AN population responses to an arbitrary acoustic input that is sent to the auditory ascending pathways. Fig. 6.16 shows an illustrative example of the model's capabilities to mimic

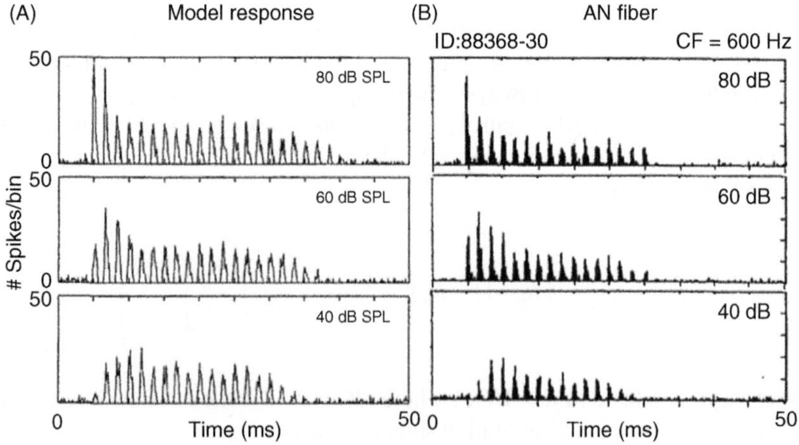

FIGURE 6.16 Model responses to 25 ms tone pips, presented at $cf = 600$ Hz (A) at three different sound levels, compared to recorded data from cat AN (B). Note the good correspondence in phase-locking behavior for the different conditions. *(Source: After Zhang et al., 2001, with kind permission.)*

recorded AN data. The fiber has a f_{cf} = 600 Hz and shows clear phase locking. Note, however, that the histograms also change their shape with changing sound level; this property is due to synaptic adaptation.

6.7 MULTITONE RESPONSES OF THE AUDITORY NERVE

The sensitivity of the tuning curves of AN fibers appears to closely resemble the BM motion (Section 6.1 and chapter: The Cochlea). However, the relatively low-frequency spiking patterns of the AN are inappropriate to phase-lock with high-frequency sounds, and together with their limited dynamic range (of about 20 dB) preclude a full reconstruction of the high-frequency motion patterns of especially the basal and mid-frequency sections of the BM over a larger dynamic range of sounds. Moreover, although AN fibers of the apical cochlea do show phase locking, there are hardly any mechanical data of BM motion patterns at the lower frequencies, because the apical portion is inaccessible to measurement.

Van der Heijden and Joris (2003) developed an interesting method that relies on the nonlinear input-output characteristics of neurons, and on the resulting nonlinear interactions (like second-order distortions) between the components of particular tone complexes. This could in principle allow for a precise amplitude and phase reconstruction of BM motion across the entire length of the cochlea. How does this method work?

The underlying idea is that the envelope of a tone complex contains low-frequency interactions among the primary high-frequency components, and that the AN can phase lock to these low-frequency amplitude modulations. For example, in case of two simultaneous tones, with frequencies ω_1 and ω_2, the amplitude of the summed waveform is modulated with the difference beat frequency, $\Delta\omega = |\omega_1 - \omega_2|$ (eg, chapter: The Nature of Sound, binaural beats, Eq. 2.39). The difference frequency is also manifest as a second-order interaction in a nonlinear system. To understand why, consider the following amplitude-modulated tone

$$s(t) = \left[1 + a_{\text{mod}} \sin(\omega_{\text{mod}})\right] \sin\left(\omega_{\text{carr}} t\right) \tag{6.19}$$

which can be rewritten (Exercise) as follows:

$$s(t) = \sin(\omega_{\text{carr}} t) + \frac{a_{\text{mod}}}{2}\left[\sin(\omega_{\text{carr}} - \omega_{\text{mod}})t + \sin(\omega_{\text{carr}} + \omega_{\text{mod}})t\right] \tag{6.20}$$

where ω_{carr} is the signal's carrier frequency, a_{mod} the modulation amplitude, and ω_{mod} the modulation frequency ($\omega_{\text{mod}} \ll \omega_{\text{carr}}$).

This modulated signal therefore contains three spectral components:

$$\left[\left(\omega_{\text{carr}}-\omega_{\text{mod}}\right),\quad \omega_{\text{carr}},\ \left(\omega_{\text{carr}}+\omega_{\text{mod}}\right)\right] \tag{6.21}$$

In case the carrier frequency < 3 kHz, the AN fiber will respond to this stimulus by phase-locking to the half-cycles of the amplitude modulated envelope only (due to the rectification by the IHC). Suppose that the response of the AN fiber is described as follows:

$$r(t)=\max\left\{0,\ \sin\left(\omega_{\text{carr}}t\right)+b\left[\sin\left(\omega_{\text{carr}}-\omega_{\text{mod}}\right)t+\sin\left(\omega_{\text{carr}}+\omega_{\text{mod}}\right)t\right]\right\} \tag{6.22}$$

Note that the spectrum of the AN response (Exercise) contains the following frequency components:

$$\left[0,\ \omega_{\text{mod}},\ \left(\omega_{\text{carr}}-\omega_{\text{mod}}\right),\quad \omega_{\text{carr}},\ \left(\omega_{\text{carr}}+\omega_{\text{mod}}\right)\right] \tag{6.23}$$

In other words, the envelope's modulation frequency is represented in the response of the AN fiber, despite the fact that there is *no* acoustic energy for that frequency in the acoustic input! This interesting nonlinear property of AN fiber responses to modulated sounds is illustrated in Fig. 6.17.

The amplitude of the AN fiber's response to the sinusoidal envelope modulation varies systematically with ω_{mod}. The modulation transfer function (MTF) quantifies the sensitivity of AN fibers to such amplitude modulations. The MTF has a low-pass amplitude characteristic, with a cut-off that varies linearly with the neuron's characteristic frequency [and hence, with the fiber's bandwidth of its frequency tuning curve, Eq. (6.9)].

The same principle holds when a tone complex contains $N(>2)$ frequency components, all of which may have different amplitudes and phases:

$$s(t)=\sum_{n=1}^{N}A_n \exp[\mathrm{j}(\omega_n t+\phi_n)] \tag{6.24}$$

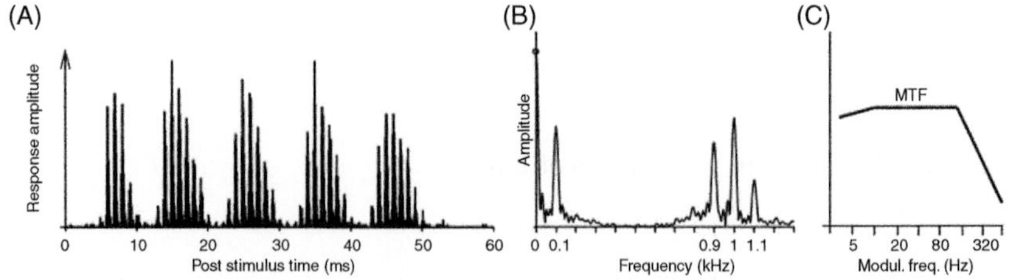

FIGURE 6.17 Response of an AN fiber (f_{cf} = 1 kHz) to the modulated tone of Eq. (6.12), where f_{mod} =100 Hz and f_{carr} = 1 kHz. Due to rectification in the phase-locked responses (A), the spectrum (B) also contains a strong component at the difference frequency of 100 Hz. The dc (*open circle*) equals the mean firing rate. (C) Schematic MTF. *(Source: After Joris et al., 2004, with kind permission.)*

fcf

The squared amplitude of the signal's envelope is then given by (see Exercises):

$$|s|^2 = \sum_{n=1}^{N}(A_n)^2 + \sum_{n=1}^{N}\sum_{m=n+1}^{N} 2A_nA_m \cos\left[(\omega_n - \omega_m)t + \phi_n - \phi_m\right] \tag{6.25}$$

In other words, all possible low-frequency differences between the primary components determine the beat frequencies of the envelope:

$$\Delta\omega_{nm} = (\omega_n - \omega_m), \text{with } n \in \{1 \cdots N\}, \text{ and } m \in \{(n+1) \cdots N\} \tag{6.26}$$

and AN fibers could phase lock to each of these $\binom{N}{2}$ interaction components. In particular, the "*zwuis*" (beat–noise) tone complex consists of a set of irregularly spaced tones, for which all low-frequency differences, $\Delta\omega_{nm}$, are unique, and their corresponding amplitudes and phases are given by the right-hand part of Eq. (6.25):

$$\begin{cases} A_{mn} = 2A_nA_m \\ \phi_{nm} = \phi_n - \phi_m \pmod{2\pi} \end{cases} \tag{6.27}$$

After extracting all (or most) beating responses from an AN fiber, the amplitudes and phases of the primary components of the tone complex, A_n, ϕ_n, can be readily determined, because for all $N > 3$ there will be more independent equations than unknowns [as Eq. (6.27) is redundant]. For example, a complex with 5 irregularly spaced primaries yields 10 equations for beat amplitudes as well as beat phases, whereas there are only 5 unknowns for each to be solved.

Fig. 6.18 shows an example of an irregular multitone stimulus (with 7 randomly selected components) and its associated spectrum for the high-frequency fine structure (between 2 and 3 kHz) and the low-frequency (<1 kHz) envelope with its 21 components.

By shifting different sets of tone complexes across the full tuning curve of the AN fiber, meanwhile preserving sufficient overlap among adjacent complexes, the complete amplitude and phase characteristics of the BM vibrations at the site of the neuron's characteristic frequency can be reconstructed at a high spectral-temporal resolution, and over a remarkable dynamic range. By ensuring that the tone complexes overlap, the different reconstructed segments of the underlying BM tuning curve can be glued together. Fig. 6.19 shows an example of the full reconstruction procedure for a single cat high-frequency AN fiber, with f_{cf} = 14.4 kHz.

Note that the primary components of the complex can still be faithfully reconstructed if the AN fibers display a significant nonlinearity, for example,

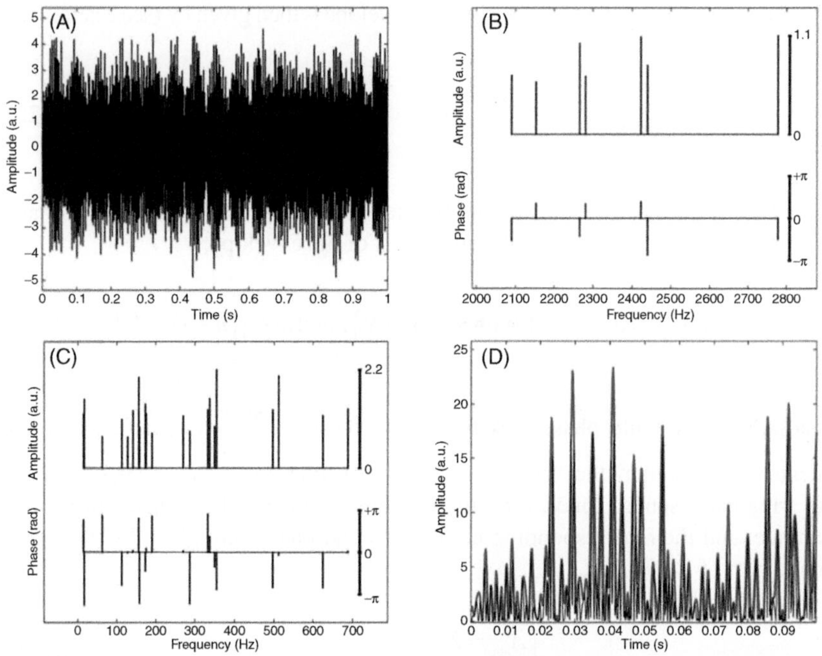

FIGURE 6.18 (A) Irregular multitone stimulus ($N = 7$); frequencies (between 2 and 3 kHz), amplitudes, and phases were randomly selected. (B) Amplitude and phase spectrum of the stimulus in (A). (C) Amplitude and phase spectrum (21 components between 0 and 700 Hz) of the squared envelope of the sound. (D) Squared envelope of the first 0.1 s of the stimulus in (A) (*black*), and the calculated envelope from Eq. (6.25) (*red*).

a strong compression in the overall gain transduction process. The reason for this spectral robustness lies in the constant (dc) term of Eq. (6.17) (its left-hand part), which will strongly dominate the (nonlinear) response when N is sufficiently large. As a result, the interaction components will survive the nonlinear transduction stage. Fig. 6.20 illustrates this interesting property for a compressive nonlinearity of the form:

$$f(\textbf{Envelope}) = \text{Envelope}^{0.3} \tag{6.28}$$

For low-frequency (apical) AN fibers, the phase locking will also be preserved for the primaries, ω_n, so that the power spectrum of neural responses from LF fibers will contain both linear (ie, primary) and, due to nonlinearities in the transduction process, second-order (frequency differences) components (even third-order contributions, such as $2f_n - f_m$ may be expected). This property allows for a *linear* analysis of the apical fiber responses to the irregular "zweus" complexes, which are constructed such that all primaries differ from

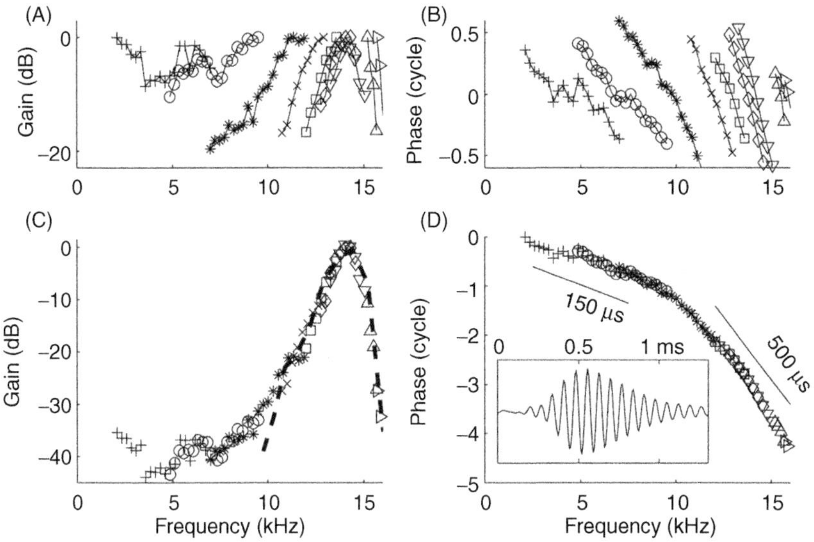

FIGURE 6.19 (A) Amplitudes of the different (overlapping) primaries obtained from different "zweus" complexes, presented to a cat AN fiber with f_{cf} = 14.4 kHz. (B) The associated phase curves of the primaries. (C) Total amplitude tuning curve, obtained by matching the different amplitude segments of (A). The dashed line is the fiber's (inverted) tuning curve obtained with classical pure-tone stimulation. (D) Total phase response. Inset shows the corresponding impulse response function of the transfer characteristic. Oblique line segments indicate average group delays. *(Source: After Van der Heijden and Joris, 2003, with kind permission.)*

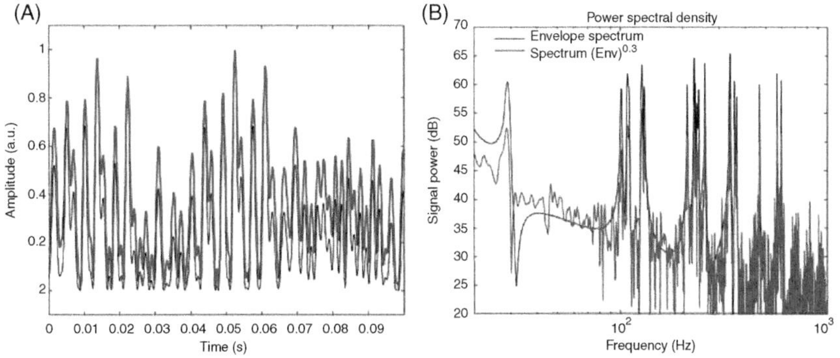

FIGURE 6.20 (A) Envelope of an irregular tone complex ($N = 7$; primaries between 2 and 3 kHz) in *black*, and its compressed version [applying Eq. (6.20)] in *red*. Signals have been normalized for better comparison. (B) Power spectra of the signals in (A). The spectrum of the compressed waveform is shifted by 37 dB to allow for direct comparisons. Note that all dominant spectral peaks of the original envelope survive the nonlinear compression.

the second- and third-order distortion components. This technique allows for a reconstruction of the entire BM motion along the cochlea, also for the apical locations, for which mechanical data cannot be obtained. See Van der Heijden and Joris (2006), for further details.

6.8 EXERCISES

Problem 6.1: Derive Eq. (6.8).

[*Hint:* First calculate the FT of Eq. (6.7). Note that the FT of the cosine term gives two delta functions: $S(f)=(1/2)[e^{-j\phi}\delta(f+f_{cf})+e^{j\phi}\delta(f-f_{cf})]$. Then, show that the FT of the prefactor, $g(t)=t^{n-1}e^{-at}$ is given by: $G(f)=(n-1)!/(a+j2\pi f)^n$. Now combine the two FTs: $H(f) = S(f)*G(f)$, with * convolution. Finally, note that one of the two terms arising from the cosine transform may be neglected relative to the other term.]

Problem 6.2: Calculate the Fourier spectrum (frequency, amplitude, phase) of the rectified (positive only) modulated sinusoid, which describes the firing of a phase-locked AN fiber, as given by Eq. (6.14) (see also chapter: The Nature of Sound).

Problem 6.3: Verify Eq. (6.13).

Problem 6.4: Get a feeling for the validity of Eq. (6.25), by deriving the expressions explicitly for $N = 2$ and $N = 3$:

(a) Calculate A and Ψ from:

$$ae^{i(\omega_1 t+\varphi_1)}+be^{i(\omega_2 t+\varphi_2)}=Ae^{i(\psi)}$$

(b) Use the result from (a) to extend to three components:

$$ae^{i(\omega_1 t+\varphi_1)}+be^{i(\omega_2 t+\varphi_2)}+ce^{i(\omega_3 t+\varphi_3)}=Ae^{i(\psi)}$$

Problem 6.5: Verify (analytically, or through a numerical simulation) that Eq. (6.16) lets the time constant $\tau(t)$ vary between τ_{Narrow} and τ_{Wide}.

REFERENCES

Arthur, R.M., Pfeiffer, R.R., Suga, N., 1971. Properties of two-tone inhibition in primary auditory neurons. J. Physiol. 212, 593–609.

De Boer, E., De Jongh, H.R., 1978. On cochlear encoding: potentialities and limitations of the reverse correlation technique. J. Acoust. Soc. Am. 63, 115–135.

Evans, E.F., 1972. The frequency response, and other properties of single fibers in the guinea pig cochlear nerve. J. Physiol. 226, 263–287.

Hartmann, W.H., 1998. Signals, Sound, and Sensation. Springer, New Jersey, USA.

Heil, P., Neubauer, H., Irvine, D.R.F., 2011. An improved model for the rate-level functions of auditory nerve fibers. J. Neurosci. 31, 15424–15437.

Joris, P.X., Smith, P.H., 2008. The volley theory and the spherical cell puzzle. J. Neurosci. 154, 65–76.

Joris, P.X., Schreiner, C.E., Rees, A., 2004. Neural processing of amplitude-modulated sounds. Physiol. Rev. 84, 541–577.

Liberman, M.C., Kiang, N.Y., 1978. Acoustic trauma in cats. Cochlear pathology and auditory nerve activity. Acta Otolaryngol. Suppl. 358, 1–63.

Palmer, A.R., Russell, I.J., 1986. Phase-locking in the cochlear nerve of the guinea-pig and its relation to the receptor potential of inner hair cells. Hear. Res. 24, 1–15.

Price, R., 1958. A useful theorem for nonlinear devices having Gaussian inputs. IRE Trans. Inf. Theory 4, 69–72.

Ringach, D., Shapley, R., 2004. Reverse correlation in neurophysiology. Cogn. Sci. 28, 147–166.

Ruggero, M.A., Robles, L., Rich, N.C., 1992. Two-tone suppression in the basilar membrane of the cochlea: mechanical basis of auditory-nerve rate suppression. J. Neurophysiol. 68, 1087–1099.

Ruggero, M.A., Shyamla Narayan, A., Temchin, A.N., Recio, A., 2000. Mechanical bases of frequency tuning of the cochlea: Comparison of basilar membrane vibrations and auditory-nerve-fiber responses in chinchilla. Proc. Natl. Acad. Sci. 97, 11744–11750.

Sachs, M.B., Abbas, P.J., 1974. Rate versus level functions for auditory nerve fibers in cats: tone-burst stimuli. J. Acoust. Soc. Am. 56, 1835–1847.

Van der Heijden, M., Joris, P.X., 2003. Cochlear phase and amplitude retrieved from the auditory nerve at arbitrary frequencies. J. Neurosci. 23, 9194–9198.

Van der Heijden, M., Joris, P.X., 2006. Panoramic measurements of the apex of the cochlea. J. Neurosci. 26, 11462–11473.

Zhang, X., Heinz, M.G., Bruce, I.C., Carney, L.H., 2001. A phenomenological model for the responses of auditory-nerve fibers: I. Nonlinear tuning with compression and suppression. J. Acoust. Soc. Am. 109, 648–670.

Zilany, M.S.A., Bruce, I.C., Nelson, P.C., Carney, L.H., 2009. A phenomenological model of the synapse between the inner hair cell and auditory nerve: long-term adaptation with power-law dynamics. J. Acoust. Soc. Am. 125, 2390–2412.

Chapter 7

Acoustic Localization Cues

7.1 INTRODUCTION

The foregoing chapters: The Cochlea; The Auditory Nerve, describe how the cochlea and auditory nerve encode sounds in terms of their spectral and temporal acoustic properties. For now we will ignore the vast complexities that arise when multiple sound-sources are present in the environment (see chapter: A Brief Introduction to the Topic), and instead focus on the seemingly simple problem of just a single sound-source in the environment. We also assume that the subject's head is stationary. The sound's spectral-temporal (Fourier-like) representation in the auditory nerve firing uniquely refers to its *identity*, useful to solve the *source-identification* (what?) problem (see chapter: A Brief Introduction to the Topic). However, the firing patterns of auditory nerve fibers do not provide a direct spatial representation of the sound, so that the *source-localization* task (where?) has to be derived at later stages in the auditory pathway.

In the double-pole coordinate system used to describe sound-source directions, the *azimuth* angle is defined within parallel horizontal plane(s) through the head, where straight ahead (the mid-sagittal plane) corresponds to azimuth, $\alpha = 0$ degrees, locations leftward from the mid-sagittal plane yield negative azimuth angles, and rightward positions correspond to positive azimuth angles. Likewise, *elevation angles* are measured in parallel medial plane(s): the horizontal plane through the ear canals and inion separate the positive (upward) elevation angles from the negative (downward) elevation directions. The relationship between the polar coordinates of a target, given by $T = [E,\Phi]$, where E is the angle corresponding to the target's eccentricity (taken positive, between $[0,\pi]$) relative from straight ahead, and Φ is the target direction (between $[-\pi,+\pi]$) with respect to the horizontal plane, and the target's corresponding azimuth–elevation angles, $T = [\alpha,\varepsilon]$, is given by (Fig. 7.1):

$$\begin{aligned}\alpha &= \arctan\left[\tan(E)\cos(\Phi)\right]\\ \varepsilon &= \arctan\left[\tan(E)\sin(\Phi)\right]\end{aligned} \tag{7.1}$$

Note, that although azimuth and elevation angles can in principle run between $\alpha = [-90,+90]$ degree, and $\varepsilon = [-180,+180]$ degree, not all $[\alpha,\varepsilon]$ combinations are possible when sounds are presented from all possible directions at a fixed distance from the subject. For example, for a target at an azimuth far

The Auditory System and Human Sound-Localization Behavior. http://dx.doi.org/10.1016/B978-0-12-801529-2.00007-6

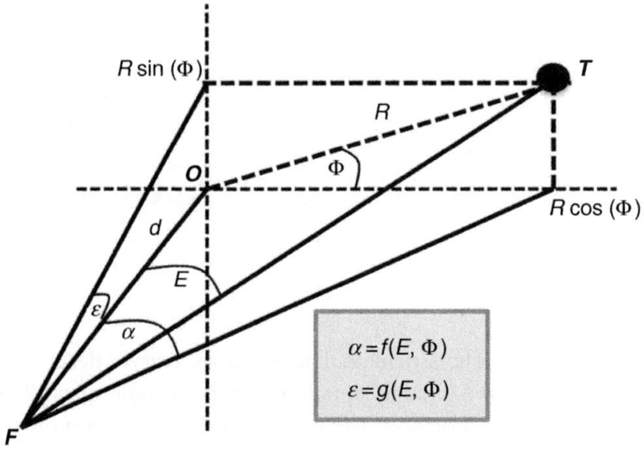

FIGURE 7.1 **Geometrical relationships between the polar angles, (E,Φ), of a target and the double-pole azimuth–elevation coordinates, (α,ε).** As an aid to the reader, the target is projected on a flat screen at distance d from the center of the head at F. On the screen it thus obtains coordinates (R,Φ). The straight-ahead direction is found at $O = (\alpha,\varepsilon) = (E,\Phi) = (0,0)$.

right or left (ie, at $\alpha = \pm 90$ degree, respectively, ie, at the azimuth poles), the elevation angle can only be zero. Similarly, at the zenith or nadir, $\varepsilon = \pm 90$ degree (the elevation poles), the azimuth angle can only be zero (Fig. 1.7, and Exercise 7.1, and Section 7.6, where this feature leads to an interesting possibility for localization in the medial plane).

Sound localization relies on the physical interaction of sound waves with torso, head, and ears, giving rise to specific, frequency- and direction dependent changes in the sound-evoked activities in the auditory nerves for either ear. These interactions involve interference in the ear canal from reflections of the sound waves at the different body parts (for high frequencies) and from diffraction of sound waves around these body parts (for the lower frequencies). These interactions lead to frequency-dependent and sound-direction dependent amplifications and attenuations of the different spectral components of the sound (see also chapter: The Nature of Sound, for background). These acoustic interactions may provide valid *sound-localization cues* when the patterns of acoustic changes correspond in a unique way to a particular location in space. For example, although cues may be *ambiguous* with regard to the sound-source direction when looking at their values at *either* ear, the *binaural differences* between these cues (which then have to be derived in the brain through a binaural integration of signals from both ears) may to a large extent disambiguate the spatial information. This binaural principle was already known at the end of the 19th century (eg, Rayleigh, 1896), and was important for military applications (Fig. 7.2).

Humans require three acoustic cues to unambiguously point to the sound-source direction in azimuth and elevation. These cues are interaural time differences, or ITDs, interaural level differences, or ILDs, and spectral shape cues

FIGURE 7.2 **A Czech sound localizer.** Such systems were used in World Wars I and II to locate enemy artillery and cavalry. Many years before the barn owl's binaural sound-localization cues were discovered by Knudsen and Konishi (1978), military engineers had figured out a way to assess the direction of distant sound sources in azimuth and elevation. These passive systems predate the active radar systems of later years. *(Source: Photo kindly made available by TNO Museum Waaldorp, The Netherlands.)*

provided by the complex filtering of (broadband) sounds by torso, head, and pinna (the so-called head-related transfer functions, or HRTFs). In the following sections we will discuss these different cues, and their potential neurobiological implementations.

7.2 INTERAURAL TIME DIFFERENCES

As sound propagates through the air at a constant, frequency-independent velocity of about $c = 343$ m/s (chapter: The Nature of Sound), the arrival times of the sound wave to the left versus the right ear will differ when a sound emanates from a particular azimuth angle. The resulting time difference depends in a systematic way on the size of the head, the incident azimuth angle of the sound wave front, and the sound velocity. If for simplicity we assume a spherical head without pinnae, radius r, a positive source-azimuth angle of α_0 degree, and a planar wave front (which is a reasonable approximation when the distance of the sound source (typically a small loudspeaker) from the ears exceeds 1 m), the path-length difference between the right and left ear is (Fig. 7.3)

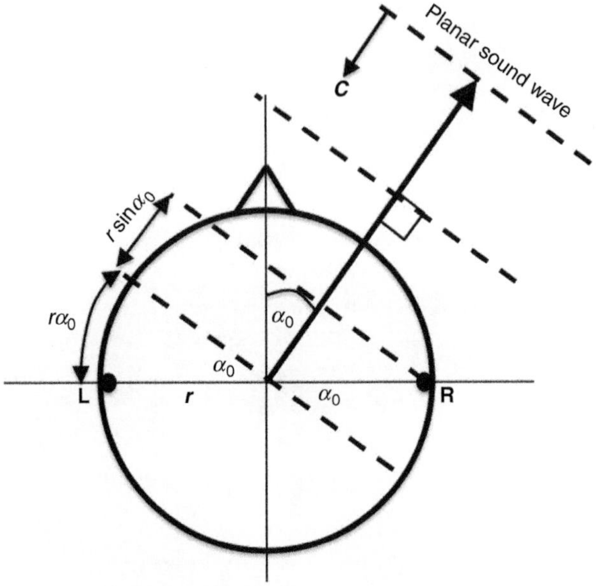

FIGURE 7.3 Geometrical approximation of the interaural path length difference for a planar sound wave front at incident azimuth angle α_0 degree. R, right eardrum; L, left eardrum; r, effective radius of the head; c, the speed of sound.

$$\Delta x = r(\alpha_0 + \sin\alpha_0) \quad \text{for} -\frac{\pi}{2} \leq \alpha_0 \leq +\frac{\pi}{2} \tag{7.2}$$

and therefore the interaural time difference is determined by

$$\Delta T = \frac{\Delta x}{c} = \frac{r}{c}(\alpha_0 + \sin\alpha_0) \tag{7.3}$$

To get an idea of the size of this effect, here are some numbers: for a typical human head, $r \sim 10$ cm, so that the prefactor in Eq. (7.3) amounts to about 0.000292 s. Per degree azimuth angle around straight ahead Eq. (7.3) may then be linearized to the following estimate:

$$\Delta T \approx 0.000292 \frac{\pi}{180} 2\alpha_0 \approx 10\ \mu\text{s/degree}$$

This is an extremely short time difference, even way shorter than the duration of a single action potential, but psychometric and behavioral evidence (see also later) of normal-hearing human listeners indicate a spatial resolution in the horizontal plane of approximately one degree! As a result, the binaural sound-localization system may be sensitive to time differences in the order of 10 μs. The largest ITDs in this simple model occur for the far-lateral azimuth angles, and amount to approximately $\Delta T \simeq 750$ μs.

The ITD resolution, expressed in μs/rad, for the total range of azimuth angles is obtained by taking the derivative of Eq. (7.3) with respect to α:

$$G_{\mathrm{ITD}}(\alpha) \equiv \frac{d(\Delta T)}{d\alpha} = \frac{r}{c}(1 + \cos\alpha), \tag{7.4}$$

which has a maximum at $\alpha = 0$ (straight ahead), and decreases to a minimum at $\alpha = \pm\pi/2$. According to this simple model, ITD perception for front and back locations will be exactly the same. However, psychoacoustic measurements indicate clearly better performance for the frontal versus the rear hemifield.

Note that the ITD cue for a pure tone, say $s(t) = A\sin(\omega_0 t)$, starts to point at ambiguous (multiple) azimuth angles, as soon as the tone's wavelength, $\lambda = c/f$, exceeds the maximum path-length difference,

$$\Delta x_{\max} \sim \frac{2.57}{c} < \lambda_0 \quad \text{that is for} \quad f_0 < \frac{c}{2.57 \cdot r} \approx 1335\,\mathrm{Hz} \tag{7.5}$$

Indeed, for pure tones the ITD effectively becomes an interaural *phase* difference, because there is no tone onset, and is hence constrained to IPD $\in [0,2\pi]$ (Exercise 7.2).

Because pure tones yield phase differences, rather than interaural time differences (as the onset of a pure tone is ill-defined), it's interesting to consider the following. For a given, fixed, time difference ΔT, the interaural phase difference for a tone at frequency ω_0 is:

$$\Delta\Phi(\alpha_0) = \omega_0 \Delta T(\alpha_0) \tag{7.6}$$

As a result, the phase difference varies linearly with frequency, so that neurons with different characteristic frequencies, and sensitive to the phase difference would fire at different phases of the stimulus.

How can the system determine the real acoustic time difference without having to worry about frequency-dependent phase differences? A possible answer is given by taking the cross-correlation between the two inputs at the left and right ear. For the situation that one of the signals is a delayed version of the other, say $x_L(t) = x_R(t - \Delta T)$, the cross-correlation function is simply the autocorrelation function of $x_L(T)$, shifted by ΔT.

Suppose a tone of frequency ω_0 and amplitude A stimulates the right ear, and the left ear with an interaural delay of ΔT. The cross-correlation between the left and right signals is then:

$$\Phi_{\mathrm{LR}}(\tau) = A^2 \int_0^{T_0} \sin(\omega_0 t)\sin\left[\omega_0 (t + \tau - \Delta T)\right] dt = \Phi_{\mathrm{RR}}(\tau - \Delta T) \tag{7.7}$$

Clearly (Exercise 7.3), the cross-correlation function reaches its first maximum at $\tau = \Delta T$, which is indeed independent of the tone's frequency.

The precedence effect and summing localization: The precedence effect is also called the "*law of the first wave front*," and plays a role in the perception of sound location in the presence of multiple sounds, usually reflections (Wallach et al., 1949; Brown et al., 2015). When two sounds arrive at the ears within about 1–5 ms from each other, they are typically merged into one azimuth percept that is dominated by the first-arriving sound source (provided that the intensity of the lagged sound does not exceed the first by more than 10–15 dB). In these conditions, the location of the delayed source has only a small effect on the localization percept. The precedence effect is most dominant for low frequencies (for which the ITDs are important), and for stimuli with a long duration, and is explained by assuming that the auditory system determines the ITD only over the first few ms of the acoustic signal. In this way, robust localization is still possible in the presence of reverberations (echoes) in confined spaces. When the delay exceeds about 40 ms, two sounds are heard, whereby the second sound is perceived as an echo of the first. For longer delays both sounds are separated and accurately localized near their veridical locations.

When the time difference is very small (<1 ms) the localization percept resembles a weighted average of the locations of the two sources, with relative sound level and time delay as weighting factors. This effect has been termed "*summing localization*" (see also chapter: Assessing Auditory Spatial Performance):

$$\alpha_{\text{Perceived}} = a\alpha_{\text{ITD}} + (1-a)\alpha_{\text{ILD}} \quad \text{with} \quad 0 \le a \le 1 \tag{7.8}$$

7.2.1 Jeffress' Delay-Line Model

How could the auditory system detect the time differences associated with the azimuth-dependent interaural delays, as predicted by Eq. (7.3)? And through which mechanism could the system map these measured time differences to inferred azimuth angles? Already as early as 1948, Lloyd Jeffress proposed that the auditory system could respond reliably to tiny interaural timing differences, by using axonal delay lines that project to coincidence detectors (Fig. 7.4A). The idea is as elegant as it is simple: while action potentials travel at a finite speed along the auditory nerve, it takes time to reach its target nucleus. When the axons have a fixed diameter, d, the spike-propagation speed will be the same for all fibers (proportional to $\sqrt{d}$), and the total axonal length determines the time needed to reach a particular projection neuron in the target nucleus.

Thus, if the axonal path lengths from the left and right ear to the target cell differ, the action potentials traveling along these axons will arrive at exactly the same time, only when the azimuth-dependent sound delay exactly matches the delay from the difference in axonal path lengths.

If the target cells act as coincidence detectors, for which the probability of generating a spike:

$$p(Spike) = \begin{cases} \text{High when } \Delta T_{\text{LR-spks}} < \delta \approx 0 \\ \text{Low when } \Delta T_{\text{LR-spks}} > \delta \end{cases} \tag{7.9}$$

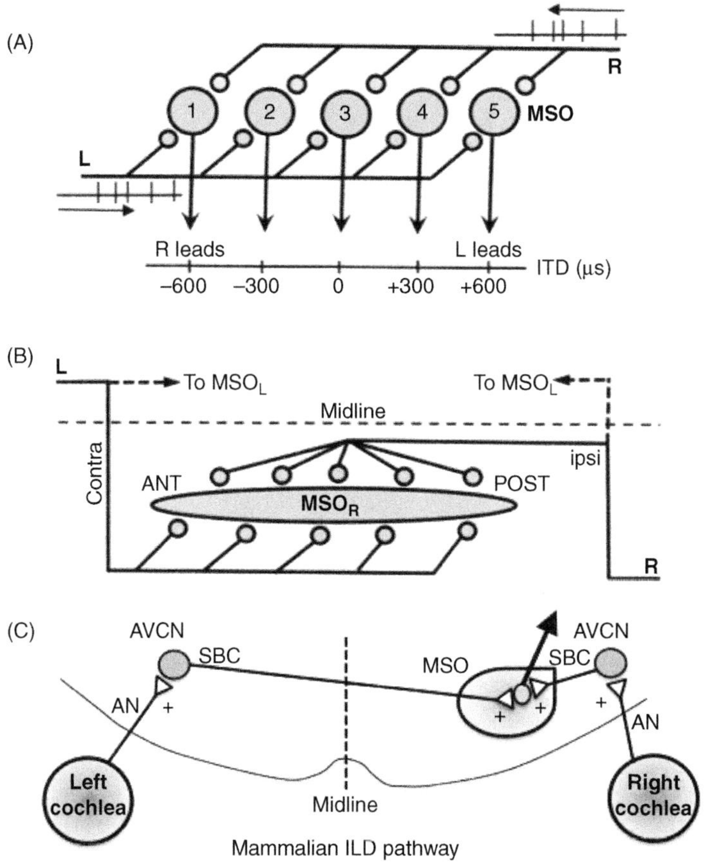

FIGURE 7.4 (A) Jeffress' model generates a topographic azimuth map through ladder-like dual delay lines that project to coincidence detectors in the MSO. (B) Neuroanatomical implementation of delay lines in cat. The contralateral projections branch at different portions in the MSO, creating systematically different path lengths; the ipsilateral projection lacks this systematic organization; yet, the joint projections onto each MSO cell create a systematic progression of interaural delay along the antero posterior dimension of the nucleus for each characteristic frequency. (C) The ITD neural pathway. AN, auditory nerve; SBC, stellate bushy cell; AVCN, anteroventral cochlear nucleus. Right: schematic tuning characteristic of an MSO neuron to ITD. Note that the characteristic delay corresponds to a contralateral azimuth location.

the target cells effectively operate as cross-correlators on the incoming spike trains from both sides [as Eq. (7.7)].

The binaural coincidence detectors reside in the Medial Superior Olive (MSO), where cells receive bilateral excitatory inputs from phase-locked stellate bushy cells with the same characteristic frequency (SBC; Fig. 6.10) in the anteroventral cochlear nuclei (AVCN).

Anatomical tracer studies in cats and birds (Carr and Konishi, 1990) have suggested that the dual delay line organization of Jeffress' original proposal

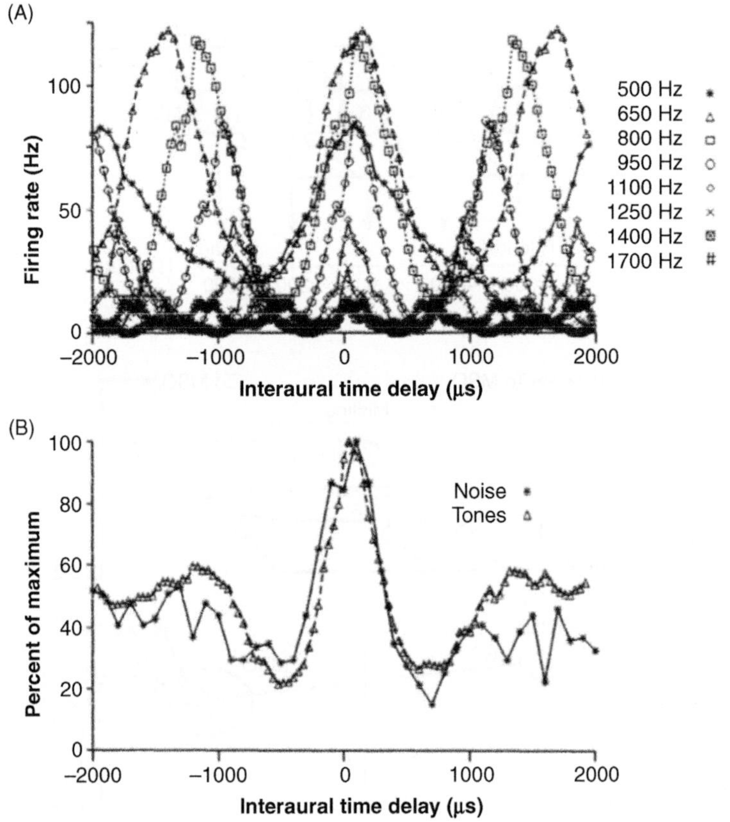

FIGURE 7.5 (A) Response of a single MSO neuron to tones of different frequencies within its response field. In line with the cross-correlation prediction of Eq. (7.7) the responses all peak at a fixed interaural delay, which is the cell's characteristic delay (CD). (B) Response of same neuron to broadband noise peaks at the same CD. The dashed line is the linear summed response for the tones shown in (A). *(Source: After Yin and Chan, 1990, with kind permission.)*

(Fig. 7.4A) is not implemented as such in the MSO: while the contralateral axonal branches from the AVCN seem to follow Jeffress' idea of systematically organized delays along the MSO from anterior (short) to posterior (long) locations, the ipsilateral branches follow a different, forked-like pattern that does not topograpically organize into particular delays (Fig. 7.4B).

Fig. 7.5 shows an example of a single-unit recording from an MSO cell (Yin and Chan, 1990). The cell was recorded during binaural stimulation with tones of different frequencies that all belonged to its response field (here: 0.5–1.7 kHz), while systematically varying the interaural delay between −2.0 ms (ipsilateral ear leading) and +2.0 ms (contralateral leading). The cell's firing rate is nicely modulated with the ITD, at a modulation period that matches the inverse of the tone frequency. As a result, the multiple peaks in the firing rates do typically not coincide, except at the characteristic delay for this cell, which is at an ITD = 33 μs. Indeed, after summing the responses to the eight different tones only a single peak emerges at the characteristic delay (Fig. 7.5B, dashed curve). Similarly, when the cell is stimulated by binaural broadband noise, the response curve nicely coincides with the summed sinusoid responses (solid curve). This response behavior is reminiscent of a cross-correlation between the left- and right-ear inputs.

7.2.2 Modeling Coincidence Detection

A neuron can act as a coincidence detector whenever the epsp's from spikes arriving on its synapses from one side (say, from the contralateral AVCN) are not sufficient to get the neuron above its firing threshold, unless joint inputs from ipsi and contralateral synapses arrive within a narrow time window, δ [Eq. (7.9)].

Gerstner et al. (1996) proposed a self-organizing computational model of how a neuron can become a precise coincidence detector that is tuned to a particular ITD (Fig. 7.6A). They assumed a simple Integrate-and-Fire model neuron for the MSO cells, for which the membrane potential, $v(t)$, is described by:

$$\begin{aligned}\frac{dv}{dt} &= -\frac{v(t)}{\tau_m} + \frac{1}{\tau_s}\sum_{j=1}^{N} w_j \exp\left(-\frac{t-t_j^f}{\tau_s}\right)U\left(t_j^f - t\right)\text{for } v(t) < \Theta \\ &\text{for } v\left(t^{\text{spk}}\right) = \Theta : v\left(t^{\text{spk}}\right) = \text{Spike and } v(t) = 0 \text{ for } t^{\text{spk}} \le t \le t^{\text{spk}} + t^{\text{REFR}}\end{aligned} \tag{7.10}$$

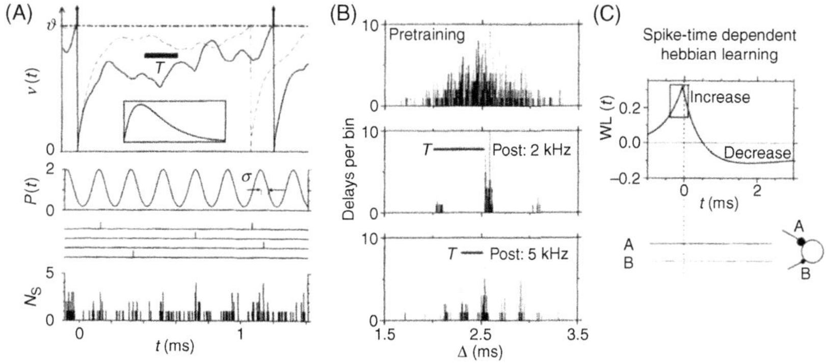

FIGURE 7.6 Self-organizing model of MSO cells that learns precise tuning to interaural timing differences. (A) Membrane potential and spikes of the integrate-and-fire neuron [Eq. (7.10)] are phase-locked to the sound (period $T = 0.2$ ms) after training; when synapses are homogenously distributed, this periodicity is absent (*dashed line*). Input neurons fire sparsely according to Poisson statistics (center panel, 4 input lines; spike jitter $\sigma = 40$ μs), and the population of all inputs yields jittering spike arrivals (bottom). Inset: epsp. (B) Top: Initial transmission delays of input spikes, Δ, have a broad random distribution, but after training to tones of 2 kHz (center), or 5 kHz (bottom), they self-organize precisely to phase-locked inputs. (C) The spike-time dependent Hebbian learning rule. Bottom: two example lines show increase (A) and decrease (B) of the weights. *(Source: After Gerstner et al., 1996, with kind permission.)*

The right-hand term of Eq. (7.10) describes the total input current to the neuron. The neuron's membrane time constant, τ_m, and the synaptic time constant, τ_s, are both supposed to be very short, about ~0.1 ms; the neuron's refractory time $t^{REFR} = 0.5$ ms, and $U(t)$ is the unit step function. As the neuron's threshold, Θ, was taken to be 36 times larger than the epsp amplitude, a significant volley of input spikes is needed to bring the neuron to spiking. The presynaptic input spike trains (spike times of input line j are described by t_j^f ,and transmitted and effective synaptic weights, w_j) are generated by a Gaussian process around a certain phase of the pure tone input sound, with a temporal jitter of 40 μs and transmission delay Δ_j (Fig. 7.6A, center).

A spike-time dependent Hebbian learning mechanism trained the synaptic weights (all with starting values $w_j = 1$). The learning rule is specified in the following way (Fig. 7.6C):

$$\Delta w_j = 0.002 \sum_f \left[0.1 + \sum_n \mathrm{LW}\left(t_j^f - t^n\right) \right] \tag{7.11}$$

$$\text{with } \mathrm{LW}(t) = \begin{cases} 0.3 \exp\left(\dfrac{t + 0.05}{0.5}\right) & \text{for } t < -0.05 \\ 0.5 \exp\left[-(t + 0.05)/0.5\right] - 0.2 \exp\left[-(t + 0.05)/5\right] & \text{else} \end{cases}$$

LW(t) is the spike-time dependent learning window, where t_j^f are the presynaptic input spikes, and t^n the postsynaptic firings of the model neuron of Eq. (7.10). The synaptic weight increases most strongly when a presynaptic spike occurs slightly before the postsynaptic spike (Fig. 7.6C, line A). If, on the other hand, the presynaptic spike *follows* the postsynaptic spike, the weight will be decreased (line B).

As all synaptic weights are constrained to values between 0 and 3, the learning rule prunes the ineffective synapses. After exposing the cell to a 2 kHz input tone, the number of effective synapses reduced from $N = 600$ at the start of the training to about $N = 150$ after about 3 s of sound stimulation (Fig. 7.6B).

Interestingly, after learning the model neuron was better phase locked to the input sound than its input spike trains from the cochlear nucleus, as the temporal jitter reduced from 40 to about 25 μs (Fig. 7.6B). Note that when the synapses are divided evenly as inputs coming from the left and right ear, respectively, and these inputs are given a particular interaural delay, only the synapses whose spikes correspond closely to that delay will survive. In this way, a population of MSO neurons can self-organize to tune to the entire range of interaural time delays across the range of relevant frequencies.

7.2.3 Timed Inhibition

Despite its elegance, and its neurophysiological and neuroanatomical support across different species, serious doubt has been cast on whether the Jeffress model could also be applicable as a neural mechanism for sound-localization in (small) mammals. As pointed out by McAlpine and Grothe (2003), although MSO cells in birds as well as mammals respond in a phase-locked manner to ITDs, alternative mechanisms for localization in the horizontal plane should not be excluded. This becomes especially evident in (small) mammals like, for example, gerbils, for which ITD sensitivity is restricted to low frequencies (<1.5 kHz). For all animals, the maximal ITD, which determines the animal's so-called *physiological range*, is determined by the size of the head [Eq. (7.3)]:

$$\mathrm{ITD_{MAX}} \cong 2.57\frac{r}{c} = 0.0075 \cdot r \text{ ms} \tag{7.12}$$

with r (the head radius) expressed in millimeters. For example, in gerbils the interaural distance ($2r$) is about 30 mm, leading to a maximum ITD of about ±110 μs, which is roughly the same as for the barn owl. However, as the frequency range to which gerbils respond to ITDs is <1500 Hz, the corresponding phase differences (for pure tones) span a frequency-dependent range of (Fig. 7.6):

$$\mathrm{IPD_{MAX}} = \pm \mathrm{ITD_{MAX}} \cdot f_0 \text{ cycles} \tag{7.13}$$

For gerbils, at 1500 Hz, the IPD range is limited to 0.165 cycles, while at 500 Hz the IPDs are confined to only ±0.055 cycles. In other words, when relying on ongoing IPDs, determined by the delays within the physiological range, the gerbil auditory system would span a range of IPDs that is far from optimal (as it potentially could cover ±0.5 cycles; see Fig. 7.7). For barn owls ITD sensitivity runs over frequencies from 3–10 kHz, rendering the IPDs to cover the full range of relevant frequencies.

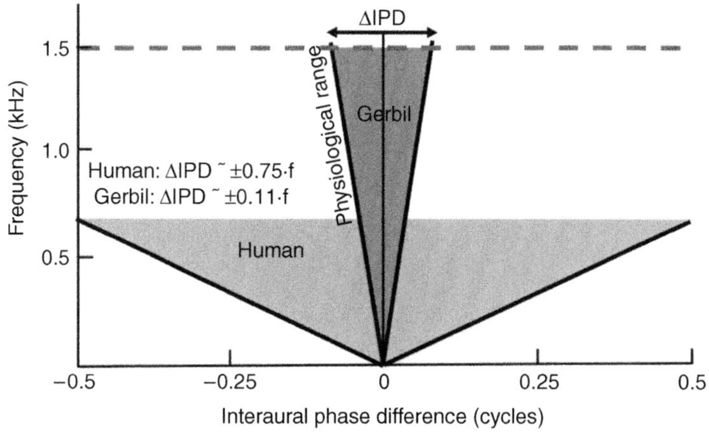

FIGURE 7.7 Physiological range of ITD detection of gerbil versus human [based on Eq. (7.13)]. The maximum phase differences detectable (*solid black lines*) depend on head size and frequency.

Humans are sensitive to ITDs over roughly the same frequency range as gerbils. However, as the human head has a radius $r \sim$ 100 mm, the ITDs vary over ±750 μs, so that at 500 Hz the IPDs span a range of ±0.375 cycles, and already at ~ 670 Hz the full IPD range is covered.

Interestingly, however, gerbils (and other small mammals) tend to be sensitive to ITDs that extend to *beyond* their physiological range, and this sensitivity is not readily explained by Jeffress' delay-line mechanism.

McAlpine and Grothe (2003) therefore proposed that in mammals a prominent role for sound localization encoding could be provided by a very different mechanism, which relies on precisely timed contralateral *inhibition* on the MSO cells. This alternative mechanism contrasts with the bilateral excitatory processes from the AVCN in Jeffress' model, needed for the cross-correlation analysis (Fig. 7.4C). Indeed, the lateral nucleus of the trapezoid body (NTB) provides glycinergic, inhibitory inputs to the MSO cell bodies in mammals. Through this mechanism, the peak of the ITD tuning curve of MSO cells can shift as a function of frequency and interaural delay, such that the ITD will fall beyond the animal's physiological range (Fig. 7.8B; Brand et al., 2002). When the inhibition is

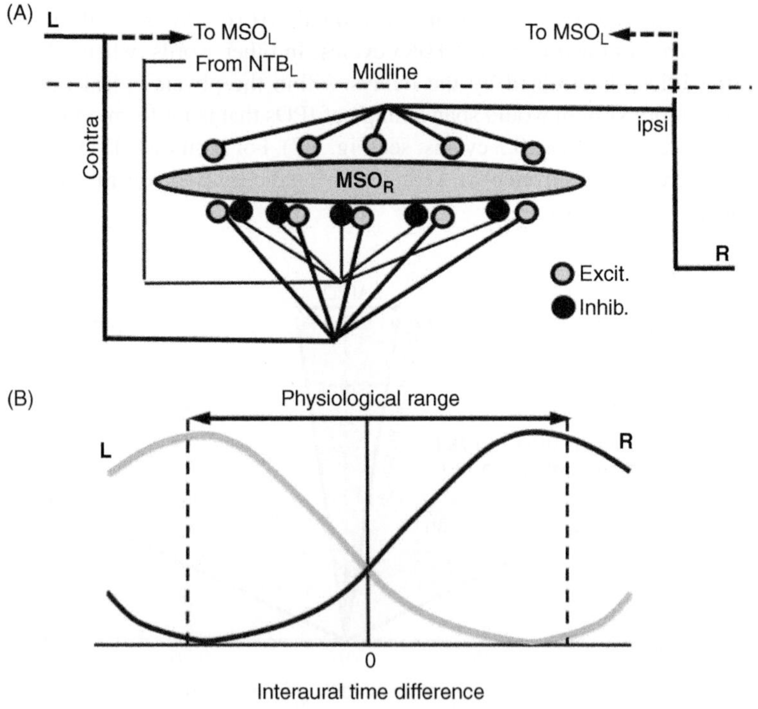

FIGURE 7.8 (A) Alternative model for ITD processing in mammals, that relies on tuned inhibition from the contralateral NTB to MSO cells. In the absence of inhibition MSO cells are tuned to ITD ~ 0. (B) Due to the timed inhibition, each ITD leads to a differential activation of the left- and right MSO cell populations.

removed by applying strychnine (the antagonist of glycine), the ITD curves return to the physiological range, close to zero ITD. With the mechanism of precisely timed inhibition, the entire population of MSO cells can thus provide a monotonic code for IPD, and hence for source azimuth, over the entire spatial range.

Which of the two ITD mechanisms dominates the encoding of source location in the horizontal plane may depend on three factors: the sound's frequency, the animal's head size, and the ITDs typically encountered in the animal's biotope (represented by the IPD probability distribution). Possibly, within a single animal species both mechanisms may be operational for different frequency ranges. For example, in cats (with an intermediate head size, and maximum ITDs around 400 μs), and possibly also in humans, the delay-line mechanism can ensure an appropriate azimuth encoding for the higher frequencies (>750 Hz for cats, >400 Hz for humans), whereas the timed-inhibition mechanism would encode source locations for the low frequencies. In barn owls, however, the very high-frequency sensitivity to ITDs (3–10 kHz) can be exclusively covered by Jeffress' delay-line mechanism (Harper and McAlpine, 2004).

7.2.4 Timed Inhibition?

However, it has been pointed out that timed inhibition may only be effective for detecting the ITD at sound *onsets*, rather than for the ongoing IPD for tones. Indeed, the data of Brand et al. (2002) seem to suggest that after application of strychnine, the *first* peak of the ITD tuning curve shifts to approximately zero ITD, but *without affecting* the side peaks (eg, Fig. 7.5A). If the inhibitory mechanism would be responsible for encoding the ongoing IPD, all peaks should have to shift by the same amount when the inhibition is abolished (Joris and Yin, 2007).

At present, the discussion has not been settled yet. The current Jeffress' model cannot readily account for the finding in gerbils that the ITD peak shifts as a function of frequency (as for a cross-correlation it should not, see earlier). Yet, the neural evidence in favor of the timed inhibitory mechanism is not conclusive either, given that it is so far based on a relatively small population of recorded neurons. Furthermore, the idea that the ITD is represented by a weighted distribution of activity in the two hemispheres (a two-channel scalar weighting model) is incompatible with unilateral lesion studies (eg, of the Inferior Colliculus), which show that the resulting localization deficit is confined to the contralateral hemifield. The two-channel population model would predict a localization deficit for both hemifields.

Finally, a possible alternative mechanism to create very short neural delays in the peripheral auditory system could be implemented by small, systematic asymmetries in the projections from the ipsi- and contralateral cochlea to an MSO neuron: as the traveling wave takes about 5 ms to reach the apex (see chapter: The Cochlea), projections from slightly different cochlear sites to an MSO cell will automatically create a small delay in the arrival of action potentials. Note, however, that such asymmetric projections need to be very systematic in order to account for the required change in delays for the full range of contralateral azimuth angles (Joris and Yin, 2007).

7.3 INTERAURAL LEVEL DIFFERENCES

A second localization cue in the horizontal plane arises as a consequence of the head-shadow effect (HSE). For high frequencies, the head effectively attenuates (part of the) incident acoustic energy for the far ear, as it is reflected back from the head whenever the sound wavelength, λ, is smaller than the size of the head (~0.2 m diameter). As a lower bound for the reflecting sound frequencies we thus estimate:

$$f_{MIN} = \frac{c}{\lambda_{MAX}} \approx \frac{343 \text{ m/s}}{0.2 \text{ m}} \sim 1.7 \text{ kHz} \tag{7.14}$$

The head-shadow effect depends on three factors: (1) sound frequency, f, (2) head size, r, and (3) source azimuth, α. For broadband noise, the HSE of an adult human head could be approximated to reasonable accuracy by the following function (Van Wanrooij and Van Opstal, 2004; Fig. 7.9):

$$\text{HSE}(\alpha) = 10\sin(0.13\alpha + 0.3)\text{dB} \tag{7.15}$$

with the azimuth angle expressed in degrees.

The HSE is obtained by subtracting the absolute sound-source level at the position of the ear without the subject in the setup from the measured proximal sound levels at the recorded (eg, right) ear, while moving the speaker to different azimuth positions:

$$\text{HSE}(\alpha) = I_{PROX}(\alpha) - I_{SOURCE} \tag{7.16}$$

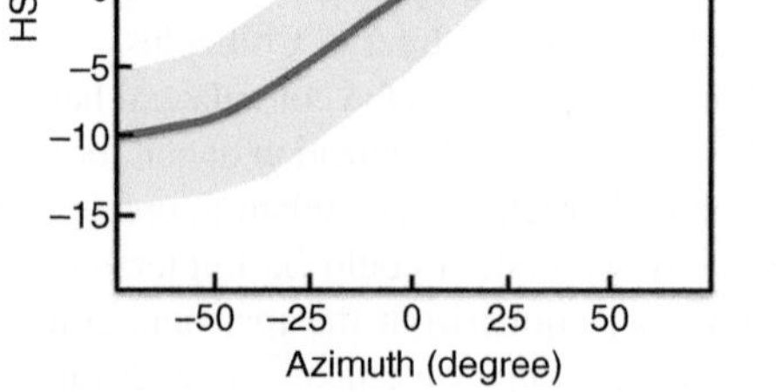

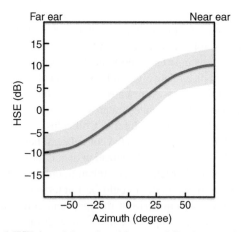

FIGURE 7.9 Measured HSE for adult male subjects, and fitted curve through the data according to Eq. (7.15) (*solid line*). The total range of proximal sound level changes across azimuth in the frontal hemifield covers about 20 dB SPL. Negative azimuths: sound presented on the side of the far (nonrecorded) ear. Positive azimuths: sound presented on the side of the near (recorded) ear.

As a result of Eq. (7.16), the HSE is an *ambiguous cue* for sound localization, as in principle infinitely many combinations of proximal sound levels (and hence source azimuths) and actual source intensities can produce the *same* value for the HSE. This ambiguity disappears, however, when taking the *interaural level difference*, or ILD: the I_{PROX} curves for the right and left ears are mirror images with respect to the mid-sagittal plane of the head, and the contribution of the absolute source level in Eq. (7.16) is removed. The ILD thus exclusively depends on the sound's azimuth and frequency in a monotonic way:

$$\mathrm{ILD}(\alpha) = I^{R}_{\mathrm{PROX}}(\alpha) - I^{L}_{\mathrm{PROX}}(\alpha) \approx A(f)\cdot\sin(\alpha) \tag{7.17}$$

with $A(f) \approx 20$ dB for broadband noise.

Interestingly, the lower bound of Eq. (7.17) for the potential use of head-shadow almost seamlessly matches the highest frequencies usable for the detection of IPDs [Eq. (7.5)]. This fact of acoustics was already described by Rayleigh (1896), and is often formulated as the *duplex theory*:

> Rayleigh's Duplex Theory: Horizontal sound localization of low ($\lesssim$1.5 kHz) frequencies relies on interaural time (phase) differences, and of high ($\gtrsim$2–3 kHz) frequencies is based on interaural level differences.

The two mechanisms thus appear to nicely complement each other for sound localization in the horizontal plane across the entire audible frequency range.

The dependence of the HSE (and therefore, the ILD) on the incident angle, the sound frequency and head geometry can be quite complex, as it requires solving the wave equation with the appropriate boundary conditions at the head's surface (Fig. 7.10). The sound field at all points in space can be considered as a superposition of the incident (planar) sound wave and the scattered sound wave arising from the surface of the head, which is usually approximated by a sphere. The properties of the scattered wave then depend on the size of the sphere and on the wavelength of the incident sound, and involve an approximation with spherical Bessel functions. A full analysis of this problem is described in Rayleigh's (1896) seminal work on acoustics. Briefly, the wave equation in 3D is given by:

$$\nabla^2 p = \frac{1}{c^2}\frac{\partial^2 p}{\partial t^2} \tag{7.18}$$

and because of the linearity of the problem, we can consider harmonic solutions of the form:

$$p(\vec{r},t) = p(\vec{r})\exp(-j\omega t) \tag{7.19}$$

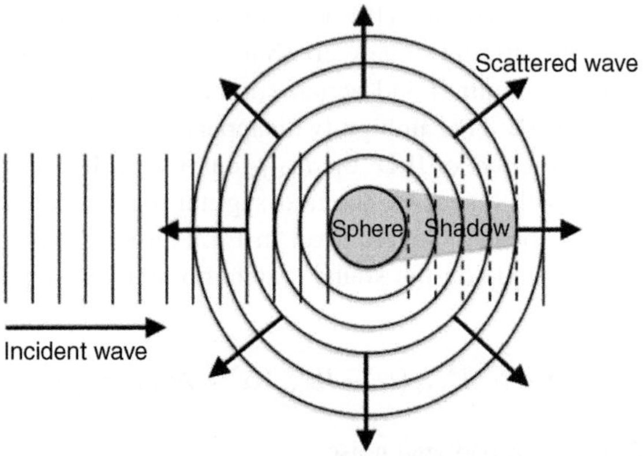

FIGURE 7.10 The sound field in the presence of a spherical obstacle is composed of the linear superposition of the incident planar sound wave, and the scattered wave emanating from the surface of the sphere.

Substitution of this solution into Eq. (7.18) yields the so-called Helmholtz equation:

$$\left(\nabla^2 + \frac{\omega^2}{c^2}\right)p(\vec{r}) = 0 \tag{7.20}$$

which has to be solved by applying the appropriate boundary conditions on the sphere. For example, if the sphere may be reduced to a point source (ie, the large wave length approximation, when there is no head shadow), the total solution is described by the sum of the incident planar wave (traveling in the x-direction), and a spherical wave (radiating in all directions from $r = 0$):

$$p(\vec{r},t) = p_{\text{inc}} \exp\left[i(kx - \omega t)\right] + p_{\text{scat}} \frac{\exp\left[i(\vec{k}\cdot\vec{r} - \omega t)\right]}{r} \tag{7.21}$$

At low frequencies, for which the head is acoustically nearly transparent, an intensity difference also arises for sources that are close to the head (ie, within about 1 m from the ears; Shinn-Cunningham et al., 2000). The reason for this is that the intensity of a (point) source [$I(t) \sim p^2(t)$] decreases with distance r from the source as $1/r^2$, so that for ears, separated by about 15 cm, the different distances to the source can create noticeable intensity differences (Exercise 7.5; see also Section 7.6).

Instead of solving the acoustic problem analytically, which becomes quite tedious for frequencies (wave lengths) that are neither very low, nor very high (especially when calculating the near-field solutions at the ear canal entrance), we can look at the results of measurements as function of frequency and source

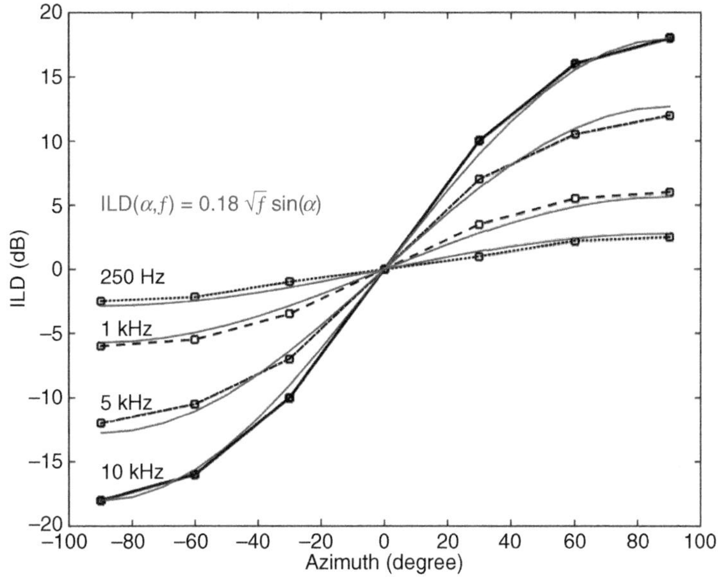

FIGURE 7.11 Frequency-dependence of the ILD for an adult human subject, recorded at four different frequencies (*open dots*), and approximated by the heuristic function of Eq. (7.22) (*red lines*).

azimuth, for sources that are far from the head. Fig. 7.11 shows such a measurement for four frequencies, which here were approximated over the audible range by the following heuristic, separable, function:

$$\text{ILD}(\alpha, f) = 0.18\, g(f) \sin\alpha \quad \text{with} \quad g(f) \sim \sqrt{f} \tag{7.22}$$

7.3.1 ILD Encoding at the LSO

Cells in the lateral superior olive (LSO) are tuned to the ILD for sounds around their characteristic frequency. These cells receive binaural input: a monosynaptic connection from spherical bushy cells in the ipsilateral AVCN, and a disynaptic connection with the contralateral AVCN, where globular bushy cells project to the medial nucleus of the trapezoid body (MNTB; see also Section 7.2.3). The latter provides an *inhibitory* projection to the LSO, which effectively inverts the sign of the GBC signal. The LSO cells thus receive excitatory–inhibitory (EI) inputs from the cochlear nuclei. The number of spikes of an LSO cell has a monotonic, saturating relationship with the ILD corresponding to ipsilateral azimuth angles for the frequencies within its response field. Moreover, each cell has its own recruitment threshold. The response of an LSO neuron (its mean

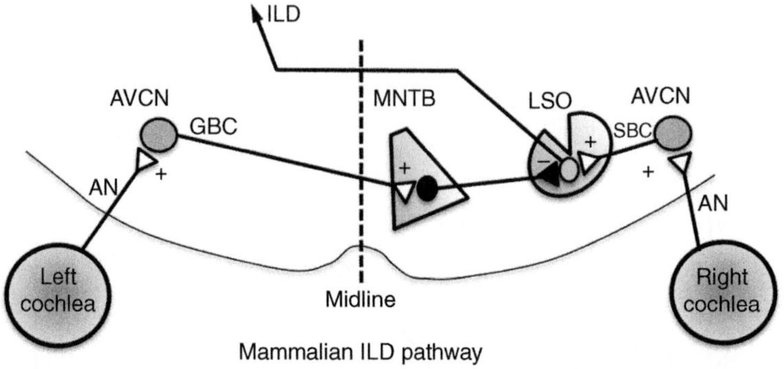

FIGURE 7.12 **Neural pathway that implements the encoding of a frequency-specific interaural level difference in the LSO.** Same format as Fig. 7.4C. GBC, globular bushy cell; MNTB, medial nucleus of the trapezoid body. Right: schematic response of an LSO neuron to ILD: the cell fires when the ipsilateral sound level is higher than the contralateral sound level.

firing rate, or number of spikes) to ILD at its best frequency can thus be approximated by a sigmoid characteristic:

$$\bar{f}^{\mathrm{LSO}}(\mathrm{ILD}) = \begin{cases} \dfrac{2f_{\mathrm{MAX}}}{1+\exp\left[-\beta\left(\mathrm{ILD}-\Theta_{\mathrm{ILD}}\right)\right]} & \text{for} \quad \mathrm{ILD} > \Theta_{\mathrm{ILD}} \\ 0 \quad \text{otherwise} \end{cases} \tag{7.23}$$

with β the neuron's sensitivity (slope at the curve's midpoint), f_{MAX} the neuron's maximum activity, and Θ_{ILD} the cell's recruitment threshold. Given a range of different recruitment thresholds across neurons, the summed population output of ILD-tuned cells in the LSO can faithfully represent the azimuth angle of a sound-source in a nearly linear way, despite the saturating nonlinearity of Eq. (7.23) (Zwiers et al., 2003):

$$\sum_{n=1}^{N_{POP}} \bar{f}_n^{LSO}(\mathrm{ILD}) \cong \kappa\alpha_T \tag{7.24}$$

Fig. 7.12 illustrates the neuroanatomical circuit involved in ILD processing in the mammalian auditory brainstem. As the ILD cells project to the contralateral Inferior Colliculus, cells there become responsive to *contralateral* azimuth locations.

7.4 THE CONE OF CONFUSION

Clearly, for a symmetric head all locations in the midsagittal plane (elevations) give rise to the same ITDs (namely zero) and ILDs (also zero), as the path length differences for these locations to either ear are zero, and the head shadows for the left and right ear will be identical too. Thus, on the basis of ITDs and ILDs alone, none of the locations in the median plane can be dissociated, leading to a

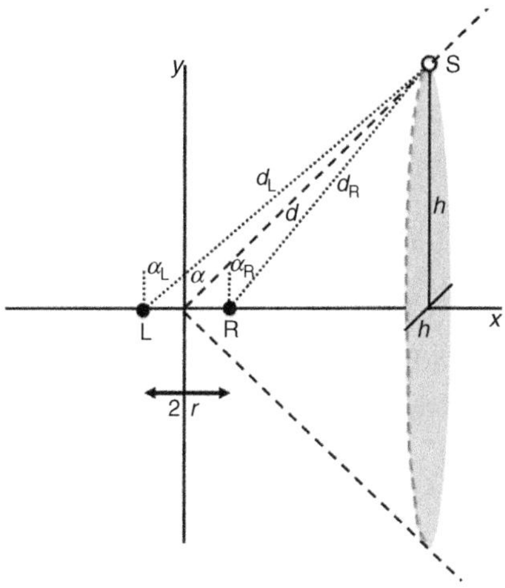

FIGURE 7.13 **The cone of confusion describes the surface of points that yield the same interaural difference (either ITD, or ILD), determined by the path length difference, $\Delta = d_L - d_R$.** Here, source S is assumed to be far away from the head, that is, $d \gg r$, so that $\alpha \simeq \alpha_R \simeq \alpha_L$ (planar waves). The dashed lines that protrude from the origin at angle α delineate the cone's surface.

serious ambiguity problem for sound localization if the head would contain only two ear canals as it acoustic input channels.

Moreover, this ambiguity is not confined to the midsagittal, median plane: many more locations share the property that they yield identical (nonzero) ITDs and ILDs. These locations are described by the so-called "*cone of confusion*," which is an interesting concept in the field of sound localization. When considering distant sound-sources (planar wave approximation), each azimuth angle α yields its own cone of confusion, which is obtained by rotating the line $y = \text{cotan}(\alpha x)$ around the interaural (x-)axis, with α the azimuth angle (in Fig. 7.13, with the y-axis). Consider the situation in Fig. 7.13 where the sound is presented at a distance d, from the center of the (spherical, acoustically transparent) head, for which $d \gg r$ (r is the head's radius).

The path length difference for the sound waves, given by $\Delta = d_R - d_L$, yields an ITD $= \Delta/c$. From Fig. 7.13, note that $\Delta = 2r \cdot \sin\alpha$. All sources on the surface of a cone with its base given by $h = d \cdot \cos\alpha$ will yield the *same* ITD values. For sources near the head, where the approximation $\alpha_R \approx \alpha_L \approx \alpha$ no longer holds, the surface will deviate from the cone to a hyperbolic curve, but the idea remains the same: the ITD cannot uniquely specify the source location in 2D (or 3D) space. The surface for which the ITDs are constant is described by:

$$\frac{4x^2}{\Delta^2} - \frac{4(y^2 + z^2)}{4r^2 - \Delta^2} = 1 \tag{7.25}$$

where (y,z) are the directions perpendicular to the interaural axis (Shinn-Cunningham et al., 2000).

7.5 SPECTRAL PINNA CUES

Due to the cone of confusion, the direction of a distant sound source cannot be uniquely determined from the ITD and ILD cues alone. As a consequence, a third acoustic cue will be required that constrains all potential source directions to a unique pair of azimuth–elevation coordinates. This cue is provided by direction-dependent spectral filtering properties of torso, shoulders, head, and most notably, the pinnae (Fig. 7.14).

The different cavities within the pinna form an asymmetric acoustic aperture into which sound waves diffract, resonate, and reflect in a direction-dependent way. Interestingly, the directional sensitivity of the pinna is predominantly (arguably, almost exclusively) confined to the elevation direction, and because the geometries of the left and right ear are typically very similar, the subtle acoustic differences between the pinnae are probably of minor importance for sound localization. Thus, the acoustic cues provided by the pinna are essentially monaural.

Let us first look at the spectral cues, and then discuss briefly how they arise. First, it is important to realize that the acoustic transformation of the sound input by the pinna is entirely described by the linear wave equation [ie, the Helmholtz equation, Eq. (7.20)]. As a result, the pinna (plus the head and torso) can be modeled as a linear, direction-dependent filter with an impulse response and its associated frequency transfer characteristic:

$$h_{\text{pinna}}(\tau;\, \varepsilon) \overset{\text{FFT}}{\Leftrightarrow} H_{\text{pinna}}(\omega;\varepsilon) \tag{7.26}$$

This filter has become known in the literature as the head-related transfer function, or HRTF (Wightman and Kistler, 1989). The acoustic signal at the entrance of the ear canal for an arbitrary sound, $s(t)$, at elevation angle ε is

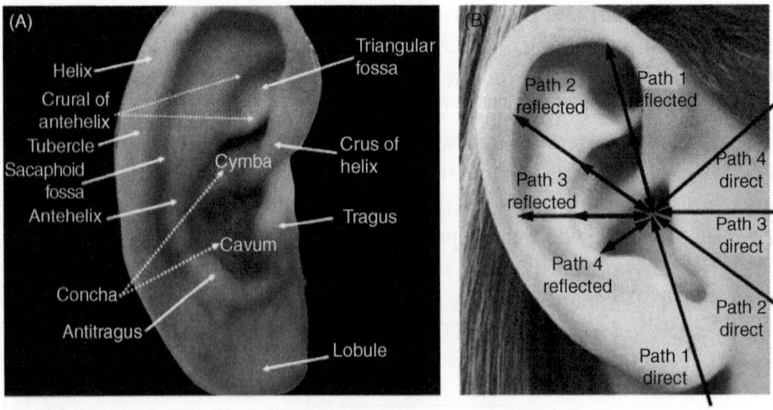

FIGURE 7.14 The human pinna acts as a direction-dependent acoustic aperture. (A) Anatomical features. (B) Acoustic waves from a particular direction reach the ear canal directly, and via reflections at the rims of the cavities, here simplified by a single reflection from the helix. Note that the reflections have different path lengths.

determined by the convolution between the sound and the direction-dependent pinna filter (see chapter: The Nature of Sound):

$$y_{\text{canal}}(t;\varepsilon)=\int_0^{\infty} s(t-\tau)\,h_{\text{pinna}}(\tau;\,\varepsilon)\,d\tau \tag{7.27}$$

In the frequency domain, the acoustic signal at the oval window is formed by the serial cascade of pinna filter, ear canal, and middle ear transfer (eardrum, ossicles, and the acoustic impedance of the cochlea), respectively:

$$Y_{\text{sens}}(\omega;\varepsilon)=H_{\text{pinna}}(\omega;\varepsilon)H_{\text{canal}}(\omega)H_{\text{mid}}(\omega)S(\omega) \tag{7.28}$$

The ear canal effectively functions as a half-closed resonator with a length of about $L = 3.5$ cm, which produces resonances for wavelengths of

$$\lambda_{\text{resonance}}=\frac{4L}{2n+1}\quad\text{with}\quad n=1,2,3,\cdots \tag{7.29}$$

The transfer of the middle ear is independent of the sound's direction, is not readily accessible for direct acoustic measurement, and is usually discarded in the description and analysis of acoustic localization cues. Thus, the HRTF is usually described as the sensory spectrum at the eardrum, which follows from the lumped frequency characteristic:

$$Y_{\text{sens}}(\omega;\varepsilon)=H_{\text{HRTF}}(\omega;\varepsilon)\cdot S(\omega) \tag{7.30}$$

Fig. 7.15 shows an example of human HRTFs for a restricted elevation range in the frontal hemifield as measured in the lab with a small probe microphone positioned deep in the ear canal near the tympanic membrane. At around 2.5 kHz there is a strong amplification that is equal for all sound directions, and caused by the first, low-frequency resonance within the ear canal [Eq. (7.29), $n = 1$]:

$$\lambda_1=4L\Leftrightarrow f_{\text{res}}=\frac{c}{\lambda_1}=\frac{343}{0.14}=2.5\,\text{kHz}$$

The HRTFs for different elevations start to deviate from each other for frequencies above about 4 kHz. There, each elevation angle produces a HRTF with a unique pattern of peaks and notches (Wightman and Kistler, 1989; Kulkarni and Colburn, 1998; Hofman and Van Opstal, 1998). Note that the sound wavelengths for 4–12 kHz correspond to about $\lambda \sim 8.5-2.5$ cm, so that the spectral patterns, visible in the HRTF, can be expected to arise from geometrical features (path lengths, cavities) that are indeed part of the pinna.

Already as early as 1967, Batteau proposed a simple model on how the pinna might function as a direction-sensitive acoustic antenna (Exercise 7.6). According to this model (Fig. 7.14B), the incident sound wave enters the ear canal directly as well as indirectly from delayed reflections within the cavities of the pinna. Let's suppose there is one such reflection (path 4), arising at Δx cm from the ear canal, and that, for the sake of simplicity, this reflection is complete (ie, no absorption; hard walls). The total sound pressure in the ear canal is then given by:

$$p_{\text{tot}}(t)=p_{\text{in}}(t)+p_{\text{in}}(t-\Delta T) \tag{7.31}$$

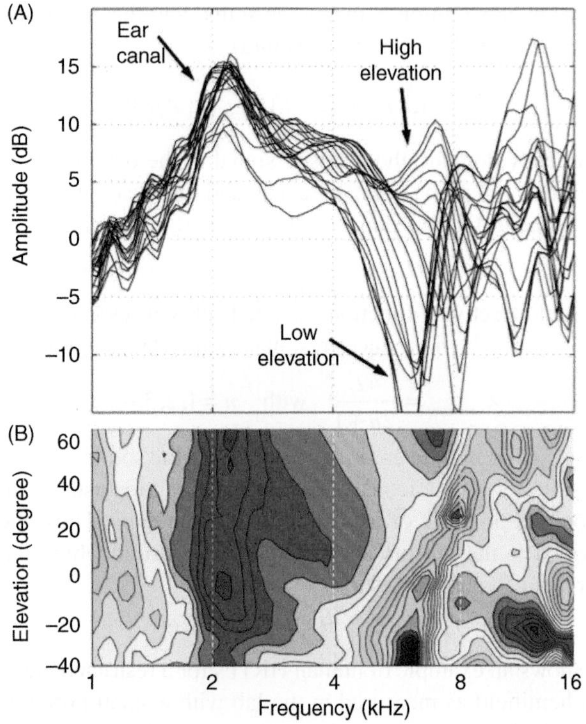

FIGURE 7.15 Measured head-related transfer functions of a male adult human subject. Data are shown on log–log scale. (A) Amplitude spectra (in *dB* re. subject's head absent) for GWN stimuli presented in the midsagittal plane at elevations between −40 to +60 degree in 5 degree steps. Up to about 3 kHz the curves all coincide, and show no direction dependence. The amplification around 2–2.5 kHz is due to the first resonance in the ear canal. Curves start to separate at about 4 kHz. (B) Same data, plotted in color scale. Note the typical elevation-dependent notch (dark blue), running from about 6 kHz at −40 degree, to about 12 kHz at +60 degree.

What is the transfer characteristic of such a process? According to what we've seen in chapter: Linear Systems, the frequency response of Eq. (7.31) is given by:

$$H(\omega) \equiv \frac{\text{Output}}{\text{Input}} = \frac{P_{\text{TOT}}(\omega)}{P_{\text{IN}}(\omega)} = [1 + \exp(-i\omega\Delta T)] \tag{7.32}$$

and the amplitude characteristic can then be written as (but please check):

$$G(\omega) = \|1 + \cos\omega\Delta T - i\sin\omega\Delta T\| = \sqrt{2 + 2\cos\omega\Delta T} \tag{7.33}$$

Note, however, that the delay of the reflection is supposed to depend on the incident elevation angle of the planar sound wave:

$$\Delta T = \frac{2\Delta x}{c} = \frac{2\Delta x(\varepsilon)}{c} \tag{7.34}$$

with c the speed of sound (Fig. 7.14). The HRTF is hence described by

$$G(\omega;\varepsilon)=\sqrt{2+2\cos\left(\omega\frac{\Delta x(\varepsilon)}{c}\right)}, \tag{7.35}$$

and the first notch of this model HRTF is encountered when the direct and reflected waves are exactly in antiphase:

$$\omega_{N1}(\varepsilon)=\frac{\pi c}{\Delta x(\varepsilon)} \quad \Rightarrow \quad f_{N1}(\varepsilon)=\frac{c}{2\Delta x(\varepsilon)} \tag{7.36}$$

Despite the blatant oversimplification, this model does capture, at least qualitatively, some essential features of real HRTFs. First, when the reflective path length varies systematically with the incident angle of the sound (which happens when the ear canal is eccentrically positioned within the asymmetrically shaped pinna), the notch will shift systematically with elevation too (like in Fig. 7.15). For example, referring to Fig. 7.14, when sound waves enter from below, the reflective path lengths with helix-fossa are longer than when sounds enter from above. As a result, the first notch for a downward elevation angle is found at a longer wave length (ie, a lower frequency) than for an upward elevation [Eq. (7.36); Fig. 7.15]. Second, multiple reflections for a given incident angle will in principle induce multiple notches (eg, paths 1, 2, and 3 in Fig. 7.14; Exercises 7.6 and 7.7).

This simple model can provide a reasonable approximation for the HRTF at the higher frequencies; say above 5 kHz, where the acoustics are dominated by reflections. Interestingly, the reflective add-and-delay model also holds that from the measured notch locations of real HRTFs (Fig. 7.15), the approximate geometry of the reflecting surfaces within the pinna can in principle be reconstructed. Spagnol et al. (2013) recently explored this idea on a more rigorous geometrical model of the pinna, which is based on contour matching of pinna features obtained from 2D images.

For a more precise treatment of the acoustics over the entire frequency domain, including the lower frequencies, however, calculating the acoustic interactions with the pinna, head, and torso involves the inclusion of sound-wave diffraction with head and pinna (Lopez-Poveda and Meddis, 1996), as well as the effects of nontrivial resonances within the complex shapes of the 3D pinna cavities (Takemoto et al., 2012). An analytic treatment of the Helmholtz equation requires extensive physical modeling of approximated geometries (like a parabolic reflector for the concha) that may capture (some of) the essentials of the pinna (Lopez-Poveda and Meddis, 1996). Alternatively, the full complexity of the 3D geometry of the pinna may be included, in which case the boundary conditions for the Helmholtz equation must be approximated by calculating the acoustic pressure at many vertices on the overlaying mesh (boundary elements methods). This leads to intensive calculations, as high frequencies will require the mesh to be very fine-grained. The full 3D pinna geometry can currently be measured and stored to sub-mm precision with the aid of magnetic resonance imaging (MRI). Recent work by Takemoto et al.

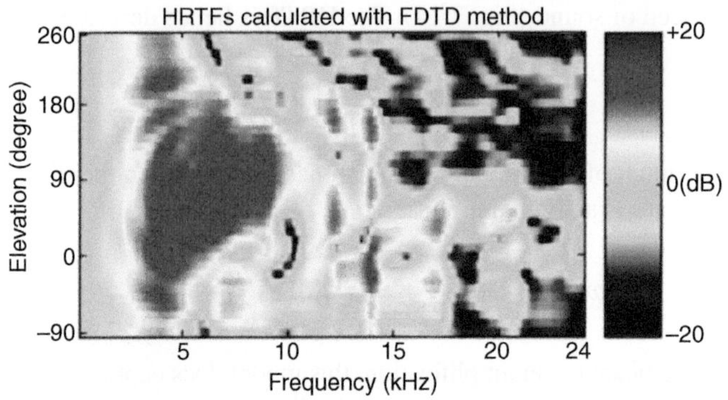

FIGURE 7.16 **Modeled HRTFs of a human ear, calculated over the full elevation and frequency domain on a MRI-generated 3D volume of head and pinna with the FDTD method.** *(Source: After Takemoto et al., 2012, with kind permission.)*

(2012) has demonstrated that the numerically less expensive finite difference–time domain (FTDT) method leads to an adequate approximation of the resulting acoustic patterns and HRTFs, even for the high frequencies (Fig. 7.16).

7.5.1 Possible Neural Correlate for Spectral Cues

There is some good neurophysiological evidence that the unique, complex neural circuitry within the dorsal cochlear nucleus (DCN) could be involved in the processing of spectral pinna cues. Most of the evidence is obtained from the cat DCN (Young and Davis, 2002). The output cells of the DCN project directly to the central nucleus of the midbrain Inferior Colliculus (ICc). The ICc is the converging stage that receives its inputs from the three auditory brainstem nuclei known, or considered, to process the three types of sound-localization cues (LSO for ILD, MSO for ITD, and DCN for HRTF). As such, the ICc play an important role in the creation of a neural representation of auditory space. Small, focal lesions that specifically target the output of the DCN produce subtle deficits only in the cat's sound-localization performance, and predominantly in its ability to point to targets in elevation.

The cat pinnae produce a particular type of HRTF in which the spectral location of a prominent notch varies in a systematic way with the azimuth and elevation coordinates of the sound source (Rice et al., 1992). Fig. 7.17A,B shows examples of cat HRTFs (measured for three azimuth angles at a fixed elevation in Fig. 7.17A, demonstrating also the head-shadow effect, and three elevation angles at a fixed azimuth in Fig. 7.17B, without a HSE). The first notch (FN) systematically varies over a frequency range between 5 and 18 kHz and, by combining the cue information from the left and right ears, unambiguously points to a specific source location in azimuth and elevation. Fig. 7.17C shows iso-FN contours for the right (solid lines) and left (bold dashed lines) ear for eight different frequencies.

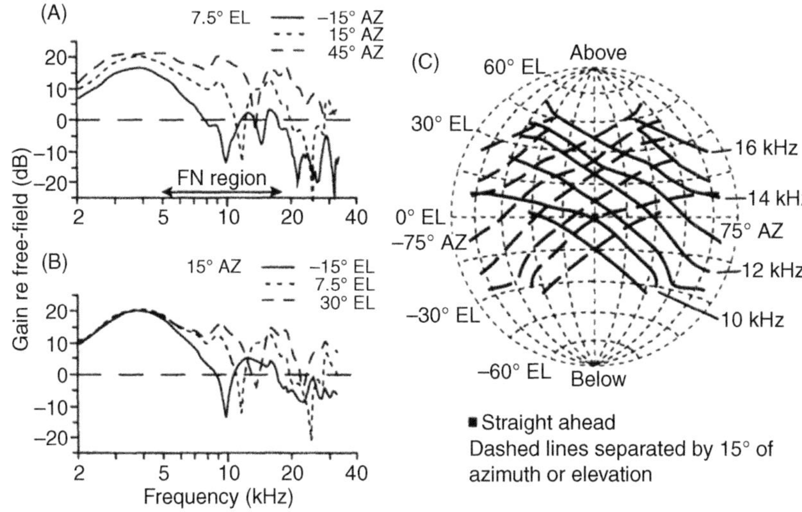

FIGURE 7.17 **Cat pinna acoustics are best described by the systematic azimuth–elevation dependence of its characteristic first notch (FN).** *(Source: Rice et al., 1992, with kind permission.)*

A brief account on how DCN output cells (so-called Type IV cells) can become tuned to a particular notch frequency is illustrated in the circuit model and response patterns of Fig. 7.18 (Young and Davis, 2002). The Type-IV unit is excited by a pure tone at its best frequency of about 8 kHz (BF; Fig. 7.18B, bottom trace). However, as the bandwidth of the stimuli is increased, the response patterns become more and more complex, ending eventually in an inhibitory response for frequencies around the cell's BF (top traces). A minimum circuit that can explain this behavior contains two types of inhibitory inputs to the Type-IV unit: a narrow-band Type-II input that is tuned to a lower frequency than its target Type-IV unit, and a wide-band inhibitory input (W) that strongly inhibits the Type-II unit, and provides a weaker inhibition to the Type-IV cell.

7.5.2 Localization in Elevation: an Ill-Posed Problem

Although the HRTFs provide a detailed spectral shape cue for sound-source elevation, which could in principle resolve the cone of confusion, the extraction of the elevation angle from the sensory spectrum [Eq. (7.30)] still poses the auditory system with an interesting problem. Note that the sensory spectrum, which is the only source of acoustic information to the brain, is composed of the product of two spectra, both of which are a priori unknown to the auditory system: the HRTF, which refers to the actual elevation angle of the sound-source, and the spectrum of the sound-source itself,

$$Y_{\text{sens}}(\omega;\varepsilon) = H_{\text{HRTF}}(\omega;\varepsilon)S(\omega)$$

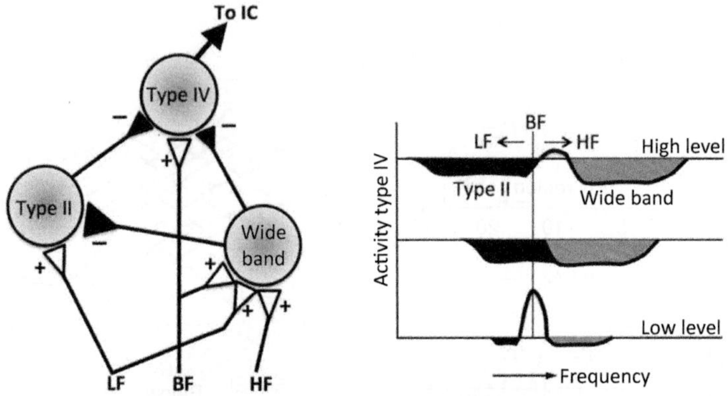

FIGURE 7.18 **Tuning properties of a Type-IV output neuron with a best frequency (BF) of about 8 kHz, from cat DCN (B) to tone bursts of various bandwidths (rows), and a schematic circuit model (A) that explains how Type IV cells become tuned to a particular notch (Young and Davis, 2002).** Type II (excited at a lower BF than its Type IV target) and wide-band (W) neurons both inhibit the Type-IV unit at and around its best frequency (BF). The differences in inhibitory strengths (symbolized by synapse sizes) and bandwidths give rise to the complex, nonlinear patterns of excitatory (*white*) and inhibitory (*black/gray*) responses around BF.

As a result, infinitely many combinations of source spectra and HRTFs can produce the *same* sensory spectrum:

> The extraction of sound-source elevation from the sensory spectrum is a fundamentally ill-posed problem.

In chapter: A Brief Introduction to the Topic we already alluded to this problem. Such problems are fundamentally unsolvable, unless the auditory system can rely on additional sources of (nonacoustic) information. Such sources of information, often called "*priors*," will constrain the infinite number of potential solutions to only a few (or ideally, only one), and may rely on reasonable (physical, or statistical) assumptions about the acoustic world in which we live (as acquired through experience and learning, or by expectation and prediction), and on assumptions about the sensory measurement (in this case, the HRTFs). Later we will describe the more general theoretical framework of this idea (called Bayesian inference in sensory perception) in more mathematical detail, but for now we will simply pose a potential solution to the ill-posed elevation problem, as it was proposed earlier by Hofman and Van Opstal (1998).

Their *spectral correlation model* hinges on the following assumptions:

1. The auditory system has an explicit representation of its own HRTFs, which has been acquired by (sensorimotor) learning.

2. All HRTFs of an individual listener are unique. This means that the spectral shapes of HRTFs do not resemble each other, so that each elevation angle refers to a unique HRTF, and vice versa.
3. The auditory system compares the incoming sensory spectrum [Eq. (7.28)] with the stored HRTFs in the auditory system. The comparison could be based on computing spectral correlations.
4. The auditory system assumes that the spectra of sound sources in the environment correlate poorly with the spectral shapes of HRTFs.
5. The HRTF that best resembles the sensory spectrum (ie, the HRTF that yields maximum correlation) then corresponds to the perceived source elevation.

Assumptions (2.) and (4.) of this model refer to the sensory environment (its relevant biotope) in which the (human) auditory system has evolved, and to the sensory measurement device itself (the pinna), respectively. Evidence for the first assumption (sensorimotor learning) will be discussed in chapter: The Midbrain Colliculus. The other two components of this model (3. and 5.) refer to specific neurocomputational implementations that are less critical.

In principle, the two priors (2. and 4.) can be tested experimentally by (1) recording a wide variety of environmental sound spectra and test how well or how poorly they resemble spectral shapes of the HRTFs of many human subjects and (2) correlate the HRTFs of an individual listener with each other to verify that they are indeed unique for each elevation angle. The test on environmental sounds is not so simple, as it may be conceivable that (in the arms race of evolution) some sound sources *do* resemble (human) HRTFs. The problem is not so much that sounds may *never* resemble human HRTFs, but that it is quite unlikely (at least uncommon) in the human biotope that they do! In the latter case, the brain would make a reasonable, statistical assumption.

The test on uniqueness of the HRTFs, however, is crucial and can be readily performed. Indeed, if certain HRTFs would resemble each other, the brain would be unable to dissociate the different elevation angles to which they refer, and the ill-posed problem would remain unresolved. Fig. 7.19 shows the mutual correlations of the HRTFs of a human subject (calculated with Eq. (7.37) over $f \in [2-16 \text{ kHz}]$) over a range of elevation angles in the frontal hemifield. The prior holds that the correlations should only be high when the HRTF is correlated with itself, that is, the correlation matrix should have its maximum values only on the main diagonal, and very low (close to zero) values for the off-diagonal correlations. This turns out to be the case. Note that HRTFs for nearby elevations still do resemble each other to some extent, which indicates that the spectral shapes are not random functions, but are due to a physical process that causes a relatively smooth and continuous change from one HRTF to the next.

Assuming that the requirements of the two priors are met, how does the spectral correlation model cope with (not "solve") the ill-posed problem? Let's assume that the HRTFs are stored in a tonotopic format, in which spectrum

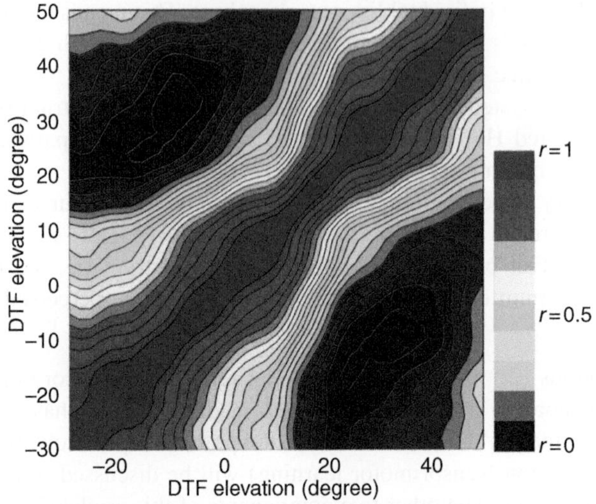

FIGURE 7.19 **Spectral correlations [Eq. (7.34)] of directional transfer functions (DTFs) in the frontal hemifield are only high for DTFs at nearby elevations (around the main diagonal).** Thus, the spectral pinna cues are indeed unique for each elevation angle.

and amplitude refer to their octave and logarithmic values, respectively (as illustrated in the Bode plot of Fig. 7.15A). The spectral correlation between two functions, *A*(*f*) and *B*(*f*), is defined as follows:

$$C\left[A(f),B(f)\right]=\left\langle\left[\frac{A(f)-A}{\sigma_A}\right]\left[\frac{B(f)-B}{\sigma_B}\right]\right\rangle \tag{7.37}$$

where <*X*> is the expectation value (the mean) of function *X*(*f*), given by

$$\langle X\rangle \equiv \int_{f_{\min}}^{f_{\max}} p(f)X(f)df \tag{7.38}$$

with *p*(*f*) a nonzero weighting function, here chosen to be constant (uniform) over the relevant frequency domain (taken from $f = 2-16$ kHz). Note that the expectation value does not depend on frequency, *f*, but may be a function of some other variable, like elevation, ε. Likewise the standard deviation of the spectral function is determined as

$$\sigma_X = \sqrt{\left\langle\left[A(f)-A\right]^2\right\rangle} \tag{7.39}$$

As the sensory periphery effectively performs a logarithmic transformation of the signal's amplitude, and physical frequency is represented in octaves, (see chapter: The Cochlea), Eq. (7.27) transforms into:

$$\hat{Y}_{\text{SENS}}(\Omega\ ;\varepsilon_S)=\hat{H}_{\text{HRTF}}(\Omega\ ;\varepsilon_S)+\hat{S}(\Omega) \tag{7.40}$$

where $\hat{X} \equiv \log_{10} X$, $\Omega \equiv \log_2(f/f_{min})$, and ε_S is the actual source elevation, to be estimated by the auditory system. According to the spectral correlation model, the brain calculates the correlations between the sensory spectrum of Eq. (7.40) and all stored HRTFs, $\hat{H}_{HRTF}(\Omega;\varepsilon)$:

$$C_{\hat{Y}}(\varepsilon;\varepsilon_S) = C\left[\hat{Y}_{SENS}(\Omega;\varepsilon_S), \hat{H}_{HRTF}(\Omega;\varepsilon)\right] \tag{7.41}$$

It can now be readily shown (Exercise 7.8) that under the assumption of a poor correlation of source spectra, $\hat{X}(\Omega)$, with stored HRTFs, $\hat{H}_{HRTF}(\Omega;\varepsilon)$, the global maximum of the correlation Eq. (7.41) is found at elevation

$$\varepsilon_{Perceived} = \text{MAX}\left[C_{\hat{Y}}(\varepsilon;\varepsilon_S)\right] \cong \varepsilon_S \tag{7.42}$$

Thus, if the perceptual decision is based on the criterion of maximum correlation between the sensory spectrum and all stored HRTFs, the auditory system will correctly localize the target's elevation angle, as long as the source spectrum does not correlate with any of the HRTFs.

A consequence of this mechanism is that in a localization experiment the actual amplitude spectrum of the sound-source would not be critical to measure localization performance (eg, it doesn't have to be flat GWN in order to be localized correctly). Conversely, if a sound's amplitude spectrum does correlate significantly with any of the HRTFs, corresponding to say a particular elevation angle, ε^*, the model predicts that the target's elevation will be severely mislocalized in the direction of ε^* (Exercise 7.9).

7.5.3 Alternative Elevation Cues?

As described in Section 7.4, the cone of confusion for a symmetrical head without pinnae prevents the auditory system from resolving the elevation angle of a sound. To determine the sound's elevation angle from the HRTFs requires a sophisticated analysis of the source spectrum. This all suggests that the elevation of a *pure tone* cannot be determined, even if the tone falls within the relevant frequency band (say, between 6 and 9 kHz) for which its perceived level would vary systematically with elevation (Fig. 7.15). The problem is, of course, that the absolute intensity of the tone is not known to the auditory system. As a result, from a perceived low sound level it cannot be decided unambiguously whether it is due to a loud sound-source positioned at a low elevation, or from a soft sound at a high elevation, as both situations (and infinitely many more!) could produce the same perceived intensity.

As will be discussed in chapter: Assessing Auditory Spatial Performance, pure tones can indeed not be localized under situations in which the head is not moving and/or sounds are presented briefly (say, 150–300 ms). But would it be possible to localize a sustained pure tone in elevation by using head movements? One could imagine that by making a vertical head movement, the perceived tone's intensity would systematically change in a way that couples to

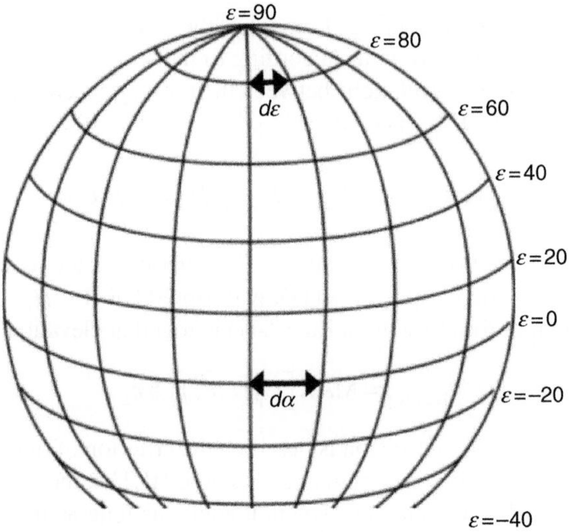

FIGURE 7.20 **Perceived change in ILD/ITD, indicated by $d\varepsilon$, of a sound source at elevation ε, after a head rotation, $d\alpha$, varies between $d\alpha$ (for elevation $\varepsilon = 0$) to 0 (for elevation $\varepsilon = 90$).** For a known head rotation, the sound's elevation can be estimated.

the HRTFs associated with the real sound position and the associated change in head position. Although the mapping from elevation-related intensity change to head-position change is far from trivial, it might be a conceivable strategy, although there are as yet no convincing data to support this idea.

Interestingly, however, it is also possible, at least in theory, to determine the elevation of a sustained pure tone when making small *horizontal* head movements! This idea was recognized as early as 1929 by Van Soest, and is described in the (Dutch) PhD thesis of Kornelis de Boer (referenced in Blauert, 1997). Here is how this might work: suppose that the elevation angle of the tone is ε degrees and that the head makes a horizontal rotation of α degrees. Then the quantity

$$D = \frac{d\alpha(\varepsilon)}{d\alpha_H} \tag{7.43}$$

can specify the tone's elevation.[a] Here, $d\alpha(\varepsilon)$ refers to a *perceived* change in azimuth of the source at elevation ε, and $d\alpha_H$ is the associated azimuth change of the head: if the sound source lies in the horizontal plane (elevation zero), $D = 1$, and if it is located at the zenith, $D = 0$ (Fig. 7.20). Intermediate elevation angles yield $0 < D < 1$. Thus, by assessing the variation in ILD (or ITD) that results from

[a]Note that this measure narrows the number or potential possibilities down to four, at mirror-symmetric up/down and front/back positions, respectively. By making small additional head movements the correct signs can be readily established (De Boer, 1940).

a (known) horizontal head movement from α_1 to α_2, the brain can in principle estimate the elevation angle of the sound:

$$\varepsilon = f\left(\int_{\alpha_0}^{\alpha_1} D d\alpha\right) \tag{7.44}$$

where the function $f(x)$ maps the perceived change in ILD/ITD to the elevation angle. Experiments in the early 1940s suggested that the brain could indeed successfully use this remarkable, highly nontrivial, strategy (De Boer, 1940). Note, however, that this procedure can only work provided the sound-source lasts sufficiently long (probably in the order of seconds, although a minimum duration has not been established), and that the brain has access to an accurate signal about head orientation.

Further complexities arise when not only the head, but also the source moves through space. In Chapter 11, we will see that orienting head movements play indeed an important, even crucial, role in our sound-localization behavior, also for very brief (down to tens of milliseconds) sounds.

7.6 DISTANCE PERCEPTION

7.6.1 The Free Far Field

Intensity: In the far free field the auditory system cannot rely exclusively on acoustic cues to determine the distance of a sound-source. Like for elevation, also the estimation of source distance turns out to be an ill-posed problem. Clearly, a sound-source that is far away will have a lower perceived intensity than when the same source is nearer. So, why couldn't the perceived sound level be a faithful measure for source distance? In the free field (total absence of reflections, only a direct, unobstructed pathway from source to listener), and at a large-enough (beyond a few meters) distance r from a point source with intensity I_0 (in Watt per meter square), the perceived sound intensity falls as:

$$I_{\text{Perceived}} \sim \frac{I_0}{r^2} = I(I_0, r) \tag{7.45}$$

(and hence the perceived sound pressure, as $p \sim 1/r$). Thus, the perceived intensity is a function of both I_0, and r. As a result, when the absolute intensity of the sound-source is unknown, the distance r remains ambiguous. Indeed, subjects overly underestimate the true distance of a sound-source when the only cue provided is an intensity cue.

Spectrum: As sound is propagated through the air, and the acoustic impedance of air varies with frequency, the perceived spectrum of the sound-source could provide an additional distance cue: high frequencies are attenuated more for a source located far away than for nearby sounds. Indeed, when a broadband noise burst is low-pass filtered, its perceived distance increases with decreasing

LP filter cut-off. This LP filtering property of air starts to play a role at distances beyond 10–15 m from the listener (4 kHz is attenuated by about 4 dB/100m). In that case, the perceived intensity is no longer proportional to $1/r^2$, but to a power that will also depend on frequency:

$$I_{\text{Perceived}}(\omega) = \sim \frac{I_0(\omega)}{r^{2+g(\omega)}} = I\left[I_0, \omega(r)\right] \tag{7.46}$$

with $g(\omega)$ a positive-only, increasing function that captures the frequency dependence of the air's acoustic impedance (which will be low for low frequencies, and high for high frequencies).

However, without knowledge of the original source spectrum, and of its absolute intensity, this cue remains ambiguous for estimating source distance. If the source spectrum is (approximately) known (eg, as prior knowledge acquired through learning), the auditory system can, and will, make educated guesses about the distance of the sound-source. For example, cars, buses, and trucks can be expected to have typical, nearly constant intensities, and they produce speed-dependent spectra that can be used to approximate their distance. It is quite conceivable that our auditory system uses this type of knowledge in making distance judgments in our real-world environments. A similar strategy could hold for familiar human voices and other relevant (and supposedly characteristic) sound-sources.

7.6.2 The Free Near Field

Note that Eqs. (7.45) and (7.46) refer to the situation of a point source (ie, planar waves) in the *far* free field. In the near field, however, (ie, within about 1.5 m from the listener), these relations are no longer valid, as now even without atmospheric attenuation the perceived sound intensity depends on frequency. Moreover, the effects of pinna filtering will also depend on source distance in a rather complex way. This nontrivial aspect of wave mechanics thus provides a potential spectral cue for (near-)distance perception, as was already recognized by Von Békésy (1938).

Use of ILDs? Shinn-Cunningham et al. (2000) demonstrated that for nearby sources and for low frequencies to which the head is acoustically transparent (no head shadow), the auditory system could use the free-field approximation of Eq. (7.45), by accounting for the difference in distance to the near and far ear for locations off the midsaggital plane. Indeed, by taking the log of the ILD (expressed in dB), the dependence of I_0 drops from the equation:

$$\begin{aligned}\widehat{\text{ILD}}_{\text{Perceived}} &\sim \log\frac{I_0}{r^2} - \log\frac{I_0}{(r+\Delta r)^2} \\ &= \log\frac{(r+\Delta r)^2}{r^2} = \widehat{\text{ILD}}\left[r, \Delta r(r, \alpha_S)\right]\end{aligned} \tag{7.47}$$

Note that the perceived ILD increases rapidly with decreasing source distance (increasing path-length difference). However, still two unknowns remain in this equation (apart from a torus of confusion): the distance r, and the path-length difference, Δr, which is related to the sound's azimuth and distance. As a result, neither the distance, nor the azimuth angle can be unambiguously dissociated on the basis of ILD information alone, despite the fact that source intensity itself no longer plays a role. However, that the behavior of the ITDs for a given source distance and azimuth angle differ from the ILD as a function of the source coordinates [Eq. (7.25); Shinn-Cunningham et al., 2000]. Thus, by combining the ITD and ILD information, the azimuth angle, and hence the source distance can both be dissociated. For higher frequencies, to which the head is not acoustically transparent, the ILDs become frequency dependent. Although this complicates the analysis, it does not affect the major conclusion that the combined use of ITD and ILD information provides useful distance estimates for nearby sound-sources.

7.6.3 Reverberant Environments

In rooms with reflecting walls, the sound reaches the eardrum both directly and after one or more wall reflections. The latter give rise to particular reverberation patterns that depend on the location of the sound-source relative to the receiving ear and reflecting walls. A simple linear model for such a process is described above for the pinna [Eqs. (7.31)–(7.35)]. When incorporating reflections, the signal at the eardrum is described by:

$$p_{\text{Ear}}(t) = p_{\text{Direct}}(t) + \sum_{n=1}^{N} p_{\text{Reflect, n}}(t) \tag{7.48}$$

where $p_{\text{Direct}}(t)$ is the sound pressure from the direct sound path, propagated to the ear as if through the free field [ie, obeying Eq. (7.45) or (7.46)], and $p_{\text{Reflect},\,n}(t)$ is the nth attenuated wave (as it propagated toward and back from the walls), delayed and filtered (by the walls) (the room's reverberation; Fig. 7.21). When incorporating only the first-arriving reflection, the two pressures relate to the source pressure, $p_0(t)$, as (Exercise 7.10):

$$\begin{aligned} p_{\text{Direct}}(t) &= \frac{p_0\left(t - \frac{r}{c}\right)}{r} \\ p_{\text{Reflect,1}}(t) &= \frac{p_0\left(t - \frac{r_{11}}{c}\right)}{r_{11}} + \frac{1}{r_{11}r_{12}}\int_0^{\infty} w_1(\tau)\,p_0\left(t - \frac{r_{11}}{c} - \frac{r_{12}}{c} - \tau\right)d\tau \end{aligned} \tag{7.49}$$

with $w_1(\tau)$ the linear acoustic absorption filter of the first wall, r_{11} the distance from the source to the nearest wall, r_{12} the distance from the wall to the listener, and c the speed of sound (Exercise 7.10). The delay of the first reflected wave, ΔT_1, is:

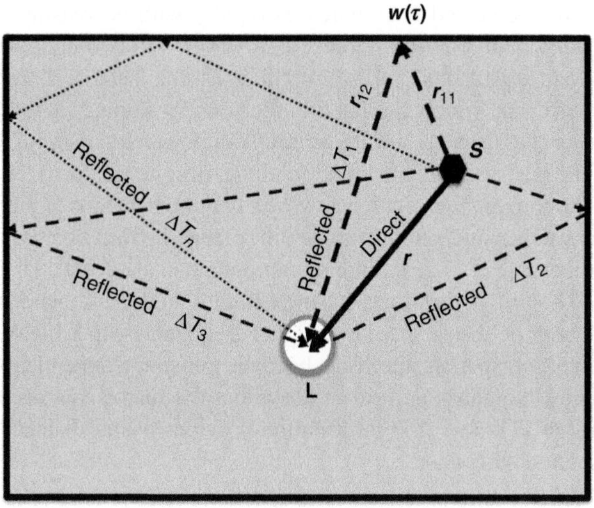

FIGURE 7.21 **The perceived distance in a reverberating room is contingent upon the relative energies associated with the direct sound wave from source (S) to listener (L), and the energies of many reverberations, each filtered by the walls [$w(\tau)$].** Just a few reflections are indicated in this 2D representation. Eq. (7.45) only incorporates the reflection with the shortest delay, ΔT_1. This delay is determined by the difference between the direct path length (the source distance, r) and the path length of the reflected wave, $r_1 = r_{11} + r_{12}$.

$$\Delta T_1 = \frac{r_{11} + r_{12} - r}{c} \tag{7.50}$$

It can be readily appreciated that the signal at the ear changes systematically, as the sound-source is located at different positions within the room.

It has long been held that the auditory system depends on the relative energies of the direct sound path versus the reverberations to estimate the perceived distance of the source in rooms (Von Békésy, 1938; Mershon and King, 1975):

$$r_{\text{Perceived}} \propto f\left[\frac{p^2_{\text{Reflected}}(t)}{p^2_{\text{Direct}}(t)}\right] \tag{7.51}$$

with $f(x)$ a (less than linear) monotonically increasing function. As the sound moves away from the listener, the direct contribution of sound energy decreases as $1/r^2$, while the indirect contribution of the reverberations will increase. As a result, the perceived distance of the source will increase too. Bronkhorst and Houtgast (1999) demonstrated, however, that the neurocomputational process involves temporal integration of the direct (first arriving) sound wave, within a window of about $\Delta T = 6$ ms:

$$r_{\text{Perceived}} = 0.14\sqrt{\frac{GV}{T}}\sqrt{\frac{\int_{\Delta T}^{T} p_{\text{Reflected}}^{2}(t)\,dt}{\int_{0}^{\Delta T} p_{\text{Direct}}^{2}(t)\,dt}} \tag{7.52}$$

where $t = 0$ is the arrival time of the direct sound at the ears. The prefactor depends exclusively on the room acoustics, with G the directivity of the sound source, V the room volume, and T the reverberation time of the room. The $\Delta T = 6$ ms integration window for the direct sound (found by fitting psychophysical results) corresponds nicely with the temporal integration window that is thought to underlie the *precedence effect*, which holds that the first few ms of a sound strongly dominate the extraction of binaural cues for azimuth localization.

Note that as room acoustics can vary substantially, the auditory system should be able to quickly adapt to the acoustics of new, unfamiliar spaces, as Eq. (7.49) depends heavily on the properties of the room (its dimensions, reverberation time, and the frequency-dependent absorption of the walls). The computational model of Eq. (7.52) accounts for this requirement, as it includes the parameters that specify the room acoustics. Experiments indicated that subjects indeed rapidly adapt their judgments when the room acoustics change (Bronkhorst and Houtgast, 1999).

7.7 EXERCISES

Problem 7.1: Suppose that sound sources are attached to the surface of a sphere with diameter R, with the subject's head in its center (ignore head size). We constrain the azimuth of the sounds to α_0 degree. Show that the range of elevations to which the sound sources are constrained in such a stimulus system is given by

$$\varepsilon = \pm \arcsin \cos \alpha_0$$

Problem 7.2: For frequencies above about 1350 Hz [Eq. (7.5)], the IPD points to multiple azimuth angles. Analyze how the number of potential azimuth angles depends on the tone's frequency.

Problem 7.3:

(a) Show that the cross-correlation between a signal and its delayed version is equal to the delayed autocorrelation of that signal.
(b) Calculate explicitly the shape of $\Phi_{LR}(\tau)$.
(c) Now adapt the cross-correlation of Eq. (7.7) in the following way: in chapter: The Auditory Nerve we saw that a phase-locked auditory nerve can represent the half-wave rectified tone of the neuron's best input frequency quite well. Calculate the cross-correlation function for the half-wave rectified tone and its delayed version.

Problem 7.4: Make an educated estimate of the axonal path-length difference needed to encode an azimuth direction of α_0 degree. To that end, first derive (or find information about) the relationship between axonal diameter and signal propagation speed, and values for the relevant parameters. Assume an average axonal diameter of 3 μm.

Problem 7.5: Assuming that the head is acoustically transparent (low frequencies), estimate the intensity difference (in dB) for a sound source at a distance of 0.7 m from the center of the head, at an azimuth angle of 45 degree. Assume that the ears are at positions $R = (0.075, 0)$ (in meters) and $L = (-0.075, 0)$, respectively.

Problem 7.6: Calculate the HRTF when the pinna reflections arise from two different cavities: the helix (a long path) and the conch (a shorter path). See, for example, Fig. 7.13, path 1. Again assume that the pinna walls are perfect reflectors (no attenuation).

Problem 7.7: In this exercise we assume that the reflection is not complete, but filtered and attenuated by a linear filter with impulse response $r(\tau)$. Rewrite Eq. (7.27) for this situation, and derive an extended expression for the HRTF.

Problem 7.8: Insert Eq. (7.36) into the definition of the spectral correlation of Eq. (7.34) to show that

$$C_Y(\varepsilon;\varepsilon_S) = \frac{\sigma_{\hat{H}_S}}{\sigma_{\hat{Y}}} C\left[\hat{H}_{\text{HRTF}}(\Omega;\varepsilon_S), \hat{H}_{\text{HRTF}}(\Omega;\varepsilon)\right] + \frac{\sigma_{\hat{X}}}{\sigma_{\hat{Y}}} C\left[\hat{H}_{\text{HRTF}}(\Omega;\varepsilon_S), \hat{X}(\Omega)\right]$$

Thus, under the validity of the prior assumptions on sound sources, X, and HRTFs the perceived elevation, obtained where C_Y reaches its global maximum, will always equal the actual sound-source elevation.

Problem 7.9: The spectral correlation model predicts that if the source spectrum does correlate with a stored HRTF, the perceived elevation will point to the wrong location. Show that it is possible to design a stimulus spectrum such that the perceived sound elevation can point to any elevation, ε^*, regardless the actual stimulus elevation.

Problem 7.10: Expand Eq. (7.45), by incorporating all first-order reflections in the 2D room (ie, the primary reflections, such as paths 1, 2, and 3, shown in Fig. 7.20); if you dare, also incorporate a number of second-order reflections (like path n).

REFERENCES

Batteau, D.W., 1967. The role of pinna in human localization. Proc. R. Soc. London Ser. B. 168, 158–180.

Blauert, J., 1997. Spatial Hearing: The Psychophysics of Human Sound Localization, second ed. MIT Press, Cambridge, MA.

Brand, A., Behrend, O., Marwuardt, T., McAlpine, D., Grothe, B., 2002. Precise inhibition is essential for microsecond interaural time difference coding. Nature 417, 543–547.

Bronkhorst, A.W., Houtgast, T., 1999. Auditory distance perception in rooms. Nature 397, 517–529.

Brown, A.D., Stecker, C., Tollin, D.J., 2015. The precedence effect in sound localization. JARO 16, 1–18.

Carr, C.E., Konishi, M., 1990. A circuit for detection of interaural time differences in the brain stem of the barn owl. J. Neurosci. 10, 3227–3246.

De Boer K., 1940. Stereofonische Geluidsweergave. PhD thesis (in Dutch) of TH Delft, The Netherlands.

Gerstner, W., Kempter, R., Van Hemmen, J.L., Wagner, H., 1996. A neuronal learning rule for sub-millisecond temporal coding. Nature 383, 76–78.

Harper, N.S., McAlpine, D., 2004. Optimal neural population coding of an auditory spatial cue. Nature 430, 682–686.

Hofman, P.M., Van Opstal, A.J., 1998. Spectro-temporal factors in two-dimensional human sound localization. J. Acoust. Soc. Am. 103, 2634–2648.

Jeffress, L.A., 1948. A place theory of sound localization. J. Comp. Physiol. Psychol. 41, 35–39.

Joris, P., Yin, T.C.T., 2007. A matter of time: internal delays in binaural processing. Trends Neurosci. 30, 70–78.

Knudsen, E.I., Konishi, M., 1978. Mechanisms of sound localization in the barn owl (*Tyto alba*). J. Comp. Physiol. 133, 13–21.

Kulkarni, A., Colburn, H.S., 1998. Role of spectral detail in sound-source localization. Nature 396, 747–749.

Lopez-Poveda, E.A., Meddis, R., 1996. A physical model of sound diffraction and reflections in the human concha. J. Acoust. Soc. Am. 100, 3248–3259.

McAlpine, D., Grothe, B., 2003. Sound localization and delay lines – do mammals fit the model? Trends Neurosci. 36, 347–350.

Mershon, D.H., King, L.J., 1975. Intensity and reverberation as factors in the auditory perception of egocentric distance. Percept. Psychophys. 18, 409–415.

Rayleigh, J.W.S., 1896. The theory of sound, second ed. vols. I and II, Dover Publications, New York (first American edition 1945).

Rice, J.J., May, B.J., Spirou, G.A., Young, E.D., 1992. Pinna-based spectral cues for sound localization in cat. Hear. Res. 58, 132–152.

Shinn-Cunningham, B.G., Santarelli, S., Kopco, N., 2000. Tori of confusion: binaural localization cues for sources within reach of a listener. J. Acoust. Soc. Am. 107, 1627–1636.

Spagnol, S., Geronazzo, M., Avanzini, F., 2013. On the relation between pinna reflection patterns and head-related transfer function features. IEEE Trans. Audio Speech Language Process. 21, 508–519.

Takemoto, H., Mokhtari, P., Kato, H., Nishimura, R., 2012. Mechanism for generating peaks and notches of head-related transfer functions in the median plane. J. Acoust. Soc. Am. 132, 3832–3841.

Van Wanrooij, M.M., Van Opstal, A.J., 2004. Contribution of head shadow and pinna cues to chronic monaural sound localization. J. Neurosci. 24, 4163–4171.

Von Békésy, G., 1938. Über die Enstehung der Entfernungsempfindung beim Hören. Akust. Zeitschrift 3, 21–31. Available in: Von Békésy, G., 1960. Experiments in Hearing. McGraw Hill, NY, pp. 301–313.

Wallach, H., Newman, E.B., Rosenzweig, R., 1949. The precedence effect in sound localization. Am. J. Psychiatr. 62, 315–336.

Wightman, F.L., Kistler, D.J., 1989. Headphone simulation of free-field listening. I. Stimulus synthesis. J. Acoust. Soc. Am. 91, 1648–1661.

Yin, T.C.T., 2002. Neural mechanisms of encoding binaural localization cues in the auditory brainstem. In: Oertel, D., Fay, R.R., Popper, A.N. (Eds.), Integrative Functions in the Mammalian Auditory Pathway. Springer, Heidelberg, pp. 99–159.

Yin, T.C.T., Chan, J.C.K., 1990. Interaural time sensitivity in Medial Superior Olive of cat. J. Neurophysiol. 64, 465–488.

Young, E.D., Davis, K.A., 2002. Circuitry and function of the dorsal cochlear nucleus. In: Oertel, D., Fay, R.R., Popper, A.N. (Eds.), Integrative Functions in the Mammalian Auditory Pathway. Springer, Heidelberg, pp. 160–206.

Zwiers, M.P., Van Opstal, A.J., Paige, G.D., 2003. Plasticity in human sound localization induced by compressed spatial vision. Nat. Neurosci. 6, 175–181.

Chapter 8

Assessing Auditory Spatial Performance

8.1 INTRODUCTION

The spatial perception of a sound's location can be measured in different ways. As the following chapters will deal predominantly with perception and dynamic localization behavior in the free-field auditory space, it is worthwhile to also briefly consider other methods that assess different aspects of spatial performance in audition.

It is useful to distinguish two different hearing conditions: listening to sounds presented in external space (the laboratory room, or the free field) versus listening over headphones. In a natural listening condition sounds would be presented in the free field, which is characterized by the $1/r^2$ law of sound intensity as a function of source distance (see chapter: Acoustic Localization Cues). This natural hearing condition in an ideal open environment (say, the desert) is approximated in the laboratory by covering all walls and reflecting objects in the experimental room by high quality–sound absorbing foam wedges. In such environments the acoustic waves from the sound source reach the ears practically unobstructed while preserving all natural acoustic localization cues that are present in a real free field (chapter: Acoustic Localization Cues). Note that in such environments the acoustic inputs to the two ears are typically highly correlated (so-called *diotic* hearing), as the *same* sound waves reach (possibly somewhat attenuated, or slightly delayed) either ear.

When listening over headphones the stimuli to either ear can be manipulated independently, creating binaural sensations that may, or may never, occur in natural acoustic environments. Now it is possible, for example, to decorrelate the inputs to the two ears (this is called *dichotic* hearing). This may be advantageous when studying the subject's sensitivity to a particular cue for spatial hearing in isolation, without the potentially confounding contributions from other cues, or to present entirely different sounds to each ear altogether.

In a particular application of headphone stimulation the technical challenge is to simulate the external acoustic input to the ears as faithfully as possible. In this hearing condition, called *virtual acoustics*, the listener is unable to distinguish real free-field sound stimulation from headphone acoustic input.

The Auditory System and Human Sound-Localization Behavior. http://dx.doi.org/10.1016/B978-0-12-801529-2.00008-8

In contrast to these different forms of binaural listening, it is also possible to present sound over only one ear (monaural stimulation, or *monotic* hearing), either through the use of headphones, or by plugging the nonstimulated ear.

These different methods of acoustic stimulation address different aspects of spatial hearing, that can for example relate to the auditory system's *sensitivity* to changes in spatial location, to the system's *representation* of space for navigation and orientation in the environment, or to the system's *discriminative* power of sounds as a function of their spatial or acoustic parameters. In the next sections we will briefly describe these different aspects of spatial perception, and how they are usually measured, but before we do that we first briefly introduce the powerful analysis technique of signal-detection theory, and how it can be applied in psychoacoustics.

8.2 SIGNAL-DETECTION THEORY

In psychophysics and psychoacoustics, signal-detection theory (SDT) plays an important role. The technique has a wide range of applications, as the *signal* to which it refers can be virtually anything: the presence or absence of a sound, the difference between sound locations, the frequency of a tone, the sound intensity, the spectral shape of a sound, sound duration, etc. It is particularly useful in the detection, lateralization, and discrimination tasks described earlier.

The following example, which refers to the problem of interpreting the information from a radar system at the end of WOII, well illustrates the background of SDT (Fig. 8.1). An operator looks at the radar screen and briefly sees a number of echoes at each radar sweep. One of the dots may be an enemy bomber, others mere atmospheric disturbances. The operator will have to decide whether the enemy is present (and approaching) in the cloud of dots, or not. In SDT, this problem is transformed into the decision matrix in Table 8.1.

In Table 8.1, the hits and correct rejections are appropriate responses of the operator, associated with a positive gain, or reward, whereas the false alarms (also called *Type I* errors) and misses (*Type II* errors) are incorrect responses that potentially lead to costs, or losses.

Clearly, the radar example is a difficult task, as the dots are only briefly visible, and their spatial–temporal behavior is difficult to assess. In situations like

TABLE 8.1 There are Four Possible Outcomes in the Radar Decision Task

		Signal: enemies present?	
		YES	NO
Decision: enemies present?	YES	Hit	False alarm
	NO	Miss	Correct rejection

Each outcome is associated with either a gain (good) or a cost (bad).

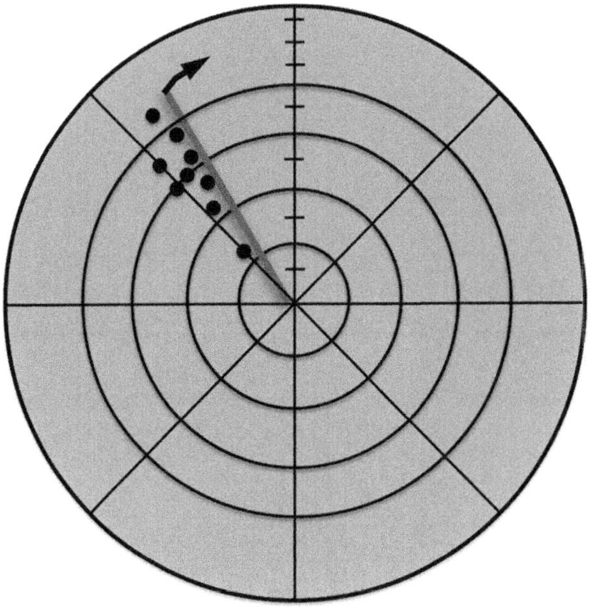

FIGURE 8.1 **Radar signals where the gray diagonal line indicates the direction of view of the rotating antenna, and the dots are measured reflections from potential objects, only briefly visible during each sweep.** Many reflections may be mere weather related (clouds, rain), and therefore are noise in the signal. These dots are indistinguishable from enemy aircraft, but have different spatial–temporal behavior. Is there an enemy aircraft among the dots, or not?

this, the challenge is to maximize the number of hits and correct rejections (as these are good decisions), while at the same time minimize the number of false alarms (these are costly decisions, as the army will be alerted and recruited in vain), and misses (leading to potentially deadly mistakes!).

SDT starts from the central idea that perceptual decisions are fundamentally *probabilistic*. This idea is supported by the typical large variability found in perceptual experiments. The stochastic nature of the process is modeled in psychophysics as *noise* (Fig. 8.2). The noise can be part of the acoustic input received by the decision-making system (as in Fig. 8.1), and/or it can be internally generated in the brain and sensory systems.

The basic assumption is that the signal is only *added* to the noise (no multiplicative interactions), so that the probability distributions of signal plus noise and noise alone only differ in their means, while having the same variance (Fig. 8.2):

$$P(x|\text{no } S) = \frac{1}{\sigma_N\sqrt{2\pi}}\exp\left[-\frac{(x-\mu_N)^2}{2\sigma_N^2}\right]$$
$$P(x|S) = \frac{1}{\sigma_N\sqrt{2\pi}}\exp\left[-\frac{(x-\mu_{S+N})^2}{2\sigma_N^2}\right] \quad (8.1)$$

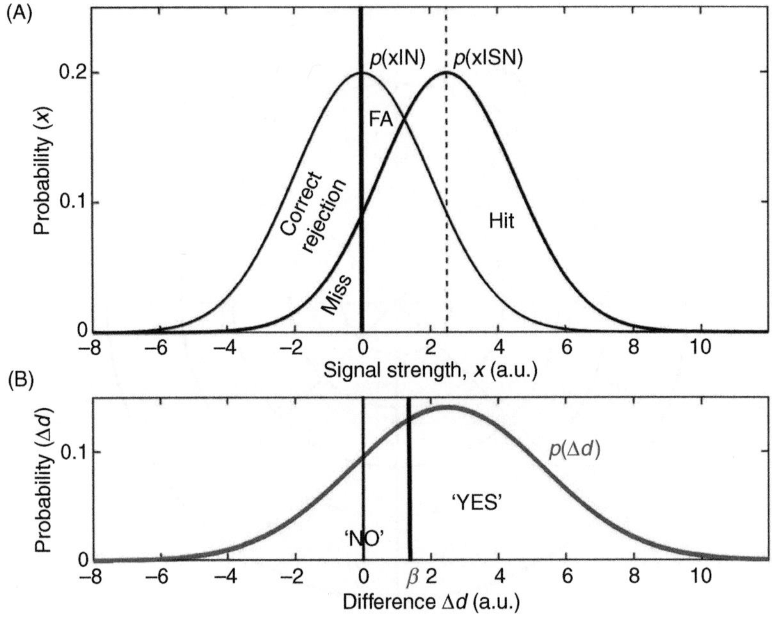

FIGURE 8.2 (A) Gaussian noise is present in the responses of a psychophysical experiment. The deterministic signal shifts the internal response to a higher mean, with the same variance as the noise. The amount of overlap between the two distributions sets task difficulty, leading to correct and incorrect responses (Table 8.1). (B) The distribution of differences between x_{SN} and x_N has a larger variance than the noise. SDT assumes that the perceived difference has to exceed a certain criterion, β, to elicit a "YES" response.

Further, the two random variables in Eq. (8.1) are supposed to be statistically independent. Somehow, the listener has to determine whether or not the response (in her brain) contained the signal. Here is how it is usually modeled: the subject is assumed to have an internal representation of the *difference* response signal, $\Delta r = x_{SN} - x_N$, which is based on a comparison of the different stimuli in the experiment. When the two underlying probability distributions for these alternatives are independent, the distribution of differences is also a Gaussian, for which the mean and variance are given by (Exercise 8.1):

$$\begin{aligned} \mu_D &= \mu_{SN} - \mu_N = \mu_{SN} \\ \sigma_D &= \sqrt{\sigma_{SN}^2 + \sigma_N^2} = \sigma_N \sqrt{2} \end{aligned} \tag{8.2}$$

Note that a *correct response* (hit, or correct rejection) is identified as follows:

$$\begin{aligned} \Delta r &\equiv x_{SN} - x_N > 0 \quad \text{"hit"} \\ \Delta r &\equiv x_{SN} - x_N < 0 \quad \text{"correct rejection"} \end{aligned} \tag{8.3}$$

However, this cannot be the complete story. What is still missing is the following important question: How certain is the subject that the difference signal indeed differs from zero? This seemingly subjective aspect of the psychophysical experiment is quantified by an internal *decision criterion* that reflects the subject's evaluation of costs and benefits associated with making an error (false alarm, or miss) or correct response (hit or correct rejection). As this criterion may vary from trial to trial it is in itself a stochastic variable too. Therefore, in SDT two quantities summarize this inherently stochastic aspect of signal detection: the *d-prime*, and the *decision criterion.*

The crucial outcome of an analysis based on SDT is the d' ("d-prime"). This is the (dimensionless) ratio of the size of the response to the signal plus noise, μ_{SN}, to the standard deviation in the responses to noise alone, σ_N:

$$d' \equiv \frac{\mu_{SN}}{\sigma_N} \tag{8.4}$$

The larger d', the more the two probability distributions are separated, and the easier it is for the subject to detect the signal. As such, the d' measures the subject's sensitivity to the signal.

The second important quantity in the decision task is the *criterion*, β, on which the subject bases a decision (Fig. 8.2B). If the perceived difference exceeds β, the subject reports "YES"; otherwise, the answer will be "NO."

$$\begin{array}{ll} \Delta r > \beta & \text{"YES"} \\ \Delta r < \beta & \text{"NO"} \end{array} \tag{8.5}$$

Note that if β differs from zero, and the distributions overlap (which is typically the case), the subject will unavoidably make errors. For example, if β is positive, there will be misses for $0 < \Delta r < \beta$. Conversely, if β is negative, there will be false alarms for $\beta < \Delta r < 0$. If a "miss" would be potentially fatal (like in the radar example of Fig. 8.3), the decision criterion will probably be lowered (leading to many false alarms). Conversely, if the cost for a "false alarm" would be gigantic, β will likely be increased (leading to more "misses").

The "YES/NO" task of the radar example of Fig. 8.1 (and its possible results in Table 8.1) is a popular application of these ideas in psychophysics. It's essentially a two-alternative forced choice task (2AFC). In an acoustic version of this task, the listener will hear a sound and will have to decide whether it contained the target signal ("YES") or just the noise ("NO"). Clearly, scores on the 2AFC task should lie between 50% (pure chance performance; signal was not detected) and 100% correct.

The *psychometric function* quantifies the percentage correct responses (or, alternatively, the percentage of "YES" responses) in the experiment as function of signal strength (tone intensity, sound location left vs. right from straight ahead, etc.). The subject's sensitivity to the signal, d, and the internal criterion, β, are directly related to this psychometric function.

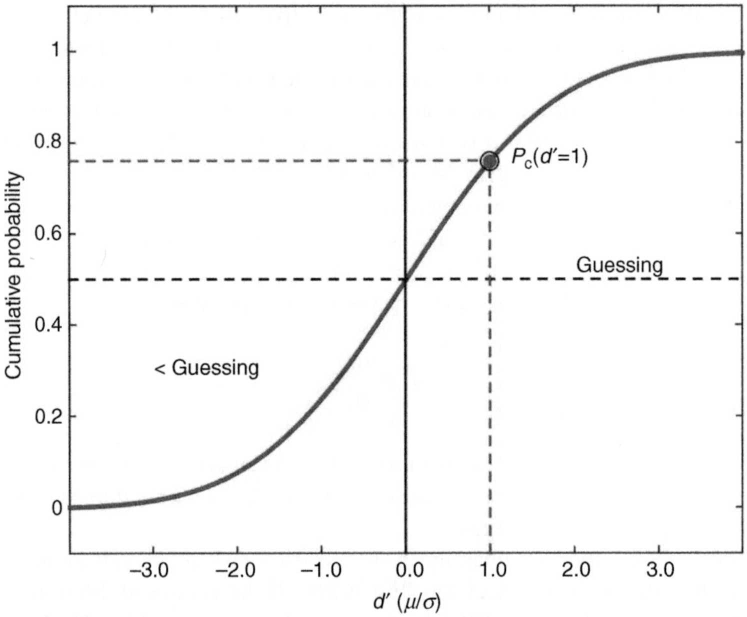

FIGURE 8.3 Normalized Gaussian CDF for the probability of a correct response, $P_C(\Delta r > 0)$ in the 2AFC (or YES/NO) task. The horizontal axis depicts the values of d', which are found by the z-scores of Eq. (8.10). Eg, $d' = 1$ is found at $\Delta r = \sigma_N$.

The probability for a correct response (Fig. 8.2B) is by definition

$$P_C(\Delta r > 0) = \frac{1}{\sigma\sqrt{4\pi}} \int_0^{\infty} \exp\left[-\frac{(\Delta r - \mu)^2}{4\sigma^2}\right] d(\Delta r) \tag{8.6}$$

which can be expressed into standard form of a Gaussian with mean zero, and variance one as the cumulative distribution function (CDF; Exercise 8.2).

$$P_C(\Delta r > 0) = \Phi_G\left(\frac{\mu}{\sigma\sqrt{2}}\right) \equiv \Phi_G\left(\frac{d'}{\sqrt{2}}\right) \tag{8.7}$$

The standard Gaussian probability distribution and its cumulative integral are defined as follows:

$$\begin{aligned} \phi(x) &\equiv \frac{1}{\sqrt{2\pi}} \exp\left(-\frac{x^2}{2}\right) \\ \Phi_G(x) &\equiv \frac{1}{\sqrt{2\pi}} \int_{-\infty}^{x} \phi(s)\, ds \end{aligned} \tag{8.8}$$

and the CDF can be numerically approximated by

$$\Phi_G(x) \approx 0.5 + \frac{\exp\left(-\frac{x^2}{2}\right)}{\sqrt{2\pi}} \cdot \left[x + \frac{x^3}{3} + \frac{x^5}{3\cdot 5} + \cdots + \frac{x^{2n+1}}{3\cdot 5\cdots(2n+1)}\right] \tag{8.9}$$

The common problem in an experimental context is to find d' ("how sensitive is our subject to the stimulus?") from P_C. This measure follows from Eq. (8.7):

$$d'(P_C) = \sqrt{2}\Phi_G^{-1}(P_C) \tag{8.10}$$

where the inverse of $\Phi_G(x)$ gives us the z-score: $z = x/\sigma$. In other words, when $d' = 1$, the cumulative probability for a correct response is obtained at 0.76 (ie, a stimulus strength yielding 76% correct responses; Fig. 8.3).

In the 2AFC task the subject should decide whether or not the signal was present. Note that a subject could answer with a series of "YES" responses, which could be due to a very strong signal, a real freak occurrence in the randomization of *SN*/*N* trials, or to a bias (preference, or otherwise) of the subject to answer "YES," rather than "NO."

How does Signal Detection Theory allow us to dissociate these different possibilities? Clearly, the four conditional probabilities of Table 8.1 and Fig. 8.2 are not all independent, as in a given trial the subject must always answer with either a "YES" or a "NO" to identify the perceived stimulus. As a result:

$$\begin{aligned} P_H + P_M &= 1 \quad \text{and} \\ P_{FA} + P_{CR} &= 1 \end{aligned} \tag{8.11}$$

and two independent measures suffice to fully characterize the experimental results. The most common choices in SDT are the hit rate, P_H, and the false-alarm rate, P_{FA}.

In SDT, the subject's "bias" to answer "YES" is modeled as a criterion. If the criterion is low, the subject ensures that she almost never misses a signal, so that the hit rate is almost 100%. However, in that case, the false-alarm rate will also be high. An optimal strategy could be to set the criterion such that P_H is maximized, while at the same time P_{FA} is minimal. Fig. 8.2 shows the situation in which the subject has an internal, subjective, criterion set at β.

It can be readily demonstrated (Exercise 8.3) that

$$d' = \Phi_G^{-1}(P_H) - \Phi_G^{-1}(P_{FA}) \tag{8.12}$$

The hit rate P_H and false alarm rate P_{FA} relate directly to the subject's internal criterion β, which is also known as the *likelihood ratio*, *LR* (see Fig. 8.2). The larger β, the more likely that the subject will answer "Yes," whereas for small β the subject is reluctant to say "Yes." Using Eq. (8.12) it can now be shown that

$$\beta \equiv \frac{P(x=\beta|SN)}{P(x=\beta|N)} = \exp\left(-\frac{1}{2}\{[\Phi_G^{-1}(P_H)]^2 - [\Phi_G^{-1}(P_{PA})]^2\}\right) \tag{8.13}$$

ROC: When plotting the cumulative probability of the hit rate as a function of the cumulative probability of the false-alarm rate for different decision criteria, we obtain a curve that is known as the *Receiver Operating Characteristic*, or *ROC* of the experiment. When the criterion is at β, the two cumulative distributions for FA and H, respectively, are determined as follows:

$$\begin{aligned} &\text{False Alarms:} \, \Phi_{FA}(\beta) = \int_\beta^\infty P_N(x)dx \\ &\text{Hits:} \, \Phi_H(\beta) = \int_\beta^\infty P_{SN}(x)dx \end{aligned} \tag{8.14}$$

The ROC plots $\Phi_H(\beta)$ as a function of $\Phi_{FA}(\beta)$ (Fig. 8.4). Clearly, as the number of hits versus false alarms depends on the subject's criterion, so does the ROC value. In addition, the ROC curve depends on d' (the detectability of the signal). When a subject responds purely at chance level, the two cumulative distributions are equal, and the ROC curve coincides simply with the main diagonal. Thus, the more the ROC curve deviates from the main diagonal, the easier it was to detect the signal, and the better the subject's performance can be. Fig. 8.6 depicts a couple of different ROC curves, obtained for different signal strengths, while varying the detection criterion over a large range. The red curve highlights one point of a simulated experiment for which the signal had a strength of $d' = 1.25$, and the detection criterion was set at $\beta = 1$. When $d' = 2$ (signal strength is two standard deviations away from the mean noise level), the number of false alarms is much lower than for the red curve and the hit rate rapidly reaches ceiling performance of 100%.

Note that in the absence of noise the hit rate will always be 100%, and there will be no false alarms. This situation corresponds to point [0,1] in Fig. 8.4.

From the ROC curve the best possible performance for a given signal strength (ie, the optimal decision criterion) is found by finding the point of largest deviation of the curve from the main diagonal. Referring to Fig. 8.4, and Eq. (8.14), the distance of each point of the ROC curve to the diagonal is (Exercise 8.4):

$$\begin{aligned} &D(\beta) = \sqrt{[\Phi_{FA}(\beta)]^2 + [\Phi_H(\beta)]^2} \sin[\varphi(\beta) - \pi/4] \\ &\text{with } \varphi(\beta) = \arctan\left[\frac{\Phi_H(\beta)}{\Phi_{FA}(\beta)}\right] \end{aligned} \tag{8.15}$$

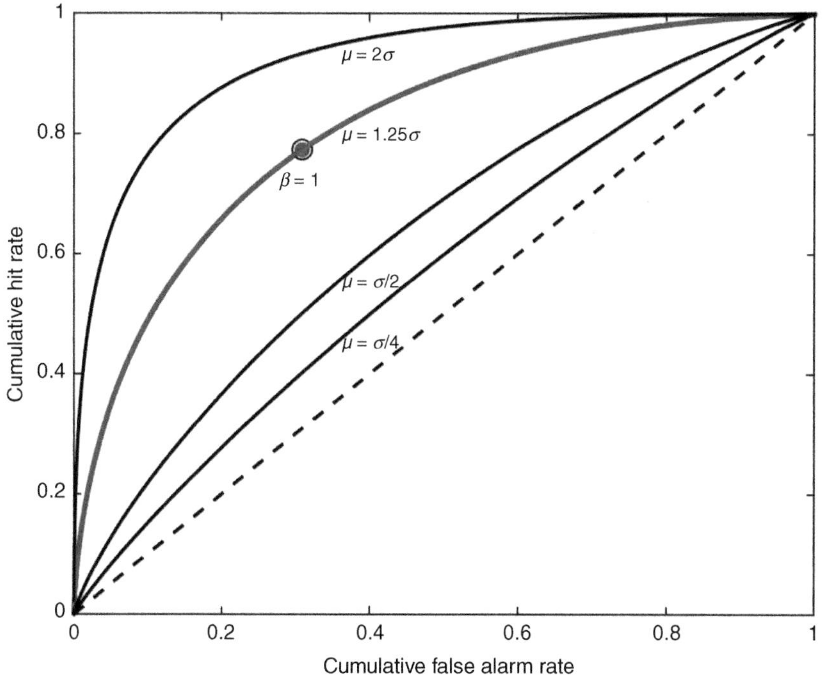

FIGURE 8.4 **ROC curves of the 2AFC task for four different d' values (d' = 0.25, 0.5, 1.25, and 2).** The curves were generated by calculating Eq. (8.14) for a range of criterion values between −8 and +8. The *dashed line* indicates chance performance (d' = 0). The *red curve* (at d' = 1.25) highlights a single point for which the criterion was $\beta = 1$. The optimal criterion (highest hit rate vs. lowest false alarm rate and misses) of a given stimulus is found where the ROC curve deviates furthest from the diagonal.

from which

$$\beta_{\mathrm{OPT}}(d') = \mathrm{MAX}\left[D(\beta)\right] = d' \tag{8.16}$$

As shown in Fig. 8.5A, the optimal distance increases monotonically, nearly linearly, with d' over the range shown, although for $d' \rightarrow \infty$ the distance approaches the asymptotic limit of $1/2\sqrt{2}$. The maximal distance is obtained at the criterion that equals d' (Fig. 8.5B). So, the best possible criterion in a decision task that maximizes the probability for a hit, while at the same time minimizes the number of false alarms, is d'. As d' is a property of the stimulus (ie, the external world), and β is an internal decision criterion of the subject, it is an interesting question whether subjects (learn to) find the optimal criterion in a real decision task, and whether they appropriately adapt their criterion when the signal-to-noise conditions (and hence d') change.

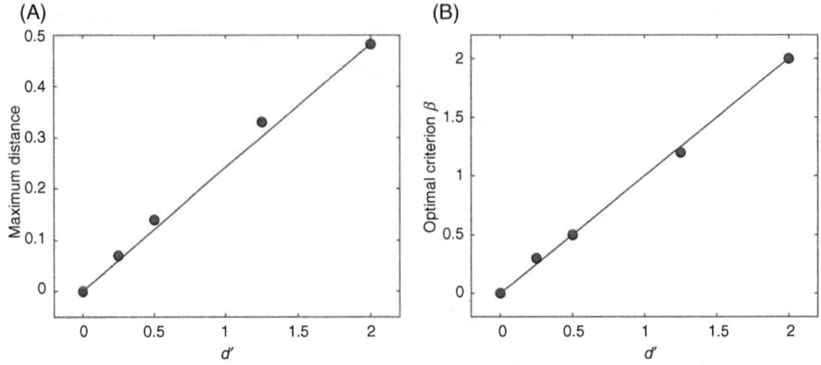

FIGURE 8.5 (A) Maximum distance of the ROC curve as function of d' [Eq. (8.16)]. The point at (0,0) reflects the pure noise condition (chance performance). The relationship is nonlinear as it asymptotes at $D \sim 0.71$ for very large d' (the pure signal condition). (B) The optimal criterion, β, equals d'.

8.3 DETECTION, LATERALIZATION, DISCRIMINATION, AND LOCALIZATION

Detection: Arguably the simplest auditory task is a *detection* task. In such an experiment the subject just has to indicate whether she heard a particular sound (which may be hidden in noise), without having to identify or spatially localize the sound source. The experiment can be carried out under all monaural and binaural hearing conditions described in Section 8.1. The result of the experiment could be a simple button press for "YES," and (when recorded) the associated reaction time of the subject to the sound's onset. No reaction needs to be given when no sound is perceived. In principle, sounds can be presented at randomized moments within a single, continuous experimental recording session, or within a series of fixed time windows that separate individual test trials.

Although in detection experiments sounds may be presented from different locations (eg, to measure the effect of sound direction on the reaction time, or on the detection threshold), the subject does not directly report or point to the sound's location. The potential effects of a spatial manipulation are thus indirectly obtained as a statistical estimate from a series of psychophysical measurements that typically require many sound presentations (Section 8.2).

Lateralization: Sound *lateralization* refers to the ability of the auditory system to discriminate left from right with respect to the head. In this context it is useful to relate sounds to the *midsagittal median plane* of the head, either physically (in the room, or the free field, where the reference may be the straight-ahead direction), or perceptually (usually with headphones, with respect to the estimated center of the head). In the latter case the *auditory median plane* (AMP) is defined as the point in acoustic space (eg, a binaural level or time difference) for which the subject responds 50% to the right with respect to the center of the head (Fig. 8.6).

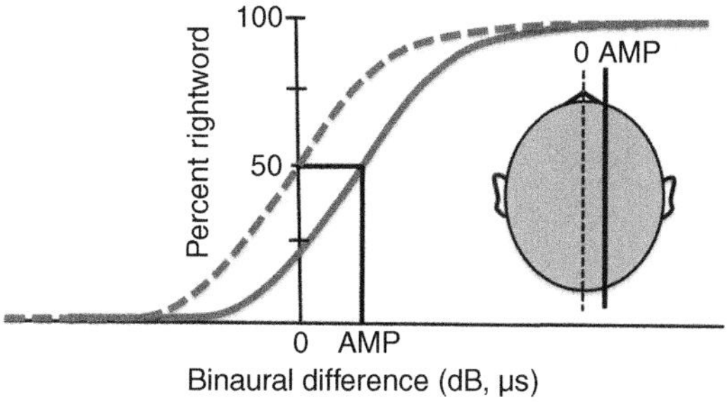

FIGURE 8.6 **The AMP is defined at the 50% point in a left/right decision task.** The acoustic stimulus is a binaural difference (BD) in intensity or time. With the head upright (*dashed curve*) the AMP corresponds to the center of the head. In case of a passive head tilt the AMP shifts to a BD > 0 (*solid curve*). This means that to perceive the sound in the center of the head (*dashed line*), the BD should be −AMP dB or μs.

With the head upright and under normal hearing conditions it is expected that the AMP will be close to the center of the head, that is, at ILD = 0 dB, or at ITD = 0 μs (Fig. 8.6, dashed psychometrical curve). However, under certain experimental conditions the AMP may be at a different location. For example, it has been demonstrated in numerous studies that the AMP shifts away from the center when the head is passively tilted with respect to gravity. When the head is tilted with the right ear downward, the AMP shifts in the direction of head rotation, which in this case would be rightward in the head (ie, AMP > 0 dB). This effect, which is supposed to reflect a computational mechanism for neural integration of vestibular and acoustic spatial information, has been called the *audiogravic illusion* (Lewald and Karnath, 2001; Van Barneveld et al., 2010). Also under hearing-impaired conditions (eg, in case of a unilateral hearing loss) the AMP may shift in a particular direction.

In the L/R decision task the subject never reports the actual perceived position of the stimulus. She indicates, for example, with a button press, whether the sound was heard left or right of the reference location. Note that with headphone stimulation perceived stimulus locations seem to shift along a line inside the head with changes in ILD or ITD. These changes are not directly related to degrees in external space. Yet, from data such as in Fig. 8.6 (see also Section 8.2) several parameters can be derived that relate to spatial hearing:

- The *spatial* (*or ILD/ITD*) *sensitivity* (or *resolution*) around the central reference position is related to the slope of the function at the AMP;
- Asymmetries for the perceived left versus right hemifield arise when the AMP is offset from zero;
- A subjective bias (a preference to choose left over right, or vice versa, irrespective of stimulus strength) becomes apparent when the percentage right never equals zero, or 100%.

A psychometric function that captures these properties is based on the cumulative Gaussian (Section 8.2), and is described by

$$\psi(x;\mu,\sigma,\gamma,\lambda) \equiv \gamma + (1-\gamma-\lambda)\cdot F_G(x;\mu,\sigma) \tag{8.17}$$

with γ and λ the so called–stimulus independent *lapses* (these are usually small, and are indicative of biases for rightward and leftward responses, respectively), μ the mean, σ the standard deviation, and $F_G(x;\mu,\sigma)$ the cumulative Gaussian distribution. When $\lambda = \gamma$, the AMP is found at $x = \mu$, and the standard deviation is a direct measure for the width of the psychometric function, and hence the subject's resolution to stimulus variable x.

Discrimination: In the lateralization task the research question relates to the percept of a spatial (or cue) discrimination of left versus right with respect to a (head- or body-) fixed landmark, typically the estimated center of the head. In a more general spatial discrimination task one studies the sensitivity of the auditory system to changes in spatial displacements (or changes in acoustic cue strengths) across the auditory field. In such experiments, the subject may hear two sounds in succession, and has to judge whether the second stimulus is to the left or right, up versus down, or different versus the same, with respect of the first stimulus (hence, a 2AFC task). In a diotic headphone task the first stimulus may be presented at any ILD or ITD, and is therefore not confined to the center of the head. Discrimination experiments thus measure the *resolution* of the spatial localization cues across the auditory domain as function of the reference cue strength. Also in this task the subject never indicates the real perceived location (or cue) of the (first or second) stimulus, only whether or not she perceived a change in a given direction with respect to the first sound.

Minimal audible angle: In the free-field variation of the discrimination task, the subject has to differentiate spatially separated acoustic stimuli (as "same/different," "left/right," "up/down," etc.), with the separation angle between subsequent stimuli as experimental variable. The result of this experiment is quantified as the *minimum audible angle* (MAA) and can be measured as function of spatial location in azimuth (Fig. 8.7A), elevation, or both (Fig. 8.7B), and of acoustic parameters such as sound frequency (Fig. 8.7A) or bandwidth (Mills, 1958; Perrott and Saberi, 1990). The MAA is a measure for the spatial resolution of the auditory system across the auditory field. The early experiments of Mills (1958) indicated that the MAA is at a minimum of about one degrees for locations near straight ahead, and that this high resolution remains relatively robust for different sound directions with respect to the horizontal plane, even up to about 70 degrees orientation of the stimulus array (Perrott and Saberi, 1990). For locations in the medial plane (at 90 degrees), however, the MAA sharply increases to about 3.5 degrees (Fig. 8.7B).

Hartmann and Rakerd (1988) argued that the MAA is actually smaller by a factor $1/2\sqrt{2}$ than the reported values of Mills (1958) and others (Fig. 8.7). Their reasoning goes as follows: the original MAA paradigm utilizes three, instead

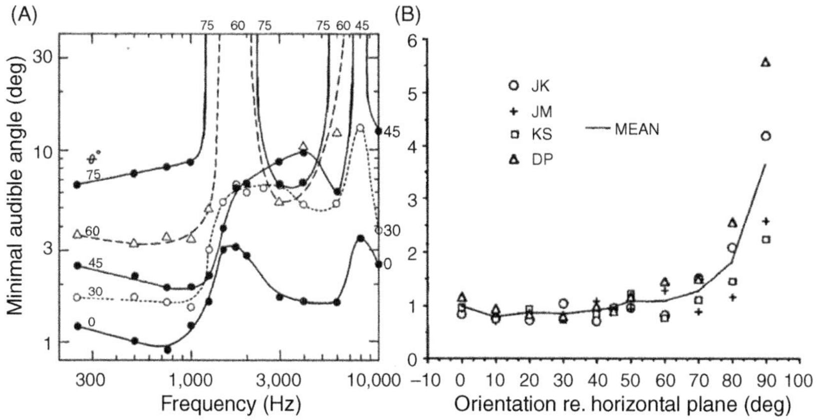

FIGURE 8.7 (A) Minimum audible angles (MAA) in the horizontal plane as function of azimuth (different curves) and frequency. Note the prominent decrease in spatial resolution around 1.5 kHz (and 8 kHz), and for lateral stimulus locations. (B) MAA around straight ahead as function of the orientation of the speaker test array. Spatial resolution remains stable at about 1.0 degree for angles up to 70 degrees, but increases to 3.5 degrees for the elevation direction near 90 degrees. *(Source: Part A: After Mills, 1958. Part B: After Perrott and Saberi, 1990, with permission.)*

of two speaker positions, namely at $x_0 \pm \Delta x$, with x_0 the reference location, and Δx the test locations at either left or right of x_0 in randomized trials. They then showed that an SDT model, in which subjects discriminated $2\Delta x$ (thereby comparing Δx vs. $-\Delta x$ *between* trials, while ignoring x_0), rather than Δx (corresponding to a comparison *within* a trial, as assumed in the MAA literature), provides a better explanation for the MAA results of Fig. 8.7. Therefore, around straight ahead the spatial resolution for azimuth would be ~0.7 degree, and for elevation ~2.5 degrees.

Localization: In a sound-localization task the subject is asked to point to the perceived spatial location of the sound source. Many different pointing methods have been used in the literature that differ in speed, precision, range of pointing, and cognitive load, but overall yield qualitatively similar results.

- Head orienting movements (eg, Middlebrooks and Green, 1991; Van Wanrooij and Van Opstal, 2004; Perrott et al., 1987);
- Arm (finger) pointing (Zwiers et al., 2001);
- "Gun" pointing (Oldfield and Parker, 1984), and laser pointing with a joystick (Zwiers et al., 2003);
- Pointing on a model sphere with an azimuth–elevation coordinate grid (Wightman and Kistler, 1989b);
- Pointing with fast eye movements (called *saccades*; Jay and Sparks, 1984; Frens and Van Opstal, 1995; Hofman and Van Opstal, 1998);
- Pointing with combined saccadic eye–head movements (*gaze shifts*; Whittington et al., 1981; Goossens and Van Opstal, 1997; Populín, 2006; Van Grootel et al., 2012).

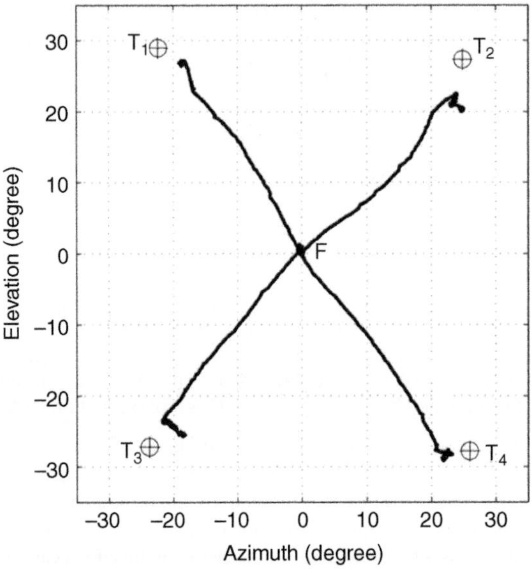

FIGURE 8.8 Individual trials of evoked eye movements from the central fixation position (F) to four different target locations in the frontal hemifield. Responses were evoked in darkness (open loop). *(Source: After Hofman and Van Opstal, 1998, with kind permission.)*

Fig. 8.8 illustrates 4 example responses of saccadic eye movement responses toward broadband sounds (selected from a larger data set of 98 targets, shown in Fig. 8.9). Stimuli (GWN at 58 dBA, duration 500 ms) had been randomly presented in complete darkness. As eye movements do not provide any acoustic feedback to the listener, and subjects cannot see the speaker, this paradigm exemplifies a fully *open-loop* experiment, in which participants can only rely on acoustic input to plan and generate their localization response.

Pointing methods that rely on fast eye-, head-, or combined eye–head movements typically yield responses with very short reaction times (150–200 ms). Arguably, such responses reflect the early stages of acoustic spatial processing, as they leave little time for cognitive interference. We will expand further on these pointing methods later, and in the following chapters.

Sounds are usually presented in a free-field environment, although localization experiments can also be performed in reverberant rooms. In some cases, localization is tested with headphones, with an experimental technique that is called *virtual acoustics* (Section 8.6).

The psychoacoustic tasks described in the previous section require *relative* judgments about the stimuli, which result in binary responses (eg, "YES/NO") that lack information about absolute spatial processing. In contrast, the sound-localization pointing task yields responses that can be measured and calibrated on a continuous, *absolute* spatial scale, and quantitatively evaluated in terms of speed, accuracy, and precision (see later). It is thus possible to evaluate the

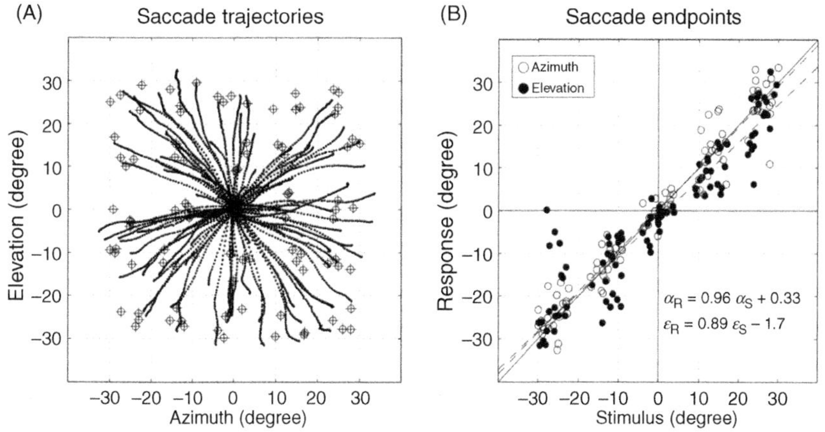

FIGURE 8.9 Sound-localization responses in the frontal hemifield using saccadic eye movements as a pointer. Only the first saccade in a trial is taken as localization response. Correction saccades are not included. (A) Spatial trajectories to 98 targets randomly presented within the oculomotor range. (B) Stimulus–response relationships for the azimuth (*open symbols*) and elevation (*closed symbols*) saccade components. Linear regression coefficients [Eq. (8.19)] are shown at the bottom-right. Note lower gain for elevation (0.89 vs. 0.96), and slightly higher variability, than for azimuth components. *(Source: After Hofman and Van Opstal, 1998, with kind permission.)*

responses with linear regression models: perfect sound-localization behavior would require the listener to exactly point (with response, R) to the target (T); hence, in two dimensions (azimuth and elevation) the ideal localizer yields:

$$\vec{R} = \vec{T} \tag{8.18}$$

Linear regression: If stimulus location is the only variable in the experiment, the data can be conveniently modeled by two linear regression equations, which can be separated for the azimuth and elevation components, respectively (Fig. 8.9):

$$\begin{aligned} R_{\mathrm{AZI}} &= aT_{\mathrm{AZI}} + b \\ R_{\mathrm{ELE}} &= cT_{\mathrm{ELE}} + d \end{aligned} \tag{8.19}$$

with $[a,c]$ the dimensionless slopes, and $[b,d]$ (in degrees) the offsets of the regression lines. Often, the slope is called the *response gain*, and the offset the *response bias*.

Azimuth/elevation: The major reason for separating the regressions into azimuth and elevation components is the remarkable independence of the neural processing stages of acoustic target components within the auditory system. As we have seen in chapter: Acoustic Localization Cues, the azimuth component relies on binaural difference cues in intensity and timing, which are in turn extracted by two different, independent neural pathways (passing through LSO and MSO, respectively) in the auditory brainstem. In contrast, the elevation

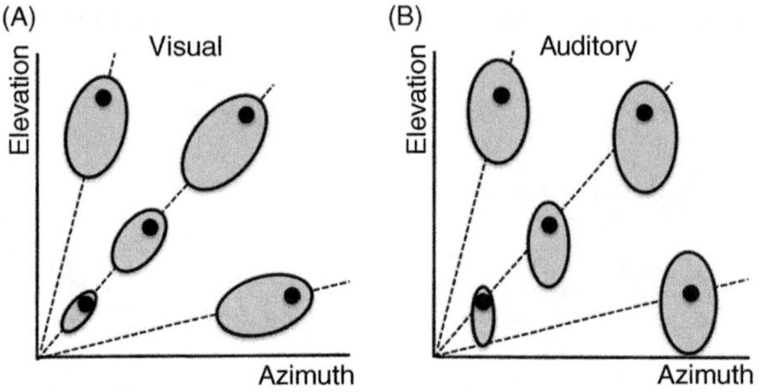

FIGURE 8.10 Different endpoint scatter patterns for saccade responses to visual (A) versus auditory (B) targets reflect fundamental differences in sensory processing and reference frames.

component is extracted from the spectral pinna cues (HRTFs); this is essentially a monaural process (but also see further), carried out by a third neural pathway, possibly embodied within the DCN.

The organization in independent Cartesian coordinates of sounds differs profoundly from *vision*, in which spatial coordinates reflect the circular-symmetric organization of the retina around the fovea. The organization of the visual system is therefore better described in *polar coordinates* (amplitude, R, and direction, Φ) than in Cartesian (horizontal, vertical) coordinates. The fundamental difference in reference frames: head-centered Cartesian coordinates for sounds, versus eye-centered polar coordinates for vision, can be readily appreciated from observed distributions of eye-movement end points to either sounds (Hofman and Van Opstal, 1998) or lights (Van Opstal and Van Gisbergen, 1989).

These patterns are schematically illustrated in Fig. 8.10. Eye movement endpoint scatter evoked by flashed light stimuli reflects a polar coordinate organization, as the endpoints vary independently along the eccentricity (amplitude, R), and angular (direction, Φ) dimensions. Amplitude scatter increases linearly with target eccentricity (and hence, eye-movement amplitude), while the angular variance remains constant (Van Opstal and Van Gisbergen, 1989). In contrast, the endpoint variance for sound-evoked eye movements varies independently along azimuth-elevation coordinates, leading to vertically oriented scatter ellipses in which the elevation variance exceeds azimuth variance (Hofman and Van Opstal, 1998). This latter property nicely reflects the observed differences in MAA, described earlier.

Some caution: It can become a bit tricky to compare different pointing methods in terms of their absolute (in)accuracy, because different factors play a role in the mapping from *acoustic cues* → *spatial percept* → (*motor*) *response* to the perceived location. Let's highlight an obvious pitfall to illustrate this point.

Suppose that one only records head movements as a pointer for perceived sound locations, but that the eye orientation in the head is not constrained. A natural orienting response consists of a combined eye–head movement, called a gaze shift, which quickly directs the fovea to a point in space (the gaze direction, or gaze, for short). Eye–head coordination involves a division of labor between the eye-in-head movement (E_H) and the head-in-space movement (H_S, which, in this example, is our pointer). As a result, gaze in space is defined as

$$G_S = E_H + H_S \tag{8.20}$$

So, if the eye points to a location 80 degrees away from straight ahead, it could be achieved by a head rotation of 80 degrees (which is the aim of the paradigm), by a head rotation of 60 degrees, and an eye rotation of 20 degrees, by $H_S = 70$ degrees and $E_H = 10$, etc. In short, unless we constrain E_H, we cannot dissociate these different possibilities. But if the natural contributions of eye and head to a gaze shift would be fixed (eg, 80% head movement, 20% eye movement), it would not be a fundamental problem to use the head as a reliable pointer, albeit that the gain of the responses could underestimate the real response gain.

However, from gaze control studies we know that vertical gaze shifts are typically associated with smaller head movements (and thus larger eye movements), than horizontal movements (eg, Goossens and Van Opstal, 1997). This means that Eq. (8.20) should actually be rewritten to

$$G_S = a(\phi)E_H + [1 - a(\phi)]H_S \text{ with } 0 \le a(\phi) \le 1 \tag{8.21}$$

with Φ the direction of the gaze shift with respect to horizontal. Moreover, the coefficient $a(\phi)$ strongly depends on the total gaze amplitude. Small gaze shifts hardly involve a head movement, and are carried out exclusively by the eyes [$a(\phi) = 1$], whereas large gaze shifts, say >50 degrees, necessarily involve a substantial head movement, as the eye will reach its oculomotor range around an eccentricity of 35–40 degrees.

One could thus find that the head-movement responses to targets in elevation ($\phi = 90$) have a much lower gain than those in azimuth ($\phi = 0$). But because subjects (perhaps unintentionally) may have pointed with their eyes (gaze), rather than with their nose (head alone, as intended by the experimenter), the actual gaze responses may have been much more accurate, even dead-on target! Thus, to ensure that the pointer is appropriate for the task, the eye position should be constrained, for example, to $E_H = 0$, in order to force the head to make the entire movement [$a(\phi) = 0$].

Similar potential problems apply to hand or arm pointing (then, subjects use eye–hand coordination to do the task!). As a result, the absolute localization error in a particular pointing experiment may not always reflect real perceptual misjudgments of the spatial location, but could merely be part of a particular sensorimotor strategy of the system that was used as a pointer.

Multiple regression: If multiple (continuous) variables play a role in the experiment, like stimulus duration, D; bandwidth, B; level, L; etc. the regression model should be extended (let's suppose first-order effects, for simplicity) to include these variables. For example, for the azimuth response components the multiple-linear regression model could read:

$$R_{\mathrm{AZI}} = aT_{\mathrm{AZI}} + bT_{\mathrm{DUR}} + cT_{\mathrm{BW}} + dT_{\mathrm{LEV}} + e \tag{8.22}$$

and coefficients can be found by standard least-squares-error techniques.

However, because the response (in degrees) and the stimulus variables (in degrees, millisecond, Hertz, decibel, respectively) are measured in fundamentally different quantities, the values of the regression coefficients [*a*,*b*,*c*,*d*,*e*] cannot be directly compared as they are expressed in different dimensions [*dimensionless, degree per millisecond, degree per Hertz, degree per decibel, degree*]. It is therefore not easy to decide from Eq. (8.22) whether, for example, the effect of stimulus duration (slope b, in degree per millisecond) on the localization accuracy is stronger than the effect of sound level (slope d, in degree per decibel), and by how much. To avoid this problem it is customary to express the variables of Eq. (8.22) as dimensionless *z*-scores:

$$\hat{x} \equiv \frac{(x-\mu_x)}{\sigma_x} \tag{8.23}$$

where μ_x, σ_x are mean and standard deviations of variable x, computed over the entire data set. After this transformation Eq. (8.22) reads:

$$\hat{R}_{\mathrm{AZI}} = p\hat{T}_{\mathrm{AZI}} + q\hat{T}_{\mathrm{DUR}} + r\hat{T}_{\mathrm{BW}} + s\hat{T}_{\mathrm{LEV}} \tag{8.24}$$

where [*p*,*q*,*r*,*s*] are the *partial correlation coefficients* (all dimensionless). Now, the contributions of different stimulus variables to the subject's responses can be directly compared. The method can also be used to determine the power of different models, by letting each of the models contribute to the predicted response. Below, we apply this method to quantify potential binaural interactions in spectral cue extraction for elevation. In later chapters we will apply multiple regression to address sound-localization performance under monaural hearing conditions, to eye–head orienting responses to sounds, and for localization responses to rapid stimulus changes.

Precision, accuracy, error: A report of a subject's localization performance should preferably include several quantities, as different performance measures capture different aspects of the task. Fig. 8.10 illustrates the difference between accuracy and precision of responses to a target.

Accuracy refers to the systematic error of the responses: it equals zero when responses are on the target, on average. The mean response inaccuracy (*IA*) is computed as the mean signed difference between target location and response location across trials:

$$IA(\alpha) = \frac{1}{N}\sum_{n=1}^{N} T_n^{\alpha} - R_n^{\alpha} \quad \text{and} \quad IA(\varepsilon) = \frac{1}{N}\sum_{n=1}^{N} T_n^{\varepsilon} - R_n^{\varepsilon} \tag{8.25}$$

The coefficients of the linear regression line (slope and offset) together provide direct measures for response accuracy: the *gain* measures the *sensitivity* of the subject's responses to changes in stimulus eccentricity, whereas the *offset* (in degrees) quantifies the subject's bias, irrespective of target location. If the gain = 1 and the offset = 0, the accuracy will be perfect (mean error = inaccuracy = 0). Any deviation of the gain or offset from their optimal values leads to a particular pattern of inaccuracies in the responses:

If the offset = 0, and the gain differs from 1, the inaccuracy increases linearly with target eccentricity; the relative inaccuracy, however, (response amplitude divided by stimulus eccentricity) will be constant, and is determined by the gain. Conversely, if the offset is nonzero, but the gain is one, the mean error is constant, and so the inaccuracy is independent of stimulus eccentricity. In this case, the relative inaccuracy decreases inversely with stimulus eccentricity. When both regression coefficients differ from their ideal values the error patterns form a combination of these behaviors.

Precision is a measure for response *reproducibility*, and quantifies the variability of responses. It is usually defined as:

$$J \equiv \frac{1}{\sigma^2} \tag{8.26}$$

The *mean absolute error* across trials (MAE; separately calculated for azimuth and elevation) is determined by

$$\mathrm{MAE}(\alpha) = \frac{1}{N}\sum_{n=1}^{N}\left|T_n^{\alpha} - R_n^{\alpha}\right| \text{ and } \mathrm{MAE}(\varepsilon) = \frac{1}{N}\sum_{n=1}^{N}\left|T_n^{\varepsilon} - R_n^{\varepsilon}\right| \tag{8.27}$$

In general, the MAE differs from the inaccuracy measure of Eq. (8.25), as the former also differs from zero when the average response position coincides exactly with the target. For example, in Fig. 8.11 the accuracies of panels 2 and 4 are equal (zero), but the MAE of panel 2 is larger than that of panel 4. Furthermore, the MAE of the data in panel 2 is much larger than for panel 1, although the variances are the same.

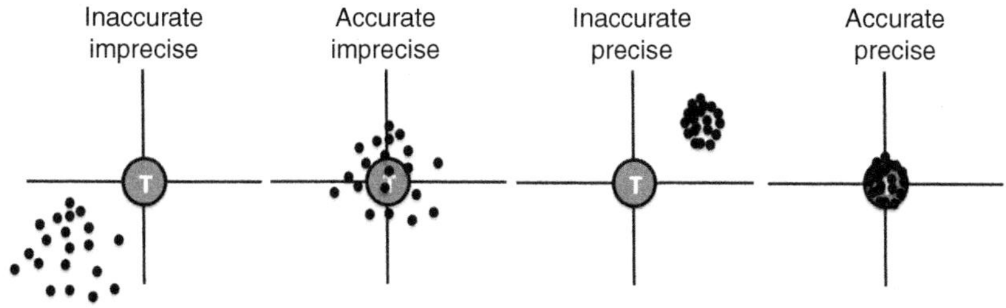

FIGURE 8.11 Accuracy (quantified by mean response error) and precision (the inverse of response variance) refer to fundamentally different aspects of localization performance. Accuracy is the same and high in panels 2 and 4, precision is the same and low in panels 1 and 2, and high in panels 3 and 4. The MAE is different for all four cases.

The *goodness of fit* for a regression model is usually quantified by the squared correlation coefficient between data and model prediction, r^2 (this is called the *coefficient of determination*). It specifies the amount of variance in the data that is explained by the model. For example, when $r^2 = 0.9$ it means that 90% of the response variation over the stimulus range is explained by the regression model, leaving only 10% unexplained (and considered noise, for lack of a better explanation). Note that a correlation coefficient of $r = 0.8$ seems (and often is …) quite high, but still leaves 36% of the variation in the data unexplained.

8.4 SOUND LOCALIZATION: PREY VERSUS PREDATOR

The ability to detect, identify, and localize objects in the natural environment is of paramount importance for survival, predators, and prey alike. However, the behavioral response to the detection of a brief sound, or of a subtle motion detected in the visual periphery, will be quite different for prey versus predator. Interestingly, the different ecological requirements may have had a strong impact on the development of their *visual and auditory systems*.

For prey animals like rabbits it is crucial to detect the presence of a predator in the periphery as quickly as possible, with relatively little pressure on absolute precision and accuracy. As long as it can reliably decide in which direction it's safe to run away, the animal will survive. Thus, there is an evolutionary pressure on prey animals to be particularly good at the detection, discrimination, and lateralization of sensory stimuli.

The predator has other issues to consider: it needs to know exactly where its (moving) food source is, so that it can attack and catch the prey with high accuracy and precision. If it misses its goal, it has wasted a lot of energy in its hunt. For its survival, it should better be good at absolute localization of stimuli.

As a result of these different requirements, the visual systems for prey animals and predators are organized quite differently: prey animals have their eyes positioned *laterally* on their head, providing them with a panoramic visual field of almost 360 degrees with little binocular overlap. Moreover, these animals typically lack a fovea. The fovea is the small circular area in the central retina with a very high photoreceptor density, which results in a high spatial resolution within a narrow field of view. Instead, prey animals often have a horizontal *visual streak* (a strip of higher density photo receptors along the horizontal meridian of the eye's retina) that allows them to discriminate the movement of predators in the environment over a wide visual angle (often more than 90 degrees), without the need to make conspicuous movements.

In contrast, predators are frontal eyed with large binocular overlap that allows for depth vision, and their retinae contain a fovea (area centralis). Because the photoreceptor density in the peripheral retina of foveate animals rapidly decreases with eccentricity the eyes need to make accurate gaze shifts in order to allow the visual system to make a high-resolution inspection of the stimulus. The evolutionary pressure to generate these goal-directed gaze shifts *as fast and as accurately as possible* has led to the saccadic eye–head-(body) orienting system as a primary sensorimotor system for orienting.

Similar considerations may hold for the auditory system: prey animals versus predators may have developed different sound-localization strategies that might optimally suit their ecological requirements.

Indeed, in an interesting comparative study, Heffner and Heffner (1992) tried to answer the following question: *which factors determine the auditory spatial resolution (say, the MAA) of a given species*? One might suppose that head size would be a strong determinant. Indeed, according to Eqs. [(7.2) and (7.3)] head size sets the path length difference for sound waves and has an immediate effect on the absolute interaural time (and level) differences. Larger heads create longer time differences, resulting in a higher spatial resolution (because of the better signal-to-noise ratio) in the azimuth direction. Surprisingly, however, by comparing different animal species, ranging from small rodents to cows and humans, the explanatory power of head size to the animal's MAA turned out to be insignificant. Also other potentially important factors, like the animal's lifestyle (eg, nocturnal vs. diurnal activity, or their trophic level, ranging from pure predator to pure prey) could not explain the observed MAAs. The single factor that did strongly correlate with the MAA, however, was the *size of the fovea or visual streak* ($r = 0.91$)! Additional factors like the amount of binocular overlap (which is large for frontal-eyed animals) or the animal's trophic level (often associated with lateral vs. frontal eyes), despite correlating well with the MAA when analyzed in isolation, added only 13% to the total variance accounted for in a multiple regression analysis of the studied sample of species. The first-order correlative relationship of MAA versus size of the field of best vision is shown in Fig. 8.12A, which contains 13 different animal species, including humans.

An earlier study by Harrison and Irving (1966) had demonstrated that the number of cells in the medial superior olive (encoding interaural time differences; chapter: Acoustic Localization Cues) and the abducens nucleus (whose cells innervate the lateral rectus eye muscle for ipsilateral eye movements; chapter: The Gaze-Orienting System) were strongly correlated (Fig. 8.12B), both for diurnal animals (which rely on *cone* vision), and for nocturnal animals (relying on *rod* vision). This is quite a remarkable finding, as it suggests that the integrity of the sound-localization and oculomotor (ie, visuomotor) systems may be functionally related.

> These independent lines of evidence both support the notion that an important function for the sound-localization system in predators is to direct the eyes (in particular, the field of best vision) to the perceived sound location.

An auditory-evoked eye movement will thus allow the animal to quickly acquire accurate visual detail for full identification of the stimulus. It is for

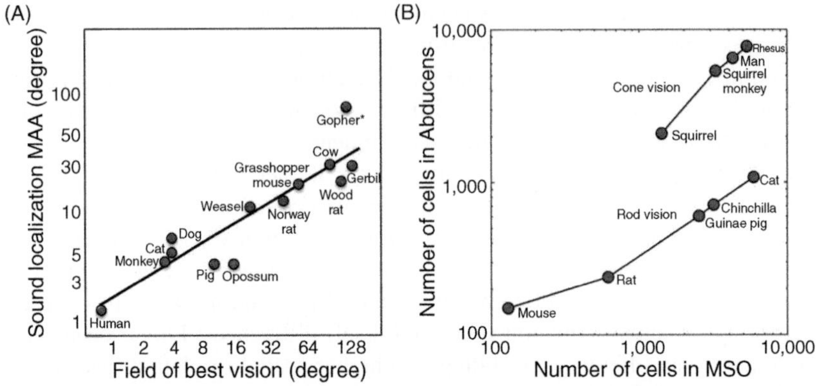

FIGURE 8.12 (A) Sound-localization thresholds for a range of mammalian species are best predicted by the width of the field of best vision (the fovea, or visual streak). Note log–log scales. (B) Number of cells in the MSO and oculomotor nucleus, n.VI (Abducens), correlates well for diurnal animals (cone vision), as well as nocturnal animals with good vision (rods). *(Source: Part A: From Heffner and Heffner, 1992, with kind permission. Part B: After Harrison and Irving, 1966.)*

this reason that several research groups have employed eye movements (the "audio–ocular response"; Zahn et al., 1979) as a natural pointer to study the sound-localization system (Whittington et al., 1981; Zambarbieri et al., 1982; Jay and Sparks, 1984; Frens and Van Opstal, 1995; Populín, 2006; Van Grootel et al., 2012; Fig. 8.9).

To appreciate the role of eye movements (or better: the coordinated orienting gaze shift of body, head, and eyes) in sound localization, we will go into some more depth of the gaze-orienting system in the next chapter. Furthermore, because of their mutual interdependence, the auditory and visual systems also strongly interact through a process called *multisensory integration* (discussed further in chapter: Impaired Hearing and Sound Localization). In the remainder of this chapter, we describe some fundamental findings on sound localization as revealed by eye movements as a pointer.

8.5 SOUND LOCALIZATION: EFFECTS OF SPECTRAL CONTENT, DURATION, AND LEVEL

In a typical eye-movement experiment the subject is seated in a dark, sound-attenuated, echo-poor laboratory room. To avoid complete disorientation of the subject in such an environment, and to control the initial conditions of the experiment, a central visual fixation spot in front of the subject indicates the start of a sound-localization trial. When this spot disappears, a peripheral sound appears, and the task of the subject is to orient the eyes as fast and as accurately as possible toward the perceived sound location. In a pure eye-movement task

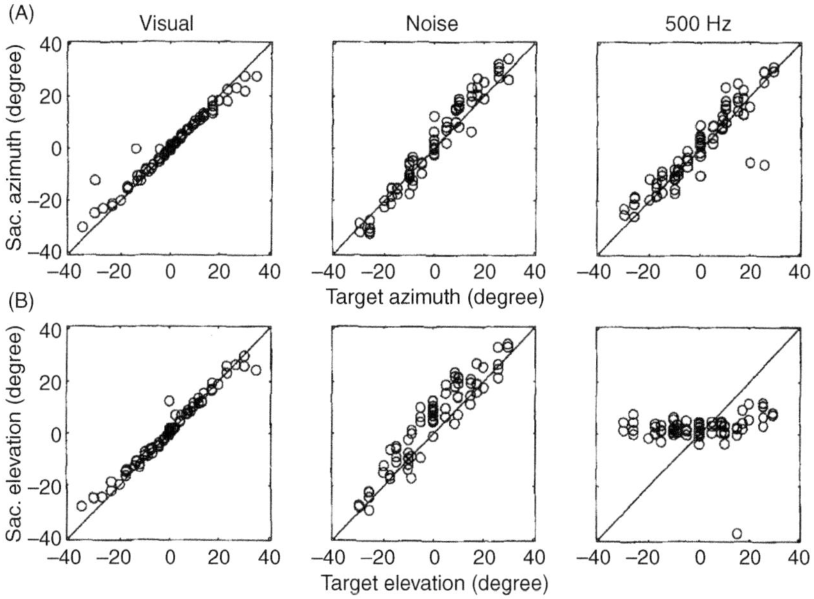

FIGURE 8.13 Saccadic eye-movement components in azimuth (A) and elevation (B) elicited by visual stimuli, Gaussian white noise (500 ms duration), and pure tones (500 Hz) presented within the 2D oculomotor range (35 degrees around the central fixation point) in complete darkness. Note the high accuracy of noise-evoked responses in all directions, comparable to vision, and the absence of a significant elevation response for the tone. *(Source: From Frens and Van Opstal, 1995, with kind permission.)*

the head is not allowed to move, which can be achieved, for example, by a neck rest or a bite bar.

To study the acoustic factors that determine sound-localization performance, the sound can be given certain spectral or temporal properties, or varying intensities, which may all be systematically manipulated.

Spectral content: As an example, Fig. 8.13 compares the results of an eye-movement localization experiment when the acoustic manipulation was the sound–source spectrum. When the sound is a broadband noise of sufficiently long duration (central panels), response accuracy is almost as high as for visual-evoked eye movements toward brief light flashes at the same locations (response gains are close to one for all response components). The major difference between the sound-evoked responses and the visual saccades is the larger variability of the former. Another interesting feature to observe is that the elevation components of the sound-evoked saccades (center-bottom panel) have a larger variability than their azimuth components, which reflects the unique Cartesian azimuth-elevation organization of the auditory system; no such difference is observed for the visual saccades.

When the sound spectrum is reduced to a pure tone (in this case a low-frequency tone of 500 Hz, right-hand panels) it is immediately evident that the auditory system extracts the azimuth and elevation components of sound locations through independent neural pathways. As the azimuth components for the tone responses are indistinguishable from the broadband noise-evoked saccades, the elevation components show that the auditory system has no clue about the real target elevation: both the gain and the stimulus-response correlation are indistinguishable from zero, while the bias is a few degrees upward. This bias turns out to be frequency (and subject) specific (Frens and Van Opstal, 1995):

$$R_{\mathrm{AZI}}^{\mathrm{tone}} = aT_{\mathrm{AZI}} + b \quad \text{and} \quad R_{\mathrm{ELE}}^{\mathrm{tone}} = d\left(f_{\mathrm{tone}}\right) \tag{8.28}$$

The explanation for the latter finding is, of course, that a pure tone cannot provide a unique cue for sound–source elevation, as any tone (amplitude A_0, frequency f_0), can only probe a single point of its associated HRTF (chapter: Acoustic Localization Cues). In the frequency domain, this operation amounts to:

$$\mathrm{HRTF}\left(f;\, \varepsilon_0\right) A_0 \delta\left(f - f_0\right) = A_0 \mathrm{HRTF}\left(f_0;\, \varepsilon_0\right) \tag{8.29}$$

Although the perceived intensity of the tone may vary for different elevations, ε_0 (Fig. 7.15), that potential cue is entirely ambiguous because the perceived intensity is always an entangled mixture of the tone's own intensity, A_0, and the relative amplitude of the HRTF for that frequency and elevation [Eq. (8.29)].

It might be conceivable, however, that by adding more tones their joint variation in perceived intensity, *PI*, could be uniquely related to their change in elevation:

$$PI\left(f_0, f_1\right) = A_0 \mathrm{HRTF}\left(f_0;\, \varepsilon_0\right) + A_1 \mathrm{HRTF}\left(f_1;\, \varepsilon_0\right) \Rightarrow f\left(\varepsilon_0\right) \tag{8.30}$$

How many probing points will be needed to allow for appropriate localization of a sound's elevation remains an interesting question that so far has yielded relatively little attention (see, eg, Kulkarni and Colburn, 1998). We have observed in our own lab (unpublished results) that when higher harmonics (2*f*, 3*f*, …) are added to a tone, even at very low relative intensities compared to the ground frequency, the auditory system immediately uses the added piece of information to improve its zero-order (constant bias) localization estimate. The result is that saccade elevation responses start to correlate significantly with the target coordinates, albeit that the response gain will still be very low (below 0.2–0.3).

The frequency dependence and idiosyncrasy of the bias in Eq. (8.28) is thought to be a manifestation of the way in which the auditory system extracts elevation from (idiosyncratic) HRTFs. We will come back to these issues in chapter: Impaired Hearing and Sound Localization, where we discuss (Bayesian) models of sound localization.

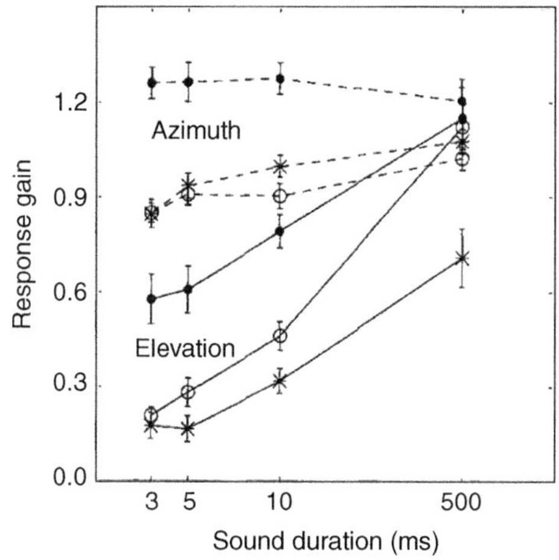

FIGURE 8.14 **The elevation gain of eye-movement responses depend strongly on the duration of the broadband GWN stimuli.** In contrast, saccade azimuth components do not systematically vary with sound duration. *(Source: After Frens and Van Opstal, 1995.)*

Sound duration and level: Sound-localization accuracy to broadband stimuli depends systematically on the *duration* of the sound source (Frens and Van Opstal, 1995; Hofman and Van Opstal, 1998), and on the *sensation level* (perceived, or proximate, intensity) of the stimulus (Macpherson and Middlebrooks, 2000; Vliegen and Van Opstal, 2004). Also, in this case, the effects are quite different for azimuth versus elevation response components.

The accuracy and precision of sound–source azimuth are virtually insensitive to varying stimulus durations (from as short as 1–3 ms, and up) and sound levels: whenever the sound intensity exceeds the detection threshold, its azimuth is accurately determined.

The accuracy of the elevation components, however, depends strongly on these acoustic parameters. Frens and Van Opstal (1995) and Hofman and Van Opstal (1998) showed that the elevation gain started to decrease systematically with decreasing duration of (60–70 dBA) broadband noise bursts, from about 20–30 ms downward (Fig. 8.14). Similar effects of the temporal structure of broadband sounds result when varying the duty cycle of burst trains, and the sweep duration of broadband FM sweeps. Hofman and Van Opstal (1998) proposed that the estimate of stimulus elevation, apart from a neural correlation stage {as described in chapter: Acoustic Localization Cues, Eqs. [(7.41)–(7.42)]}, also involved a neural *integration* stage in which the auditory system acquires acoustic (spectral) evidence by averaging across multiple short time windows of about 5 ms duration. The neural representation of the stimulus-associated HRTF would thus gradually improve over time, because the signal-to-noise ratio resulting from each window in itself would be insufficient to provide the fine spectral detail needed to estimate the elevation angle.

TABLE 8.2 Summary of The Effects of Sound Duration (Short/Long) and Sound Level (Low/High) on Elevation Gain, as Predicted by the Neural Integration and Cochlear Adaptation Models

	Effect of duration/level			
Model	Short		Long	
Neural integration	Low: LG	High: LG	Low: HG	High: HG
Cochlear adaptation	Low: HG	High: LG	Low: HG	High: HG

LG, low elevation gain; HG, high elevation gain; box highlights conditions for which the model predictions differ. See also Fig. 8.15.

Macpherson and Middlebrooks (2000) forwarded an alternative theory for the "multiple view," neural integration model. They argued that very brief and loud noise bursts would generate unstable cochlear responses that hamper the spectral estimation process. They thus predicted that cochlear adaptation (and the associated low gains) would be absent for low sound levels.

Table 8.2 summarizes the predictions of the effects of simultaneous manipulation of sound duration and sound level for the neural integration (NI) and cochlear adaptation (CA) models. The condition for which the models differ in their predictions is a short-duration sound burst presented at low levels: for such sounds the NI model still predicts a low gain, whereas the CA model predicts (near-)normal, high gain values.

A subsequent experimental test of these predictions by Vliegen and Van Opstal (2004) showed that both mechanisms (cochlear adaptation and neural integration) may actually be needed to explain the results (Fig. 8.15). The effect of sound level on the elevation gain resulted to be more complex than accounted for by either model: the data showed a nonmonotonic relationship: for both low and high sound levels the gain is low, reaching a maximum at intermediate sound levels, ~50 dBA. This nonmonotonic level effect on the elevation gain (possibly attributable to cochlear processes) combines with the effects of the sound's temporal structure (explained by neural integration).

Fig. 8.16 presents a neurocomputational model that explains the experimental data for the localization of sound–source elevation described so far. The model contains a nonmonotonic cochlear stage that shows adaptation at high sound levels, as well as a neural integration stage that averages the multiple short-term measurements (symbolized by the on–off switch) of the cochlear spectral signal to build a gradually improving spectral estimate. However, when the signal-to-noise level of the cochlear output decreases (because of the low responses for both very soft and loud sounds), the spectral estimate has a lower reliability too, resulting in a lower elevation gain for both situations.

Reaction times: Reflexive saccadic eye movements can provide interesting insights into the neural processes underlying sound localization when analyzing their onset reaction times to different types of sounds. The idea is that the more

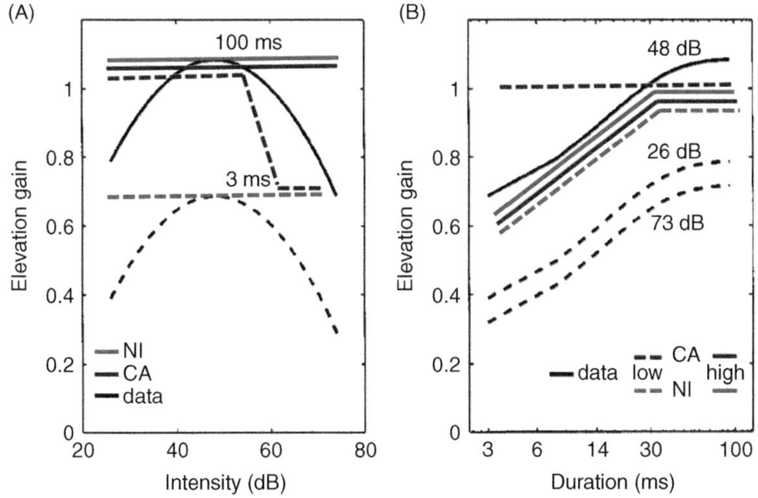

FIGURE 8.15 Summary of the effects on elevation gain of sound-evoked eye movements to noise bursts of varying intensity and duration. *Black*, data; *red*, NI model; *blue*, CA model. Neither model accounts for all data: the observed gain effect of sound level is nonmonotonic. (A) Elevation gain as function of sound level for different stimulus durations. (B) Gain as function of sound duration for different levels. *(Source: After Vliegen and Van Opstal, 2004.)*

difficult the sound-localization task, the longer the neural processing time, and hence the reaction time of the eye movement that is evoked by the sound. For example, if the target sound is barely audible within the environmental background noise, localization is more difficult, and the reaction time will be longer than when the sound has a high signal-to-noise ratio (SNR) with respect to the background.

But this effect of task difficulty on reaction times can also become apparent in quite subtle acoustic manipulations, such as the ones described above (spectral content, temporal properties, or spectral–temporal modulations). When the gain of the stimulus-response relation for a certain sound is lower than for some standard hearing condition, it may be expected that the task is more difficult (has a higher cognitive load, or requires more processing stages) than for the

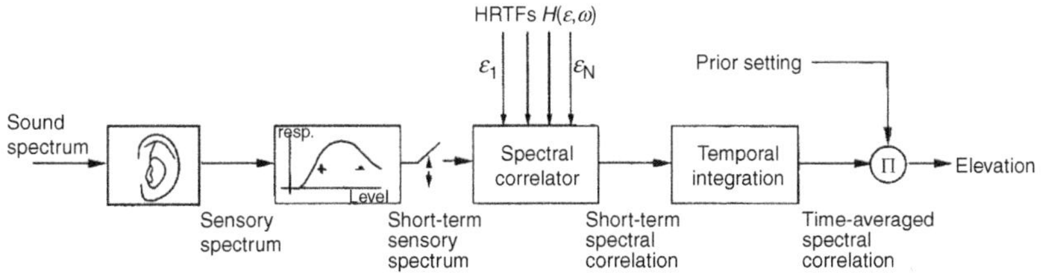

FIGURE 8.16 Neurocomputational model that explains the neural estimation of sound elevation from spectral correlations with stored HRTF representations. It accounts for the influence of sound level (CA) and duration [short-term, few milliseconds "look" windows (switch), and long-term temporal integration, NI]. "Prior setting" includes nonacoustic factors that affect localization (eg, expectation, bias). *(Source: After Vliegen and Van Opstal, 2004.)*

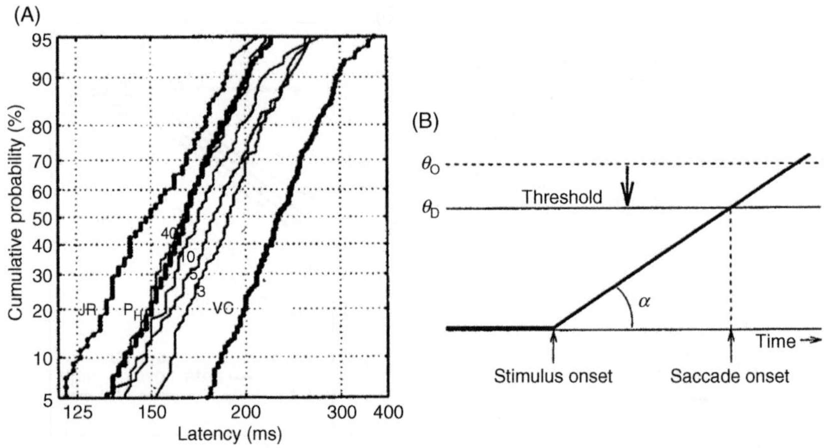

FIGURE 8.17 (A) Effect of noise-burst duration on the cumulative distributions (on a probit scale) of saccade reaction times (RT). *Solid lines*: 500 ms noise bursts for three subjects. Note the systematic increase of RT with decreasing burst duration (data from P_H are highlighted), which suggests an increased task difficulty for $D = 10$, 5, and 3 ms. (B) A diffusion model with diffusion rate, α, can explain the distributions of saccade RT as a stochastic variable under the influence of a varying criterion (the threshold, $\theta_0 \rightarrow \theta_D$). The latter depends on the acoustic manipulation. *(Source: After Hofman and Van Opstal, 1998.)*

standard. Indeed, Hofman and Van Opstal (1998) demonstrated robust systematic effects on the distributions of saccade reaction times when varying the temporal properties of broadband sounds: the duration of noise bursts, the duty cycle of burst trains, and the speed of fast frequency sweeps. Fig. 8.17A illustrates the results of the eye-movement localization task (see also above, for the effects on gain) in so-called *probit scale* format, for which a sigmoid Gaussian cumulative distribution of reaction times would be transformed into a straight line.

8.5.1 Binaural Spectral Weighting

So far, we described and modeled the mapping of spectral HRTFs into an elevation estimate as an entirely *monaural* process. Yet, the sound energy from a lateral sound source enters both ears. This is particularly obvious for sources on the midsagittal plane, which stimulate the two ears with equal power. How do we know whether the elevation estimate of a listener is due to a monaural process, or not? For typical conditions (the normal, healthy, binaural subject) the HRTFs of the two pinnae are very similar, with only minor differences in their spectral detail. In this situation it is not possible to distinguish whether the elevation estimate of the listener results from a monaural (ear-dominance) or a binaural (weighting) neurocomputational process.

To study potentially monaural versus binaural aspects of sound localization in elevation it would be necessary to impose acute and considerable asymmetries in the pinnae (eg, by inserting a mold in one of the pinnae, which fills the pinna cavities and removes, or radically changes, the original spectral cues of that ear) and measure the effects of this manipulation on localization performance in elevation as function of the lateral position of the sound source. Suppose that the mold renders localization with the altered pinna impossible. Then, if the subject can only listen with the manipulated ear, her responses will be uncorrelated with source elevation (like the performance for a pure tone, shown in Fig. 8.13).

We can readily envisage the different predictions for monaural versus binaural spectral estimation mechanisms of such an experiment, as there are three possible scenarios to consider in an acute localization experiment:

1. *Lateralized-specific ear dominance.* Localization responses on the normal hearing side would be unaffected (normal response behavior), and responses on the manipulated side would be as bad as under monaural hearing of that ear, as it remains the dominant ear with or without manipulation.
2. *Full ear dominance.* The elevation percept is then determined by only one of the ears, regardless the laterality of sounds. The results will be quite dramatic: no localization is possible at either side when the dominant ear was manipulated, or completely normal localization results on both sides when the nondominant ear contained the mold.
3. *Binaural weighting.* When the spectral information of both ears is centrally integrated to decide on the final elevation estimate, the manipulated ear affects elevation responses also on the contralateral side (leading to a decreased gain). In turn, the normal, nonmanipulated ear improves localization responses on the manipulated side (increased gain).

Studies by Morimoto (2001), Hofman and Van Opstal (2003), and Van Wanrooij and Van Opstal (2005) have provided strong evidence for the third mechanism (Fig. 8.18). When sounds are presented in the midsagittal plane, the interaction is strongest, thereby gradually decreasing for lateral positions on either side. The interaction effect on the elevation estimate extend to almost 60 degrees into the contralateral hemifield, in which the azimuth angle of the sound serves as a weighting factor. To quantitatively describe these effects, the gain and bias of the elevation regression equation [Eq. (8.19)] can be taken to depend on azimuth (Hofman and Van Opstal, 2003):

$$R(\varepsilon)=c(\alpha)T(\varepsilon)+d(\alpha) \tag{8.31}$$

When we assume a first-order dependence of the regression coefficients on azimuth, Eq. (8.31) becomes (Exercise 8.6) a multiple linear regression model. It is called a *"twist" model*, as it contains a second-order term of the 2D target coordinates:

$$R(\varepsilon)=p+qT(\varepsilon)+rT(\alpha)+sT(\alpha)T(\varepsilon) \tag{8.32}$$

Parameters r and s express the azimuth dependence of the regression coefficients c and d. To allow for a direct quantitative comparison of the

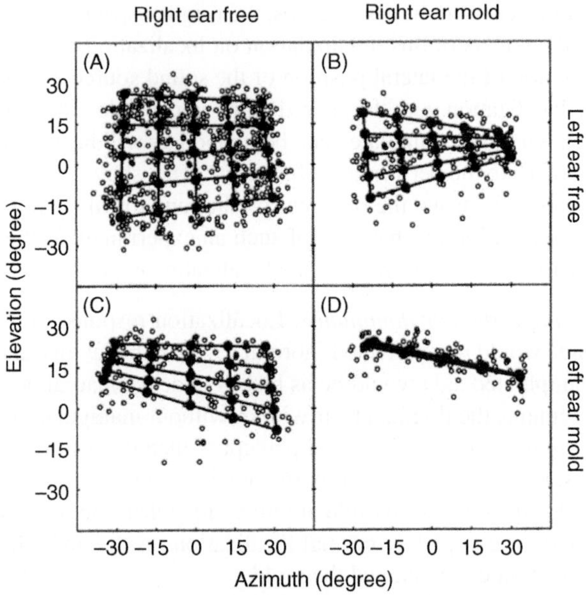

FIGURE 8.18 Binaural weighting of spectral cues. Eye-movement localization data (*open dots*) to broadband and high-pass filtered (>3 kHz) GWN, in which a listener was equipped with a mold in either ear (B, C), or in both ears (D). Data in (A) are the control localization data (no molds). *Filled dots* with connecting lines: least squares fits of Eqs. (8.33) and (8.34) to the data. *(Source: From Hofman and Van Opstal, 2003, with permission.)*

elevation-independent (p, r) versus the elevation-dependent regression parameters (q, s), Eq. (8.32) is transformed into the following normalized form:

$$R(\varepsilon)=\left[p+\Delta r\frac{T(\alpha)}{T(\alpha_{\mathrm{MAX}})}\right]+\left[q+\Delta s\frac{T(\alpha)}{T(\alpha_{\mathrm{MAX}})}\right]T(\varepsilon) \tag{8.33}$$

with α_{MAX} the maximum azimuth eccentricity of targets used in the experiment (which would be about 30 degrees for eye movements, but can extend over 90 degrees for eye–head gaze shifts). In Eq. (8.33), Δr and Δs are the so-called "*bias drift*" and "*gain drift*" parameters, respectively, which express the elevation-independent and elevation-dependent azimuth effects of the subject's responses. A similar normalized relation can be used for the binaural azimuth components of the responses.

$$R(\alpha)=\left[k+\Delta m\frac{T(\varepsilon)}{T(\varepsilon_{\mathrm{MAX}})}\right]+\left[l+\Delta n\frac{T(\varepsilon)}{T(\varepsilon_{\mathrm{MAX}})}\right]T(\alpha) \tag{8.34}$$

Fig. 8.18 shows the results of applying this 2D binaural integration model on eye-movement localization data within the oculomotor range, when the ears were left untouched (normal hearing; panel A), or when molds were applied that perturbed the spectral cues of either (panels B, C), or both (panel D) ears (see also Exercise 8.5).

The results of acute monaural spectral cue perturbations are quite clear, and consistent among different studies: there is considerable binaural integration of the spectral cues in order to estimate the elevation of a sound source. The obvious follow-up question is how this integration process might be implemented in the human auditory system. There is no obvious answer to this question when we consider the (monaural) functional model of Fig. 8.16. Where in the binaural extension of that model would the binaural weighting have to take place? Here it will be useful to distinguish two potential conceptual schemes for estimating sound–source elevation in a binaural model (Exercise 8.6):

1. The WM scheme, in which first the spectral cues from either ear are binaurally integrated into a weighted spectrum (W), and subsequently mapped (M) into a spatial estimate of elevation.
2. The MW scheme, in which first each of the monaural spectra are mapped into a left- and right-ear spatial estimate (M), after which a spatial weighting is performed (W).

Interestingly, both stages (spectrum to space mapping, and binaural weighting) are *nonlinear*, so that changing the order in which these processes are implemented in the auditory brain (WM versus MW) yields essentially different results. Therefore, in principle, it will be possible to dissociate these schemes experimentally by performing clever input–output measurements (see chapter: Nonlinear Systems, which deals with nonlinear systems analysis) (Fig. 8.19).

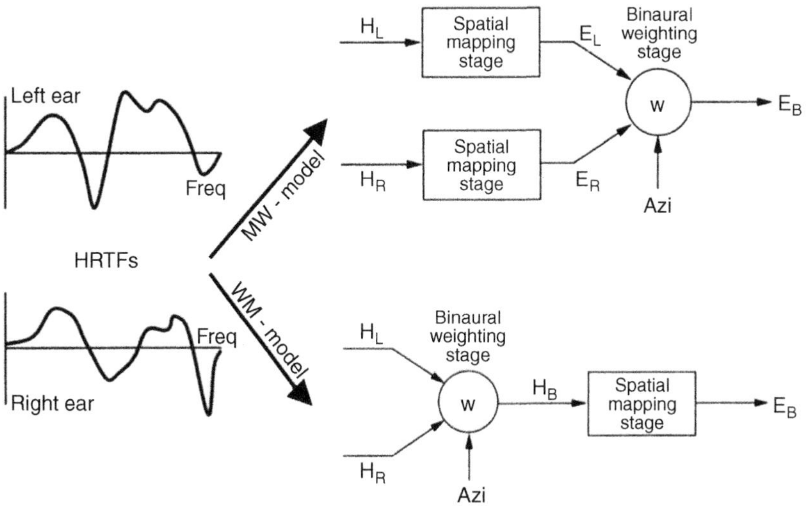

FIGURE 8.19 Two conceptual models may potentially explain the binaural weighting of spectral cues. The MW model (first spatial mapping, then spatial weighting) versus the WM model (first spectral weighting, then spatial mapping). H, Spectral cues; E, elevation estimate; W, binaural weighting. *(Source: After Hofman and Van Opstal, 2003.)*

The results of Fig. 8.18 do not yet allow for a firm dissociation of these models, although monaural lesion studies at the level of the DCN output in cat (May 2000) may have provided some support for the MW scheme. If so, a potential candidate for the monaural spectral-to-space mapping might be speculated to be embedded in the DCN to ipsilateral inferior colliculus (IC) projections (further discussed in chapter: Coordinate Transformations), whereas the bilateral spatial weighting stage could potentially be implemented at the level of the midbrain superior colliculus (SC). The latter is a gaze-control structure that is known to be organized in spatial sensory coordinates, and it occupies a pivotal role in the programming and generation of eye–head gaze shifts to sensory stimuli, including sounds (the SC is described in more detail in chapter: The Midbrain Colliculus).

8.6 VIRTUAL ACOUSTICS

Up to now we have described sound-localization experiments in which sound sources are presented in the external space (usually the free field). Under such hearing conditions, all acoustic cues for azimuth and elevation (and potentially for distance, see chapter: Acoustic Localization Cues) are always intrinsically coupled and consistent. For example, a given binaural time difference measured within the low frequency spectrum of the sound source, will point to the same azimuth angle as the binaural level differences obtained for the high-frequency part of the sound spectrum, and also the (binaural) spectral cues will point to a unique location in the elevation direction.

In virtual acoustics, the experiment uses headphones to study sound localization over the full 2D directional space. For a valid simulation of the natural free-field hearing condition, the natural coupling of acoustic cues should be faithfully mimicked. Once this is achieved, it is then possible to systematically dissociate (and arbitrarily change) the mutual cue relationships in order to study their importance in isolation. Such experiments are not possible in free-field hearing. So how is natural hearing simulated over headphones? For this to happen, it suffices to apply the concepts of linear systems theory (chapter: Linear Systems) because all relevant acoustic and electric transformations between the signal and the eardrum can be described by linear filters (Wightman and Kistler, 1989a).

Fig. 8.20 illustrates the virtual acoustics procedure. The idea is the following: in the free field, a signal $x_1(t)$ is produced to drive a speaker, which leads to acoustic signals, $y_{L1}(t)$ and $y_{R1}(t)$, at the eardrums of the left and right ear. When using headphones, signals $h_{L2}(t)$ and $h_{R2}(t)$ drive the two speakers, which produce eardrum signals, $y_{L2}(t)$ and $y_{R2}(t)$, respectively. The challenge is, to program the headphone signals such that the headphone's eardrum signals are identical to the free-field listening condition:

$$\begin{aligned} y_{2R}(t) &= y_{1R}(t) \\ y_{2L}(t) &= y_{1L}(t) \end{aligned} \tag{8.35}$$

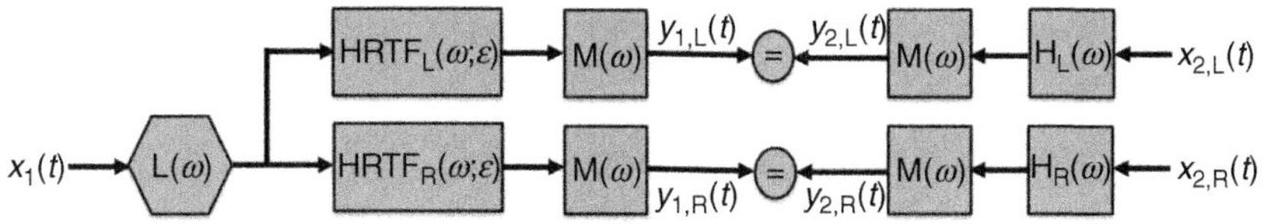

FIGURE 8.20 Procedure for generating a realistic headphone simulation of a free-field acoustic stimulus. The goal is to generate headphone signals $x_{2,R}$ and $x_{2,L}$ such that the measured signals at the eardrums, $y_{2,R}$ and $y_{2,L}$, are the same as $y_{1,R}$ and $y_{1,L}$ that result from free-field stimulation with sound x_1. M(ω) is the transfer characteristic of the probe microphone system.

In Exercise 8.7 the reader may verify (thereby referring to Fig. 8.20) that the required transfer of free field to headphone signals in the frequency domain is then given by the following equation:

$$T_{L,R}(\omega) = \frac{L(\omega)\mathrm{HRTF}_{L,R}(\omega;\varepsilon)}{H_{L,R}(\omega)} \tag{8.36}$$

where $L(\omega)$ is the free-field speaker transfer characteristic, $\mathrm{HRTF}_{L,R}(\omega;\varepsilon)$ are the HRTFs for the left and right ear, and correspond to the sound's azimuth-elevation coordinates, and $H_{L,R}(\omega)$ are the headphones-to-eardrum speaker characteristics.

Validation of the virtual acoustics method, in which subjects localize either virtual or real free-field acoustic stimuli, demonstrates that when implemented correctly, there is no difference in localization performance (Wightman and Kistler, 1989b). Subjects also report realistic externalized percepts of the virtual acoustics stimuli, provided that their own HRTFs were used to produce the virtual sounds.

Note that the central idea of virtual acoustics is that the only signal on which the auditory system will base its spatial judgments is the acoustic pressure at the eardrums. Whether these signals are produced by a free-field sound source or by any other (technical) means would be immaterial for the sound-localization system. Whenever the signals are identical, the spatial percept will be identical too.

However, this assumption may only be partially correct: when the auditory system would also rely on other, nonacoustic signals, the assumption breaks down, and the simulation will fail. We will see in later chapters that there are many such signals. Think, for example, of eye- and head position signals, eye- and head movement signals, visual signals, prior information (expectation, biases), etc.

As a result, virtual acoustics will typically work well when nonacoustic signals can somehow be excluded from the perceptual task (eg, by preventing eye- and head movements, or by eliminating possible visual interference). If, however, subjects can freely move around in their environment, realistic virtual acoustics according to the procedure of Fig. 8.20 will definitely not work.

8.7 EXERCISES

Problem 8.1:

(a) Show that the pdf, $P(x)$ of the sum (and difference) of two independent random variables, x_1 and x_2, each with pdf's $P_1(x_1)$ and $P_2(x_2)$, respectively, is given by the convolutions

$$x = x_1 + x_2 \Rightarrow P(x) = \int_{-\infty}^{\infty} P_1(s) P_2(x-s)\,ds$$

$$x = x_1 - x_2 \Rightarrow P(x) = \int_{-\infty}^{\infty} P_1(s) P_2(x+s)\,ds$$

Hint: Use the property of independence (probabilities multiply), and the constraint on x_1 and x_2 with respect to x.

(b) Apply this result to obtain the result of Eq. (8.2) for two Gaussian random variables with the same variance, σ, but with means of 0 and μ, respectively.

(c) Same as in (b.), but now for two Gaussians with different variances and means.

Problem 8.2: Transform the general Gaussian (mean μ, standard deviation σ) into the standard form of (mean zero, and standard deviation one) to arrive at Eq. (8.7) for the cumulative distribution.

Problem 8.3:

(a) From Fig. 8.2B, and the definition of the standard cumulative normal distribution, $\Phi_G(x)$, derive Eq. (8.12).

(b) Derive Eq. (8.13)

(c) Evaluate at $\beta = 1$ the relation between hit rate and false alarm rate.

Problem 8.4:

Derive Eq. (8.15).

Problem 8.5:

(a) Verify Eq. (8.32), and extend the twist model to include the azimuth response components.

(b) Identify the regression parameters of Eqs. (8.33) and (8.34) with the fit results (the connected filled dots) of the data in Fig. 8.18. In particular, how do the bias drift parameters for elevation (Δr) and azimuth (Δn) become apparent in the fit, and what about the gain drifts (Δs and Δn)?

(c) How would the response grids in Fig. 8.18 change when the drift gain parameters were both zero? What if only the bias parameters were zero?

Problem 8.6: Extend Fig. 8.16 to binaural models for which the two different weighting schemes (WM and MW) are incorporated. Where would azimuth-dependent spatial information enter the system in either scheme?

Problem 8.7: Make use of Fig. 8.20 to derive Eq. (8.36).

REFERENCES

Frens, M.A., Van Opstal, A.J., 1995. A quantitative study of auditory-evoked saccadic eye movements in two dimensions. Exp. Brain Res. 107, 103–117.

Goossens, H.H.L.M., Van Opstal, A.J., 1997. Human eye-head coordination in two dimensions under different sensorimotor conditions. Exp. Brain Res. 114, 542–560.

Harrison, J.M., Irving, R., 1966. Visual and nonvisual auditory systems in mammals. Science 154, 738–743.

Hartmann, W.M., Rakerd, B., 1988. On the minimum audible angle – a decision theory approach. J. Acoust. Soc. Am. 85, 2031–2041.

Heffner, R.S., Heffner, H.E., 1992. Visual factors in sound localization in mammals. J. Comp. Neurol. 317, 219–232.

Hofman, P.M., Van Opstal, A.J., 1998. Spectro-temporal factors in two-dimensional human sound localization. J. Acoust. Soc. Am. 103, 2634–2648.

Hofman, P.M., Van Opstal, A.J., 2003. Binaural weighting of pinna cues in human sound localization. Exp. Brain Res. 148, 458–470.

Jay, M.F., Sparks, D.L., 1984. Auditory receptive fields in primate superior colliculus shift with changes in eye position. Nature 309, 345–347.

Kulkarni, A., Colburn, H.S., 1998. Role of spectral detail in sound-source localization. Nature 396, 747–749.

Lewald, J., Karnath, H.O., 2001. Sound lateralization during passive whole-body rotation. Eur. J. Neurosci. 13, 2268–2272.

Macpherson, E.A., Middlebrooks, J.C., 2000. Localization of brief sounds: Effects of level and background noise. J. Acoust. Soc. Am. 108, 1834–1849.

May, B.J., 2000. Role of the dorsal cochlear nucleus in the sound-localization behavior of cats. Hearing Res 148, 74–87.

Middlebrooks, J.C., Green, D.M., 1991. Sound localization by human listeners. Ann. Rev. Psychol. 42, 135–159.

Mills, A.W., 1958. On the minimum audible angle. J. Acoust. Soc. Am. 30, 237–246.

Morimoto, M., 2001. The contribution of two ears to the perception of vertical angle in sagittal planes. J. Acoust. Soc. Am. 109, 1596–1603.

Oldfield, S.R., Parker, S.P., 1984. Acuity of sound localization: a topography of auditory space. II. Pinna cues absent. Perception 13, 601–617.

Perrott, D.R., Saberi, K., 1990. Minimum audible angle thresholds for sources varying in both azimuth and elevation. J. Acoust. Soc. Am. 87, 1728–1731.

Perrott, D.R., Ambarsoom, H., Tucker, J., 1987. Changes in head position as a measure of auditory localization performance: auditory psychomotor coordination under monaural and binaural listening conditions. J. Acoust. Soc. Am. 82, 1637–1645.

Populín, L.C., 2006. Monkey sound localization: head-restrained versus head-unrestrained orienting. J. Neurosci. 26, 9820–9832.

Van Barneveld, D.C.P.B.M., Van Opstal, A.J., 2010. Eye position determines audio-vestibular interactions during whole-body rotation. Eur. J. Neurosci. 31, 920–930.

Van Grootel, T.J., Van der Willigen, R.F., Van Opstal, A.J., 2012. Experimental test of spatial updating models for monkey eye-head gaze shifts. PLoS One 7 (10), e:47606.

Van Opstal, A.J., Van Gisbergen, J.A.M., 1989. Scatter in the metrics of saccades and properties of the collicular motor map. Vision Res. 29, 1183–1196.

Van Wanrooij, M.M., Van Opstal, A.J., 2004. Contribution of head shadow and pinna cues to chronic monaural sound localization. J. Neurosci. 24, 4163–4171.

Van Wanrooij, M.M., Van Opstal, A.J., 2005. Relearning sound localization with a new ear. J. Neurosci. 25, 5413–5424.

Vliegen, J., Van Opstal, A.J., 2004. The influence of duration and level on human sound localization. J. Acoust. Soc. Am. 115, 1705–1713.

Whittington, D.A., Hepp-Reymond, M.C., Flood, W., 1981. Eye and head movements to auditory targets. Exp. Brain Res. 41, 358–363.

Wightman, F.L., Kistler, D.J., 1989a. Headphone simulation of free-field listening. I: Stimulus synthesis. J. Acoust. Soc. Am. 85, 858–867.

Wightman, F.L., Kistler, D.J., 1989b. Headphone simulation of free-field listening. II: Psychophysical validation. J. Acoust. Soc. Am. 85, 868–878.

Zahn, J.R., Abel, L.A., Dell'Osso, L.F., Daroff, R.B., 1979. The audio-ocular response: intersensory delay. Sens. Proc. 3, 60–65.

Zambarbieri, D., Schmid, R., Magenes, G., Prablanc, C., 1982. Saccadic responses evoked by presentation of visual and auditory targets. Exp. Brain Res. 47, 417–427.

Zwiers, M.P., Van Opstal, A.J., Cruysberg, J.R.M., 2001. Two-Dimensional sound-localization behavior of early-blind humans. Exp. Brain Res. 140, 206–222.

Zwiers, M.P., Van Opstal, A.J., Paige, G.D., 2003. Plasticity in human sound localization induced by compressed spatial vision. Nat. Neurosci. 6, 175–181.

Chapter 9

The Gaze-Orienting System

9.1 INTRODUCTION

In the previous chapter we introduced the saccadic eye movement as a reliable continuous pointer to study human sound localization behavior. We argued, on the basis of comparative behavioral and neurophysiological studies in the literature, that the gaze-orienting system, apart from subserving visuomotor control (eye movements, combined eye–head movements, and eye–head-body orienting movements), might also be tightly connected to the sound-localization system, as these systems share a common goal:

- either to orient the fovea as fast and as accurately as possible to the peripheral target of interest for further visual inspection (a potential moving food source, a prey)
- or to get away as fast as possible from a potential threat (predator, danger)

In this chapter we will go into more depth for the predator's (cat, monkey, or human) gaze-orienting system, in order to identify the problems for which it evolved, see (and model) how it is functionally organized, and therefore better understand how gaze-orienting and spatial percepts of auditory and visual targets are intrinsically integrated in our localization behaviors.

So what seems to be the major problem for the visual systems of foveate animals like humans? The answer is that their retina is highly inhomogeneous, with a small area centralis, where the density of (color sensitive) photoreceptors (cones) with small visual receptive fields is extremely high (in the center of the fovea about 200.000 cones/mm^2; Fig. 9.1). The cell density rapidly decreases toward peripheral retinal locations, and at the same time the receptive field size of their target cells (retinal ganglion cells) increases. If the density of photoreceptors in the central fovea is σ_0 cells/mm^2, then the cell density, $\sigma(r)$, appears to roughly depend on retinal eccentricity, r (in mm), relative to the fovea (at $r = 0$), according to (Fig. 9.1):

$$\sigma(r) = \frac{n_0}{(\beta r^2 + \delta)} \quad \text{with} \quad \frac{n_0}{\delta} \equiv \sigma_0 \tag{9.1}$$

At $r \sim 0.5$ mm (corresponding to $\sim$1.75 degrees visual angle, given that the human eye has a diameter of about 2.4 cm) the cone density has already dropped

The Auditory System and Human Sound-Localization Behavior. http://dx.doi.org/10.1016/B978-0-12-801529-2.00009-X

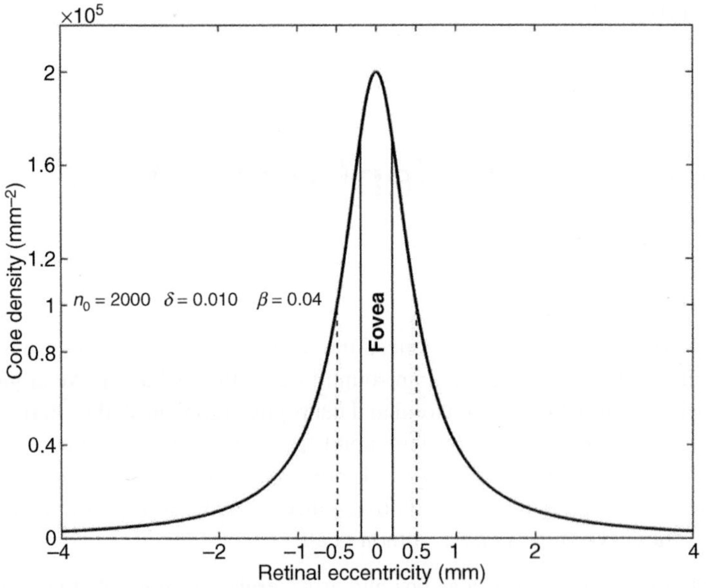

FIGURE 9.1 Cone density of the human retina decreases according to Eq. (9.1) with retinal eccentricity. In humans, the fovea extends to about 1 degree of visual angle, which corresponds to a circular area with a diameter of approximately 0.4 mm (solid lines; eye diameter ~2.4 cm). At 0.5 mm from the center (1.75 degrees) cell density has decreased by 50% (dashed).

by about 50% and 4 mm away from the fovea (at about 10 degrees visual angle), the density is <5000 mm^{-2} (Curcio et al., 1990). From these numbers we may estimate the parameter values of Eq. (9.1) at $\delta = 0.01$, $n_0 = 2000$, and $\beta \sim 0.04$ [δ only prevents $\sigma(r) \rightarrow \infty$ at $r = 0$, and can safely be ignored for $r > 0.1$ mm].

In addition, the radius of visual receptive fields of cone-sensitive retinal ganglion cells increases approximately linearly with their retinal eccentricity:

$$R_{\mathrm{RF}}(r) = \alpha r, \tag{9.2}$$

so that the area of a receptive field is proportional to r^2. As a result, the number of cones that define the receptive field of a retinal ganglion cell is approximately constant:

$$\left(\pi\alpha^2 r^2\right)\frac{n_0}{\left(\beta r^2 + \delta\right)} \approx \frac{\pi\alpha^2 n_0}{\beta} = N_0 \tag{9.3}$$

With the parameter values given previously, and $\alpha \sim 0.02$, the number of convergent photoreceptor cells on each ganglion cell is in the order of $N_0 = 50$–60.

Why saccades? The need to rapidly identify objects in the visual periphery requires that the visuomotor system has to direct the fovea as fast and as accurately as possible to the target, as beyond a few degrees of visual angle the retina cannot discriminate fine spatial detail. The saccadic eye-movement system fulfills this function: saccades are extremely fast (in a monkey, saccade peak velocities can reach 1300 degrees/s), accurate and precise. They are generated involuntarily at a rate of 3–4 saccades/s (>10,000/h). Each saccade provides the visual system with a brief "*snapshot*" of approximately 200 ms (the intersaccadic interval) of a small piece of high-acuity visual input. Our visual perception thus results from "gluing" many different snapshots into a coherent visual image of the environment. Our visuomotor behavior operates within this internally constructed visual world.

Yet, to program an accurate saccade, and to make the system this fast, is not trivial. Systems theory (see chapters: Linear Systems and Nonlinear Systems) helps us to identify a number of interesting problems, and to understand the mechanisms by which the central nervous system has tackled them. To mention a few:

1. The peripheral retina has poor spatial resolution (Fig. 9.1). Hence, there is considerable uncertainty (*sensory noise*) about the exact location of a peripheral target.
2. A target overshoot requires the system to program a saccade in the opposite direction, which involves a (time consuming) transfer of signals to the contralateral hemisphere.
3. The oculomotor plant (eye muscles, globe, and surrounding fatty tissues) is a very sluggish mechanical system with considerable viscosity at high saccadic speeds. Its effective time constant (about 200 ms) by far exceeds the duration of a typical saccade (<70 ms).
4. Due to the high saccadic speeds, visual feedback will be far too slow to guide their execution. Saccades may thus be either *ballistic* (open loop), or controlled by internal feedback.
5. In motor control, the requirements to be both *fast and accurate* are mutually exclusive. Speed-accuracy trade off should optimize these opposing constraints.
6. Each saccade induces an equally fast and opposite shift of the retinal image. Despite the rapid jumps of visual input, we perceive a stable visual environment. This phenomenon (*transsaccadic integration*) calls for online eye-movement feedback, and fast coordinate transformations.
7. Since the visual periphery is uncertain and noisy, the visuomotor system has to make clever decisions on where to make the next saccade (and prevent returning to locations already explored).

The saccadic system implements several strategies to efficiently tackle these problems, and to execute saccades with remarkable speed, efficiency, accuracy, and precision. In the later section we describe the mechanisms by which the system achieves such near-optimal behavior.

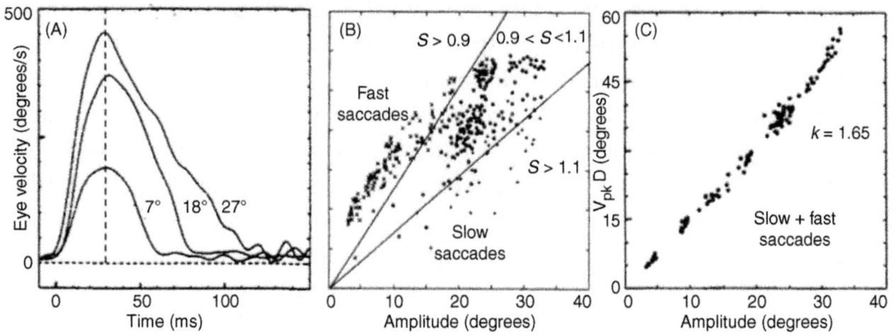

FIGURE 9.2 **Kinematic properties of saccadic eye movements.** (A) Three example velocity profiles for $R = 7$, 18, and 27 degrees, respectively. Note constant acceleration time [dashed line; Eq. (9.4)c], and an increase of skewness with saccade amplitude. (B) Main sequence for fast and slow saccades. Eq. (9.4)b is violated for slow saccades. The two lines demarcate three different skewness sectors [Eq. (9.5)]. (C) All saccades obey the very strict relation of Eq. (9.4)d. *(Source: After Van Opstal and Van Gisbergen, 1987.)*

9.2 SACCADIC EYE MOVEMENTS

Let's first consider the defining property of a saccadic eye movement, which is based on its kinematic behavior, as described by the "*main sequence*" characteristics (Fig. 9.2). The main sequence [Eq. (9.4)a,b] relates saccade duration (D) and peak eye velocity (V_{pk}) to its amplitude (R). However, saccade kinematic properties are further specified by the *shape* of saccade velocity profiles. This property can be quantified by the *skewness* parameter, S, which is determined by the third-order moment of the velocity function. (the nth-order moment of function $f(t)$ on $t \in (0, \infty)$ is defined by $\mu_n = \int_0^{\infty} (t - \mu_1)^n f(t)dt$). Taken together, the following five relationships parameterize the saccade kinematics:

$$\begin{array}{ll} (a) & D = aR + b \\ (b) & V_{pk}(R) = V_{max}\left[1 - \exp(-\mu R)\right] \\ (c) & T_{pk}(R) = \text{constant} \\ (d) & V_{pk}D = kR \\ (e) & S = cD + d \end{array} \tag{9.4}$$

Eq. (9.4)a,b state that saccade duration increases in a straight-line fashion with its amplitude, while peak velocity will saturate at V_{max} for large saccades. Eq. (9.4)c expresses the observation that the time to peak-velocity (the saccade acceleration phase) is approximately constant (about 20–25 ms), irrespective of saccade amplitude. Eq. (9.4)d describes a remarkable fact of saccades that directly relates to the shape of their velocity profiles. This relation is very tight (with correlations $r \sim 0.95$, and higher; Fig. 9.2C), and even holds for abnormally slow saccades (due to fatigue or drugs, or made to remember target locations), when the main sequence [Eq. (9.4)a,b] breaks down (Fig. 9.2B; Van Opstal and Van Gisbergen, 1987). For human saccades: $k \sim 1.6$–1.7 (dimensionless; Exercise 9.2). Finally, Van Opstal and Van Gisbergen (1987) found that saccade skewness related better to saccade duration, D, than to saccade amplitude, R [Eq. (9.4)e].

Combining Eq. (9.4)d,e yields a summary relation for the three kinematic quantities that unifies all saccades, slow and fast, in the amplitude–velocity plane (oblique lines in Fig. 9.2B):

$$V_{\mathrm{pk}} = \frac{kc}{S-d} R \tag{9.5}$$

Now that we have seen some properties of saccadic eye movements (the system's output), it's time to apply the concepts of *linear systems theory* (see chapter: Linear Systems) to gain some further insight into this system. The first thing to note is that a saccade is evoked by a rapid target jump from the fovea to a peripheral retinal position. Thus, in systems–theoretical lingo:

A saccade is the position step response of the visuomotor system.

The second important thing to note is that Eq. (9.4)a,b,d,e [but not Eq. (9.4)c] (…) immediately demonstrate that the saccadic system must be a *nonlinear system*! For a *linear system*, the saccade main sequence relations should have been (Exercise 9.3):

$$\begin{array}{ll}
\text{(a)} & D = D_0 \\
\text{(b)} & V_{\mathrm{pk}}(R) = V_{\mathrm{pk},\,(R=1)}\; R \\
\text{(c)} & T_{\mathrm{pk}} = T_0 \\
\text{(d)} & V_{\mathrm{pk}} D = \left[D_0 V_{\mathrm{pk},(R=1)}\right] R \\
\text{(e)} & S = S_0
\end{array} \tag{9.6}$$

The crucial question to answer is, of course, what causes the nonlinear behavior of the system? Although extraocular muscles clearly do possess nonlinear properties (they can only pull, not push, so they operate as rectifiers, and at some point their stretch will surely saturate) one might suspect that the main sequence could be due a nonlinear plant. However, there are a number of reasons to refute this hypothesis:

1. The eye muscles are organized in antagonistic push–pull pairs, which effectively remove a rectifying nonlinearity.
2. The transfer characteristic of the plant is well approximated by a second-order *linear* system (further analyzed below) over its main working range.

3. The activity of oculomotor neurons (and premotor burst neurons, see further section) reflects the fact that saccade duration (*burst duration*) increases with amplitude, and that peak velocity (*burst peak firing rate*) saturates for large amplitudes.

These findings all demonstrate that the main sequence properties are likely generated within the central nervous system. Let's find out how, by first looking at the properties of the oculomotor plant.

The oculomotor plant. The overall transfer characteristic of the oculomotor plant is well described by a second-order linear model, which contains two viscoelastic elements in series (Fig. 9.3; Goldstein, 1983).

Let's analyze this model by determining its transfer function in Laplace notation (see chapter: Linear Systems). The force $F(t)$, generated by a change in firing rate of the oculomotor neurons [$\Delta R(t)$], is applied to the two first-order viscoelastic (Voigt) elements:

$$\begin{gathered} F(t) = k_1 x_1(t) + r_1 \frac{dx_1}{dt} = k_2 x_2(t) + r_2 \frac{dx_2}{dt} \,, \\ E(t) = x_1(t) + x_2(t) \end{gathered} \tag{9.7}$$

where k_n are the elasticities, and r_n the viscosities of the Voigt elements. $E(t)$ is the output eye position of the model in response to the force input.

The reader will show in Exercise 9.4 that Laplace transformation on these equations yields for the overall transfer function of the plant:

$$H_{GR}(s) \equiv \frac{E(s)}{F(s)} = C \frac{(sT_z + 1)}{(sT_1 + 1)(sT_2 + 1)}, \tag{9.8}$$

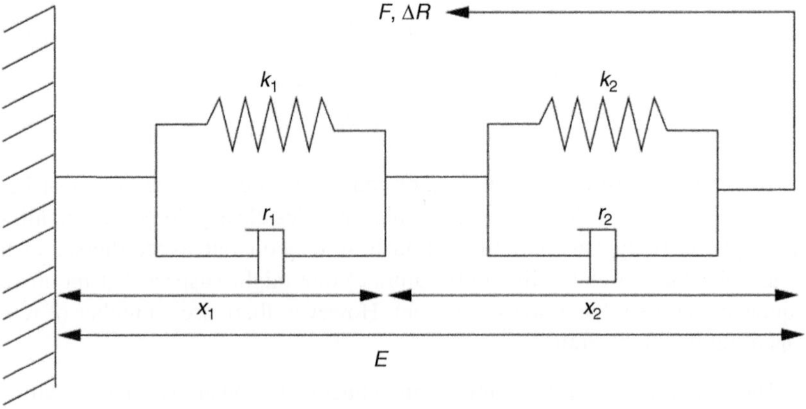

FIGURE 9.3 Second-order linear model as proposed by Goldstein (1983) for the horizontal oculomotor plant. Input force F (expressed by a net change in motorneuron firing rate, ΔR) activates two Voigt elements in series. The resulting output, $E(t)$, is the sum of x_1 and x_2.

which has one zero, and two poles. Further, $T_2 < T_z < T_1$ (about 20, 70, and 200 ms, respectively). The longest time constant, $T_1 \sim 200$ ms, specifies the minimum duration of the model's impulse response (an exponential decay; Exercise), and the *step response* of this model is:

$$S(s) = \frac{H_{\mathrm{GR}}(s)}{s} \tag{9.9}$$

The step response in the time domain can then be readily found by inverse Laplace transformation of Eq. (9.9), which is left as an exercise for the reader.

Also in this case, systems theory helps us make the following observations: First, as the impulse response decays to zero after about $3T_1$, which is due to the plant's elastic restoring forces, a *saccade* must involve a *tonic* motorneuron input signal (ie, a DC force) to keep the eye at the final eye position without a fast drift back to the neutral position. Second, the step response of the plant is much too slow to account for the saccade itself (being the step response of the total visuomotor system), as its minimum movement duration until steady state would then be about 200–400 ms.

The inevitable conclusion is therefore that the oculomotor innervation of the plant is neither an impulse, nor a step. *But what is it*?

Pulse–step innervation. Here, we will show how linear system's analysis can provide surprisingly deep insights into the neurocomputational secrets of saccade generation! We will illustrate this first for a highly simplified plant model, but the reader will work this out for the more elaborate Goldstein–Robinson plant model in Exercise 9.5. Suppose that the plant can be reduced to the following first-order model:

$$H_1(s) = \frac{1}{sT + 1} \tag{9.10}$$

Instead of asking the question: what is the response of this model to some input signal, we will ask the reverse: *what type of input signal will produce a given output*? Here, we make use of the concept that the *ideal saccade* (which brings the eye as fast and as accurately as possible on the target) would have to be a *step*! So, let's assume that the output of the plant is indeed a step of amplitude R, then in Laplace notation:

$$E_1(s) = \frac{R}{s} \tag{9.11}$$

The oculomotor input that is responsible for such an output is computed by:

$$M_1(s)H_1(s) = E_1(s) \Leftrightarrow M_1(s) = \frac{E_1(s)}{H_1(s)} = \frac{R}{s}(sT + 1) = (RT) + \frac{R}{s} \tag{9.12}$$

The last step in the calculation brings the Laplace terms to well known invertible Laplace functions, because:

$$M_1(s) = (RT) + \frac{R}{s} \Leftrightarrow m_1(t) = RT\delta(t) + RU(t) \tag{9.13}$$

This interesting result shows that in order to generate a perfect saccade (a step), the oculomotor neurons should fire a *pulse–step input signal.* The weighted pulse kicks the sluggish eye at the highest possible speed (in this case, infinite…) to the final location, *R*, while the step will keep it there forever.

The reader will verify in Exercise 9.5, that the input signal for the more realistic oculomotor plant of Eq. (9.8) is given by the following equation:

$$m_{GR}(t) = R\left[\frac{T_1T_2}{T_z}\delta(t) + U(t) + \frac{(T_1 - T_z)(T_z - T_2)}{T_z^2}\exp\left(-\frac{t}{T_z}\right)\right] \tag{9.14}$$

The signal contains three components: a *weighted pulse*, a *unity step*, and an *exponential slide* (*decay*). Interestingly, all three components are actually observed in the firing of real oculomotor neurons (Fig. 9.4; Goldstein, 1983).

The PSG. Systems theory has so far led us to the conclusion that the step input of the visuomotor system is somehow transformed by a nonlinear saccadic system into a pulse–slide–step innervation of the extra-ocular muscles. We make yet another observation: the step and the pulse of the pulse–step in Eqs. (9.13) and (9.14) are intrinsically connected, because

> The step component of the oculomotor pulse–step innervation is the time-integral of the pulse.

This tight relation ensures that the eye stays exactly at the position where the saccade has brought it. If the step is too small, the eye will drift backward immediately after the saccade (with a 200 ms time constant) to its new equilibrium position, and if the step is too large there will be a forward drift. Clearly, both scenarios are not in line with an optimally functioning saccadic system.

This line of reasoning prompted Robinson (1975) to propose that a *neural integrator* generates the eye-position step that follows a saccade by *temporal integration* of a centrally generated eye–velocity pulse:

$$S(t) = \int_0^t P(\tau)d\tau \tag{9.15}$$

Both signals, the pulse, *P*(*t*), appropriately scaled by the plant's time constant(s), and the step, *S*(*t*), are linearly added at the motorneuron to produce the pulse–step innervation of the muscle, and both then automatically have

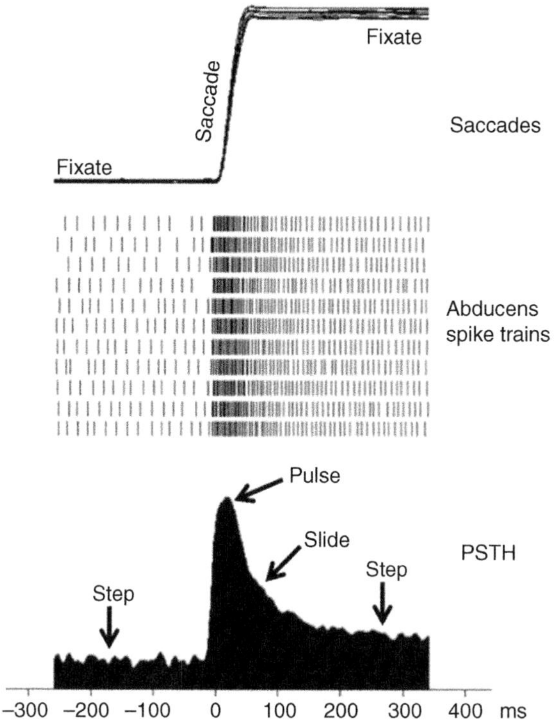

FIGURE 9.4 Recording of an Abducens oculomotor neuron of the right LR muscle in macaque monkey for 11 rightward saccades. Note the pulse–slide–step innervation in the PSTH (postsaccadic time histogram) of the cell [predicted by Eq. (9.14)], and the different tonic firing rates for pre- and postsaccade fixations.

the correct amplitudes. Based on the highly simplified, first-order linear plant model of Eq. (9.10), this systems-theoretical analysis already leads to a quite detailed model of how saccades are generated (Fig. 9.5).

The Neural Integrator. The neural integrator concept is one of the great success stories of the application of systems analysis in neuroscience. After it was proposed, it still took more than a decade to identify the neural substrate of the horizontal oculomotor NI. It is distributed over the medial vestibular nucleus (MVN) and nucleus prepositus hypoglossi (NPH; Cannon and Robinson, 1987). In addition, the vertical/torsional integrator is found in the mesencephalic interstitial nucleus of cajal (INC; Crawford et al., 2003). The NI increases the time constant of the oculomotor system's viscoelasticity to about 20 s. In darkness, without visual feedback, the eyes can be seen to drift slowly at about this rate (eg, the gaze trace in Fig. 9.9; Van Grootel and Van Opstal, 2010).

A simple model of how (linear) cells with membrane time constants of only 5 ms could build a neural integrator with a time constant of 20 s is analyzed in Exercise 9.6. This exercise points at an interesting problem: the signals that feed

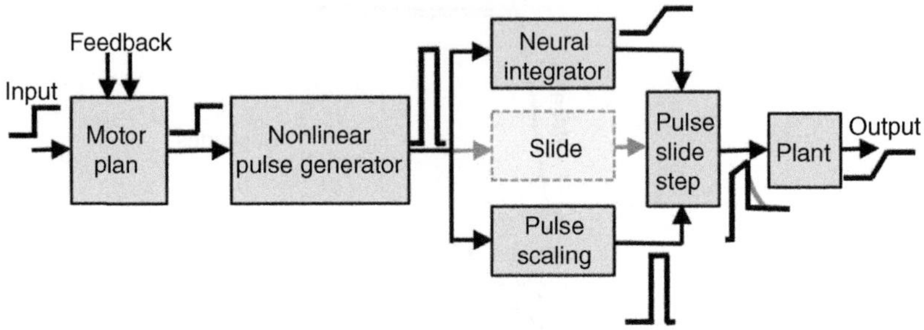

FIGURE 9.5 **Functional scheme of the saccadic system.** The step input on the retina is transformed (combined with feedback signals) into a motor plan for the nonlinear PG. The pulse–(slide–) step generation stage contains parallel pathways with a neural integrator and scaling of the pulse. The oculomotor neurons innervate the plant to produce the goal-directed saccade. The slide-generation stage is shown shaded.

into the NI often contain a combination of bursts (rapid changes in firing) and tonic activity (spontaneous background firing). However, the NI should only integrate the neural *signal* (the *change* in firing rate), and not the constant (DC) spontaneous firing rates. If it would also integrate the latter, its output would linearly increase with time and rapidly drive the eyes to the limits of the OMR. The exercise shows that this problem can be solved by symmetric cross-inhibitory feedback connections between neurons from the left and right hemispheres. It thus suggests that the oculomotor NI is built by distributed neural networks.

Inverse model. According to the analysis of the first-order linear plant model [Eq. (9.10)], the pulse–step generator is a parallel pathway that consists of the neural integrator (gain 1) and a direct path with a gain that exactly matches the plant's time constant, T. At this point it is worthwhile to think a bit further about this, and ask: what does this mean? What is the exact functional significance of this parallel pathway, and how critical are its parameters?

To answer these questions we have to determine the total transfer characteristic of the system in Fig. 9.5, from the pulse generator (PG; considered to be the input to the PSG), to the output of the oculomotor plant. This analysis can best be performed in the Laplace domain. The output of the PG is indicated by $P(s)$, the eye movement by $E(s)$, and note that the Laplace representation of the neural integrator is given by 1/s. Following the signal flow from left to right, we thus obtain:

$$P(s)\left(\frac{1}{s}+T\right)\frac{1}{sT+1}=E(s) \tag{9.16}$$

from which the overall brainstem transfer function, $B(s)$, is obtained as:

$$B(s)\equiv\frac{E(s)}{P(s)}=\left(\frac{1}{s}+T\right)\frac{1}{sT+1}=\frac{1}{s} \tag{9.17}$$

In other words, the functional significance of the parallel path of the PSG is to render the total transfer of the brainstem final common pathway completely *independent* of the plant's sluggish mechanics (with its long time constant), and let it functionally operate as a *perfect pulse integrator* (ie, a perfect step generator!). Or, expressed in the terminology of systems control theory:

The PSG embodies an inverse model of the oculomotor plant.

So, no matter how sluggish the oculomotor plant, or whatever happens to the eye–muscle mechanics, the brainstem pathway can (at least in principle) readily adapt its synapses (weights) to fully compensate for the mechanical changes. It can now be appreciated what happens to eye movements, if the gain of the direct pathway is either increased, or decreased, while keeping the plant's time constant at T (Exercise 9.7). The reader can extend this analysis also to the Goldstein–Robinson plant model in Exercise 9.8.

Note that the pathway from PG output to oculomotor plant output is entirely modeled by linear transfer functions, so that this total system can be considered a *linear system* (and when tuned appropriately: *a pure neural integrator*). The saccade nonlinearity that underlies the main sequence kinematics [Eq. (9.4)] is therefore implemented upstream from the PSG, and most saccade models assume that it is located at the level of the brainstem PG (however, see chapter: The Midbrain Colliculus, where this classical view will be challenged).

Indeed, the saccadic burst cells in the pons (described later) appear to encode the major nonlinear hallmarks of the saccade main sequence: increase of burst duration with saccade duration, and saturation of their peak firing rates.

Pulse Generation; internal feedback. The cells that embody the PG for horizontal saccades are so-called medium-lead burst neurons, or MLBs, as their lead time to the saccade is only 8–10 ms. They are found in the brainstem paramedian pontine reticular formation (PPRF). A total PPRF lesion abolishes all ipsilateral horizontal saccades: it's over and out, and it will never recover. For a given saccade (and only for saccades) these cells fire a high-frequency burst of spikes that correlates well with the instantaneous saccade velocity (Van Gisbergen et al., 1981; Cullen and Guitton, 1997). Conversely, the number of spikes in their burst encodes the saccade amplitude. It is commonly thought that these cells are driven by a *dynamic eye–motor error signal*, $m_e(t)$, that is constructed by a dynamic, local feedback circuit. Precisely how this error signal is produced, however, is still a matter of debate, which has not been resolved so far. Here, we highlight two important models: the Robinsons model and the Scudder model.

In the Robinson model (Van Gisbergen et al., 1981) the motor error is the difference between the planned end position of the saccade, E_0, and feedback about current eye position, $e(t)$:

$$MLB_{\text{In}}^{\text{Rob}}(t) \equiv m_e(t) = E_0 - e(t) \tag{9.18}$$

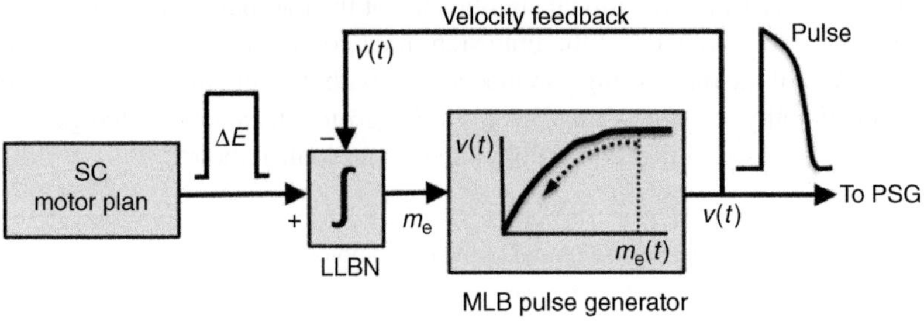

FIGURE 9.6 Scudder's model (Scudder, 1988; simplified version) of the nonlinear feedback PG in the brainstem. The pulse is fed to the PSG of Fig. 9.5. Input to the MLBs is the integrated difference between the SC burst (encoding the desired eye displacement), and the PG (current eye displacement). LLBN, long-lead burst neurons (comparator). The MLB nonlinearity, which accounts for the main sequence [Eq. (9.4)], is shown in the box. Omnipause neurons are omitted for clarity.

The feedback eye-position signal is derived from the output of the neural integrator in Fig. 9.5. The model of Eq. (9.18) is therefore a *head-centered* saccade model, since orbital eye positions are referenced in absolute head-centered coordinates.

A more recent model of the PG is Scudder's proposal (Scudder, 1988; Fig. 9.6), in which the dynamic motor error is determined by the integrated difference between the output of the midbrain superior colliculus (see chapter: The Midbrain Colliculus) and feedback of instantaneous eye velocity:

$$MLB_{\text{In}}^{\text{Scud}}(t) \equiv m_e(t) = \int_0^t \left[SC(\tau) - v(\tau)\right] d\tau \tag{9.19}$$

Note that since Eq. (9.19) uses *eye–velocity* feedback, the driving input signal from the SC represents mean eye velocity. The integrals of these velocity signals thus correspond to desired and actual eye *displacements*, respectively. These are *relative* signals, as they are measured with respect to the initial state of the saccade (displacement 0), irrespective of orbital eye position. The Scudder model is therefore an *oculocentric* model.

One additional feature in the saccade models that hasn't been touched upon so far is the *gating* of MLBs through the omnipause neurons (OPNs). These cells fire at a tonic rate between saccades, and fully inhibit the MLBs during fixation. Once the OPNs are silenced (presumably by an inhibitory trigger from the SC), the MLBs are released and can respond to the dynamic motor error signal. During the saccade the OPNs remain silent, as they are in turn inhibited by the MLBs. This inhibition stops as soon as the saccade is over, and the OPNs will immediately resume their tonic firing. This mechanism leads to the following change of Eq. (9.19):

$$MLB_{\mathrm{In}}^{\mathrm{Scud}}(t)=\begin{cases}\int_{0}^{Trig} SC(\tau)d\tau & \text{for } t<Trig\\ \int_{Trig}^{t}(SC(\tau)-v(\tau))d\tau & \text{for } t>Trig\end{cases} \tag{9.20}$$

The two models [Eqs. (9.20) and (9.18)] are mathematically equivalent: they have the same input–output relationships, and therefore cannot be dissociated through psychophysical experiments. The major differences reside in the putative role of the Superior Colliculus (not specified in Robinson's model), and in the nature of the feedback signals (position vs. velocity or displacement feedback).

In Exercise 9.9 the reader will analytically solve the simplified scheme of Fig. 9.6 when the trigger is given after the SC burst has finished.

Oblique Saccades. This far, we have discussed horizontal saccades only, but eye movements can be generated in any direction. Each eye is equipped with six extraocular muscles, organized in a nearly perfect orthogonal Cartesian coordinate system that aligns with the activation planes of the semicircular vestibular canals (see Section 9.3). The six muscles comprise three antagonistic pairs that enable rotations around a fixed axis in the positive (eg, rightward) and negative (eg, leftward) direction.

- The lateral and medial rectus muscles (LR and MR) pull the eye laterally and medially, respectively, to rotate the eye around the vertical axis.
- The superior rectus (SR), inferior rectus (IR), and the superior and inferior oblique muscles (SO and IO), together make the eye rotate in either a purely vertical direction (ie, around the interaural, horizontal, axis), or torsional direction (ie, around the visual axis).

How should we extend the model of Fig. 9.6 for oblique saccades? Also here, the concepts of systems analysis will provide some further insights into the saccadic system.

To simplify matters, we will ignore the control of torsional movements (3D eye rotations) and constrain our description to two dimensions (horizontal/vertical) only. Although a gross simplification, we assume a vertical oculomotor plant with essentially the same properties as the horizontal system (it's dynamics will be compensated, anyway…). Two brainstem saccade generators drive the plants: horizontal/vertical PGs, and horizontal/vertical pulse–slide–step generators, respectively.

We need to consider one more issue: the target location has been mapped onto a desired saccade vector (the motor plan, in Fig. 9.5). It is well established that the superior colliculus represents this motor plan by a population of neurons within a motor map that is organized in *polar coordinates* (see chapter: The

Midbrain Colliculus for detailed description of this map). Clearly, to appropriately drive the horizontal and vertical pulse–step generators with horizontal and vertical velocity signals, this planned motor vector should be decomposed into its horizontal and vertical displacement coordinates:

$$\text{Vector decomposition:} \quad \begin{array}{l} \Delta H = \Delta E \cos \Phi \\ \Delta V = \Delta E \sin \Phi \end{array} \tag{9.21}$$

Thus, the transformation from motor plan into eye movement involves two computational processes: (1) vector decomposition (polar to Cartesian), and (2) pulse generation (motor error to eye velocity). The latter is described by the static nonlinearity that we've introduced earlier, and is reiterated here for the more general case:

$$\textit{Pulse generation}: \dot{c} = C_{\max}\left(1 - \exp(-\alpha[\Delta C - c(t)])\right), \tag{9.22}$$

where $\dot{c} \equiv dc/dt$ represents the time derivative (velocity) of the ongoing vectorial eye displacement, $\Delta e(t)$, or for each of the saccade components, $\Delta h(t)$, or $\Delta v(t)$, respectively. ΔC is the desired vectorial (ΔE), or component ($\Delta H,\Delta V$) displacement.

Due to the assumed nonlinearity of the PG, the order in which the two stages are implemented in the system matters! Van Gisbergen et al. (1985) therefore proposed two possible ways to generate oblique saccades, which distinguish two different models (Fig. 9.7):

1. *Common Source* model: vectorial pulse generation (motor error vector to vectorial velocity) is followed by vector decomposition of the velocity command (Fig. 9.7A).

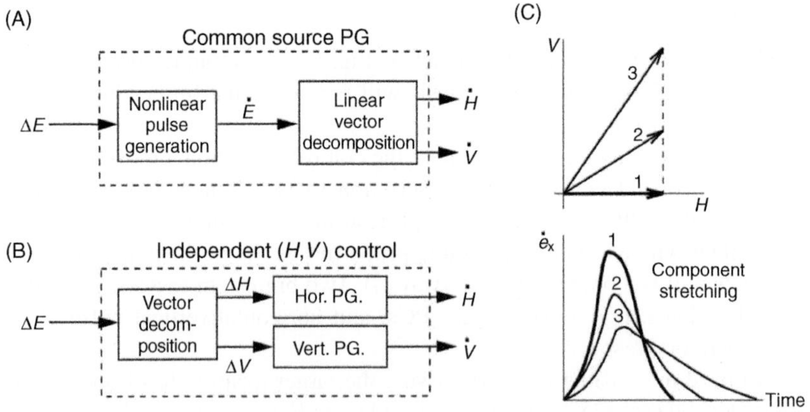

FIGURE 9.7 (A) Common Source model with a single vectorial PG stage. (B) Independent nonlinear horizontal and vertical PGs. (C) Component stretching in oblique saccades (here, fixed horizontal component). Component duration increases and peak velocity decreases with increasing angle.

2. *Independent PG* model: vectorial decomposition of the planned eye-displacement vector is followed by horizontal and vertical PGs (Fig. 9.7B).

Since the two models make different predictions about the behavior of horizontal and vertical components of (large) oblique saccades, they can be readily dissociated through psychophysical experiments.

The independent PG model predicts that the horizontal and vertical durations and velocities of the saccade vector only depend on their size, irrespective of saccade direction. For example, a 30 degrees amplitude saccade in a direction of 60 degrees (nr. 3 in Fig. 9.7C) has a horizontal component of 15 degrees and a vertical component of 26 degrees. Since the horizontal component has a shorter duration than the vertical component, the saccade trajectory has concave curvature. For a saccade direction of 30 degrees curvature will be convex.

In contrast, according to the Common Source model the saccadic system contains a single *vectorial PG*, as a common velocity drive to the brainstem. As a consequence, the horizontal and vertical velocities remain *scaled versions* of each other, and oblique saccades will be straight (Exercise 9.10). The durations of the saccade components are *equal*, and match the duration of the total saccade vector. Thus, a given 10 degrees horizontal saccade component will have its shortest duration for a purely horizontal ($\Phi = 0$ degrees) 10 degrees saccade, but its duration will systematically increase (and hence, its velocity decrease!) for oblique saccades ($\Phi \neq 0$ degrees). This phenomenon has been called *component stretching* (Fig. 9.7C).

Experiments clearly demonstrate the existence of component stretching in oblique saccades (Van Gisbergen et al., 1985; Smit et al., 1990), thereby clearly refuting the independent PG hypothesis. Although alternative schemes exist, in which the horizontal and vertical PG are coupled to produce stretching and straight saccades (Grossman and Robinson, 1988), the common source model is by far the simplest scheme to explain the observed behavior of saccades. Importantly, in the common source model component stretching and straight trajectories are *emerging properties*. In cross-coupling models these features have to be explicitly tuned by the coupling strengths, which involves many free parameters (Smit et al., 1990). In chapter: The Midbrain Colliculus we will argue that the vectorial PG may be embedded in the population response of the midbrain superior colliculus.

Double-stimulus responses. In the case of a single target, be it visual, auditory, or multisensory, there is a close correspondence between the target coordinates and the coordinates of the saccade vector, which can be well modeled by linear regression on the horizontal/vertical coordinates (see chapter: Assessing Auditory Spatial Performance; Fig. 9.8, top row). However, in natural environments, the gaze-control system is typically confronted with many potential targets, so that a selection has to be made on where to direct the eyes next. Rather than discussing gaze control in natural scenes (a research field in itself), it is already illuminating to study responses of the saccadic system to two competing

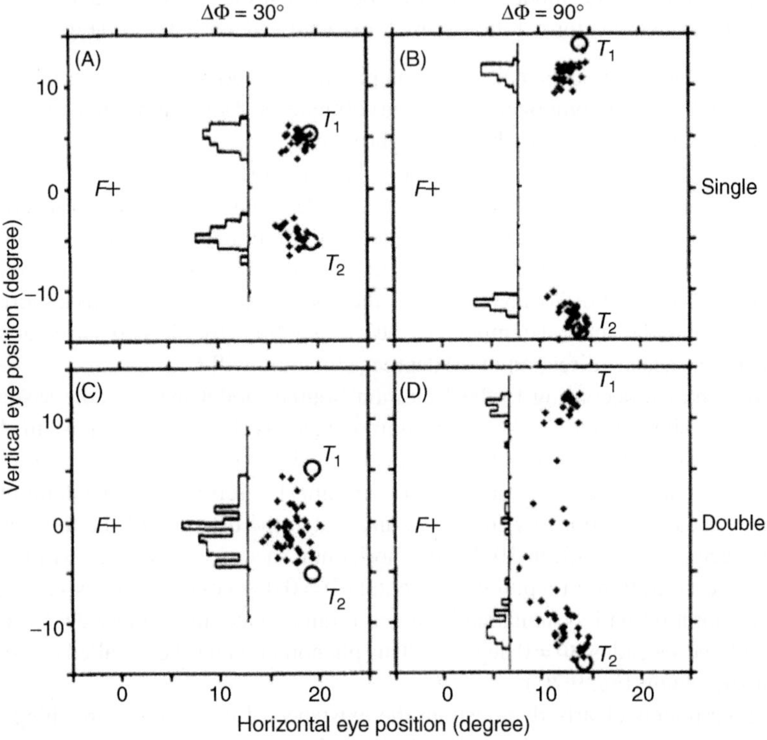

FIGURE 9.8 (A,B) Single-target first-saccade end point distributions to either T1 or T2. (C,D) Double-target responses. Saccade averaging responses (C) versus bistable responses (D) to simultaneous double stimuli depends on angular target separation. Insets: response histograms (not to scale). (F) Initial fixation point. *(Source: After Ottes et al., 1984.)*

stimuli, simultaneously presented at different locations in the visual field, say at $\vec{T}_1$ and $\vec{T}_2$. The results will also be of importance when we look at localization responses to competing sources in chapter: Multisensory Integration.

Fig. 9.8 shows what happens for two different situations: target separation relatively small (30 degrees angle; panels A,C), versus a large target separation (90 degrees, panels B,D). When the targets are close together, the first-saccade endpoints do not land on one of the target locations. The response endpoint distribution (inset in Fig. 9.8C) has a single peak, which is located at an intermediate location. The peak location (and hence the target weights) depends on visual (eg, size, luminance, contrast, spatial separation) and nonvisual factors (eg, task, like target/nontarget), but also on the saccade reaction time. The mean of the distribution is usually described by a weighted average of the single-target responses:

$$\vec{S}_{\mathrm{Avg}} = \frac{a\vec{S}_1 + b\vec{S}_2}{a+b}, \tag{9.23}$$

with a and b the stimulus weights. This response phenomenon has been termed the *global effect* (Findlay, 1982), or target averaging (Ottes et al., 1984). Eq. (9.23) hints at another *nonlinearity* in the visuomotor system because of the division by weighting factors $a + b$.

Note that in a typical double-target trial, a second corrective saccade will be directed toward one of the two stimuli. Supposing that the secondary saccade is made to $\vec{T}_2$, its motor coordinates should be:

$$\vec{S}_2 = \vec{T}_2 - \vec{S}_{\mathrm{Avg}} \tag{9.24}$$

In chapter: Sound Localization Plasticity, we propose and discuss an alternative, *linear* formulation for the averaging phenomenon that incorporates the planning of both saccades of Eqs. (9.23) and (9.24) in this paradigm.

When the two targets are widely separated (beyond 60 degrees, see Fig. 9.8C,D), the response endpoints are typically described by a *bimodal* response distribution. The system then chooses a response to *either* location $\vec{T}_1$, *or* to $\vec{T}_2$:

$$\vec{S}_{\mathrm{Bim}} = \left(\vec{S}_1 \quad \text{or} \quad \vec{S}_2\right) \tag{9.25}$$

In the scheme presented in Fig. 9.5 the saccade endpoint would be implemented at the *motor planning* stage. Neurophysiological studies have implicated the superior colliculus (discussed in chapter: The Midbrain Colliculus) in this process (Robinson, 1972; Van Opstal and Van Gisbergen, 1990), but also upstream visuomotor centers, like the frontal eye fields and posterior parietal cortex, may be involved (see chapter: Sound Localization Plasticity).

Double steps. Similar averaging phenomena are observed when the two targets are not presented simultaneously, but in rapid succession. In such a paradigm (the *double-step paradigm*), $\vec{T}_1$ first appears as a brief flash (20–50 ms) on the retina, and after it is extinguished it is immediately followed by $\vec{T}_2$, which may or may not remain present for the rest of the trial (Ottes et al., 1984). This paradigm has been of paramount importance for the study of visuomotor planning, of spatial representations and reference frames, and of (visual) localization performance. Due to its importance also for understanding sound-localization behavior, this paradigm will be discussed in more depth in Chapters 11. Here it suffices to briefly mention the main results from visuomotor studies.

Also in double steps, visual and task-related factors can modulate the averaging responses, as described earlier. However, in a double-step trial the endpoints of the first saccade systematically depend on the saccade *reaction time* with respect to the onset of the second (final) target:

- For short RTs the response will end close to the first target, for long RTs the saccade will be directed near the final target.
- For intermediate RTs, however, one obtains either averaging (small target separations), or bi-stable responses for larger target separations.

Note that a second corrective saccade usually carries the eye rapidly to $\vec{T}_2$. Sometimes, the intersaccadic interval can be as short as 30–50 ms, which strongly suggests that the corrective saccade of Eq. (9.24) is programmed in parallel of the first saccade, because any visual feedback after the first saccade will be far too slow to be of any use (Goossens and Van Opstal, 1997). In chapter: Sound Localization Plasticity we will see how predictive mechanisms in the brain could achieve this remarkable behavior.

Subjects are typically unaware that they generated two saccades in rapid succession, and they often do not perceive the presentation of two brief targets, when double steps are randomly interleaved with single-target trials.

9.3 SACCADIC EYE–HEAD GAZE SHIFTS

Eye movements are limited to their oculomotor range (OMR), which in humans and monkeys is about 40 degrees in all directions (and in cats only about 20–25 degrees). Beyond this range, the eye muscles cannot pull the eye any further. When an orienting response is planned to a target beyond the OMR (which often happens when responding to a sound source located in the periphery of the visual field) a combined eye–head movement (or: gaze shift) will be generated. The head will rotate in the direction of the target with respect to the body, or world (supposing that the body is restrained from moving), and carries the eye along with it. In the mean time, the eye makes a rapid saccade within the head (Fig. 9.9).

Clearly, the two motor systems should be precisely coordinated to let the eye foveate the target as fast and as accurately as possible. Fig. 9.9 shows how the system achieves this for a typical human eye–head gaze shift (amplitude about 36 degrees) toward an auditory target presented in total darkness. Three systems have to closely work together in a gaze shift: the oculomotor (saccadic) system, the head–motor system, and the vestibular system. The latter ensures that the eye stays on the target, despite the ongoing head movement, once the gaze shift has ended (at gaze-shift offset, G_{OFF}). Control of the VOR has to occur with millisecond precision to prevent unwanted retinal slip of the target.

The gaze shift consists of an eye-in-head movement, ΔE_{H} and a head-in-space movement, ΔH_{S}. Both movement components are measured at the gaze-shift offset, so that:

$$\Delta\vec{G}_{\mathrm{S}} \equiv \Delta\vec{E}_{\mathrm{H}} + \Delta\vec{H}_{\mathrm{S}} \tag{9.26}$$

How could this tight interplay between the three systems be implemented in neural circuitry? The model shown in Fig. 9.10 provides a simple account for the experimental data described later. It is an adapted version of Scudder's saccade model (Fig. 9.6), and includes the full repertoire of eye–head gaze shifts. We extended Scudder's model in the following ways (where it should be noted that all kinematic signals represent vectorial, 2D, quantities):

1. The number of spikes in the SC burst represents the desired *gaze shift* (ΔG); the SC firing rate is then a desired *gaze-velocity* signal.

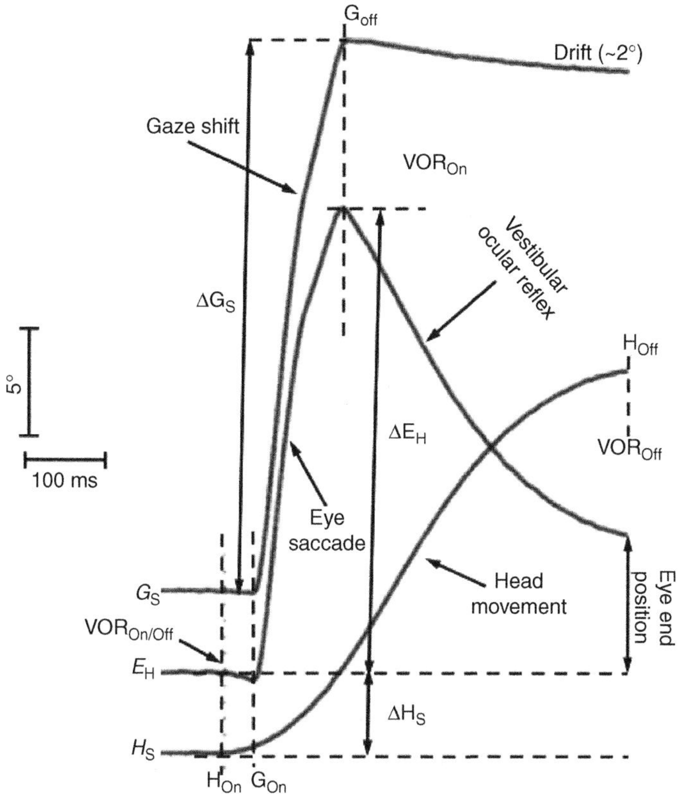

FIGURE 9.9 Human eye–head gaze shift toward an auditory target. Only the horizontal movement components are shown, for clarity. Note the tight interplay between the VOR and the gaze-control system. For auditory-evoked gaze shifts, the head often starts to move earlier (at H_{ON}) than the eye (at G_{ON}) (Section 9.4). The VOR then keeps the eye stable at the fixation point, but has to be suppressed as soon as the gaze shift starts ($VOR_{On/Off}$). At the gaze-shift offset the VOR is reactivated to compensate for any ongoing head movement, until H_{Off}. At that moment, the eye is at an eccentric orbital position. Note the slight drift of gaze (about 2 degrees in 4.5 s), as the subject is in complete darkness. Initial positions of eye, head, and gaze are shown vertically shifted for illustrative purposes.

2. The SC signal is compared (by integrating the differences) with actual eye- and head velocity signals to determine a dynamic gaze motor error. This signal will be the common drive for the eye-head-motor systems:

$$GM(t) = \int_0^t \left[SC(\tau) - EV_H(\tau) - HV_S(\tau)\right] d\tau \tag{9.27}$$

3. Although $GM(t)$ drives eyes and head, the eyes have to be limited by their physical OMR (inset in Fig. 9.10). To that end, the system first needs to know where the intended gaze shift will (hypothetically) bring the eye, by

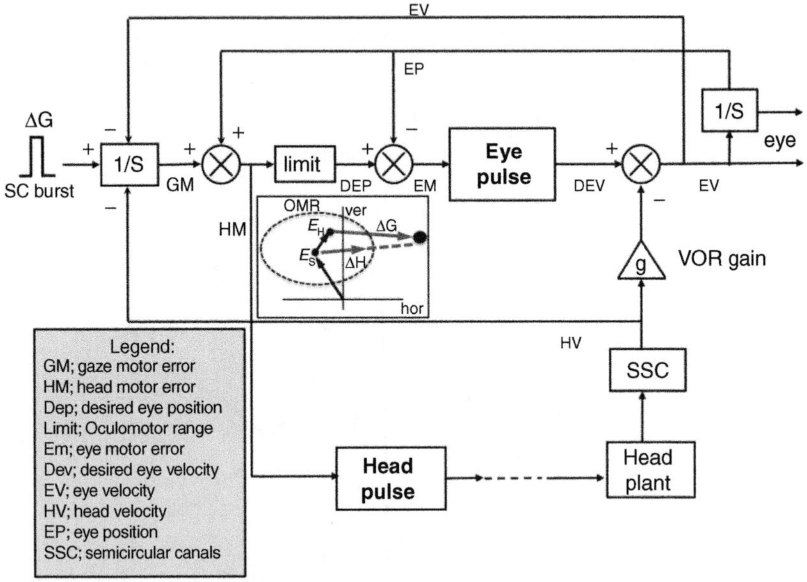

FIGURE 9.10 **Quantitative, schematized model for eye–head gaze control.** The OPNs that gate eye- and head motor systems are omitted, for clarity. Inset: vectorial relationships between initial conditions of eye and head regaze target (large solid dot). H_S, initial head orientation in space; E_H, initial eye orientation in the head (small solid dot represent fovea); OMR, oculomotor range. See text for explanation.

adding the current eye position (EP), which is derived from the neural integrator (1/s). By limiting this estimate to the OMR, the actual desired eye position for the gaze shift (Dep) can be specified:

$$\begin{aligned} Dep(t) &= Max\left[OMR,\ GM(t) + EP(t)\right] \\ HM(t) &= g_{A,V}[E_H(0)]GM(t) \end{aligned} \tag{9.28}$$

Experiments show that the head-movement contribution depends on the initial position of the eye in the orbit, and on the sensory stimuli (auditory vs. visual), which we here simplify by including a simple linear gain (Goossens and Van Opstal, 1997). For example, if the eye already looks rightward at the start of a large rightward gaze shift, it can only contribute a small eye-in-head saccade. Thus, the planned head movement is relatively large, and the resulting gaze shift will be relatively slow, as the eye contribution (by far the fastest motor system) is limited. In contrast, when the eye looks leftward at the start of a large rightward gaze shift, the head contribution can be much smaller. As the gaze shift is now predominantly carried by the eye, it will be much faster.

Note that in the gaze-control model of Fig. 9.10 the eye PG is driven by a desired *eye-position* signal, just like in the original Robinson model of Eq. (9.18):

$$EM(t) = Dep(t) - EP(t) \tag{9.29}$$

An important difference with Eq. (9.19), however, is that the desired eye position signal, *Dep*(*t*), is now a *dynamic* signal, as it changes rapidly during the gaze shift.

4. The output of the eye's PG is modified by the vestibular ocular reflex (VOR), which signals the head velocity. However, this latter signal is gated by gaze–motor error: when GME is large, the VOR is turned off to allow for a fast saccade in the direction of the target, irrespective of head motion; when GME approaches zero, the VOR gain turns on, to compensate for the ongoing head movement (Fig. 9.9). Therefore, the output of the eye PG is a *desired* eye velocity, not the real eye velocity:

$$\begin{aligned} EV(t) &= EV_{\max}\{1-\exp[-\beta EM(t)]\} - g_{\text{VOR}}[GM(t)]\dot{h}(t) \\ EP(t) &= \int_0^t EV(\tau)d\tau \end{aligned} \tag{9.30}$$

Simulations with this 2D model generate realistic gaze shifts to auditory and visual targets in all directions, and for arbitrary initial conditions (Goossens and Van Opstal, 1997).

9.4 AUDITORY AND VISUAL-EVOKED GAZE SACCADES

In line with the common-source model described earlier, also the horizontal and vertical components of gaze shifts are tightly coupled.

This leads to the interesting hypothesis that *a central vectorial gaze-PG controls gaze shifts*. Chapter: Coordinate Transformations will discuss the possibility that the SC could be blamed for this. It should be realized that a *gaze-velocity* signal is in fact an *abstract* signal! It cannot directly drive either the oculomotor plant, or the head plant, and should rather be considered as a *desired* (perhaps *optimal*) motor plan. At downstream levels, involving brainstem, spinal cord, and cerebellar circuitry, this desired motor plan is distributed across the different motor systems, in ways that depend on a number of factors: initial conditions of eye and head orientation, sensory conditions (auditory, visual, or otherwise), motor constraints, motivation, task, etc.

As predicted by a common-source controller, gaze trajectories of visual and auditory-evoked responses are approximately straight over the entire range of movement amplitudes and directions. Fig. 9.11 shows some examples for visual (Fig. 9.11A) and auditory (Fig. 9.11B) gaze, eye, and head trajectories, when the eye and head started from the same, straight-ahead position (initial condition with

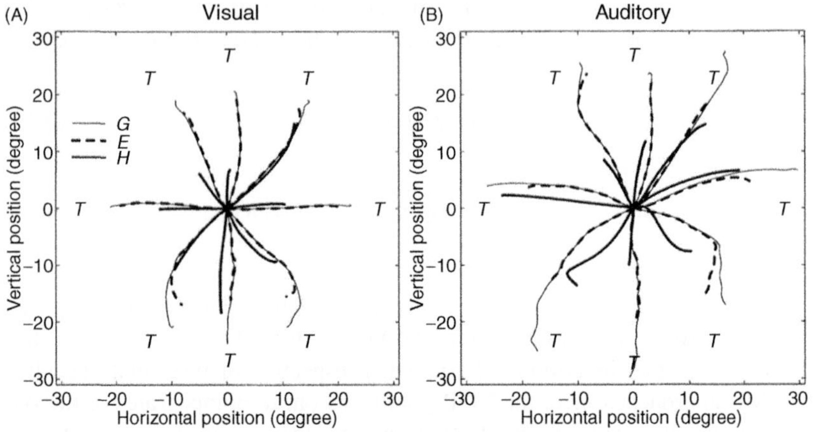

FIGURE 9.11 **Eye–head gaze shifts to visual (A) and auditory (B) targets. Aligned initial conditions. Note that eye and head both move toward the goal.** *(Source: From Goossens and Van Opstal, 1997.)*

the eyes and head aligned). It can be observed that the three components all move in the same direction, toward the perceived target location (Goossens and Van Opstal, 1997; Populín, 2006). (see also later, when the initial conditions differ).

There are some interesting differences in the control of visual and auditory-evoked gaze shifts:

- Auditory gaze shifts tend to be accompanied with larger head movements than visual gaze shifts.
- The head-movement onsets for auditory gaze shifts fall, on average, earlier than for visual gaze shifts, and can often even *precede* the onset of the gaze shift (like in the example of Fig. 9.9). This never occurs for visual gaze shifts.

The latter observation suggests the existence of independent trigger mechanisms for the eye- and head-motor systems that may be modulated by task constraints and sensory conditions.

Initial conditions. There is another interesting difference for visual versus auditory gaze control, which is due to the essentially different reference frames in which the sensory signals are represented. Since in humans the ears are placed in a fixed orientation on the head, the acoustic cues are referenced in *head-centered (craniocentric) coordinates* (see chapter: Acoustic Localization Cues). That is, each time the sound source or the head moves through space, the acoustic cues change appropriately. Similarly, visual signals are referred to in *eye-centered (oculocentric) coordinates*: whenever the visual stimulus, or the eye, moves through space, the retinal input changes accordingly. As a result, programming a head movement toward a sound source may directly use the head-centered cues, while generating an eye movement to a visual stimulus can

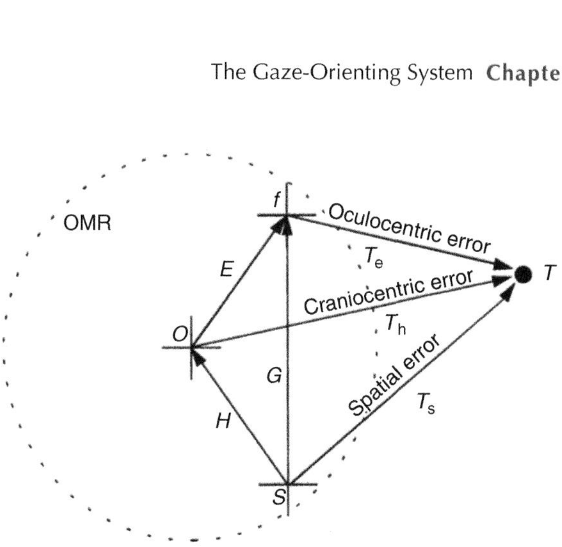

FIGURE 9.12 **The different reference frames for eye (fovea, f), head (O) and body (or world, S) specify different coordinates for their motor commands (T_e, T_h, and T_s, respectively) to target *T*.** *H*, head orientation rebody; *E*, eye orientation rehead; *G*, eye orientation rebody (gaze in space, $G = H + E$); OMR, oculomotor range.

employ its retinal coordinates relative to the fovea (Fig. 9.12). In Chapter 11 we will modify this statement when we discuss eye–head orienting in dynamic situations where eyes and head may have made intervening movements prior to the localization response.

Stated in a slightly different way: to generate an eye movement to a target requires a motor command expressed in eye-centered coordinates, and to drive a head movement will require a motor command in head-centered coordinates. As long as the eyes and head are aligned, the reference frames coincide, and both systems can be driven by the same command. However, when they are not aligned (like in Fig. 9.12), the eye-centered (T_e) and head-centered (T_h) vectors will differ. So how does this work for eye–head coordination to visual and auditory targets? Table 9.1 summarizes the requirements:

Thus under general initial conditions (the typical situation is that eyes and head do not point in the same initial direction) the appropriate motor commands

TABLE 9.1 Required Transformations for Visual and Auditory Targets Into Oculocentric and Craniocentric Coordinates, Respectively. T_V = Visual Target on the Retina and T_{AC} = Auditory Target with Respect to the Head From Acoustic Cues (Fig. 9.12)

Eye in $\vec{E}_H$	Visual target	Auditory target
Eye	$\vec{T}_E = \vec{T}_V$	$\vec{T}_E = \vec{T}_{AC} - \vec{E}_H$
Head	$\vec{T}_H = \vec{T}_V + \vec{E}_H$	$\vec{T}_H = \vec{T}_{AC}$

for eyes and head differ by the amount of eye-orientation in the head. The question is whether or not the gaze-orienting system should bother at all with these coordinate transformations.

A classical model for gaze control is that eyes and head are both driven by the *same* gaze-error command (Guitton, 1992). Indeed, when eyes and head are in aligned initial conditions, this is an appropriate driving signal for both motor systems. As originally gaze-control studies were entirely confined to visual-motor behavior, the hypothesis was that this common driving signal was automatically an *oculocentric* signal. An alternative version of this idea could be that the gaze-error signal would become craniocentric when it concerns an auditory target, and oculocentric when it's a visual target. In that case, the eyes would miss an auditory target by an amount that is determined by the initial eye position, while the head would miss a visual target by the same amount, but in the opposite direction. Since the head doesn't see, and the eyes don't hear, one might argue that perhaps the system wouldn't care so much... But what would the system do in case of an audiovisual target?

Indeed, given the important functional links between vision and audition (explained in chapter: Assessing Auditory Spatial Performance), a better strategy would be to carry out the coordinate transformations of Table 9.1, in which case the eyes and head would have to move in different directions to acquire the target. Fig. 9.13 shows that this is indeed what happens, both for visual and for auditory gaze shifts. To quantify these motor strategies Goossens and Van Opstal (1997) described the head displacements in these unaligned visual and

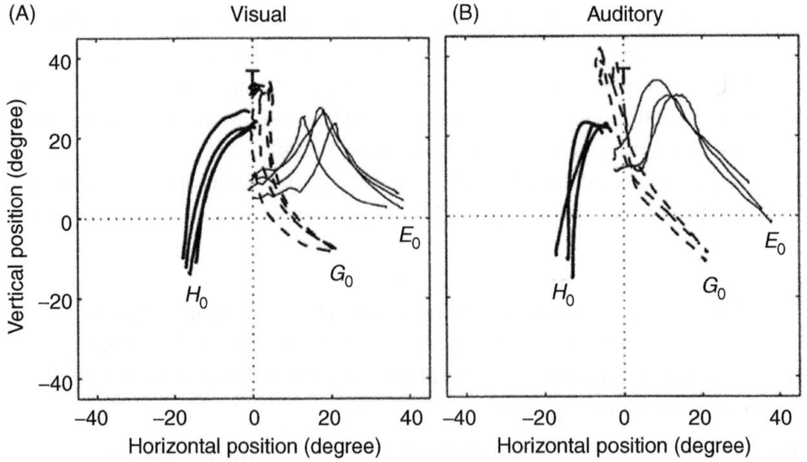

FIGURE 9.13 Eye–head gaze shifts toward a visual (A) and auditory (B) target when eyes and head are in different initial pointing directions (eyes point about 30 degrees rightward in the head). Note that gaze (the eye in space), and head both move toward the target, and hence in different directions. *(Source: After Goossens and Van Opstal, 1997, with kind permission.)*

auditory trials by multiple linear regression, with the initial gaze error, T_E, and initial head motor error, T_H, as independent variables:

$$\begin{aligned} \Delta\vec{H}_V &= a\vec{T}_E + b\vec{T}_H + c \\ \Delta\vec{H}_A &= c\vec{T}_E + d\vec{T}_H + e \end{aligned} \tag{9.31}$$

For all conditions (visual, auditory, and varying eye positions) the head displacement vector was best described by the head–motor error (ie, *b* and *d*, much larger than *a* and *c*). Note that since the head movement typically does not cover the entire distance to the target, gains were typically much lower than 1.0 (range *b*,*d*: 0.5–0.9 vs. *a*,*c*: 0.0–0.08). Furthermore, the head-movement contributions to auditory gaze shifts exceeded those for visual gaze shifts (averages: d ~ 0.77 vs. b ~ 0.67).

9.5 EXERCISES

Problem 9.1: From Eq. (9.1) it is possible to estimate the total number of cones in the human retina. Assume that the eye is a sphere with a radius of 1.2 cm, and that the retina covers the entire backside of the spherical surface. Assume that cells are distributed in a rotation-symmetric way around the fovea. Write down and solve the (2D) integral equation for the total number of retinal cones [*note: number of cells = cell density × surface; use polar coordinates*, (r,φ)].

Problem 9.2:

(a) Approximate a saccade velocity profile by a simple triangle (base = duration, height = peak velocity, at fixed T_{pk}). Comment on Eq. (9.4)d.
(b) Also try to derive Eq. (9.4)d from Eq. (9.4)a,b.
(c) When the duration increases according to Eq. (9.4)a, predict $V_{pk}(R)$ from the simple triangular approximation.

Problem 9.3: Refer to the superposition principle described in chapter: Linear Systems, to argue why Eq. (9.6) have to apply for *any* linear system.

Problem 9.4:

(a) Apply the LT on Eq. (9.7) to demonstrate Eq. (9.8), where

$$C = \frac{k_1 + k_2}{k_1 k_2}, \quad T_z = \frac{r_1 + r_2}{k_1 + k_2}, \quad T_1 = \frac{r_1}{k_1}, \quad T_2 = \frac{r_2}{k_2}$$

(b) By applying inverse LT, show that the temporal behavior of the impulse response in the time domain is independent of T_z, and given by:

$$h(\tau) = \frac{1}{T_1 - T_2}\left[\left(\frac{1 - T_z}{T_1}\right)\exp\left(-\frac{\tau}{T_1}\right) + \left(\frac{T_z}{T_2 - 1}\right)\exp\left(-\frac{\tau}{T_2}\right)\right]$$

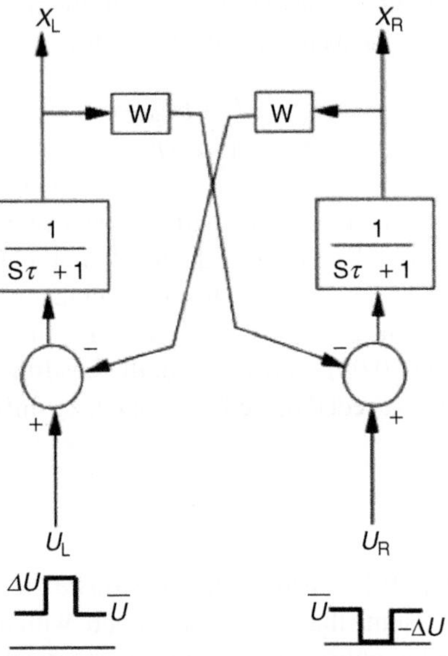

FIGURE 9.14 **Cross-inhibitory neural network to model the neural integrator.**

Problem 9.5 Apply the inverse reconstruction technique to the Goldstein–Robinson model to derive Eq. (9.14).

Problem 9.6 Here we analyze the cross-inhibitory neural network of Fig. 9.14, and show that it performs a perfect neural integration on the *difference* signal between the left and right inputs, thereby getting rid of the DC components in the input. Suppose that the neurons have the following input–output relationships (in Laplace notation):

$$X_{\mathrm{R,L}}(s) = \frac{IN_{\mathrm{R,L}}(s)}{s\tau + 1} \quad \text{with } IN_{\mathrm{R,L}}(s) = U_{\mathrm{R,L}}(s) - WX_{\mathrm{L,R}}(s) \text{and } \tau = 5 \text{ ms}$$

$U_{\mathrm{L,R}}(s)$ is the push–pull input to the neurons (when U_L goes up, U_R goes down, and vice versa), and is described by

$$U_{\mathrm{L,R}} = \bar{U} \pm \Delta U$$

Now show (use an appropriate change of variables to uncouple the pair of equations) that as W→1, the output of the neurons can be written in the following way:

$$X_{\mathrm{L,R}}(s) = \frac{\bar{U}/2}{s(\tau/2)+1} \pm \frac{\Delta U}{s}$$

In other words, the DC disappears at the output with a time constant of 2.5 ms, but the change in firing is integrated perfectly ($T \to \infty$)!

Problem 9.7 Repeat the analysis of Eq. (9.17) for an arbitrary gain in the direct pathway of the PSG. Determine the brainstem transfer characteristic, and analyze (in the time domain) what happens to the eye movements when $T' = T \pm \Delta T$.

Problem 9.8 Include the slide pathway in the brainstem Pulse–slide–step Generator, and determine the transfer characteristics of each of the three paths, such that the total model from PG to eye movement implements a pure integrator.

Problem 9.9 Here we will solve the Scudder model of Fig. 9.6 analytically. The SC burst is described by a rectangular pulse: $SC(t) = P$ for $0 \le t \le D$ and 0 elsewhere, and $PD = \Delta E$ (the desired saccade amplitude). The MLB nonlinearity is $v(t) = v_{\max}\{1 - \exp[-\beta m(t)]\}$, with $m(t)$ the dynamic motor error. Assume that the omnipause trigger is given at $t = D$.

(a) Show that the dynamic motor error obeys the following differential equation:

$$\frac{dm}{dt} = \Delta E - v_{\max}\left[1 - \exp(-\beta m)\right]$$

(b) Solve this equation for $m(t)$.
(Hint: use substitution: $s = \exp(-\beta m) - 1$ and first solve for s).

(c) Use the definition of dynamic motor error to determine the instantaneous eye displacement of this model, $\Delta e(t)$, and show that for $t \to \infty$ the eye reaches the desired displacement amplitude.

Problem 9.10 In the "common source" model, the brainstem is driven by a vectorial PG that transforms the amplitude of a 2D motor error vector, Δevec, into a vectorial eye–velocity command according to:

$$\dot{e}_{\text{vec}}(t) = v_{\max}\left[1 - \exp\left(-\frac{|\Delta \vec{e}_{\text{vec}}|}{m_0}\right)\right]$$

with $|x|$ the magnitude of the vector. Give expressions for the velocity profiles of identical horizontal saccade components for the situations sketched in Fig. 9.7C (top panel; ie, the component amplitude is fixed, but the vector rotates over angle φ with respect to the horizontal direction).

REFERENCES

Cannon, S.C., Robinson, D.A., 1987. Loss of the neural integrator of the oculomotor system from brainstem lesions in monkeys. J. Neurophysiol. 57, 1383–1409.

Crawford, J.D., Tweed, D.B., Vilis, T., 2003. Static counterroll is implemented through the 3-D neural integrator. J. Neurophysiol. 90, 2777–2784.

Cullen, K.E., Guitton, D., 1997. Analysis of primat IBN spike trains using system identification techniques. I. Relationship to eye-movement dynamics during head-fixed saccades. J. Neurophysiol. 78, 3259–3282.

Curcio, C.A., Sloan, K.R., Kalina, R.E., Hendrickson, A.E., 1990. Human photoreceptor topography. J. Comp. Neurol. 292, 497–523.

Findlay, J.M., 1982. Global visual processing for saccadic eye movements. Vision Res. 22, 1033–1045.

Goldstein, H.P., 1983. The neural encoding of saccades in the rhesus monkey. Thesis, Johns Hopkins University. Baltimore, MD, USA.

Goossens, H.H.L.M., Van Opstal, A.J., 1997. Human eye-head coordination in two dimensions under different sensorimotor conditions. Exp. Brain Res. 114, 542–560.

Goossens, H.H.L.M., Van Opstal, A.J., 1997. Local feedback signals are not distorted by prior eye movements: evidence from visually-evoked double-saccades. J. Neurophysiol. 78, 533–538.

Grossman, G.E., Robinson, D.A., 1988. Ambivalence in modeling oblique saccades. Biol. Cybern. 58, 13–18.

Guitton, D., 1992. Control of eye-head coordination during orienting gaze shifts. Trends Neurosci. 15, 174–179.

Ottes, F.P., Van Gisbergen, J.A.M., Eggermont, J.J., 1984. Metrics of saccade responses to visual double stimuli: two different modes. Vision Res. 24, 1169–1179.

Populín, L.C., 2006. Monkey sound localization: head-restrained versus head-unrestrained orienting. J. Neurosci. 26, 9820–9832.

Robinson, D.A., 1972. Eye movements evoked by collicular stimulation in the alert monkey. Vision Res. 12, 1795–1808.

Robinson, D.A., 1975. Oculomotor control signals. In: Lennerstrand, G., Bach-y-Rita, P. (Eds.), Basic Mechanisms of Ocular Motility and their Clinical Implications. Pergamon Press, Oxford, pp. 337–374.

Scudder, C.A., 1988. A new local feedback model of the saccadic burst generator. J. Neurophysiol. 59, 1455–1475.

Smit, A.C., Van Opstal, A.J., Van Gisbergen, J.A.M., 1990. Component stretching in fast and slow oblique saccades in the human. Exp. Brain Res. 81, 325–334.

Van Gisbergen, J.A.M., Robinson, D.A., Gielen, S., 1981. A quantitative analysis of generation of saccadic eye movements by burst neurons. J Neurophysiol 45, 417–442.

Van Gisbergen, J.A.M., Van Opstal, A.J., Schoenmakers, J.J., 1985. Experimental test of two models for the generation of oblique saccades. Exp. Brain Res. 57, 321–336.

Van Opstal, A.J., Van Gisbergen, J.A.M., 1987. Skewness of saccadic velocity profiles: a unifying parameter for normal and slow saccades. Vision Res. 27, 731–745.

Van Opstal, A.J., Van Gisbergen, J.A.M., 1990. Role of monkey superior colliculus in saccade averaging. Exp. Brain Res. 79, 143–149.

Van Grootel, T.J., Van Opstal, A.J., 2010. Human sound-localization behavior accounts for ocular drift. J. Neurophysiol. 103, 1927–1936.

Chapter 10

The Midbrain Colliculus

10.1 INTRODUCTION

In chapter: The Gaze-Orienting System we extensively discussed how brainstem neural circuitry is thought to program, start, and control a saccadic eye–head gaze shift. We argued that behavioral responses of saccadic eye movements, and of eye–head gaze shifts, provide strong support for a common nonlinear vectorial pulse–generator that drives the horizontal and vertical/torsional brainstem and spinal motor systems of eyes and head (the *common source* model). This elegant organization automatically leads to strong mutual coupling of the respective response components, without having to impose additional constraints or complex interacting connection schemes. As an emerging property, responses are automatically planned as straight (ie, shortest path) trajectories, and the components undergo the appropriate amount of *stretching* through a simple linear (sine and cosine) scaling. Although an attractive model, because of its obvious simplicity, a long-standing problem with the common-source scheme has been lack of convincing neurophysiological evidence for a nonlinear vectorial pulse generator in the system. In Scudder's model (Fig. 9.6) the brainstem is driven by a desired eye-displacement command from the midbrain superior colliculus (SC), but this signal represents at best a *mean eye–velocity* command in its firing rates. If the OPN trigger of Eq. (9.20) falls late, the SC firing rates will have no influence on the saccade kinematics whatsoever, and this has long been the accepted view on the function of the collicular control signal:

> Classical view: "the SC provides a desired eye-displacement vector with little (or no) influence on the saccade kinematics."

If true, there would be little room for a vectorial pulse generator, as at the next synapse downstream from the SC the signal is distributed right away. In this chapter, we will challenge this classical view of SC control by recent recording evidence indicating a more prominent role of the SC in the control of gaze shifts, which includes their kinematics.

The Auditory System and Human Sound-Localization Behavior. http://dx.doi.org/10.1016/B978-0-12-801529-2.00010-6

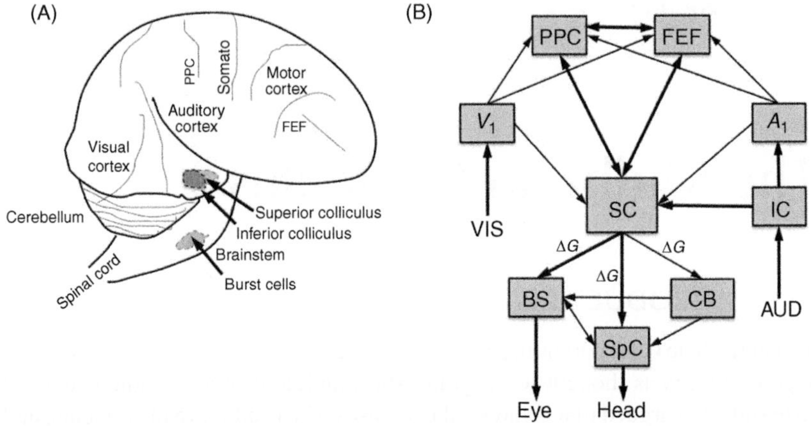

FIGURE 10.1 (A) Schematic side view of the monkey brain, showing the most important areas for visual and auditory gaze control: brainstem (BS) and cerebellum (CB), midbrain superior (SC) and inferior colliculus (IC); in cortex primary visual (V_1) and auditory (A_1) areas, frontal eye fields (FEF), and posterior parietal cortex (PPC). (B) Schematic control diagram of the major pathways involved in visual and auditory-evoked eye–head gaze shifts. SpC, spinal cord; ΔG, the desired gaze displacement vector is distributed across multiple systems.

Before doing so, we first describe the relevant neurophysiological findings and classical (nonkinematic and static) models of the SC in more detail. This includes a description and mechanistic explanation for its topographical organization into a motor map, results from single-unit recordings in eye and eye–head orienting (the movement–field concept), and of microstimulation experiments. We then discuss how SC activity patterns could embed a vectorial pulse generator. At the end of this chapter we also propose a potential role for the midbrain *inferior colliculus* (IC) in auditory-evoked eye–head orienting, through its direct projections to the SC.

Schematic: We start with the simple schematic overview of Fig. 10.1, which highlights the most important cortical and subcortical structures involved in visual and auditory-evoked gaze shifts. In cortex, cells in frontal eye fields (FEF) and posterior parietal cortex (PPC) are recruited during the planning phase and generation of voluntary saccadic gaze shifts. Both areas have reciprocal, monosynaptic connections to the midbrain SC and with each other, and both receive inputs from primary sensory areas like visual, auditory, and somatosensory cortices. The circuit that encompasses SC–FEF–PPC thus plays a crucial role in the sensory-evoked gaze behaviors described in the previous chapter, and in the chapters to follow. In chapter: Coordinate Transformations we will discuss the potential involvement of PPC and FEF in programming gaze shifts during more complex and dynamic behaviors. In this chapter, we focus exclusively on the role of the midbrain in preparing and generating visual and auditory-evoked gaze shifts to single targets. As may be clear from Fig. 10.1B, the SC appears to be a major player, as it occupies a central hub in this process: all relevant

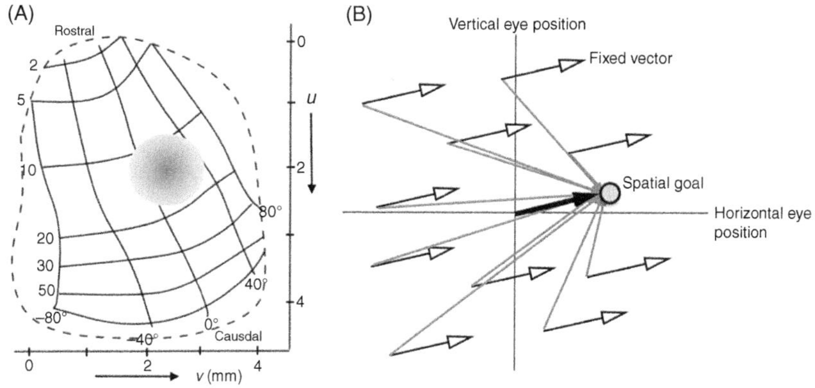

FIGURE 10.2 (A) The microstimulation map of the SC by Robinson (1972) reveals an inhomogeneous map of saccadic eye movements in polar coordinates. The map is shown superimposed as isoamplitude [$R = (2°, 50°)$] and isodirection [$\Phi = (-80°, +80°)$] lines on the anatomical coordinates from the recording chamber: v (medial–lateral, in millimeters) and u (rostral–caudal, in millimeters). (B) The motor map could represent fixed vectors (oculocentric map), or spatial goals (vectors depend on initial eye position; gray arrows). Thick arrow: saccade evoked from straight-ahead fixation at the site indicated in (*A*) by the circular area.

cortical areas (and many subcortical areas, like the basal ganglia, not shown here) project to it, it receives multisensory and motor planning inputs, and its motor output is directed to brainstem and spinal cord circuits, as well as to the cerebellum. Let's see how it's organized.

Stimulation map: The first quantitative neurophysiological studies in the SC of awake and trained macaque monkeys were performed in the early 1970s. In a series of seminal studies, Goldberg and Wurtz (1972) measured the visual receptive fields, saccade-related activity, and the effects of small focal lesions of the SC on saccades. They observed that receptive fields and saccade-related activity were in topographic register, and that monkeys were unable to make a saccade into the cell's receptive field after ablation of its SC saccade representation.

In the same year, Robinson (1972) published his extensive mapping study of the SC, obtained after evoking saccades through brief microstimulation trains. He revealed a topographic motor map, in which saccade vectors are arranged in polar coordinates: their amplitude is topographically organized from small to large along the rostral–caudal anatomical axis, with their direction perpendicular to amplitude along the lateral–medial axis (Fig. 10.2A). The results also demonstrated the strong *inhomogeneous* nature of this map, as small saccades (<10 degrees) occupied a much larger anatomical space than large saccades (>30 degrees). Below we show that this property emerges from the inhomogeneous organization of the primate retina (Fig. 9.1).

Robinson also made an important conceptual observation. Physiologists had argued that the saccadic system could not exclusively rely on retinal input, or on (slow) visual feedback. In normal orienting behavior, the saccadic system may postpone a goal-directed eye movement to an (extinguished) visual target, as it may first make (one or several) intervening saccades to other locations in the visual field. To compensate for such intervening eye movements, the retinal coordinates of the original saccade goal would no longer be valid, and would therefore be better expressed in an *extraretinal* reference frame (eg, head- or world-centered). This is achieved by adding the initial eye position (or eye and head positions) at the time of target presentation (Robinson model, see chapter: The Gaze-Orienting System) to the retinal target coordinates:

$$\vec{T}_H = \vec{T}_E + \vec{E}_H \quad or \quad \vec{T}_W = \vec{T}_E + \vec{E}_H + \vec{H}_W \tag{10.1}$$

Robinson argued that if the motor map would express extraretinal spatial saccade goals, instead of the target's visual coordinates, stimulation-evoked saccade vectors would have to vary with initial eye position. However, his experiments indicated, that evoked saccades did not change with initial eye position at all (*fixed–vector saccades*), and that the motor map was therefore expressed in *oculocentric polar coordinates* (Fig. 10.2B). For the Robinson model this seemed a bit problematic, however, as the collicular role in saccade generation remained elusive, despite its prominent input to the brainstem pulse generator. In the previous chapter we have seen how Scudder's proposal could remedy this problem.

Conformal mapping: Which neurophysiological mechanism would be responsible for the highly nonhomogeneous collicular motor map? Because the same principle appears to hold for the visual field mapping onto the surface of the primary visual cortex, we describe this mapping first. We recall from chapter: The Gaze-Orienting System that each retinal receptive field is innervated by an approximately fixed number of photoreceptors, while photoreceptor density decreases as $\sim 1/r^2$ with retinal eccentricity. Suppose that ΔA corresponds to a small area in visual space, which innervates ΔN retinal ganglion cells. The receptive field area, $A = \pi\sigma^2$ (where $\sigma \propto r$) stimulates equal numbers of ganglion cells, say N. Thus, a relative change in receptive field size will lead to the same relative change in the number of stimulated ganglion cells. The neural mapping should therefore obey the following constraint:

$$\frac{\Delta A}{A} = \frac{\Delta N}{N} \tag{10.2}$$

Using polar coordinates to describe visual coordinates, and the mapped neural coordinates, the differentials can be written as:

$$\Delta A = r\Delta r\Delta\phi \quad \text{and} \quad \Delta N = r'\Delta r'\Delta\phi' \tag{10.3}$$

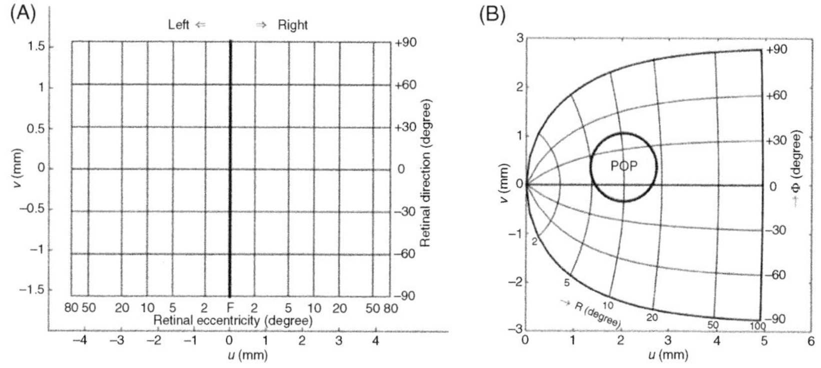

FIGURE 10.3 (A) Conformal complex–logarithmic mapping function of Eq. (10.6) that maps the entire visual field (up to eccentricity $r = 80$ degrees) onto a Cartesian (u,v) grid. (B) Afferent SC mapping function of Ottes et al., 1986 [Eq. (10.10)] for the right visual hemifield.

The mapping is then given by

$$(r',\phi') = T(r,\phi) \tag{10.4}$$

with the observation that $A = \alpha r^2$ it can now be verified that Eq. (10.2) is satisfied by the *complex–logarithmic mapping function*, which is defined by

$$\begin{aligned} &\vec{z} \Rightarrow \vec{w} = \ln \vec{z} \\ \text{where } &\vec{z} \equiv x + iy = re^{i\phi} \\ \text{and } \quad &\vec{w} \equiv u + iv = r'e^{i\phi'} \end{aligned} \tag{10.5}$$

(but this is not the only solution, see Problem 10.1). Fig. 10.3A illustrates the complex–log map $(u,v) = T(r,\varphi)$ of Eq. (10.4). Note that in visual space, isoeccentricity lines (concentric circles in visual space), become vertical lines in the map, and isodirection lines (radial spokes on the retina) are mapped as horizontal lines. In visual space circles and radii are mutually orthogonal, and their orthogonality is preserved in the complex–log map. For that reason it is called a *conformal* mapping function (a mapping that preserves angles). The neural mapping would thus be of the following form:

$$\begin{aligned} \vec{w} &= \vec{B} \ln \vec{z} \\ u &= B \ln r = B \ln \sqrt{x^2 + y^2} \\ v &= B\phi = B\,\text{atan}\left(\frac{y}{x}\right) \end{aligned} \tag{10.6}$$

with B the so-called *cortical magnification factor*:

$$\vec{B} \equiv \left(\frac{\partial u}{\partial r}, \frac{\partial v}{\partial \phi} \right) = \left(\frac{B}{r}, B \right) \tag{10.7}$$

It expresses the amount of neural space (in millimeters) traversed in the map per radian of angular displacement in visual space.

Comparing Figs. 10.3A and 10.2A one may observe striking similarities, among which are the polar–coordinate organization of the map and the logarithmic compression along the eccentricity axis. However, a noticeable difference, is that the fovea, $(R,\Phi) = (0,0)$, is a singularity in the complex–logarithmic map, as it maps as a vertical line at $u = 0$ running from $v = -\pi/2$ to $v = +\pi/2$ mm. In Robinson's microstimulation map (or in primary visual cortex) this seems not to be the case; in reality, the fovea seems to occupy a point image. A simple remedy for this problem would be to introduce a minor change in the formulation of Eq. (10.5), by adding a horizontal shift vector, $\vec{A} = (1,0)$, leading to:

$$\begin{aligned} \vec{w} &= \vec{B} \ln\left(\vec{z} + \vec{A}\right) \\ u &= B \ln\left|\vec{r} + \vec{A}\right| = B \ln \sqrt{(x+1)^2 + y^2} \\ v &= B\phi = B \,\text{atan}\left(\frac{y}{x+1} \right) \end{aligned} \tag{10.8}$$

Indeed, the fovea at $r = 0$ is then mapped to the origin at $(u,v) = (0,0)$, and the singularity is gone. In Problem 10.2, the reader can explore the effect of this change on the shape of the map. In the next section we apply the idea of a complex–log mapping to a more precise description of the afferent collicular motor map of Fig. 10.2A, and explain how this map could be understood in terms of its function in saccade control.

10.2 A MULTISENSORY MOTOR MAP FOR GAZE ORIENTING

Sensory responses of cells in the intermediate layers of the SC can be evoked by visual stimuli (Goldberg and Wurtz, 1972), but they can also be driven by auditory (Jay and Sparks, 1984) and somatosensory (Groh and Sparks, 1996) inputs. Moreover, these cells have a vigorous motor burst that is tightly locked to the onset of a saccadic eye movement into their sensory receptive field (Wurtz and Goldberg, 1972a; Sparks et al., 1976). The intermediate layers of the SC can thus be considered to contain a *multisensory motor map* (Stein and Meredith, 1993).

Microstimulation in the intermediate and deep layers of the SC in head-restrained animals generates short-latency (at a delay of 20 ms), stereotyped saccadic eye movements at low thresholds (below 10–20 μA) with a vector that

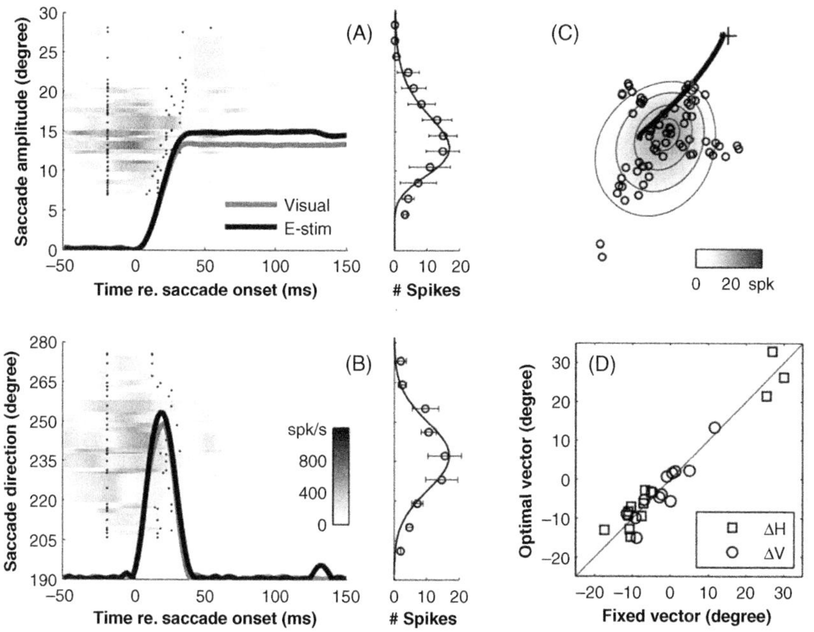

FIGURE 10.4 (A,B) Movement field of a SC neuron. Amplitude scan (A) and direction scan (B) through the movement field. The position and velocity traces for the optimal saccade (gray) and stimulation-evoked saccade (black) are also shown. Fitted curves on the right: Ottes et al. (1986) model [Eq. (10.10)] to the number of spikes across the movement field. Note that the variance of the activity increases with the number of spikes (signal-dependent noise). (C) There is a close correspondence of the stimulation-evoked saccade (solid trace) with the center of the cell's movement field. (D) Summary of stimulation versus recording results for different sites. *(Source: After Van Opstal and Goossens, 2008.)*

corresponds closely to the visual receptive field and movement field of cells near the stimulation electrode (Van Opstal et al., 1990; Fig. 10.4). Also, when the head is unrestrained, microstimulation produces perfectly normal eye–head gaze shifts (in cat: Roucoux et al., 1980; in monkey: Freedman et al., 1996) that are directed toward the receptive fields and movement fields of nearby cells.

A collicular cell in the intermediate (and deep) layers responds with a vigorous burst of spikes for a restricted range of eye (or eye–head) saccades. The set of saccade vectors for which the cell is recruited is defined as the cell's *movement field* (Wurtz and Goldberg, 1972a; Sparks et al., 1976; Ottes et al., 1986; Van Opstal et al., 1990; Goossens and Van Opstal, 2006; 2012). Although there are different ways to quantify the activity of an SC neuron (eg, the peak firing rate of the burst, or its mean firing rate during a fixed window around the saccade; Goossens and Van Opstal, 2000b), we prefer (for reasons that become clear soon) to quantify SC activity by the *number of spikes in the burst*, counted from 20 ms before saccade onset, to 20 ms before its offset.

Fig. 10.4 shows a typical example of an SC movement field. The cell responds with a maximum of 20 spikes for its optimal saccade vector, (R_{opt}, Φ_{opt}) = (14, 235) degrees, and its spike count gradually decreases the more saccades deviate from this optimum vector (Fig. 10.4C). Microstimulation at the recording site produces a normal main-sequence saccade with near-optimal metrics (black traces in Fig. 10.4A–C). Fig. 10.4D shows the close correspondence for the optimal horizontal/vertical saccade components for 13 different SC sites versus the stimulation-evoked saccades.

The movement fields (MF) across the SC motor map have the following characteristics:

- The long and short axes of a MF increase *linearly* with R_{opt}. Thus, the MF width for a cell with $R_{opt} = 20$ degrees is about 5 times larger than for $R_{\text{opt}} = 2$ degrees.
- An amplitude scan through the center of the MF is *positively skewed*: the peak is located off-center, toward the small-amplitude end of the MF (Fig. 10.4A). In contrast, direction scans are *symmetric* (Fig. 10.4B).
- The directional tuning width of a MF is roughly *constant* across sites, and amounts to about $\Delta\Phi \sim 60$ degrees (eg, Fig 10.4B).

These properties differ in essential ways from the tuning of cells in the brainstem pulse generators (in PPRF and riMLF; PPRF, paramedian pontine reticular formation, site of the horizontal saccade pulse generator; riMLF, rostral interstitial nucleus of the medial longitudinal fasciculus, site of the vertical/torsional pulse generator), the neural integrators, and the oculomotor neurons. Oculomotor brainstem cells have movement fields that are consistently described by *cosine tuning* functions (ie, these cells perform an inner vector product):

$$F(\vec{r}) = f(|\vec{r}|)(\vec{r}\,\hat{n}) = f(|\vec{r}|)\left[n_x \Delta h + n_y \Delta v\right] \tag{10.9}$$

with $\vec{r} = (\Delta h, \Delta v)$ the saccade-vector coordinates, $\hat{n} = (n_x, n_y)$ the unit vector in the cell's optimal direction, and $f(|\vec{r}|)$ a monotonically increasing (saturating) function of saccade amplitude. The directional tuning width of these neurons is therefore $\Delta\Phi$=180 degrees.

The underlying reason for cosine tuning characteristics in the brainstem is that these cells ultimately produce the *dynamic force* for saccadic muscle contractions (see chapter: The Gaze-Orienting System). Their neural activity thus represents a *temporal code* for the saccade. In contrast, the classical view given in the Introduction holds that the SC provides a *spatial saccade code* (the desired gaze-displacement vector). Thus, the SC to brainstem projections somehow perform a *spatial-to-temporal transformation* (STT).

We will provide more details on the neural implementation of the STT in Section 10.3. First, however, we model the SC motor map and SC movement fields according to the classical static view, as this will provide a basic understanding of the abovementioned properties of SC MFs. We then discuss a static population-encoding model that explains how the STT could be implemented. Thereafter, we will look at the burst profiles of SC cells to extend the static model to a dynamic STT.

The static MF: The first two steps in modeling the STT of the SC to brainstem projection is to provide an accurate description of the motor map, and of the neural population activity in the map during saccades. Ottes et al. (1986) fitted Robinson's microstimulation map (Fig. 10.2A) with an anisotropic extension of Eq. (10.7), by allowing $\vec{B} = (B_u, B_v)$. Further, $\vec{A} = (A, 0)$ was a third free parameter for the map. This extension defines the following afferent (*visual space to SC space*) mapping function (shown in Fig. 10.2B; see Problem 10.3):

$$\vec{w} = \vec{B} \ln\left(\frac{\vec{z} + \vec{A}}{\vec{A}}\right)$$

$$u = B_u \ln\left(\frac{\sqrt{R^2 + 2AR\cos\Phi + A^2}}{A}\right) \qquad (10.10)$$

$$v = B_v \operatorname{atan}\left(\frac{R\sin\Phi}{R\cos\Phi + A}\right)$$

The optimal values for the free parameters of monkey SC were found to be

$$B_u = 1.4 \text{ mm} \quad B_v = 1.8 \text{ mm/rad} \quad A = 3.0 \text{ degrees}$$

The different values for B_u and B_v make the map slightly *anisotropic*: the magnification factor (gain) for the u-direction (amplitude) is smaller than for the v-direction (saccade direction). In Problem 10.7 the reader will analyze a potential behavioral correlate for this anisotropy regarding the distribution of saccade end points to visual targets (Fig. 8.10A).

The next step in modeling the SC movement fields and the STT is the description of the neural population activity for saccades. The idea of Ottes et al. (1986) is as simple, as it is elegant: because a single SC cell is recruited for a range of saccade vectors, it follows that *each single saccade involves the recruitment of a localized population of neurons*. Thus, when mapping all saccade vectors (R,Φ) of the measured movement field of a single cell (like the one in Fig. 10.4) back onto the afferent SC motor map [through Eq. (10.10)], one obtains an estimate for the distribution of cells that was recruited for the optimal saccade of the recorded MF. This idea of *reciprocity* only assumes that the cell density in the motor map is constant. Interestingly, by doing so, the asymmetry in the amplitude scan of the MF is not present in the motor map; the SC distribution resembles a rotation-symmetric *Gaussian population profile*. The SC population is therefore conveniently modeled by:

$$N\left(u,v;u_0,v_0,\sigma_{\text{Pop}}\right)=N_0\exp\left[-\frac{\left(u-u_0\right)^2+\left(v-v_0\right)^2}{2\sigma_{\text{Pop}}^2}\right] \tag{10.11}$$

with N_0 the maximum number of spikes in the SC burst for the optimal saccade, (u_0,v_0) the SC image point [through Eq. (10.10)] of the optimal saccade vector (R_0,Φ_0), and σ_{Pop} the estimated width of the population (in millimeters). Note that Eq. (10.11) has four free parameters to fully specify a cell's MF. To find the optimal saccade vector and tuning width of the MF from hundreds of measured saccade responses, the fit is carried out on Eq. (10.11) in SC (u,v) coordinates for all measured saccade vectors. After finding $[N_0,u_0,v_0,\sigma_{\text{Pop}}]$ the inverse mapping of Eq. (10.10) is subsequently applied on (u_0,v_0) to obtain the cell's optimal saccade. The reader may verify (Problem 10.4) that the *efferent mapping function* [ie, the inverse of Eq. (10.10)] is determined by:

$$\begin{aligned} x &= A\exp\left(\frac{u}{B_u}\right)\left[\cos\left(\frac{v}{B_v}\right)-1\right] \\ y &= A\exp\left(\frac{u}{B_u}\right)\sin\left(\frac{v}{B_v}\right) \end{aligned} \tag{10.12}$$

The curves through the spike-count data in Fig. 10.4A,B show the cell's MF according to the model of Eqs. (10.10) and (10.11). In general, the model provides an excellent fit to single-unit SC data (often, $r > 0.95$ for hundreds of responses into the MF; eg, Goossens and Van Opstal, 2006; 2012). By collecting the MFs of hundreds of SC cells, the following important observations have been made (Ottes et al., 1986; Goossens and Van Opstal, 2006; 2012):

- The population-activity profile in the SC is *translation–invariant*: for every saccade the population has the same shape and size; it only shifts toward the center locus of the new optimum.
- The width of the population is approximately $\sigma_{\text{Pop}} \sim 0.5$ mm.
- The number of spikes for the optimal saccade is on average $N_0 \sim 20$ spikes. Despite considerable variability among cells (N_0 between 15 and 40 spikes), N_0 does not vary systematically across the motor map.
- Thus, assuming a *constant cell density* across the motor map, the following important corollary can be made:

> Every saccade recruits the same number of cells in the SC motor map, and the population produces a fixed total number of spikes.

In Problem 10.5 the reader may estimate this total number of spikes/saccade.

The MF model of Eqs. (10.10) and (10.11) readily explains the asymmetry observed in amplitude scans, and the linear increase of the tuning width with R_{opt}. Both properties relate to the inhomogeneous, logarithmic mapping of saccade amplitude. Let's illustrate this for three examples, say horizontal saccades at $R = 2$, $R = 10$, and $R = 20$ degrees. The horizontal meridian of the map ($y = 0; v = 0$) is given by

$$u = B_u \ln\left(\frac{x+A}{A}\right) = 1.4\ln\left(\frac{R+3}{3}\right) \Leftrightarrow R = 3\left(\exp\left(\frac{u}{1.4}\right) - 1\right) \quad (10.13)$$

The 2.0 degrees site is thus centered at $u_2 = 0.72$ mm, the 10 degrees site at $u_{10} = 2.1$ mm, and the 20 degrees site at $u_{20} = 2.86$ mm. Now suppose a $-$0.5 mm variation around these centers (corresponding to the one-sigma population width), then we can readily check how the edges of the population map to amplitude space through the efferent mapping of Eq. (10.13). For the 2.0 degrees site these edge amplitudes are 0.5 and 4.2 degrees, respectively, spanning a range of 3.7 degrees. For the 10 degrees site they are 6.4 and 16.2 degrees, with a range of 9.8 degrees, and for the site at 20 degrees we obtain 13.2 and 30.1 degrees, respectively, a range of 16.9 degrees. The ranges indeed increase linearly (slope $\sim$ 0.73) with R_{opt}. Note also that the centers are located asymmetrically with respect to the edges (see Problems 10.6 and 10.7).

The static STT: How does the total population of recruited cells encode the saccade displacement vector? Van Gisbergen et al. (1987) proposed an extremely simple, linear summation model in which only two factors determine a cell's contribution to the saccade vector:

1. It's activity: cell's that do not fire, do not contribute to the saccade.
2. It's horizontal/vertical efferent connection strengths with the brainstem, specified by Eq. (10.12).

According to this *static linear ensemble-coding model*, the saccade vector is determined by

$$\vec{S} = \eta \sum_{n=1}^{N_{\text{Pop}}} N_n \vec{w}_n \quad (10.14)$$

with N_n the number of spikes of cell n in the population, and $\vec{w}_n = (x_n, y_n)$ the efferent synaptic connection strengths of neuron n with the brainstem [Eq. (10.12)]. The fixed scaling factor η ensures that all saccades have appropriate metrics (it depends on the number of cells in the simulations, and hence on the assumed cell density). Fig. 10.5 provides a schematic overview of the static ensemble-coding model.

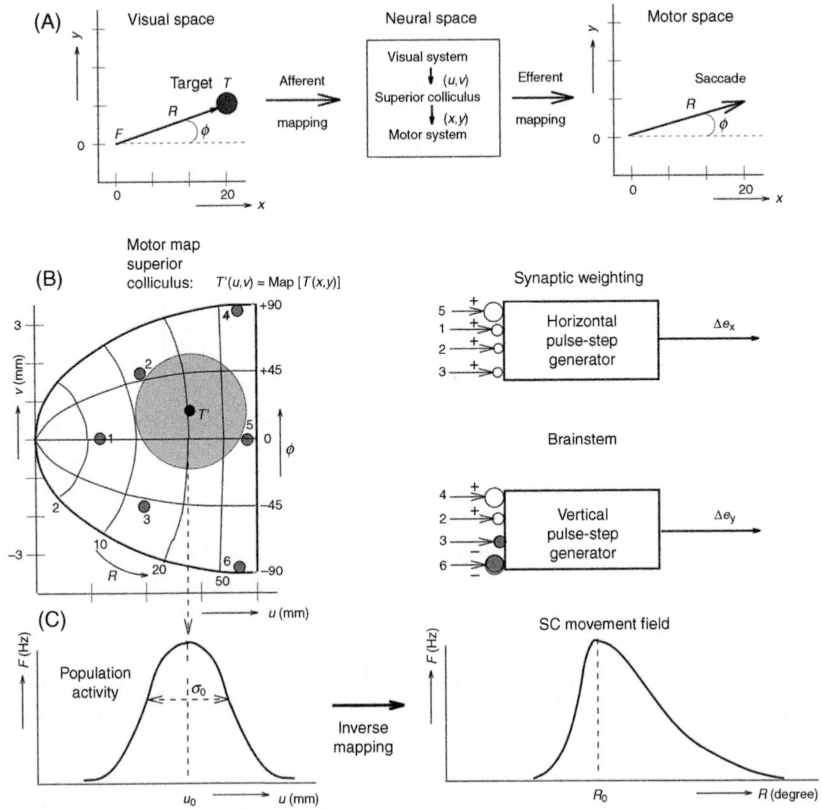

FIGURE 10.5 Linear static ensemble-coding model, after Van Gisbergen et al. (1987). (A) Conceptual relations between afferent [visual input to SC space; Eq. (10.10)] and efferent [SC space to motor output; Eq. (10.12)] mapping functions. (B) The center panels show the afferent motor map with a hypothetical neural population at the neural image of target T. The efferent synaptic connections of six example neurons [Eq. (10.12)] are shown as inputs for the horizontal/vertical oculomotor brainstem. (C) Population activity profile in SC, and the related amplitude scan of the central cell's movement field.

Note that the linear ensemble-coding model of Eq. (10.14) was not designed to generate real saccades: it is a *"static"* model because time does not play an explicit role in this scheme. There are no dynamic feedback loops, and the SC activity is only specified *after* the burst is already over. It thus provides an account on how the SC population can encode a displacement vector, but not how that vector is used as a motor command. It does, however, issue a vectorial signal to the brainstem, which is subsequently decomposed, through the efferent mapping, into horizontal/vertical displacement commands for the pulse generators. If the temporal behavior of the SC signal would be irrelevant (as held by the classical models, see Introduction), the static ensemble-coding model suggests a neural implementation for the independent and coupled PG models discussed in chapter: The Gaze-Orienting System.

However, if the temporal behaviors of SC spike trains *do* matter, the SC output signal could potentially embed a *dynamic vectorial signal*, like required by the common-source model. In Section 10.3 we discuss the evidence for this latter hypothesis.

Vector averaging? The linear summation model of Eq. (10.14) is not the only possibility of how a population of cells could encode saccadic eye-displacement vectors. Indeed, Lee et al. (1988) suggested, on the basis of focal reversible lesions in the SC motor map, that the linear model could not account for the following pattern of observed deficits in saccade metrics:

1. The saccade directed into the center of the lesion had correct metrics [this is, indeed, *not* explained by Eq. (10.14)].
2. Saccades that are not directed into the center of the lesion had their endpoints *directed away* from the lesioned site: smaller saccades were too small [this is also explained by Eq. (10.14)], but larger saccades were too large [*not* explained by Eq. (10.14)].

Lee et al. (1988) therefore proposed an alternative weighting scheme for the cells that is based on a *center-of-gravity* computation (vector averaging):

$$\vec{S} = \frac{\sum_{n=1}^{N_{\text{Pop}}} N_n \vec{w}_n}{\sum_{n=1}^{N_{\text{Pop}}} N_n} \tag{10.15}$$

Note that Eq. (10.15) is a *nonlinear weighting model*, as it relies on division by the total population activity. Although this model could indeed explain the observed deficits in saccade endpoints, a second result of this study was that the affected saccades were also much *slower* (by as much as 20%). Because in the vector-averaging model the SC is explicitly denied a dynamic role (in fact, due to the normalization, the saccade vector is already specified by the first few spikes...), Eq. (10.15) does not readily account for this strong effect on saccade kinematics, unless additional nonlinearities are incorporated (like a population-dependent gain control of the brainstem PG). In the next section we will see how a much simpler *dynamic linear ensemble-coding* model may explain both phenomena.

10.3 SC: A VECTORIAL PULSE GENERATOR

Berthoz et al. (1986, in cat), later also Lee et al. (1988), and Van Opstal and Van Gisbergen (1990, both in monkey), had provided some early evidence that activity levels in the SC might be related to saccade kinematic performance: Slow saccades were often associated with lower peak firing rates of SC cells than fast saccades of the same metrics. However, a problem with recording data, was that the variability in saccade kinematics was typically quite small when compared to the large intrinsic variability of single-cell firing rates. Finding strong causal relationships between a neuron's firing rate and saccade kinematics on single trials was therefore hard, or next to impossible (eg, Goossens and Van Opstal, 2000b).

In an attempt to vastly increase the variability of saccade kinematics, while measuring single-cell responses in the monkey SC, Goossens and Van Opstal (2000a,b) noninvasively perturbed saccades by evoking fast reflexive eyelid blinks to a brief air puff on the eye. In this paradigm perturbed saccades became heavily curved, highly variable, and slowed down by more than 50% (see Fig. 10.7 for a representative example). Interestingly, SC activity also changed dramatically in the blink-perturbation trials, peak-firing rates dropped substantially, and burst durations increased accordingly. However, despite these large behavioral and neurophysiological changes, two quantities remained unaffected:

1. Saccade-endpoint accuracy, or, in other words: the saccade *displacement vector* (Goossens and Van Opstal, 2000a).
2. The *number of spikes in the saccade-related burst*, counted from 20 ms before saccade onset, to 20 ms before saccade offset (Goossens and Van Opstal, 2000b).

Goossens and Van Opstal (2006) hypothesized that the number of spikes in the SC burst provides the key for its neural code. They proposed that the SC issues a *dynamic desired eye-displacement vector* through the *cumulative* number of spikes from the recruited neural population in the following way:

$$\vec{S}(t-\Delta T_0)=\gamma\sum_{n=1}^{N_{\text{Pop}}}\sum_{s=1}^{N_{\text{spks}}}\vec{w}_n\delta(t-\tau_{n,s}) \qquad (10.16)$$

with γ a fixed scaling constant, $\delta(t-\tau_{n,s})$ represent the occurrence of spike s of cell n at time $\tau_{n,s}$. The lead time, ΔT_0, is the fixed time delay between the SC burst onset and the saccade onset (20 ms). Each spike in the burst of each recruited cell thus adds a fixed (infinitesimal) vector contribution, $\gamma\vec{w}_n$, to the saccade.

Like in Scudder's model (see chapter: The Gaze-Orienting System), the collicular output is linearly integrated at horizontal and vertical comparators, the outputs of which (horizontal and vertical dynamic motor errors) are sent to the horizontal and vertical pulse generators. In the dynamic linear ensemble-coding model. However, the latter are taken to be entirely *linear* (with gain B). The pulse generators are positioned within two independent local feedback loops, which feed horizontal and vertical eye-velocity signals back to the comparators after a short delay, ΔT. The entire model has only two free parameters (B and ΔT), which can be determined by fitting the model's output to a small subset of selected saccades (see Problem 10.8).

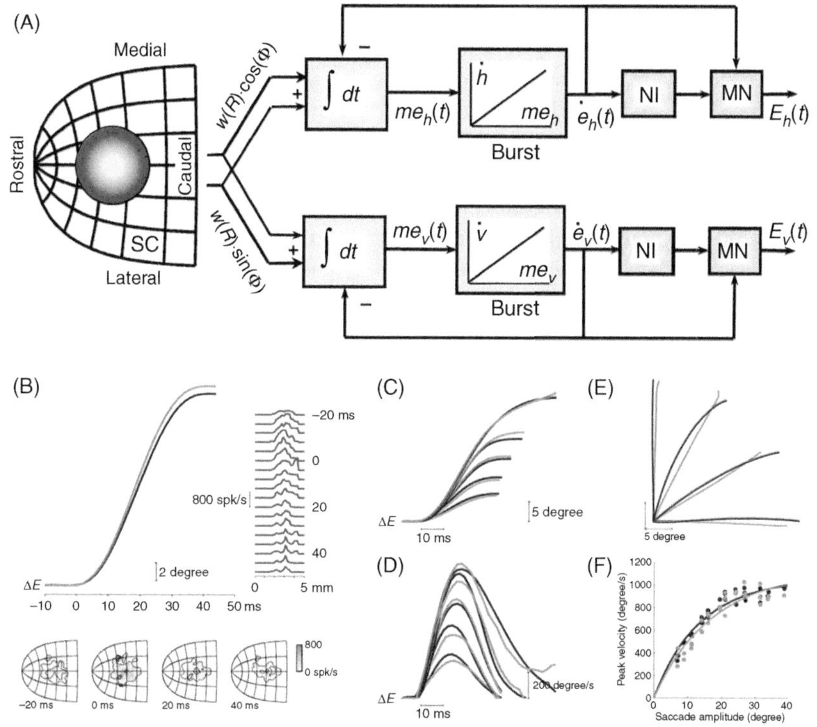

FIGURE 10.6 **Dynamic ensemble-coding model of Goossens and Van Opstal (2006).** Simulations with the model (only two free parameters, pulse–generator gain *B*, and feedback delay) yield excellent results. The model accounts for the nonlinear main sequence, straight saccades with component stretching, and skewed velocity profiles. These are all (nonlinear) emergent properties that reside in the spatial–temporal SC responses. Note that the recruited population does not move toward the rostral zone during the saccade. Black: eye-movement data; gray: model predictions. *(Source: Goossens and Van Opstal, 2006, with kind permission.)*

Model performance: Since it's almost impossible to make it any simpler than this, let us first verify the performance of this model, before going any further. Fig. 10.6A shows the model circuit, which is driven by *raw recorded* spike trains of about $N_{Pop} = 150$ cells according to Eq. (10.16). Of each cell the movement field parameters were first obtained by applying Eqs. (10.10, 10.11) to recorded spike trains 100–400 saccades across the oculomotor range. The MF then allowed us to implement the (fixed) efferent weightings for each cell.

The output of the model is a 2D eye-movement trajectory that is fully specified by the firing patterns of the recorded cells. Note that in principle the predicted eye-movement trajectory could be completely haphazard, which would occur, for example, when cells would fire erratically during each saccade. The result, however, is astonishingly close to the real measured saccadic eye movements (Fig. 10.6B–F), which reflects a high level of synchronization in the population activity (see Goossens and Van Opstal, 2012, for a detailed analysis on this aspect).

Not only does the model correctly predict the overall saccade displacement vector [Fig. 10.6B; which should be identical to the prediction of Eq. (10.14)], it also matches the full trajectory, including the shape of saccade velocity profiles: small saccades yield nearly symmetric eye-velocity profiles, larger saccades are more and more skewed, with a clear increase of saccade duration (note that these properties reflect nonlinear behavior!). Furthermore, predicted saccade trajectories are straight (Fig. 10.6E), which means that the model correctly produces component stretching. And last but not least, the model predicts saccades that obey the *same* nonlinear main sequence as the measured saccades (Fig. 10.6F). As none of these nonlinear kinematics were explicitly built into the scheme of Fig. 10.6A (which is entirely *linear* from the SC henceforth), they all *have to* emerge from the spatial–temporal firing patterns within the SC motor map! This leads to the inevitable conclusion that:

> The active population in the SC motor map acts as a nonlinear vectorial pulse generator for saccades.

Model predictions. Eq. 10.16 tells us how a large population of SC cells encodes the (desired) saccade trajectory, but the model also makes specific predictions about the detailed behavior of single neurons during saccades of any direction and amplitude, irrespective of their kinematics (slow vs. fast). Note that each spike of each neuron, n, can only contribute a *fixed* mini-vector, $\gamma \vec{w}_n$, to the saccade, but the more spikes it delivers during the burst, the larger its total vectorial contribution in its own preferred direction. How many spikes the cell contributes is determined by its MF [Eqs. (10.10) and (10.11)]:

$$\vec{s}_{n,\mathrm{Tot}} = \gamma N(R,\Phi)\vec{w}_n \tag{10.17}$$

For a single neuron the hypothesis entails that each cell encodes the current displacement of the eye along its own desired trajectory. Let's first make this explicit for a saccade that is directed into the center of its movement field (Fig. 10.7). According to the model, a neuron will fire the *same* number of spikes for an exceptionally slow, and for a fast control saccade. In addition, however, while firing, the neuron signals how far the eye has moved along the straight optimal trajectory:

$$n(t) = \alpha \Delta e(t) \tag{10.18}$$

In Fig. 10.7 the time evolutions of the spike trains and movement trajectories for a fast control and slow blink-perturbed saccade are clearly very different. However, when plotted against each other, the two curves almost perfectly overlap, as predicted by Eq. (10.18).

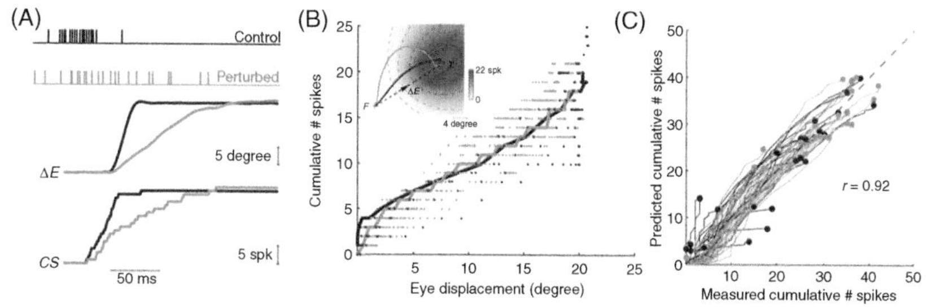

FIGURE 10.7 The cumulative number of spikes of an SC neuron (*CS*) encodes how far the eye has moved along its optimal trajectory, Δ*E*(*t*), from *F* to *T*. Here exemplified for a control trial (blue) and a blink-perturbed trial (green). (B) Inset: the cell's movement field, and a control and perturbed saccade trajectory. Both are goal directed. (A) Spike trains for the two trials. Note prolonged burst for perturbed trial; Δ*E*(*t*) is the projected eye trajectory on line FT, *CS*(*t*) is the cumulative spike count. (B) Spike count *CS*(*t*) versus Δ*E*(*t*) is identical for control and perturbed trial, as predicted by the model. (C) Predicted versus measured dynamic cumulative spike count [DMF model; Eq. (10.19)] for 400 saccades into the MF. Note the very high correlation ($r = 0.92$; $N > 15{,}000$ data points). *(Source: After Goossens and Van Opstal, 2006, with kind permission.)*

A second prediction is that the slope, α, of Eq. (10.18) varies in a specific way with saccade vector into the MF. In Fig. 10.7, the eye displacement for the optimal saccade is 20 degrees, for which this neuron fired 20 spikes. The slope in the $[n(t),\Delta e(t)]$ plot (panel B) is therefore $\alpha = 1.1$ spikes/degrees for both conditions. Now imagine a much larger saccade into the cell's movement field, say 35 degrees, for which the neuron would be expected to fire only seven spikes. In that case the model predicts that the slope of Eq. (10.18) will be reduced to $\alpha = 7/35 = 0.2$. For a much smaller saccade, for example, 10 degrees, the same number of spikes may be fired, but the slope will be $\alpha = 7/10 = 0.7$.

In other words, the expected slope α depends on both the predicted number of spikes for the particular saccade [Eq. (10.11)], and on saccade amplitude. In general, the cumulative number of spikes for the dynamic ensemble model, as function of time, amplitude, and direction is predicted to be:

$$n_{\text{DEM}}(t,R,\Phi) = \frac{N(R,\Phi)}{R}\Delta\hat{e}(t) \tag{10.19}$$

with $\Delta\hat{e}(t)$ the current length of the projection vector that runs between 0 (start of the saccade) to R (its end). Eq. 10.19 describes how the cell's movement field and firing rate evolves as function of time for *all* saccades. It therefore extends the classical, static movement field of Eqs. (10.10) and (10.11) (Ottes et al., 1986) to the concept of a *dynamic movement field* (*DMF*). Fig. 10.7C illustrates that the DMF model does a very good job in describing the full dynamics of collicular spike trains. For the population of 150 cells the obtained correlations were >0.80 for the far majority of cells.

A third crucial prediction of the dynamic linear ensemble-coding model is that the same concept should hold for *eye–head gaze shifts*. Therefore, Eq. (10.19) should actually be extended to:

$$n_{\text{Dem}}\left(t,\Delta G,\Phi\right)=\frac{N\left(\Delta G,\Phi\right)}{\Delta G}\Delta\hat{g}(t) \tag{10.20}$$

This is not a trivial extension because, as explained in chapter: The Gaze-Orienting System, a gaze shift is in fact an *abstract* motor command that consists of an arbitrary combination of eye and head contributions, and does not directly drive the eye and head motor systems. Each difference in the division of labor for eye (fast) and head (slow) in a given gaze shift leads to different gaze kinematics, but according to Eq. (10.20) this should not matter for the neural code of SC cells. The intrinsic kinematic variability of eye–head gaze shifts is automatically incorporated in Eq. (10.20).

Up to now there are no data in the published literature to either verify or refute the hypothesis, but preliminary evidence from our lab (unpublished so far) supports the prediction of Eq. (10.20), with only one extension: the predicted number of spikes for gaze shifts also depends weakly on the initial eye position in the orbit, E_0, in the following way:

$$N\left(\Delta G,\Phi,E_0\right)=\left(1+\vec{\varepsilon}\vec{E}_0\right)N\left(\Delta G,\Phi\right) \tag{10.21}$$

with $\vec{\varepsilon}$ a small eye-position tuning vector (length ~0.02–0.08 spikes/degrees). Equation 10.21 formulates a so-called *planar eye-position gain field*. It describes the modulation of neural activity for a given gaze shift in a monotonic way with changes in eye position along an optimal direction. Similar eye-position effects in SC cells were reported for head-fixed saccades (Van Opstal et al., 1995), in posterior parietal cortex neurons (Zipser and Andersen, 1988), and also in the IC (described in Section 10.4; Groh et al., 2001; Zwiers et al., 2004).

An interesting hypothesis for the function of eye-position gain fields is that they could constitute an efficient *multiplexed* population code, which would allow downstream readout systems in the brainstem to extract multiple sources of information from the same population (eg, oculocentric information, but also craniocentric coordinates, or even 3D rotational kinematic signals; Zipser and Andersen, 1988; Van Opstal and Hepp, 1995).

VPG mechanism. The linear dynamic ensemble-coding model of Eq. (10.16) successfully predicts the trajectories of oblique saccades, but it doesn't yet answer the question of how the spatial–temporal patterns of activity in the SC motor map build a nonlinear vectorial pulse generator. As the output of the pulse generator should represent the instantaneous radial eye velocity, let's first look at some saccade-related burst profiles and check for a potential velocity-related signal. Fig. 10.8 shows four averaged and normalized burst profiles that

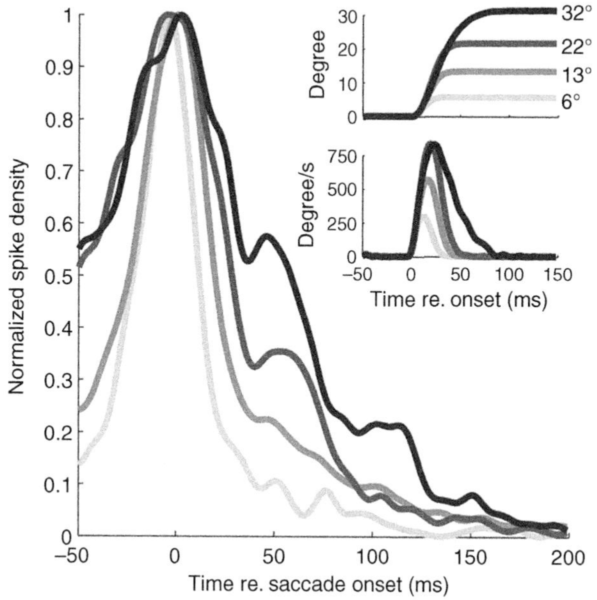

FIGURE 10.8 Normalized average saccade-related bursts for cells located at different sites (also see Fig. 10.9) encoding different amplitudes. Note similarities between burst- and saccade velocity profiles. *(Source: From Van Opstal and Goossens, 2008.)*

were calculated from four small groups of 5–10 recorded cells, each group encoding for a saccade of different amplitude (R_{opt} = 6, 13, 22, and 32 degrees, respectively). The insets show the average position traces and velocity profiles of the four saccades. Qualitative inspection suggests that these cells could indeed encode the radial eye velocity for their optimal saccade: burst asymmetry (skewness), and burst duration both change in a similar way with saccade amplitude as the respective velocity profiles.

The spatial–temporal mechanism by which the SC population implements the nonlinear vectorial pulse generator is shown in Fig. 10.9. Here we see the raw individual burst profiles of all selected cells used in Fig. 10.8, as well as their (gray encoded) averages. The most striking feature in these bursts that immediately draws attention is that the peak firing rates of the cells systematically *decrease* with the optimal saccade amplitude, while their burst duration and burst skewness increase. This seems almost counterintuitive; peak velocities of saccades are supposed to increase monotonically with saccade amplitude, they do not decrease! So what's going on here?

A quantitative analysis of the saccade-related bursts of 150 cells distributed across the SC motor map (Goossens and Van Opstal, 2012) revealed the following features:

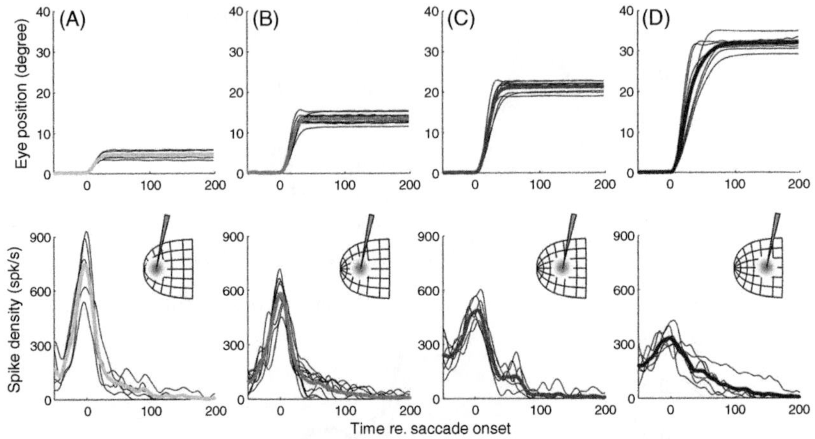

FIGURE 10.9 Mechanism by which the SC acts as a nonlinear vectorial pulse generator. Peak firing rates of SC cells systematically decrease with the optimal saccade amplitude, while burst duration increases. The total number of spikes, however, remains unaffected. *(Source: Goossens and Van Opstal, 2012, with kind permission.)*

1. Peak firing rates are inversely related to the optimal saccade amplitude, and may in first approximation be described by:

$$F_{\text{Peak}} \propto \frac{F_{\max}(R=0)}{R_{\text{opt}}+c} \tag{10.22}$$

2. The *number of spikes* in the burst for the optimal saccade does not systematically depend on a cell's location in the motor map:

$$N_{\text{Spks}}\left(R_{\text{opt}},\Phi_{\text{opt}}\right)=N_0 \approx 20 \text{ spikes} \tag{10.23}$$

3. Burst duration and burst skewness both increase with R_{opt}:

$$D_{\text{Brst}}=a+bR_{\text{opt}} \quad \text{and} \quad S_{\text{Brst}}=a+bD_{\text{Brst}} \tag{10.24}$$

4. A cross-correlation analysis of the burst profiles of all cells for all saccades indicates that for any given saccade the bursts in the recruited population *all have the same shape*. The cells in the center of the population thus seem to dictate the burst profiles of the other cells. This property suggests the presence of intracollicular connections.
5. All cells in the population reach their peak firing rates at about the same time, indicating a considerable amount of *synchronization* in the population. This ensures that the population sends its strongest possible (ie, fastest possible) signal to the brainstem.

How does the saturating main-sequence characteristic of peak eye velocity (eg, Fig. 10.6F) causally relate to these properties? Computer simulations with the model circuit of Fig. 10.6 revealed two crucial factors (Van Opstal and Goossens, 2008):

1. the decrease of the cells' peak firing rates, together with
2. a constant number of spikes in their optimal bursts

In other words, the nonlinear afferent and efferent mappings [Eqs. (10.10,10.11)] do not play any role in this problem!

A simple back of an envelope calculation to estimate peak eye velocity could go as follows: it is determined by the peak firing rate of (mainly central) cells multiplied with their efferent effect on the brainstem [which equals *R*, eg, Eq. (10.13)]:

$$V_{\text{Peak}}(R) \propto F_{\text{Peak}}(R)\left|\vec{w}(u,v)\right| = \alpha F_{\max}(0)\frac{\mathrm{R}}{R+c} \tag{10.25}$$

This result has indeed the characteristics of the main sequence: a linear increase for small *R* at slope $\alpha F_{\max}(0)/c$, while saturating at $\alpha F_{\max}(0)$ for large *R*.

Although a simplification, Eq. (10.25) also shows that when the peak-firing rate does not depend on *R*, the denominator disappears and the peak velocity increases *linearly* with saccade amplitude. In that case the entire system of Fig. 10.6A would be linear! Our simulations with realistic SC populations in the 2D complex-log map revealed that the spatial gradient of decreasing peak firing rate along the rostral–caudal amplitude axis is indeed the crucial mechanism by which the SC acts as a nonlinear vectorial pulse generator (Van Opstal and Goossens, 2008).

Why a nonlinearity? From all this we may suspect that the nonlinear saccade kinematics seem to be a *deliberate strategy* of the saccadic system, rather than a mere unavoidable (passive) effect of saturating neurons in the brainstem. Indeed, the saturation of peak-eye velocity for large saccade is *not* due to neural saturation of the vectorial pulse generator, because Fig. 10.9 clearly shows that the caudal cells fire at *much lower rates* than the rostral cells for small saccades! But why would the saccadic system go through all this trouble to *create* this nonlinearity?

Here we touch upon an interesting question, for which there is no unique, nor an obvious answer. However, given the fact that the saccadic system has to cope with conflicting demands: to direct the fovea as *fast* and as *accurately* as possible to an uncertain (noisy) peripheral location, may have prompted the system to find an *optimum speed–accuracy trade-off*.

Theoretical studies (Harris and Wolpert, 2006; Tanaka et al., 2006) have suggested that the optimal strategy for a foveate eye-movement system that has to be both fast and accurate, but at the same time has to cope with signal-dependent noise in which target uncertainty (retina) and neural noise increase linearly with eccentricity and amplitude, is *main sequence behavior*.

The saturation of eye velocity at large (more noisy) amplitudes renders the system more prudent, and therefore less vulnerable, for large and time-consuming target overshoots (see also Problem 10.7 for an analysis of the effect of noise in the motor map on saccade endpoint distributions). In chapter: Multisensory Integration we will return to the topic of optimality in sensory processing and motor control, when dealing with multisensory (audio–visual) integration.

Averaging versus summation? The model of Eq. (10.16) generates saccades by adding all spikes from the SC output, but it does not yet include a mechanism to start and stop access of SC spikes to the brainstem circuitry. A GO/STOP mechanism is, however, required as SC cells do not only fire a saccadic motor burst: often, the saccade motor burst is accompanied by a considerable prelude build-up in activity, and the burst may also be followed by additional spikes well after gaze-shift offset (Goossens and Van Opstal, 2000b). Clearly, spikes before saccade onset and after saccade offset do not contribute to the saccade proper. Since for any given saccade the population generates a *fixed* number of spikes (Problem 10.5), the OPN stop gate could be based on a simple spike-counting criterion: whenever the required number of spikes from the population is reached, the gate closes. Note that this introduces a potential nonlinearity in the model of Eq. (10.16) (a fixed cut-off).

Interestingly, this cut-off mechanism provides an alternative explanation for the remarkable pattern of saccade deficits observed in the focal lesion experiments of Lee et al. (1988), described above (Goossens and Van Opstal, 2006). Fig. 10.10 shows the results of simulated saccades generated with real spike data, by only omitting the spike contributions from a small circular area around $(R_{\text{opt}}, \Phi_{\text{opt}}) = (18, 42)$ degrees. Note that the only reason that these simulations work at all resides in the fact that SC cells tend to fire additional spikes after (normal) saccade offsets. As these later spikes are associated with low firing rates, their contribution to saccade velocity cannot compensate for the loss of vigorous burst power from the lesioned central area. As a result, the saccade will be slower too. In Problem 10.9 the reader explains why saccade vectors are consistently *directed away* from the lesioned site.

Finally, a fixed cut-off from the motor map may also explain why simultaneous microstimulation at multiple sites produces saccades that resemble a *weighted vector average* of the individual sites (Robinson, 1972). Here is how this would work for the dynamic ensemble model extended with a fixed cut-off at N_{Tot} spikes. Activity at a single site generates the saccade by

$$\vec{S}(t) = \gamma \sum_{n=1}^{N_{\text{Pop}}} N_n(t)\vec{w}_n \quad \text{with} \quad \sum_{n=1}^{N_{\text{Pop}}} N_n(t) = N_{\text{Tot}} \tag{10.26}$$

When two different sites are simultaneously activated, the resulting output is:

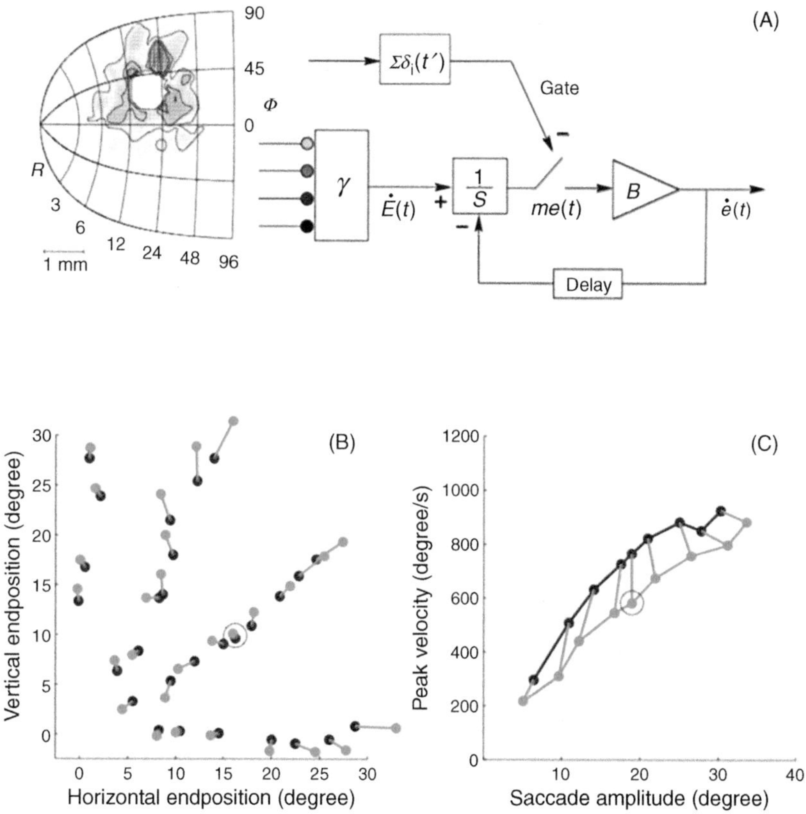

FIGURE 10.10 By adding a STOP gate to the model of Fig. 10.6A that is based on a fixed (cut-off) spike-count criterion from the SC population; becomes, the deficits on saccade metrics (B) and kinematics (C) to a small focal lesion in the motor map (the hole in (A), Lee et al., 1988) can be fully reproduced (green dots). Circle in (B,C): saccade vector at center of the lesion. *(Source: After Goossens and Van Opstal, 2006, with permission.)*

$$\vec{S}_{1+2}(t) = \gamma\left[\sum_{n1=1}^{N_1} N_{n1}(t)\vec{w}_{n1} + \sum_{n2=1}^{N_2} N_{n2}(t)\vec{w}_{n2}\right]$$
$$\text{with} \quad \sum_{n1=1}^{N_1} N_{n1}(t) + \sum_{n2=1}^{N_2} N_{n2}(t) = N_{\text{Tot}} \qquad (10.27)$$

Clearly, the contribution of each site to the saccade will be considerably smaller than for single-site activation, so that $\vec{S}_{1+2} < \vec{S}_1 + \vec{S}_2$ too. In fact, it resembles a weighted-average vector:

$$\vec{S}_{1+2}(t) = a\vec{S}_1 + b\vec{S}_2 \quad \text{where} \quad a + b = 1 \qquad (10.28)$$

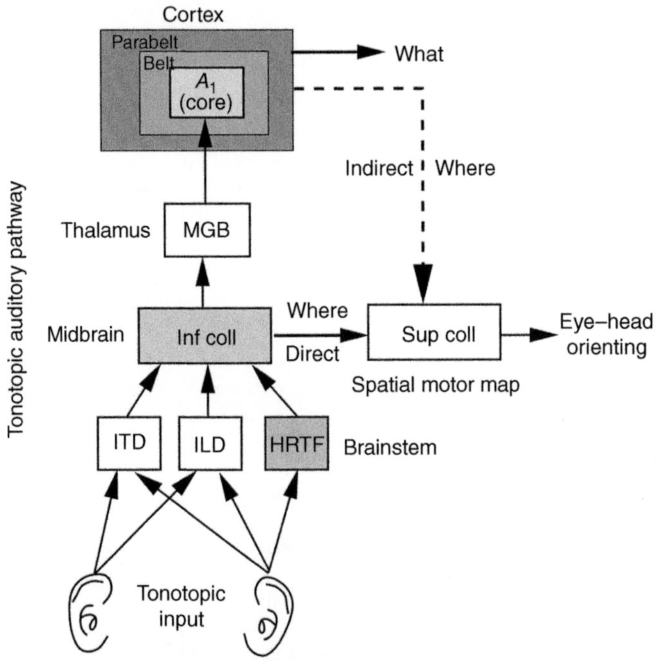

FIGURE 10.11 Overall organization of the mammalian auditory pathway for sound-localization behavior. The acoustic cues, processed in parallel brainstem pathways, converge in the auditory midbrain IC. The IC transfers its signals via the thalamus (medial geniculate body, MGB) to core auditory cortex (A1), and via an indirect path to the gaze control system (SC). Auditory cortex selects and extracts a target sound's identity (what) and location (where). There is also a direct, fast IC–SC pathway. The oculocentric spatial map of the SC drives eyes and head to the selected target.

Although also the extended dynamic ensemble-coding model has to include a nonlinearity in the system (cut-off), it is by far a much simpler mechanism than division by the entire population activity, as suggested by Eq. (10.15).

10.4 IC: PUTATIVE ROLE IN SPATIAL HEARING

The SC can be considered as a central hub for gaze-orienting behavior (Fig. 10.1B). Similarly, the IC in the auditory midbrain may be viewed as a central hub for acoustic processing, as all lower brainstem pathways converge at this stage. Interestingly, both hubs are directly and indirectly connected to each other. Here we will speculate on the potential role of the IC in mediating fast auditory-evoked gaze shifts. Fig. 10.11 provides a schematic overview of the major acoustic pathways involved in spatial hearing and in the orienting behavior to sounds. At lower brainstem levels the acoustic localization cues (ITD, ILD, and HRTF) are extracted, and passed to the IC. As a result, the IC contains (at least implicitly) all information regarding the head-centered azimuth–elevation coordinates of sound sources.

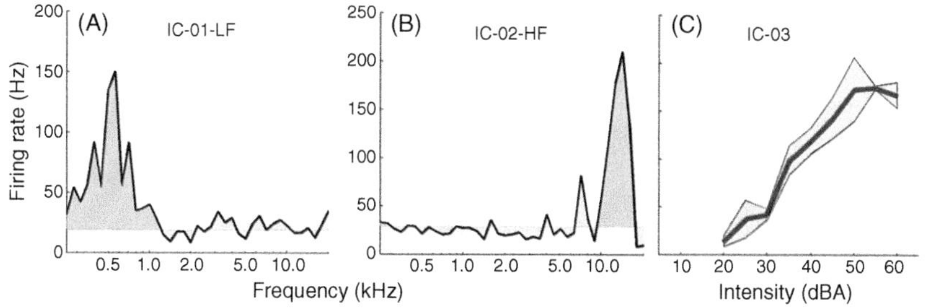

FIGURE 10.12 **Acoustic tuning of monkey IC neurons.** (A,B) Narrow frequency tuning curves for two example cells. (A) Low-frequency cell. (B) High-frequency cell. (C) Monotonic, saturating sensitivity to sound level. *(Source: After Zwiers et al., 2004.)*

All stages in the ascending auditory pathway, from cochlear nucleus to auditory cortex, appear to be *tonotopically* organized, reflecting the basilar membrane mechanics in the cochlea (fully discussed in chapters: The Cochlea and The Auditory Nerve). The mammalian IC is no exception to this principle. In mammals acoustic spatial information thus seems to be represented *implicitly* by the distributed activity of large populations of neurons, instead of by a localized neural population within a head-centered topographic map of auditory space, as found in the barn owl (Knudsen and Konishi, 1978).

For the gaze-orienting system we discussed the important concept of an STT from midbrain SC to brainstem. For the auditory system, a similar conceptual problem may be identified by the need for a tonotopic-to-spatial transformation (or TST) in the system. We will see that IC cells respond to acoustic signals (like a tone's frequency, intensity, temporal modulations, or sound localization cues), as well as to nonacoustic signals (like eye position). We speculate that the distributed population of recruited IC cells possesses all the information needed to implement the TST, and that it could thus mediate the planning of accurate eye–head orienting responses to sound sources via the SC.

Basic acoustic responses: The central nucleus of the IC (the ICc) is tonotopically organized along its dorsal (low frequencies) to ventral (high frequencies) anatomical axis. Fig. 10.12 shows some basic acoustic responses for three different ICc neurons. The frequency-tuning curves to pure tones (like in A,B) can be characterized by a best frequency at the peak (f_{BF}), and a typically narrow bandwidth. The spectral tuning curves may be described by

$$F(f - f_{BF}) = F_{max} \exp\left[-\frac{(f - f_{BF})^2}{2\sigma_{BF}^2}\right] \tag{10.29}$$

Peak firing rates of ICc cells can be as high as 250–300 spikes/s. On top of the frequency tuning, the response magnitude also depends strongly on the sound's absolute sound level. For many cells, this sensitivity can be described

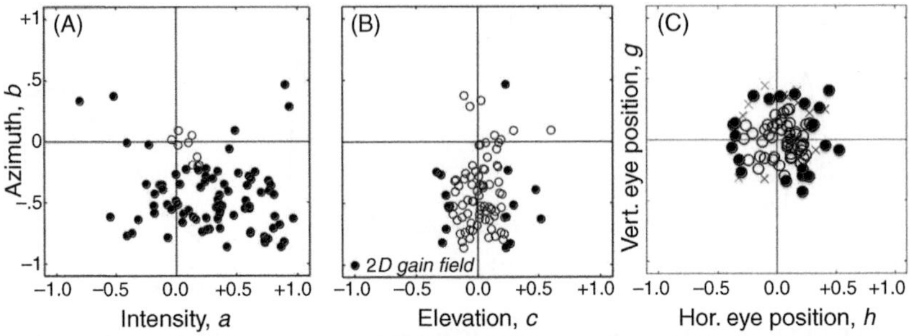

FIGURE 10.13 **Distributions of the five partial correlation coefficients of Eq. (10.29).** Black dots in (A,C): significant coefficients; open circles: nonsignificant. In (B) black dots identify cells with significant 2D spatial gain fields. Most neurons only have contralateral azimuth sensitivity (negative coefficients). *(Source: After Zwiers et al., 2004.)*

by a monotonic increase in firing rate above a cell-specific recruitment threshold, up to the peak-firing rate for high sound levels (see the example neuron in Fig. 10.12C). Since both acoustic factors, frequency and level, modulate a cell's firing rate independently, the acoustic tuning can be described by a separable function, such as:

$$IC(f, I) = F(f - f_{BF})g(I) = F(f - f_{BF})\tanh\left[a(I - I_{Thr})\right] \quad \text{for } I > I_{Thr} \tag{10.30}$$

with I_{Thr} the threshold sound level for recruiting the cell, and a (in inverse of decibel) a measure for the slope (sensitivity) of the intensity–rate curve. For a restricted range of sound intensities $g(I)$ may be linearized to:

$$g(I) \approx a(I - I_{Thr}) \quad \text{for } I > I_{Thr} \tag{10.31}$$

in which case Eq. (10.30) turns into a linear gain modulation.

Spatial sensitivity: Zwiers et al. (2004) subsequently measured the acoustic responses from ICc neurons to GWN, while varying sound location across a range of azimuth–elevation angles in the frontal hemifield. The far majority of neurons resulted to be tuned to the contralateral side of the recorded ICc, with a clear azimuth-dependent response on their firing rates (Fig. 10.13B). This response can be approximated by a linear gain modulation. Some neurons (about 27% of the recorded population) were also sensitive to changes in the *elevation* direction. For neurons with both azimuth and elevation sensitivity ($N = 15$) the neural responses could be characterized by a 2D planar gain-field modulation (Fig. 10.13B, black dots). In general, the neural tuning characteristic of Eq. (10.30) could be extended to

$$IC\left(f, I, \vec{T}_H\right) = F(f - f_{BF})g(I)m\left(\vec{T}_H\right) \approx \left[a\hat{I} + b\hat{A} + c\hat{E}\right]F(f - f_{BF}) \tag{10.32}$$

in which the regression parameters [*a*,*b*,*c*] are expressed as partial correlation coefficients, by writing their respective variables as dimensionless *z*-scores (see chapter: Assessing Auditory Spatial Performance). Fig. 10.13 shows the distributions of the coefficients for intensity (panel A) and 2D craniocentric location (panel B).

Sensitivity to eye position: Eventually, the estimate of the sound source location has to be transformed into a movement command for the gaze-orienting system at the SC output. Jay and Sparks (1984) therefore studied the acoustic responses of cells in monkey SC, when the (head-fixed) animal generated localizing eye movements to the sound source (varied only in azimuth) from different initial horizontal eye positions. The authors wondered whether *sensory* responses would remain in their original craniocentric coordinates, or whether they were transformed into a different reference frame, for example, into an oculocentric goal that could be immediately used as an oculomotor command. To study this problem one could describe the neural responses for acoustic targets by a multiple regression equation on both reference frames:

$$F = aT_{\mathrm{H}} + bT_{\mathrm{E}} + c \qquad (10.33)$$

and subsequently compare the regression coefficients *a* and *b*. Although the authors did not perform multiple regression on their data, it was clear that most cells were much better tuned to the oculocentric coordinates of sounds than to their craniocentric coordinates (ie, one expects $b \gg a$).

This highly nontrivial result suggests that sound locations have been transformed, at the level of the midbrain SC, into *oculocentric* coordinates to let the gaze-control system efficiently foveate the target. Such a remapping of sensory coordinates also enables the visual and auditory systems to spatially align within the *same* (oculocentric) reference frame. This strategy would be highly beneficial for audio–visual integration, and it also would solve the problem on establishing spatial congruence of different sensory inputs that are expressed in different reference frames (this problem is further discussed in chapter: Multisensory Integration).

Furthermore, this line of evidence provides important additional support for the arguments laid out in chapter: Assessing Auditory Spatial Performance (eg, Fig. 8.12) on the tight interplay between the sound-localization and oculomotor systems. It thus sharpens the conceptual framing of the sound-localization problem on the *TST*, mentioned previously, because what one really would like to understand is the following problem:

> The central conceptual problem in sound localization behavior concerns the neural mechanism for the tonotopic to oculocentric transformation (TOT).

Because the IC has direct projections to the SC, it is possible that a TOT already takes place *at the midbrain level* (via the fast, direct pathway in Fig. 10.11). For that reason, studies have checked, and confirmed that responses of cells in the IC (suspected to represent the craniocentric coordinates of sounds, see previous sections) are also weakly modulated by changes in eye position (Groh et al., 2001; Zwiers et al., 2004).

By letting monkeys fixate visual targets at different locations within the 2D oculomotor range, while at the same time varying the 2D craniocentric location of the sound source, Zwiers et al. (2004) could study the influence of both factors on the neural responses of IC cells. Their data indicated weak eye-position gain modulations on the firing rates, which were significant for about 22% of the recorded cells.

The results of changing sound frequency, sound level, sound location, and eye position, eventually lead to the following linearized description of IC responses:

$$IC\left(f,\, I, \vec{T}_{\mathrm{H}}, \vec{e}_{\mathrm{H}}\right) \approx \left[a\hat{I} + b\hat{A} + c\hat{E} + g\hat{e}_{\mathrm{H}} + h\hat{e}_{\mathrm{V}}\right] F\left(f - f_{\mathrm{BF}}\right) \qquad (10.34)$$

The distribution of the eye-position coefficients (g,h) is shown in Fig. 10.13C. It can be noted that the modulations are homogeneously distributed across all possible eye-position tuning directions. The recordings further indicated that the different parameter sets for sound location, eye position, frequency, or sound level were mutually uncorrelated. This suggests that the nonacoustic eye-position modulation strengths may be randomly distributed across the IC population.

The findings of Jay and Sparks (1984) suggested that the eye-position signal was used to transform the craniocentric coordinates of the sound source into an eye-centered reference frame, which is mathematically established by:

$$\Delta\vec{E} = \vec{T}_{\mathrm{H}} - \vec{E}_{\mathrm{H}} \qquad (10.35)$$

Fig. 10.14 presents a schematic for the total direct pathway of Fig. 10.11, from sound input to SC output. The scheme was implemented as a neural–network model that implements the TOT in the direct IC–SC synaptic projections. The model's output was the Gaussian population of neurons, centered at the oculocentric coordinates of the sound source [calculated by Eq. (10.35)] in the SC motor map (described in detail in Section 10.2). The input to the model was a GWN sound source of a randomly selected absolute sound level (I_0), which was presented at randomly selected head-centered azimuth–elevation coordinates. Furthermore, the initial eye position was randomly selected from any direction within the oculomotor range of 30 degrees. From the sound level, and the sound-source coordinates, the neural activity of ipsi- and contralateral tonotopic arrays of brainstem cells was calculated (ILD cues, and modeled HRTFs), and projected to the population of IC cells.

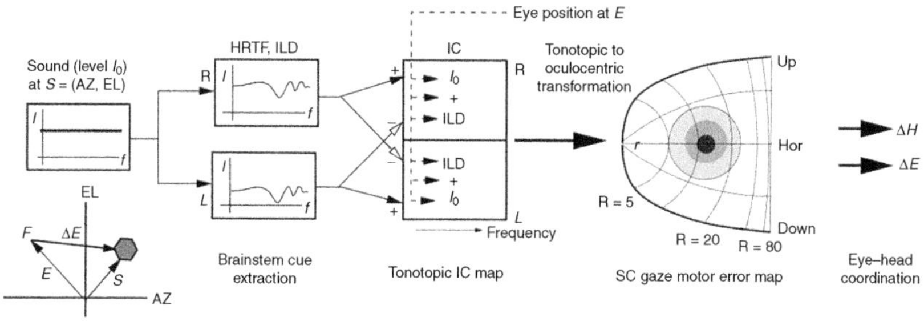

FIGURE 10.14 **Binaural and monaural processing of acoustic cues (ILDs and HRTFs) in brainstem and IC tonotopic maps.** IC activity is weakly modulated by sound–source location and by initial eye position [Eq. (10.33)]. The IC population responses are forwarded to the SC motor map (TOT), where a localized population represents the target in oculocentric coordinates. Inset at left illustrates the transformation from craniocentric azimuth–elevation to oculocentric gaze coordinates. F, fovea. *(Source: After Zwiers et al., 2004, with kind permission.)*

The bilateral IC model had 20 tonotopic maps (each IC contained 100 neurons; 5 monaural and 5 binaural frequency arrays of 10 frequencies), and randomly selected tuning sensitivities according to Eq. (10.34).

The network was trained (using the delta rule to change the synaptic IC–SC weights) on a large set of randomly selected trials, for which it was required to match the population activity profiles in the SC motor map. One such trial is shown in the inset of Fig. 10.14. The network could do so with remarkable accuracy. These simulations therefore hint at the possibility that the TOT may result from a subtle distributed population-encoding scheme at the level of the IC, in which neurons respond to multiplexed inputs, according to Eq. (10.33). In Chapter 11 we will further discuss the different mechanisms underlying the spatial transformations of sensory coordinates into appropriate eye–head motor commands.

10.5 EXERCISES

Problem 10.1: Show that the following mapping function also satisfies the constraint of Eq. (10.2):

$$T(r) \equiv r' = \sqrt{N_0^2 + \frac{2N}{\pi\alpha}\ln\frac{r}{r_0}} \quad \text{and} \quad T(\phi) \equiv \phi' = \phi$$

with N_0 and r_0 arbitrary constants, and α the proportionality constant for receptive field size. Plot the isoeccentricity and isodirection lines in a 2D representation of this function.

Problem 10.2:

(a) Draw a graph of the shifted complex-log function of Eq. (10.8). Calculate and plot the images of the vertical meridians, and of isodirection lines at

±30 and ±60 degrees, as well as for a number of logarithmically spaced eccentricities. Is this map conformal? Why (not)?

(b) Provide a prescription for the inverse mapping of this function:

$$T^{-1}(u,v)=(r,\phi)$$

Problem 10.3: Rewrite the SC afferent mapping function to (x,y) coordinates.

Problem 10.4: Verify the expression for the inverse mapping function of Eq. (10.12) (the efferent map) that relates the horizontal (x), and vertical (y) saccade components $[z_0 = (x,y)]$ to collicular neural coordinates, $w_0 = (u_0,v_0)$. Also express the efferent map in polar coordinates.

Problem 10.5: If the cell density in the SC motor map is taken to be constant, at ρ_0 cells/mm^2, show that the total number of spikes from the SC population is given by

$$N_{\text{Tot}} = 2\pi N_0 \rho_0 \sigma_{\text{Pop}}^2$$

Make an educated guess for N_{Tot}.

Problem 10.6: Use Eq. (10.13) to demonstrate that the width of the movement field (determined by the efferent mapping of 1.0 mm of SC space, symmetrically positioned around the center of a population at u_0) increases linearly with R_0. Also determine the asymmetry of the MF by taking the ratio between low-edge to peak versus peak-to-high edge.

Problem 10.7: Suppose that the center of the active Gaussian cell population in the SC determines the size and direction of the saccade vector endpoint. Assume that the center location of the population is endowed with some noise, due to retinal uncertainty, so that will scatter from saccade to saccade in response to identical target presentations. Suppose that the SC scatter is bounded by a small *circular* region around the true center, with radius ε in the collicular complex–log map of Eq. (10.10). Derive an expression for the distribution of saccade vectors as a function of saccade amplitude that results from this scatter, and show that under these assumptions the anisotropy of the motor map is directly reflected in the saccade endpoint distributions.

Hint: Place the scatter with its center on the horizontal meridian of the map and compute the resulting vectors for five points on this circle: the center, the horizontal meridian intersection points of the circle, and the two most up/down vertical points on the circle. These five points define the long and short axes of an elliptical distribution.

Problem 10.8: Since the horizontal/vertical brainstem feedback circuits (with gain B, and feedback delay, ΔT) and the downstream PSGs of Fig. 10.6 are all *linear* systems, their total function can be replaced by single feed–forward models with identical input–output characteristics.

Derive these characteristics, assuming simple, first-order plant models (time constant, *T*, and add a gain *T* in the direct path from PGs to motor neurons).

Problem 10.9: Explain the deficits observed in the simulated saccade vectors with the model in Fig. 10.10A.

REFERENCES

Berthoz, A., Grantyn, A., Droulez, J., 1986. Some collicular efferent neurons code saccadic eye velocity. Neurosci. Lett. 72, 289–294.

Freedman, E.G., Stanford, T.R., Sparks, D.L., 1997. Combined eye-head gaze shifts produced by electrical stimulation of the superior colliculus in rhesus monkeys. J. Neurophysiol. 76, 927–952.

Goldberg, M.E., Wurtz, R.H., 1972. Activity of superior colliculus in behaving monkey. I: visual receptive fields of single neurons. J. Neurophysiol. 35, 542–559.

Goossens, H.H.L.M., Van Opstal, A.J., 1999. Influence of head position on the spatial representation of acoustic targets. J. Neurophysiol. 81, 2720–2736.

Goossens, H.H.L.M., Van Opstal, A.J., 2000a. Blink-perturbed saccades in monkey. I. behavioral analysis. J. Neurophysiol. 83, 3411–3429.

Goossens, H.H.L.M., Van Opstal, A.J., 2000b. Blink-perturbed saccades in monkey. II. superior colliculus activity. J. Neurophysiol. 83, 3430–3452.

Goossens, H.H.L.M., Van Opstal, A.J., 2006. Dynamic ensemble coding of saccades in the monkey superior colliculus. J. Neurophysiol. 95, 2326–2341.

Goossens, H.H.L.M., Van Opstal, A.J., 2012. Optimal control of saccades by spatial–temporal activity patterns in monkey superior colliculus. PLoS Comp. Biol. 8 (5), e1002508.

Groh, J.M., Sparks, D.L., 1996. Saccades to somatosensory targets. III. eye-position dependent somatosensory activity in primate Superior Colliculus. J. Neurophysiol. 75, 439–453.

Groh, J.M., Trause, A.S., Underhill, A.M., Clark, K.R., Inati, S., 2001. Eye position influences auditory responses in primate inferior colliculus. Neuron 29, 509–518.

Harris, C.M., Wolpert, D.M., 2006. The main sequence of saccades optimizes speed-accuracy tradeoff. Biol. Cybernet 95, 21–29.

Jay, M.F., Sparks, D.L., 1984. Auditory receptive fields in primate superior colliculus shift with changes in eye position. Nature 309, 345–347.

Knudsen, E.I., Konishi, M., 1978. A neural map of auditory space in the owl. Science 200, 795–797.

Lee, C., Rohrer, W.H., Sparks, D.L., 1988. Population coding of saccadic eye movements by neurons in the superior colliculus. Nature 332, 357–360.

Ottes, F.P., Van Gisbergen, J.A.M., Eggermont, J.J., 1986. Visuomotor fields of the superior colliculus: a quantitative model. Vision Res. 26, 857–873.

Robinson, D.A., 1972. Eye movements evoked by collicular stimulation in the alert monkey. Vision Res. 12, 1795–1808.

Roucoux, A., Guitton, D., Crommelinck, M., 1980. Stimulation of the superior colliculus of the alert cat. II. eye and head movements when the head is unrestrained. Exp. Brain Res. 39, 75–85.

Sparks, D.L., Holland, R., Guthrie, B.L., 1976. Size and distribution of movement fields in the monkey superior colliculus. Brain Res. 113, 21–34.

Stein, B.E., Meredith, M.A., 1993. The Merging of the Senses. MIT, Cambridge, MA.

Tanaka, H., Krakauer, J.W., Qian, N., 2006. An optimization principle for determining movement duration. J. Neurophysiol. 95, 3875–3886.

Van Gisbergen, J.A.M., Van Opstal, A.J., Tax, A.A.M., 1987. Collicular ensemble coding of saccades based on vector summation. Neuroscience 21, 541–555.

Van Opstal, A.J., Van Gisbergen, J.A.M., 1990. Role of monkey superior colliculus in saccade averaging. Exp. Brain Res. 79, 143–149.

Van Opstal, A.J., Van Gisbergen, J.A.M., Smit, A.C., 1990. Comparison of saccades evoked by visual and collicular electrical stimulation in the alert monkey. Exp. Brain Res. 79, 299–312.

Van Opstal, A.J., Hepp, K., Suzuki, Y., Henn, V., 1995. Influence of eye position on activity in monkey superior colliculus. J. Neurophysiol. 74, 1593–1610.

Van Opstal, A.J., Hepp, K., 1995. A novel interpretation for the collicular role in saccade generation. Biol. Cybernet. 73, 431–445.

Van Opstal, A.J., Goossens, H.H.L.M., 2008. Linear ensemble-coding in midbrain superior colliculus specifies the saccade kinematics. Biol. Cybernet. 98, 561–577.

Wurtz, R.H., Goldberg, M.E., 1972a. Activity of superior colliculus in behaving monkey. III: cells discharging before eye movements. J. Neurophysiol. 35, 575–586.

Wurtz, R.H., Goldberg, M.E., 1972b. Activity of superior colliculus in behaving monkey. IV: effects of lesions on eye movements. J. Neurophysiol. 35, 587–596.

Zipser, D., Andersen, R.A., 1988. A back-propagation programmed network that simulates response properties of a subset of posterior parietal neurons. Nature 331, 679–684.

Zwiers, M.P., Versnel, H., Van Opstal, A.J., 2004. Involvement of monkey inferior colliculus in spatial hearing. J. Neurosci. 24, 4145–4156.

Chapter 11

Coordinate Transformations

11.1 INTRODUCTION

In the previous chapters we briefly introduced the notion of craniocentric and oculocentric reference frames, but now we will delve a bit further into this interesting topic. As it turns out, the coordinate transformations in the sensory representations, and in the different motor systems that are involved in the ensuing orienting behavior are particularly nontrivial when we consider dynamic stimulus conditions. In dynamic sensorimotor tasks, subjects orient to a sensory stimulus that had been briefly presented in midflight of a fast eye–head gaze shift that was generated toward another target. Adequate performance in this task requires the sound-localization and visuomotor systems to have access to, and use, dynamic feedback of accurate (saccadic) eye- and head-position signals. The sound-localization system in particular will allow us to peek a bit deeper into underlying neural mechanisms because of its organization in independent pathways to extract the target's azimuth and elevation coordinates. Moreover, each pathway has its own particular requirements regarding the spectral–temporal stimulus properties. This peculiar organization is unique for the auditory system, for which there is no analogue in vision. As a result, auditory- and visuomotor behaviors can be qualitatively different.

Reference frames: So, when we talk about spatial reference frames, what do we mean? Can neurons possess a "reference frame?" Do neurons, or interconnected populations of neurons, refer to a "single" reference frame? Or is the reference frame concept exclusively reserved for sensory input and motor output? How could we conceptualize the intermediate steps in the sensory-to-motor transformations within the framework of reference frames, as described for example for the gaze-control system in chapter: The Gaze-Orienting System?

Let's briefly step aside from these deeper conceptual issues, and first define the basic ideas. A reference frame consists of a reference point and a set of coordinates that together provide a unique quantitative measure (in 2D or 3D) to specify locations, distances, movements, and directions of objects in a

The Auditory System and Human Sound-Localization Behavior. http://dx.doi.org/10.1016/B978-0-12-801529-2.00011-8

given space. It is useful to distinguish two different types of reference frames (Fig. 11.1):

Egocentric reference frames measure locations of objects with respect to (an identifiable part of) the body of the observer.

Allocentric reference frames only measure relative distances and orientations of objects in the world with respect to each other, irrespective of the observer, and independent of some arbitrarily chosen reference point.

The following egocentric reference frames are relevant for eye–head orienting to acoustic and visual stimuli:

1. The *oculocentric reference frame* measures the location of targets with respect to the fovea, and specifies retinal coordinates (usually expressed in *polar-coordinate* form, $[r=\sqrt{x^2+y^2}, \phi=\text{atan}(y/x)]$). In the case of binocular eye movements, target distance with respect to the eyes also becomes relevant. It is determined by *retinal binocular disparities* (local binocular differences in the (typically horizontal) retinal coordinates of different object elements).
2. A *craniocentric reference frame* quantifies target locations with respect to a head-fixed landmark (usually the center of the head, specific landmarks on the skull, the planes of the vestibular canals, or the ear canals). It is customary (especially for its use in the auditory system) to describe target locations in *azimuth–elevation* coordinates (chapter: Acoustic Localization Cues).
3. A *body-centered reference frame* measures targets with respect to a fixed body landmark (eg, neck, chest, or belly button). Body coordinates remain stable, regardless of the motion of eyes and head.

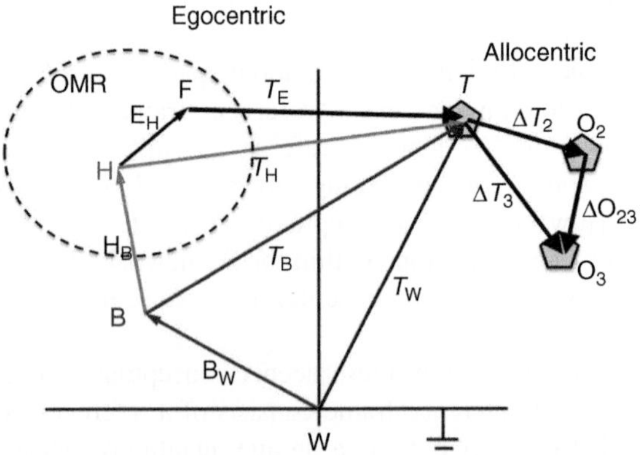

FIGURE 11.1 Egocentric and allocentric reference frames, and their mutual relationships. W, world (earth); B, body (trunk); H, head; F, eye (fovea). OMR, oculomotor range. *T* is a target in the world, and O_2 and O_3 are two other landmarks. T_E: target relative to the fovea; T_H: craniocentric target vector; T_B: body-centered target location; T_W: target in world coordinates. The allocentric reference frame labels targets as difference vectors with respect to the different landmarks.

In a typical laboratory environment the participant's body is fixed to the world (the room, the earth), unless the subject undergoes for example, whole-body vestibular stimulation. Then, also the body-centric coordinates of targets in the world will change. The egocentric reference frames are related through (Fig. 11.1):

$$\vec{T}_{\mathrm{W}} = \vec{B}_{\mathrm{W}} + \vec{H}_{\mathrm{B}} + \vec{E}_{\mathrm{H}} + \vec{T}_{\mathrm{E}} \tag{11.1}$$

with W, world; B, body; H, head; and E, eye. T_{E} specifies the retinal coordinates of the target. Let's look at what happens to the visual coordinates of a target under different initial conditions.

When the eye is fixed in space (eg, by using a central fixation spot), visual target coordinates can only change with respect to the retina. When the origins of the eyes, head, and body all point in the same world direction (eg, "straight ahead"), the reference frames are *aligned*. In that unique case, the different coordinate systems cannot be dissociated.

When head and trunk are restrained with respect to the earth, but the eye is free to rotate, targets could be expressed in head-centered coordinates, and/or in eye-centered coordinates, since:

$$\vec{T}_{\mathrm{H}} = \vec{T}_{\mathrm{E}} + \vec{E}_{\mathrm{H}} \tag{11.2}$$

with E_{H} the eye-in-head orientation. A clever eye-movement experiment could then potentially dissociate the reference frames (recall a first example for this in Fig. 10.2B). Below we make these transformations more explicit.

When the head is free to rotate (trunk remains fixed), body- (or world-) centered coordinates of the target can change (and be measured) too:

$$\vec{T}_{\mathrm{B}} = \vec{T}_{\mathrm{E}} + \vec{E}_{\mathrm{H}} + \vec{H}_{\mathrm{B}} \tag{11.3}$$

with H_{B} the head orientation on the neck. Changing eye-in-head and head-on-body orientations then potentially dissociates the reference frames.

Relative motion: For the brain, any perceived coordinate change of an object should be interpreted in terms of its own body-, head-, or eye-movement through the environment, or of object motion through the world relative to the observer. But how can the sensorimotor system distinguish these two (either potentially life-threatening, or life-saving) situations, when relativity tells us that a given change in sensory cues (like an increase of the interaural level differences) may be due to a sound source moving at a certain speed from left to right with respect to the head, of the head rotating from right to left at the same speed but listening to a stationary sound source, or of any appropriate combination of stimulus- and head movements [Eq. (11.4); Exercise 11.1]?

$$\frac{\Delta\,\mathrm{ILD}(t)}{\Delta t} = p\dot{H}_{\mathrm{W}}(\alpha) + q\dot{T}_{\mathrm{W}}(\alpha) \tag{11.4}$$

Clearly, without an additional independent signal (a *measurement*) this remains an ill-posed problem that can never be resolved! Furthermore, the brain should make educated guesses to weed out unlikely solutions. For example, it is to be expected that motion of an independent object in the world and a self-generated eye–head movement, are usually uncorrelated. It is therefore quite unlikely (except under certain laboratory conditions) that in a natural environment an object would move at exactly the same speed and direction as the head. This is why it is quite hard to produce realistic headphone simulations while moving the head, because the headphones' acoustic inputs are fixed to the head (chapter: Assessing Auditory Spatial Performance).

The major topic of this book revolves around the argument that solving real-world localization problems requires *active sensing* (active hearing and vision): in this process, knowledge about the *act of orienting* is deemed absolutely crucial. Through goal-directed eye–head orienting behavior the brain can use its own proprioceptive motor-output signals, vestibular signals from the semicircular canals and otoliths, and feedback from its own planned motor commands, to dissociate the predicted effects of self-motion from target motion. If, in the example, the auditory system would know (through its own independent measurements) that the head makes a rightward movement of 100 degrees/s, it can predict that the expected change in acoustic cues for a world-stationary sound source would be appropriate for a 100 degrees/s leftward motion. Now suppose that the auditory system measures a cue change that would correspond to a 60 degrees/s leftward movement. The system would then conclude that the target moves at a speed of 40 degrees/s leftward with respect to the head (or, equivalently, at 40 degrees/s rightward with respect to the stationary body). This knowledge can then subsequently be used to intercept the target with a goal-directed hand movement.

Clearly, this is an extremely simple example (1D, 1 target, 1 effector), but in 2D, involving multiple sensory and motor systems and relative timings, the localization problems get increasingly more complex. In what follows, we describe a series of localization experiments that studied whether the human sound-localization system processes sound locations in the appropriate reference frames under static (no movement, but potentially varying initial conditions) and dynamic conditions.

Position versus displacement feedback: In chapter: The Gaze-Orienting System, we alluded to two essentially different schemes for the control of saccadic eye movements to visual targets that relied on local feedback from the brainstem. In the Robinson model (Robinson, 1975; Van Gisbergen et al., 1981) the central idea was that the brainstem is driven by a desired head-centered goal, which is transformed into a dynamic motor error signal by subtracting the current eye-position estimate from the neural integrator:

$$\vec{m}_e(t) = \vec{T}_{\mathrm{H}} - \vec{e}(t) \tag{11.5}$$

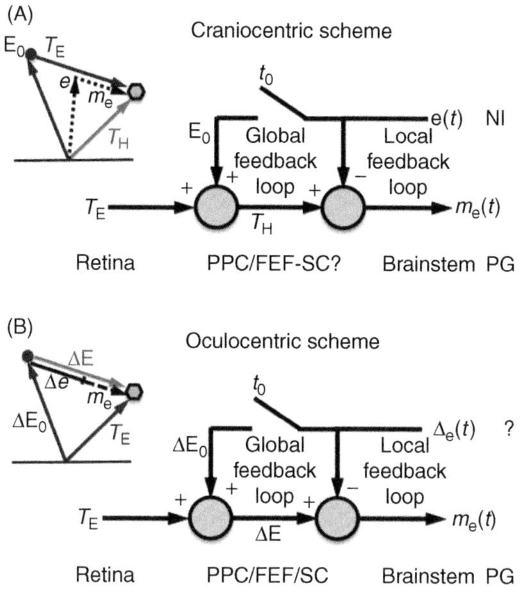

FIGURE 11.2 Different schemes for updating target locations for gaze shifts after an intervening shift of gaze to E_0. (A) The position-updating model is based on a craniocentric reference frame, and uses feedback of eye-in-head positions [E_0, $e(t)$]. It can be readily expanded to a body-centered or world reference frame, by adding initial head- and body orientations (H_0, B_0). (B) The oculocentric scheme relies on target updating through (relative) eye-displacements [ΔE_0, $\Delta e(t)$]. Putative involvement of PPC, FEF and SC is indicated. NI: neural integrator. Insets show the relevant transformations for either scheme. *Green arrows*: signal that drives the brainstem at time $t = t_0$. ?: Origin of the signal is not known, or speculative.

But how (and where) is this craniocentric goal, T_H, represented in the brain? According to Robinson's proposal, this signal is constructed prior to saccade initiation (at $t = t_0$) by storing the neural representations of the retinal error of the target, T_E, and the initial eye-in-head position, E_0. Adding these signals [Eq. (11.2)] then expresses the goal in head-centered coordinates (Fig. 11.2A; see inset). The elegance of this scheme is that the dynamic motor error is updated by just keeping track of the current eye position, regardless of how the eye got there.

However, neurophysiological recordings and microstimulation experiments have not provided direct evidence for the existence of T_H (eg, Fig. 10.2B). Instead, recordings in PPC, FEF, and SC (the three major players in the control of gaze shifts) invariantly hint at oculocentric gaze-motor signals, and microstimulation consistently produces fixed–vector saccades. Finally, the Robinson model does not provide a role for the SC motor map, which is generally considered to transmit the desired gaze-motor drive to the brainstem. Yet, recording and stimulation results indicate that this signal is oculocentric too (fixed vector).

These negative findings prompted other researchers (Jürgens et al., 1981; Scudder, 1988; Goldberg and Bruce, 1990) to propose alternative, oculocentric models, which could better account for the observed neurophysiology (Fig. 11.2B). In these schemes (which come in different versions) the retinal error, T_E, is kept in memory, but is updated by each intervening eye movement to a new oculocentric command. For example, if after target presentation the eye would have made an intervening saccade, ΔE_0, to some other location, then the updated oculocentric target coordinates become (see inset in panel B):

$$\Delta\vec{E} = \vec{T}_{\mathrm{E}} - \Delta\vec{E}_0 \tag{11.6}$$

The dynamic motor error in the local feedback loop is then determined by keeping track of the current displacement of the eyes during the saccade (Fig. 11.2B):

$$\vec{m}_e(t) = \Delta\vec{E} - \Delta\vec{e}(t) \tag{11.7}$$

Although this model provides a clear role for the SC output, a potential problem is lack of evidence for an instantaneous eye-displacement signal. Instead, the oculomotor brainstem circuitry contains an abundance of eye-position (step) and eye velocity (pulse) signals. Scudder's model remedies this problem for the local feedback loop with an integrating comparator (Fig. 9.6).

It should be noted that although the neurophysiological evidence from PPC, FEF, and SC provides by far the strongest support for oculocentric transformations for *visual-motor* control, it cannot account for *acoustic-evoked* localization responses of the eyes. In that case, the audiomotor system *must* invoke an initial eye-in-head *position* signal, as the auditory sensory cues are represented in *craniocentric* azimuth–elevation coordinates. Also, in chapter: The Gaze-Orienting System, we discussed that in head-*unrestrained* gaze shifts the eyes and head both make goal-directed movements to the stimulus (Fig. 9.13), and this behavior requires the use of a fast and accurate eye-in-head *position* signal as well, both for visual and for auditory targets.

11.2 GAIN FIELDS AND PREDICTIVE REMAPPING

It may now be useful to return to the neurophysiology and ask a different question with regard to potential neural correlates of gaze-reference frames: *Could it be that oculocentric codes (by well-identified localized neural populations in retinotopic maps) appear to dominate in visual-motor behavior merely because they reflect a preserved retinotopy in these structures?* A similar question would then relate to the preserved tonotopy of the auditory system throughout the neuraxis.

Even though such sensory-imposed topographies are almost self-evident, it does not necessarily follow that other reference frames cannot be embedded in the neural responses of these same structures, only because at single-unit level or at a local topographic organization they are not immediately visible. For example, it is conceivable that other reference frames could be represented by distributed systematic modulations of the neural population. In chapter: The Midbrain Colliculus we hinted at such a possibility for the midbrain inferior colliculus.

Response modulations by changes in eye position have also been encountered in visual responses of posterior parietal cortex cells (PPC; Andersen et al., 1985), and saccade-related responses in SC (Van Opstal et al., 1995). For the latter, these so-called *gain-field* modulations were described by:

$$N\left(u_0, v_0, \vec{E}_0\right) = N_0\left(1 + \vec{\varepsilon}\vec{E}_0\right)\exp\left(-\alpha\left[\left(u - u_0\right)^2 + \left(v - v_0\right)^2\right]\right) \quad (11.8)$$

with (u,v) the SC coordinates in the complex-log motor map [Eq. (10.9)], and $\vec{\varepsilon}$ an eye-position sensitivity vector, the amplitude and direction of which differs from cell to cell (Fig. 11.3A). In Exercise 11.2 the reader may verify that this multiplicative gain-field modulation can be understood from first-order Taylor expansion of the Gaussian tuning curve in the motor map.

Network simulations show that a population of neurons obeying Eq. (11.8) can accurately signal the target location both in head-centered and in oculocentric coordinates (Zipser and Andersen, 1988; Van Opstal and Hepp, 1995). Gain fields could thus be an efficient strategy of the system to embed multiple reference frames within the same population of neurons. Thus, a craniocentric representation of the target, T_H, which is needed to guide eye–head gaze shifts, could exist in an implicit, distributed form in the SC motor map, on

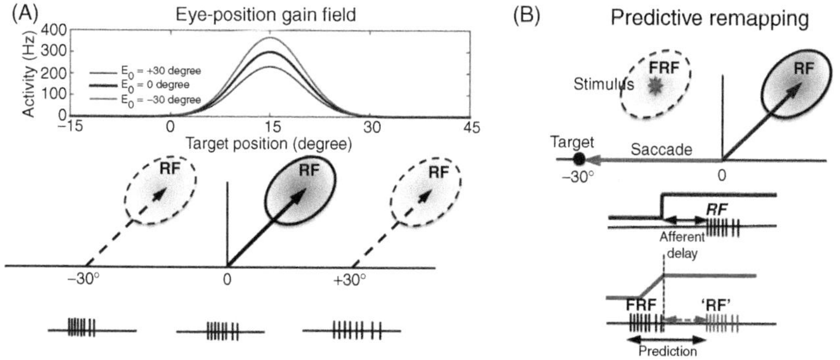

FIGURE 11.3 **Two qualitatively different response modes of PPC and SC cells.** (A) Gain fields. The cell's visual receptive field (or movement field) is modulated by static eye position. In this example, left fixation (*red*) yields a higher firing rate than right fixation (*blue*). (B) Predictive remapping. The cell responds to a visual stimulus already near the saccade onset that will bring it inside its "future receptive field" (FRF). *Blue trial*: classical receptive field mapping: the cell responds after an afferent delay. *Red trial*: a leftward saccade is made, while a stimulus (*red star*) is presented well outside the cell's RF. Yet, the cell responds near saccade onset (*black* FRF *spikes*), >100 ms before the afferent delay (*gray spikes*).

top of the oculocentric population code described in chapter: The Midbrain Colliculus.

Prediction: A subpopulation of PPC (Duhamel et al., 1992), FEF (Umeno and Goldberg 1997), and SC cells (Walker et al., 1995) demonstrate the interesting phenomenon of "predictive remapping." This response property is illustrated in Fig. 11.3B. Imagine a PPC/FEF/SC cell with a visual receptive field, RF (in the example, right and up on the retina). The cell responds to the appearance of new visual stimulus in its RF after an afferent delay of 60–80 ms. Now suppose that in a different trial a visual stimulus appears to the left and up (at the location of the red star). Obviously, the cell will not respond to this stimulus, as it falls outside its RF. The clever manipulation in this paradigm, however, is to let the monkey make a saccade to a leftward target, such that after the saccade has finished, the stimulus will have been brought into the cell's RF (the stimulus is hence located at the cell's *future receptive field*, or FRF). If the cell would be purely *visually* responsive, its neural activity would be expected 60–80 ms after saccade offset (the grey spikes, labeled RF). What happens, however, is that the cell already responds near (and sometimes even before) the saccade *onset*! Instead of an afferent delay, there is a clear *predictive* response (more than 100 ms earlier than expected) in the cell, which is appropriate for the FRF. It thus seems that these cells are part of a computational network that calculates the *new oculocentric coordinates* of the visual target based on the *predicted consequences* of an upcoming saccade, in accordance with Eq. (11.6). As this process seems to occur already *before* the saccade onset, it is not based on efferent or proprioceptive feedback from the actual saccade itself. Instead, it is hypothesized that the computation uses the *motor plan* for the *upcoming* saccade, before it actually happens. Such a signal is called a "*corollary discharge*," and its existence had already been hypothesized in the early 1980s (see further).

Goldberg and colleagues proposed that predictive remapping may be important for mediating a *stable visual percept* of the environment, despite saccades that cause the retinal image to jump around 3–4 times per second. It could thus embody the neural substrate of *trans-saccadic integration* in the visuomotor system, mentioned in chapter: The Gaze-Orienting System.

Below we will challenge the hypothesis that predictive remapping could also be the underlying mechanism for target updating in gaze-orienting behavior [Eq. (11.5)].

Remaining issues: We here briefly mention some unresolved questions and issues on the potential role of PPC/FEF/SC in target updating and spatial localization.

1. Up to now it is unclear whether gain-field cells and cells that perform predictive remapping are the same cells, or whether they are part of different (sub) populations that may be involved in different functions.
2. Despite the multitude of potential brainstem nuclei that could provide an accurate, even dynamic, eye-position feedback signal, the origin of the

eye-position signal that gain-modulates PPC, SC, (and IC) firing rates is not known. Models assume an efference copy signal from the brainstem neural integrators, but there is no evidence that this is the case.

3. Recently, Xu et al. (2012) casted some serious doubts as to whether gain fields (at least those observed in PPC) could be responsible for accurate spatial localization. They measured the influence of eye-position signals on visual responses around saccades, and observed that for the majority of cells in PPC the timing and reliability of the eye-position signal would be inappropriate and too slow for fast spatial updating. Instead, the signal seemed to reflect the initial eye position *before* the saccade, rather than during and immediately after the saccade. Only more than 150 ms later the eye-position signal reflects actual eye position. They suggest that relatively slow eye–muscle proprioceptive inputs from primary somatosensory cortex could mediate PPC gain fields (Wang et al., 2007). Their function could be to monitor and calibrate eye position, rather than to use it as a dynamic control signal.
4. There is good evidence that a corollary discharge signal of eye displacement originates from the SC, which travels via the mediodorsal thalamus (MDT) to the FEF (Sommer and Wurtz, 2002, 2004). A lesion in MDT leads to partial localization deficits (of ~19%) in the double-step paradigm described below. It remains to be tested whether this signal is also used in predictive remapping (see also chapter: The Midbrain Colliculus). Still, whatever the source and nature of the corollary feedback, as shown further, fast and accurate eye- and head *position* signals are (also) required for adequate eye–head orienting behavior.

11.3 THE STATIC DOUBLE-STEP PARADIGM

When two stimuli are briefly presented on the retina in otherwise darkness, and the subject first responds with an eye-movement to the first (extinguished) stimulus, the system can no longer rely on the retinal error of the second target (T_{E2}) to program the next saccade (Fig. 11.4). Although the coordinates for the second saccade are in principle determined by the difference vector between the two stimuli (T_{E2}–T_{E1}), a purely visual mechanism can only work if the first saccade is very accurate (Fig. 11.4A). Hallett and Lightstone (1976) introduced the *double-step paradigm* to test the idea that the saccadic eye-movement system should in fact use extra-retinal (nonvisual) signals to maintain spatial accuracy. If the actual motor response is used to update the oculocentric target coordinates, spatial accuracy is guaranteed even when the first saccade was not visually elicited (Fig. 11.4B). There is ample evidence that the upcoming goal-directed orienting response indeed compensates for the localization errors of previous saccades (Goossens and Van Opstal, 1997; Fig. 11.4C).

In their seminal series of electrophysiological studies Mays and Sparks (1980) demonstrated that the saccadic system indeed uses a corollary discharge signal to update the target coordinates, and that it doesn't even need to actively plan the intervening first saccade! In their paradigm, monkeys made saccades

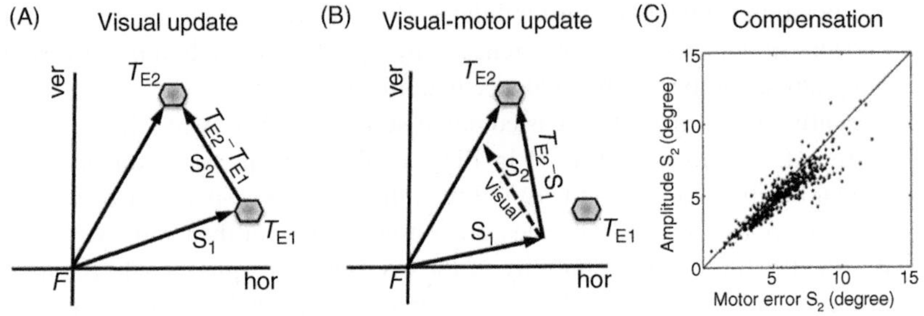

FIGURE 11.4 Static double-step paradigm. Two visual stimuli are presented in rapid succession, and are extinguished before the onset of the first saccade to T_{E1}. Clearly, the coordinates of saccade S_2 to the spatial location of T_{E2} should be updated. The response could be preprogrammed from purely retinal information (A), or it could involve feedback about the actual saccade to the first target (B). In the latter, any error of S_1 will be fully compensated. A purely visual model (*dashed arrow*) would lead to an accumulation of errors after each saccade. (C) Data show that the system compensates for its own inaccuracies in the first response. (After: Goossens and Van Opstal, 1997).

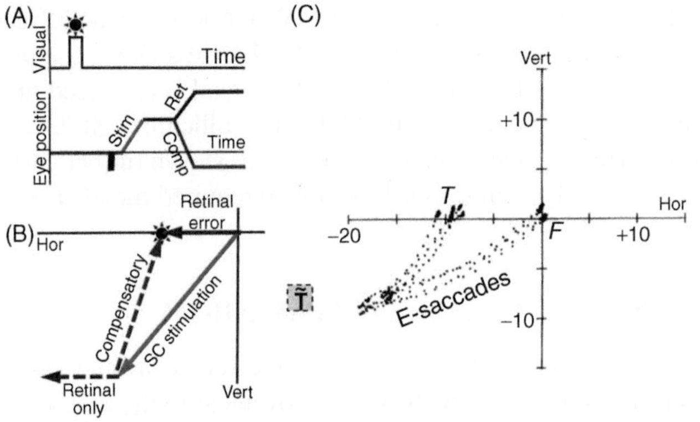

FIGURE 11.5 Collicular microstimulation paradigm (A, B) of Mays and Sparks (1980) and a typical example of their experimental results (C). (A) Temporal events (black bar: stimulation train; Stim: E-saccade; Ret: retinal saccade coordinates; Comp: compensatory saccade); (B) Spatial trajectories of the events in (A). The monkey compensates for the intervening perturbation (E-saccades). The corollary discharge signal arises at, or downstream, from the SC motor map. The first saccade need not be planned by the animal to maintain spatial accuracy.

toward briefly flashed (~20 ms) targets in otherwise darkness. In randomly selected trials they applied electrical microstimulation in the SC motor map within the saccade reaction time, in order to rapidly deviate the eyes from the initial fixation point (Fig. 11.5A). The question was whether the monkey would compensate for this sudden perturbation in eye position, and still accurately localize the target in space (dark red trace), or not. In the latter case the saccade would be driven by the target's retinal coordinates (blue). The results were very clear (Fig. 11.5C): saccades were spatially accurate, and saccade-related cells in the

SC responded to the *oculocentric* target coordinates, not to the original retinal coordinates. Correct spatial localization performance therefore uses feedback about actual motor performance, even when the animal had not planned the intervening saccade.

In a subsequent experiment they lesioned the trigeminal nerve (n. V) to remove eye–muscle proprioception, and showed that this compensation behavior was unaffected (Guthrie et al., 1983). This experiment thus provided strong support for the hypothesis that the updating behavior relied on a corollary discharge signal.[a] Moreover, when microstimulation was applied to cells in the brainstem pulse generator (PPRF; Sparks et al., 1987), the animals no longer compensated for the perturbation in eye position. The experiments therefore led to the conclusion that the saccadic system uses corollary discharge, and that this signal arises upstream from the saccadic pulse generator, possibly at the SC motor map. Furthermore, the plan for the compensatory eye-movement signal is generated at, or upstream from, the SC motor map. Together, the experiments provided strong evidence for the existence of a global motor feedback loop, as indicated in Fig. 11.2B.

11.3.1 Eye–Head Coordination

It is interesting to extend the oculomotor visual double-step paradigm to the full repertoire of eye–head coordination in response to visual and to auditory targets. In chapter: The Gaze-Orienting System we saw that in unaligned initial conditions, the eyes and head both make goal-directed movements. Even for a single target this already invokes nontrivial target remappings, as the updating computation depends on sensory modality. In the double-step this neurocomputational process is challenged even more, because of the considerable variability of the eye–head movement responses when the head is unrestrained. This remapping challenge is exemplified by the vector scheme of Fig. 11.6.

After the first gaze shift (which may be inaccurate with regard to the first stimulus) eyes and head will typically end in a different state of alignment than at the start of the trial, because the eye will have generated a net saccadic movement in the head. The eye-in-head position signal, E_1, will be quite variable from trial to trial, yet is crucial for the rapid spatial update of the target coordinates for eyes and head. Fig. 11.7 shows two examples of eye–head double-step trials to brief visual flashes and noise bursts. The gaze shifts are accurate, and both the gaze- and head-movement trajectories are goal-directed. This can be quantified by a multiple linear regression analysis, in which the measured

[a] There is, however, an alternative explanation for these results that cannot be entirely ruled out: if SC stimulation were to produce a visual "*phosphene*" (a neural, retinotopic signal), target updating could in principle be due to a visual vector calculation (Fig. 11.4A).

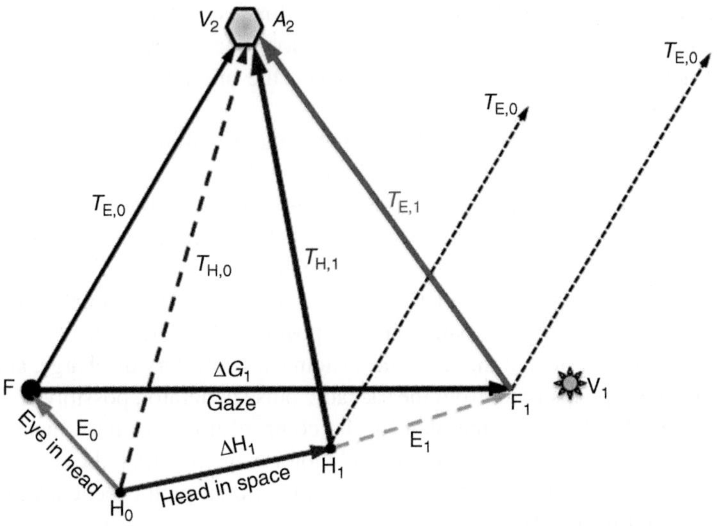

Eye-head double step	2nd eye movement	2nd eye movement
$V_1 - V_2$	$T_{E,1} = T_{E,0} - \Delta G_1$	$T_{H,1} = T_{E,0} - \Delta G_1 + E_1$
$V_1 - A_2$	$T_{E,1} = T_{H,0} - \Delta H_1 - E_1$	$T_{H,1} = T_{H,0} - \Delta H_1$

FIGURE 11.6 **Coordinate transformations required in the static eye–head double step, where the first intervening gaze shift (ΔG_1) is made to a visual target (V_1), and the second eye–head gaze shift is directed toward an auditory (A_2) or visual (V_2) target.** Both stimuli are already extinguished before the onset of the first eye–head gaze shift. Note that after the first gaze shift the eye (F_1) and head (H_1) are unaligned. The first eye-in-head saccade is given by $\Delta E = E_1 - E_0$. The second gaze shift is not indicated, only the desired target vectors for the eye (red arrow) and head (blue).

gaze- and head displacements are described as a weighted sum of the oculocentric and craniocentric motor errors:

$$\Delta\vec{G} = a\vec{T}_E + b\vec{T}_H$$
$$\Delta\vec{H} = c\vec{T}_E + d\vec{T}_H \qquad (11.9)$$

The regression results showed that the coefficients are indeed appropriate for the required transformations, shown in the table of Fig. 11.6, as $a \gg b$ and $d \gg c$, respectively (Goossens and Van Opstal, 1997; Fig. 11.9).

Two different models can explain these results: (1) Either the entire gaze shift has to be preprogrammed by the system. This would include the relative contributions of eye- and head displacements, together with the resulting changes in eye-in-head position, so that the system can already predict the required coordinate transformations for the second gaze shift near the first gaze-shift onset. This model would necessitate quite a sophisticated and extended predictive remapping mechanism. (2) A much simpler hypothesis would hold that the

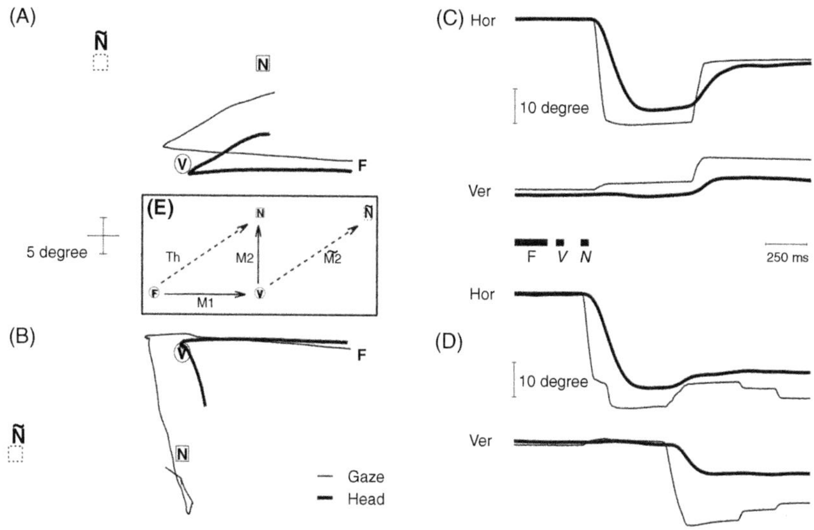

FIGURE 11.7 (A–D) Eye–head coordination in two example visual auditory–double step trials, starting with eye and head spatially aligned ($E_0 = 0$). (A,B) Spatial trajectories. (C,D) Time traces of the horizontal and vertical gaze- and head-components. The system correctly implements the coordinate transformations of Fig. 11.6. Multiple regression shows that the eye is driven by gaze-motor error and the head by a head-motor error signal (see also Fig. 11.9). *(Source: After Goossens and Van Opstal, 1997.)*

gaze-control system uses online feedback of the eye-in-head position at the end of gaze shifts (E_1 in Fig. 11.6).

Given the extremely precise requirements for a full-predictive (feed-forward, and therefore vulnerable) mechanism, and given the speed within which these transformations are carried out, we regard the predictive hypothesis as unlikely. In the next section we will provide stronger evidence for discarding predictive remapping as the mechanism for maintaining spatial accuracy in eye–head gaze shifts.

11.4 DYNAMIC DOUBLE STEP: VISUAL AND AUDITORY

What happens when the gaze-control system can't know beforehand when and where a second stimulus will appear? And that it can't predict the modality of a potential target? In those cases predictive remapping will be useless. Interestingly, a small but highly significant change in the static double-step paradigm makes it possible to study this problem in detail.

The change from a *static* double step, in which both targets were presented when eyes and head were stationary, to a *dynamic* double step introduces some crucial differences. In the dynamic double step the second target is briefly presented *in midflight* of the first eye–head gaze shift. This poses some nontrivial problems to the gaze-control system, which were not yet apparent in the static double step:

1. The system cannot know beforehand whether or not a second target will be presented. Therefore, it cannot start to preprogram its second localization responses already at, or even before, the onset of the first gaze shift.
2. The system cannot predict the sensory modality of the second target, and therefore cannot precalculate the different transformations for eyes and head, made explicit in the Table of Fig. 11.6.
3. The system cannot predict when and where the second target will be presented. Given the considerable variability in gaze shifts, depending on the relative contributions, directions, and speed of the constituent eye- and head movements, the presentation of the second target can fall anywhere within the first gaze shift.
4. Predictive remapping assumes corollary feedback about the planned first gaze shift, ΔG_1 [Eq. (11.6)]. In the static double step this is an appropriate signal for updating (Fig. 11.2B). However, in the dynamic double step the eye and head have already moved a considerable (and not predictable) part of the first gaze shift at the moment of target presentation. Let's denote these partial movements ΔG^* and ΔH^*, respectively (Fig. 11.8B). Thus, a correct

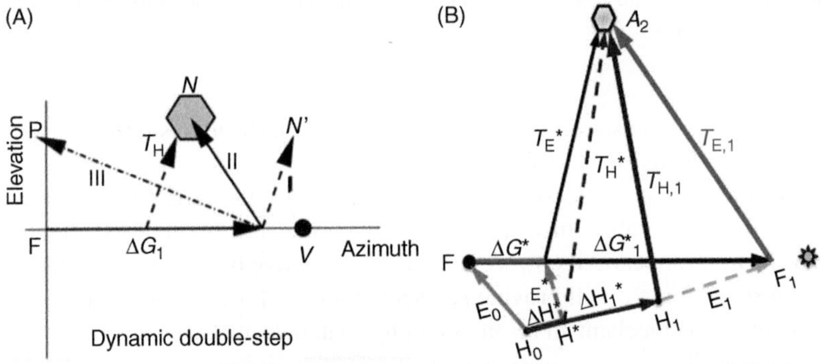

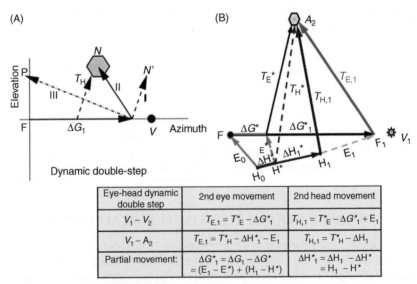

Eye-head dynamic double step	2nd eye movement	2nd head movement
$V_1 - V_2$	$T_{E,1} = T^*_E - \Delta G^*_1$	$T_{H,1} = T^*_E - \Delta G^*_1 + E_1$
$V_1 - A_2$	$T_{E,1} = T^*_H - \Delta H^*_1 - E_1$	$T_{H,1} = T^*_H - \Delta H_1$
Partial movement:	$\Delta G^*_1 = \Delta G_1 - \Delta G^*$ $= (E_1 - E^*) + (H_1 - H^*)$	$\Delta H^*_1 = \Delta H_1 - \Delta H^*$ $= H_1 - H^*$

FIGURE 11.8 (A) The dynamic double step trial (left, schematic), and three remapping models that can be dissociated by this paradigm. I, no updating model; II, dynamic updating; III, predictive remapping. Note that models I and III yield specific error patterns, in opposing directions. (B) Correct coordinate transformations should incorporate the partial movements of the first head and gaze movement, determined by the (unpredictable) moment of target presentation (at $t = t^*$). These calculations also have to compensate for the afferent delays of visual (60–80 ms) versus auditory (10–20 ms) sensory inputs.

"predictive" update should invoke only the remaining portions of the gaze- and head movements.

$$\begin{aligned} \Delta\vec{G}_1^* &= \Delta\vec{G}_1 - \Delta\vec{G}^* \\ \Delta\vec{H}_1^* &= \Delta\vec{H}_1 - \Delta\vec{H}^* \end{aligned} \qquad (11.10)$$

But, how can the system know how far it will have moved the eyes and head in its planned gaze shift without any knowledge of when and where the target is going to appear?

If the gaze-control system would nonetheless rely on the predictive remapping strategy of Eq. (11.6) (now extended to eye–head gaze shifts), the second gaze shift will miss the target by

$$\begin{aligned} &\text{Error gaze:} \quad -\Delta\vec{G}^* \\ &\text{Error head:} \quad -\Delta\vec{H}^* \end{aligned} \qquad (11.11)$$

And because eyes and head will have moved at different speeds and onsets in the first gaze shift, the directions of the second eye–head gaze shift will differ too, and no longer be directed to the same (albeit wrong) goal.

Fig. 11.8 (table) shows the required coordinate transformations for the eyes and head (in vector notation) in case of a visual versus an auditory second target. When the target is a visual flash, the sensory input is provided by a retinal signal, $\vec{T}_E^*$, with an afferent (sub)cortical delay of 60–80 ms. The motor errors for eye and head will be determined from this retinal error signal, and will at least involve the use of eye position, E_1 (at the end of the first gaze shift). A similar remapping problem occurs in case of an auditory noise burst. At the moment that the sound appears, the acoustic cues determine the head-centered target coordinates, $\vec{T}_H^*$, for example, at the level of the IC (chapter: The Midbrain Colliculus), with an afferent delay of about 10–20 ms. Now the motor errors will be based on this craniocentric input, and this again will involve the use of eye position, E_1. The biggest challenge, however, seems to be the calculation of the remaining gaze- and head motor error signals, which can have any size and direction, depending on the first gaze shift and on the exact timing of the stimulus. Let's first see whether the system indeed incorporates these dynamic coordinate transformations, before we proceed to discuss potential mechanisms, and possible neural algorithms.

From Fig. 11.9 it may become clear that the sound-localization system almost effortlessly copes with the dynamic double step. Vliegen et al. (2004, humans), and Van Grootel et al. (2012, monkeys) demonstrated that the accuracy and precision of auditory responses in the dynamic double steps were in fact indistinguishable from single-target responses to these stimuli (ie, without an intervening gaze shift), and from static double-step trials. To quantify the experimental data they performed multiple linear regression on the azimuth and elevation components of gaze- and head-movement responses, which

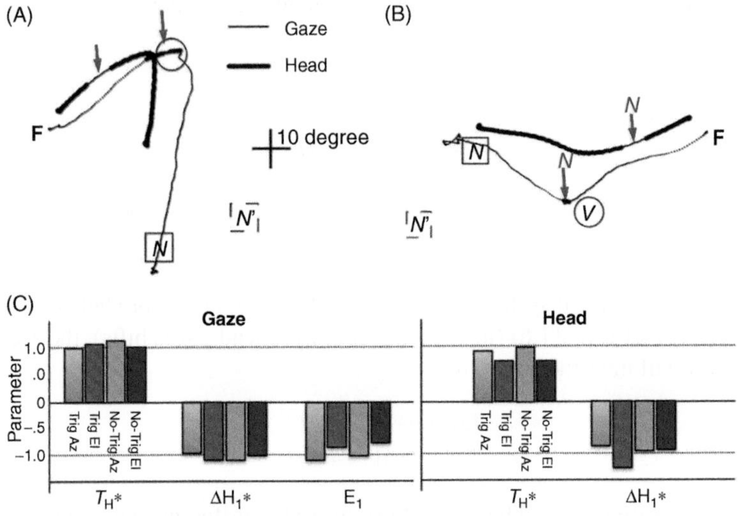

FIGURE 11.9 (A, B) Two example trials of dynamic eye–head double-steps to an auditory target, which is correctly localized by the orienting responses of gaze (*thin traces*) and head (*bold traces*). (C) Results of multiple regression on gaze- and head displacements (horizontal and vertical components) for static (nontriggered) and dynamic (triggered) double steps. All coefficients are close to the ideal values, and strongly support dynamic updating (model II in Fig. 11.8A). *(Source: After Vliegen et al., 2004.)*

incorporated all relevant signals of the vector equations shown in the table of Fig. 11.8.

$$\begin{aligned} \Delta\vec{G}_2 &= a\vec{T}_{\mathrm{H}}^* + b\Delta\vec{H}_1^* + c\vec{E}_1 \\ \Delta\vec{H}_2 &= d\vec{T}_{\mathrm{H}}^* + e\Delta\vec{H}_1^* \end{aligned} \tag{11.12}$$

Perfect localization responses in the double steps would have to yield the following values for the coefficients of azimuth and elevation response components:

$$a=[1,1] \quad b=[-1,-1] \quad c=[-1,-1] \quad \text{and} \quad d=[1,1] \quad e=[-1,-1] \tag{11.13}$$

Fig. 11.9C shows that the data approximated these ideal values quite well. In a followup study the same authors demonstrated equal accuracy for responses to brief visual flashes presented in midflight.

To summarize, the gaze-control system seems to successfully meet the computational challenges imposed by dynamic double-steps. The next question then is: *how does it do it*?

The partial gaze- and head displacements ($\Delta\vec{G}_1^*$ and $\Delta\vec{H}_1^*$) could be determined in many different ways. We here highlight two potential solutions, mathematically equivalent (since they produce the same result), but conceptually and functionally quite different.

A first observation would be that both displacement signals are in fact the *dynamic motor errors* for the eyes and head! Thus, using these signals in dynamic target updating would immediately solve the dynamic transformation problems. This seems a very elegant and novel idea indeed ... So far, dynamic motor error signals have only been proposed for the *local feedback loop* of the brainstem pulse generators, and not for the *global feedback loop* that takes care of target updating (Fig. 11.2A,B). A problem, however, is that these signals have never been recorded, neither in the brainstem [where they are supposed to be the input for the (allegedly nonlinear) pulse generators, discussed in chapter: The Gaze-Orienting System], nor in cortical areas, or in the SC. This latter idea has been vigorously defended by Choi and Guitton (2009). However, their proposed neural correlate of dynamic motor error as a caudal-to-rostral shift of the neural population across the SC motor map appears to be far too sluggish for the transformations that are required in dynamic double steps.

An alternative algorithm to solve the dynamic localization problems calculates the *world-centered coordinates* of the target as soon as it appears and is detected by the sensory systems. As illustrated previously (Fig. 11.1), a world-centered representation remains stable against any intervening movement of the eyes and head, and would therefore be extremely well suited to deal with dynamic (as well as static) localization challenges. This model does not require continuous updates of the motor error commands, as a mere sampling of the final eye- and head positions will suffice to perform the task. In short, the calculations then involve the following transformations:

$$\begin{array}{lll} & \text{Visual flash} & \text{Auditory noise burst} \\ \text{At } t = t^*: & \vec{T}_W^* = \vec{T}_E^* + \vec{H}^* + \vec{E}^* & \vec{T}_W^* = \vec{T}_H^* + \vec{H}^* \\ \text{At } t = t_1: & \Delta\vec{G}_2 = \vec{T}_W^* - \vec{H}_1 - \vec{E}_1 & \Delta\vec{G}_2 = \vec{T}_W^* - \vec{H}_1 - \vec{E}_1 \end{array} \quad (11.14)$$

Fig. 11.10 illustrates this world-centered model in a complete scheme for the audio–visual gaze-control system. It thereby extends the models presented in Figs. 9.10 and 10.14 to dynamic localization problems of the auditory and visual systems.

Note that the world-centered model represents the upcoming next gaze shift by a potentially *dynamic* signal: as the eyes and head continue to move after $t = t^*$, the subtractive feedback by current eye- and head position signals provides $\Delta G(t)$, which could be read out by the SC and brainstem circuitry at any desired moment. For example, at $t = t_1$ (end of the first gaze shift), the signal corresponds to the required goal-directed gaze shift, $\Delta G(t_1) = \Delta G_2$ (which equals the eye-centered error, $T_{E,1}$, in Fig. 11.8B).

Position signals are abundant in the oculomotor brainstem (eyes) and spinal cord (head; both as efference copies, and as proprioceptive signals), yet

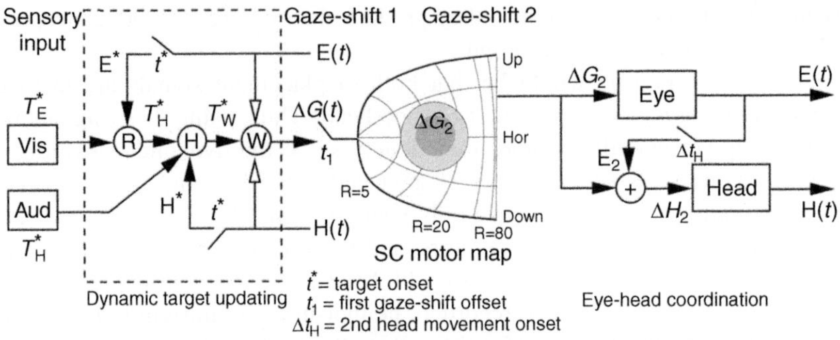

FIGURE 11.10 **Model of audio–visual gaze control that accounts for accurate performance in dynamic double-steps.** At $t = t^*$ the sensory input is transformed into a spatial reference frame by adding eye-(E*) and head-position (H*) signals. By subtracting current eye- and head positions during the first gaze shift from the world target, the dynamic coordinates of the gaze shift, $\Delta G(t)$, are continuously available to the system for read out. At first-gaze shift offset, $t = t_1$, the coordinates correspond to ΔG_2, which is represented in the SC motor map. Eyes and head both make goal-directed movements (see Fig. 9.13) by incorporating current eye position (E_2), and can be initiated at different moments (variable delay Δt_H). *(Source: After Vliegen et al., 2005.)*

the idea of having to rely on multiple time-critical samples may seem a bit too awkward for a realistic neural implementation. Note, however, that the typical time difference between the time stamps at t^* and t_1 amounts to several tens of milliseconds: the duration of a gaze shift and the accompanying head movement will be around 80–120 ms, whereas the intersaccadic interval usually adds an extra 100 ms. This allows the system ample time to carry out the calculations of Eq. (11.14) at $t = t^*$ and to transmit the signals to higher centers.

Although the scheme may suggest the presence of only one world-centered target representation in the brain, this need not be the case. It is conceivable that multiple stages within the gaze-control system represent the target in world coordinates perhaps through gain-field modulations by eye- and head-position signals, or through a different mechanism. A potential candidate would be the PPC, which has been shown to respond to visual and auditory stimuli (but see Xu et al., 2012). Yet, the FEF, and for auditory stimuli, the IC, may also be serious candidates to transmit a dynamic gaze signal to the motor SC (Walton et al., 2007).

Unfortunately, on the basis of behavioral experiments alone one cannot dissociate the different possibilities. But regardless the actual implementation of the neural algorithms, the dynamic double-step paradigm identifies several nontrivial and interesting problems for the sound-localization and visual-motor systems when they have to prepare an adequate and fast orienting response in dynamic (arguably, more natural) situations. The experimental results demonstrate that both systems are clearly up to this challenging task.

11.5 SPATIAL HEARING IS INFLUENCED BY EYE- AND HEAD POSITION

We now briefly discuss a peculiar property of the auditory system that has no equivalent in visual-motor control, and allows us to get deeper insight into the potential spatial transformations that occur in sensorimotor localization behavior. The background for this section is the idea that the auditory system is organized in independent pathways, which extract the azimuth and elevation coordinates of a target sound. As a result, certain sounds can be localized perfectly well in azimuth, and hardly (or not at all) in the elevation direction. An example of such a sound is the *pure tone* (see chapter: Acoustic Localization Cues). Although tones provide adequate ITD's (low-frequency tones, <1.5 kHz) or ILD's (high-frequency tones, >3 kHz) that accurately determine their azimuth coordinate, they cannot single out any particular HRTF to identify the sound's elevation. The auditory system relies on nonacoustic information sources (eg, statistical inference, see Parise et al., 2014, and chapter: Multisensory Integration) to determine the most likely estimate for the tone's elevation. As a result, the actual elevation coordinate of the tone has no relation whatsoever with the elevation of the orienting response (gain zero, but a frequency-dependent offset). The gaze shift to a pure-tone target in the world can thus be described by:

$$\Delta G_{\mathrm{ELE}} = aT_{\mathrm{ELE}}^{\mathrm{W}} + b(f) \tag{11.15}$$

with $a \approx 0$, $T_{\mathrm{ELE}}^{\mathrm{W}}$ the target in world coordinates, and $b(f)$ a frequency-dependent bias. Goossens and Van Opstal (1997) demonstrated that rapid goal-directed head movements do not improve the localization responses to tones (but see, however, chapter: Acoustic Localization Cues for a particular head-movement strategy that could sometimes lead to better performance).

Interestingly, the bias of Eq. (11.15) results to depend also on the initial eye-in-head and head-in-space orientations:

$$\Delta G_{\mathrm{ELE}} = aT_{\mathrm{ELE}}^{\mathrm{W}} + c(f)H_{\mathrm{S}}^{\mathrm{ELE}} + d(f)E_{\mathrm{H}}^{\mathrm{ELE}} + e \tag{11.16}$$

where $a \approx 0$ and c and d are frequency-dependent regression coefficients (Goossens and Van Opstal, 1999; Van Grootel et al., 2012), and e is a stimulus-independent bias. Fig. 11.11 shows the regression result for a typical subject in a Gaussian White Noise (top row) and a pure-tone ($f = 5$ kHz; bottom) localization experiment, where the head-in-space and eye-in-head orientations were systematically varied, and tones were presented at different world elevations. The GWN responses are quite accurate, as the responses nearly fully compensate for the changes in target (gain: $a = 0.90$), head (gain: $c = -0.91$) and eye (gain: $e = -0.75$) position. In contrast, the pure tone is localized very poorly (evidenced by a large offset at 32 degrees). However, although the target gain ($a = -0.02$) is insignificant, the contributions of the changes in head orientation (gain: $c = -0.34$) and eye position (gain: $d = -0.68$) to the responses are highly significant.

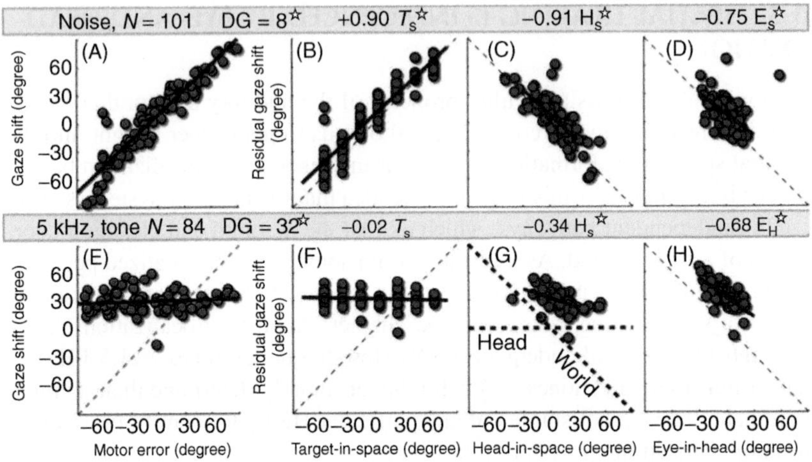

FIGURE 11.11 Gaze shifts to GWN (A–D) and a 5 kHz pure tone (E–H) presented in elevation. The panels show the overall regression (A, E), and the contributions of the spatial target location (B, F), head position (C, G), and eye-in-head orientation (D, H). For full compensatory localization the coefficients for head and eye should be −1.0. Note that although the tone is localized poorly, the eye- and head positions have significant, negative gains. These values depend on the tone's frequency. *(Source: After: Van Grootel et al., 2011, with kind permission.)*

The pure-tone localization experiments indicate that the gains for head orientation and eye position are both idiosyncratic and frequency-dependent. The frequency dependence of the coefficients hint at an interaction of eye- and head position signals *within* the auditory system. The signs of the coefficients result to be compensatory (ie, <0), but for pure tones they are typically far less than -1.0 (the value for perfect compensation). The authors forward the hypothesis that the dependence on head orientation could serve as a (distributed) signal in tonotopically organized neural populations of auditory cells, to construct a target representation in *world-centered coordinates*. This signal is subsequently transformed into the appropriate oculocentric and craniocentric gaze-shift commands, for which the initial eye position and head orientation signals are needed (Fig. 11.10). This idea is illustrated in the conceptual scheme of Fig. 11.12. The model assumes that initial head-orientation signals modulate responses in a frequency dependent way, for example, in the Inferior Colliculus. The current eye- and head positions are subsequently subtracted from T_W to define the next upcoming gaze shift. For GWN stimuli, the update to world coordinates, and subsequent recalculation into oculocentric coordinates, is nearly perfect, as all frequency channels will contribute to the gains of head- and eye position. For pure tones, however, the total population gain will be far less than -1.0, and the response will typically miss the target.

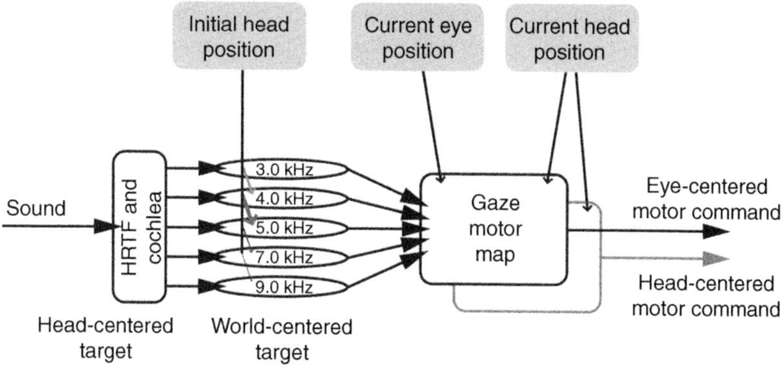

FIGURE 11.12 Conceptual model for how head- and eye-position signals interact within the auditory and gaze-control systems to enable accurate spatial performance in world centered and head centered–sound localization tasks.

11.6 LIMITS TO UPDATING?

From the previous sections it may have appeared that the gaze-control system is fully equipped to carry out the daunting task of accurately orienting to brief flashes of light, or brief sound bursts, presented during ongoing high-velocity eye–head gaze shifts, in which the sensory inputs (retinal location, or acoustic localization cues) change at a high speed. All relevant signals in the required coordinate transformations seem to be accurately represented within the system, and they all seem to be used in the control of voluntary gaze shifts. The question seems justified whether there are any limits to these sophisticated target-updating mechanisms.

For example, what would happen if the eyes and head move through space without a *planned* voluntary gaze shift to a target? Such a situation would occur, for example, under passive whole-body rotation: then, involuntary eye movements are generated as reflexive vestibular nystagmus. This ocular nystagmus is a combination of *compensatory slow-phase* eye movements that counteract the passive movement of the head through space, and *interleaved quick phases* of nystagmus, which are involuntary saccades in the direction of the head movement. The slow phase eye movement intends to keep the eyes stable in the world despite head movements, whereas the quick phases reset the eye position to around the center of the oculomotor range, in order to prevent the slow phase to drive the eyes to the mechanical limits of the OMR.

To estimate the ongoing direction of the visual axis in the world, the head movement through space has to be monitored by the semicircular canals, and continuously added to the involuntary eye-movement signals of vestibular nystagmus.

Dynamic target localization can be made even more challenging as described in the previous sections, by presenting *extremely brief* visual or auditory stimuli, with durations from less than one to only a few milliseconds. Clearly, if the sensory systems would be unable to detect and localize such brief stimuli even in a stationary situation (ie, target at a fixed location, and subject not moving), it would not come as a surprise that under dynamic passive movement conditions the system would fail too. However, if the system can localize these brief stimuli under stationary conditions, it will be interesting to introduce more complexity to the problem by imposing two different situations.

1. In the *head-fixed* condition, the target is fixed to the rotating chair (ie, the subject's head),
2. In the *world-fixed* condition, the target is nailed to the laboratory room.

Fig. 11.13 illustrates the rationale for this passive vestibular paradigm. The subject's response will be an eye movement to the perceived location of the target (and she is unaware whether the target was world- or head-fixed). The experiment can in principle dissociate four different target-updating models, regarding the compensation of intervening eye- (nystagmus) and/or (passive) head movements (Table 11.1).

In case of a world-fixed target, the appropriate response is in world-centered coordinates, but when the target moves with the chair (dashed target, *H*), the appropriate response is a craniocentric update that only accounts for the intervening eye movements. Can the gaze-control system successfully dissociate these two possibilities? And if so, which signal uniquely distinguishes the two conditions?

Van Barneveld et al. (2011b) reconstructed the visual input on the retina during the visual flashes. As the eye moves through space, the flash causes a visual streak on the retina that depends on the direction of the eye movement, and on the duration of the flash. Interestingly, because of the compensatory nature of

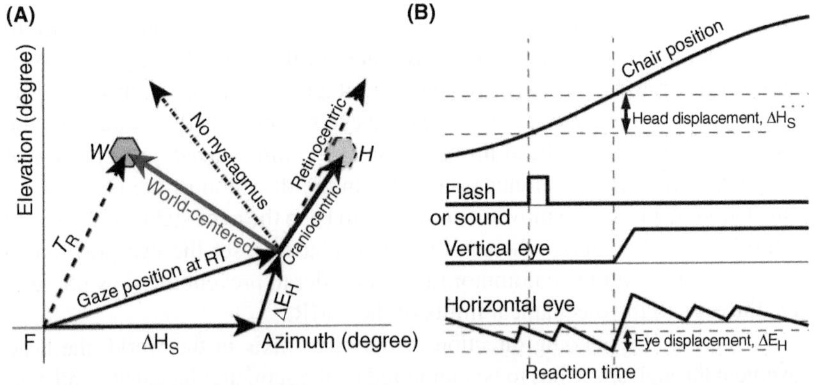

FIGURE 11.13 (A) Passive whole-body rotation (vertical axis) induces a horizontal head movement (ΔH_S), as well as a combination of ocular nystagmus and a voluntary eye movement (ΔE_H) within the reaction time of the localization response to a world-fixed (W), or chair-fixed (H) target. Four models distinguish potential updating mechanisms for the ocular localization response to a sound or light flash. (B) Temporal sequence of the different events. Lower trace exemplifies the horizontal ocular nystagmus. For target W the world-centered response is appropriate, but for condition H a craniocentric response is required. *(Source: After Van Barneveld et al., 2011a,b).*

TABLE 11.1 Four Potential Updating Models With Different Degrees of Compensation of Intervening Eye- and Head Movements

↓ Update nystagmus?	Update passive head? NO	Update passive head? YES
NO	Retinocentric model	Head-only compensation
YES	Craniocentric model	World-centered model

vestibular nystagmus, the world-centered and head-centered target conditions produce visual streaks that are in opposing directions (Fig. 11.14).

It can be seen that for the briefest flash duration, the maximal excursions of the visual streaks are about 0.05 degree, which is an order of magnitude smaller

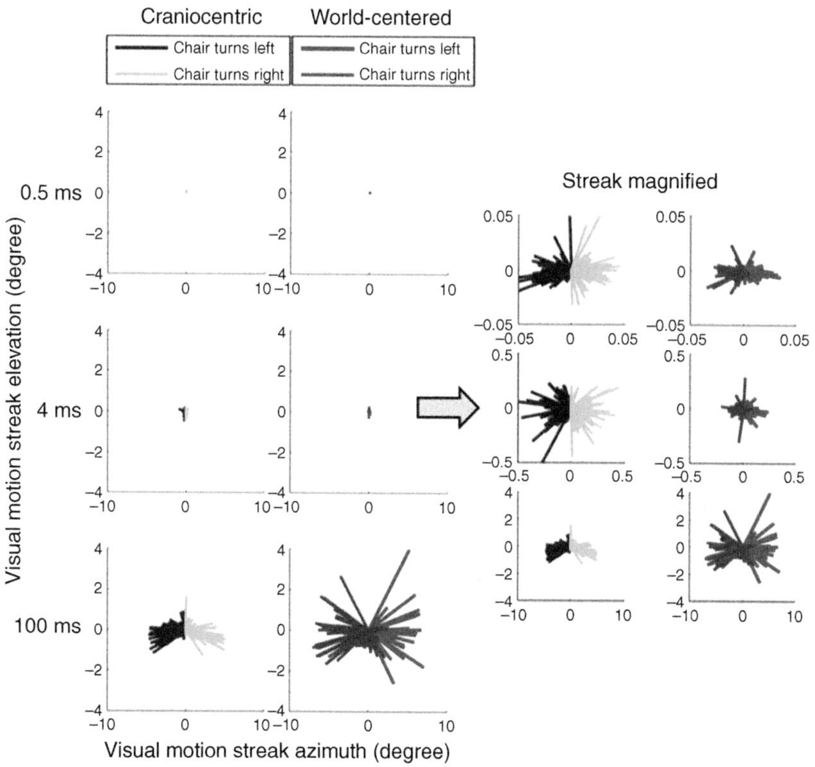

FIGURE 11.14 Retinal streaks on the retina produced by visual flashes of 0.5 ms (top) and 100 ms (bottom) for head-fixed (left) versus world-fixed (right) targets. Note the clear differences in the direction of visual streak patterns, which would allow the system in principle to dissociate the two spatial conditions. However, the differences will remain below the detection threshold for the shortest stimulus (insets: highly magnified scales).

than the radius of the fovea (0.5 degree). In other words, *there is absolutely no possibility for the visual system to tell the difference between head- and world-fixed target motion*. For longer flashes (above 10 ms) the differences are readily perceptible, as they can extend over several degrees across the retina.

The same line of reasoning holds for the pattern of interaural time- and level differences (Exercise 11.3). When sounds are too short, a world-fixed sound will produce changes in ILDs or ITDs that remain below the perception threshold. For those sound durations, the auditory system will not be able to dissociate the two different modes (Van Barneveld et al., 2011a).

As before, to differentiate the four egocentric updating models of Fig. 11.13A, the data are quantified by multiple linear regression (performed on azimuth only, since whole-body motion was restricted to a vertical axis rotation). The regression equation for visual flashes then reads:

$$\Delta G = aT_{\mathrm{E}} + b\Delta H_{\mathrm{S}} + c\Delta E_{\mathrm{H}} + d \tag{11.17}$$

where the target on the retina, T_{E}, is defined by the difference between the head-centered target location and the eye-in-head position at flash-time, $t = t^*$.

$$T_{\mathrm{R}} = T_{\mathrm{H}}^{\mathrm{Azi}} - E_{\mathrm{H}}^{*} \tag{11.18}$$

For the brief auditory bursts, the regression equation is written as:

$$\Delta G = aT_{\mathrm{H}} + b\Delta H_{\mathrm{S}} + c\Delta E_{\mathrm{H}} + d \tag{11.19}$$

with T_{H} the head-centered sound location at sound onset. For the different updating models, the following truth table holds for visual and auditory targets (Table 11.2; Van Barneveld et al., 2011a,b).

TABLE 11.2 Expected Coefficients for Multiple Regression on Eye Movements According to Different Updating Models for Visual Flashes and Brief Sound Bursts

Visual flashes				
Model [Eq. (11.17)]	*a*	*b*	*c*	*d*
World-centered	1	−1	−1	0
Craniocentric	1	0	−1	0
Retinocentric (no updating)	1	0	0	0
Sound bursts				
Model [Eq. (11.19)]	*a*	*b*	*c*	*d*
World-centered	1	−1	−1	0
Craniocentric	1	0	−1	0
No updating	1	0	0	0

So what are the results of these experiments? To summarize, when the stimulus durations are long enough for the system to perceive the two different spatial modes, the goal-directed eye movements are driven by the adequate transformations: there is (nearly) full compensation for involuntary eye- and passive head movements when targets are presented in world coordinates, and only compensation for the intervening eye movements when the targets are fixed to the head. This behavior forms another excellent example of the sophisticated neural algorithms that guide active sound- and visual localization responses.

But what happens if the stimulus duration is *too short* to distinguish world-fixed from head-fixed target locations like in the top row of Fig. 11.14? First, we have to establish that these brief targets were well localizable. Indeed, the data show that under static conditions, as well as under the vestibular rotation conditions, the target gain, $a \sim 1.0$ for both stimulus modalities. This means that these extremely brief targets were well perceived: the visual system was well aware of the target's retinal coordinates, and the auditory system had full access to the sound's azimuth coordinates. So what could we expect?

One reasonable possibility might be that the gaze-control system "assumes" that targets are by default presented in the world, because in a natural environment it would be highly unlikely that they are fixed to the head. In other words, a possible outcome would be that the responses are best described in a *world-centered* reference frame, also when the target was head-fixed. Interestingly, however, this is not the result of these experiments.

> When visual flashes are too short to create a perceptual visual streak, the visuomotor system keeps the target in its original retinal coordinates (no updating).
>
> When sounds are too short to detect acoustic motion, the audiomotor system keeps the target in head-centered coordinates, only incorporating eye movements.
>
> Thus, no spatial updating to world coordinates when there is uncertainty about stimulus motion.

This is quite a remarkable finding, because the system has access to *all signals* needed to perform the transformation of the target into world-centered coordinates: the stimulus is well localizable (T_E, or T_H), the current eye position, E_H, is known to the system, and the head motion through space, ΔH_S, is accurately measured by the vestibular system. These latter two signals are intrinsic variables, which are independent of the stimulus quality. Yet, the gaze-control system *refuses* to carry out the coordinate transformation, probably because it lacks evidence about retinal or acoustic stimulus motion when the eyes and head are moving through space. It is as if the system only accepts as a "reasonable explanation" for the lack of perceived sensory motion, that the visual target is glued to the retina, or that the auditory target is glued to the head, even though these stimulus conditions will *never* occur in natural environments. Indeed, quite surprising!

Further work is needed to understand the limits of spatial updating. For example, localization responses after self-generated active gaze shifts are accurate (Vliegen et al., 2004); even short acoustic noise bursts (down to 3 ms) are localized well in the azimuth direction (and poorer in elevation, see also chapter: Acoustic Localization Cues) for static and dynamic double steps. So far, however, these experiments have not been performed for the extremely brief stimuli as used in the vestibular paradigms.

Moreover, the experiments described in this chapter have dealt with stimuli that were either stationary in the world, or moved through space at exactly the same speed as the head. These are two extremes out of an infinite pool of potential movement conditions. Indeed, the even more challenging localization task in a dynamic double step would be to let the brief stimulus move at its own relative speed and direction with respect to the observer and the world [Eq. (11.4)].

11.7 EXERCISES

Problem 11.1: Eq. (11.4) appears to contain two free parameters (p and q), but in fact, these must be related. Simplify Eq. (11.4) to incorporate the constraint that the ILD change corresponds to a given velocity.

Problem 11.2: Gain fields [Eq. (11.7)] may follow from a first-order Taylor approximation.

(a) First approximate the Gaussian tuning characteristic in (u,v) coordinates to a first-order Taylor expansion.
(b) Then calculate the derivative for $F(u,\Delta E_0)$ by assuming an eye-position sensitivity along the u-direction only.
(c) Now express $F(u,\Delta E_0)$ as function of stimulus eccentricity, R, by applying the (1D) complex-log map, to show that

$$F(R,E_0) \approx F(R,0)(1-2\alpha AB_u \Delta E_0) \equiv (1+\varepsilon E_0)F(R,0)$$

Problem 11.3: Estimate the ILDs and ITDs produced by a stationary auditory target in the world when the burst durations are 0.5, 1.0, 3.0, 5.0, or 10.0 ms, and the chair moves at a 100 degrees/s rightward or leftward. When will it not be possible to tell the difference between head-fixed and world-fixed sounds? Use the reported resolution of the azimuth angle (chapter: Acoustic Localization Cues).

REFERENCES

Andersen, R.A., Essick, G.K., Siegel, R.M., 1985. Encoding of spatial location by posterior parietal neurons. Science 230, 456–458.

Choi, W.Y., Guitton, D., 2009. Firing patterns in superior colliculus of head- unrestrained monkey during normal and perturbed gaze saccades reveal short-latency feedback and a sluggish rostral shift in activity. J. Neurosci. 29, 7166–7180.

Duhamel, J.R., Colby, C.L., Goldberg, M.E., 1992. The updating of the representation of visual space in parietal cortex by intended eye movements. Science 255, 90–92.

Goldberg, M.E., Bruce, C.J., 1990. Primate frontal eye fields. III. Maintenance of a spatially accurate saccade signal. J. Neurophysiol. 64, 489–508.

Goossens, H.H.L.M., Van Opstal, A.J., 1997. Local feedback signals are not distorted by prior eye movements: evidence from visually-evoked double-saccades. J. Neurophysiol. 78, 533–538.

Goossens, H.H.L.M., Van Opstal, A.J., 1997. Human eye-head coordination in two dimensions under different sensorimotor conditions. Exp. Brain Res. 114, 542–560.

Goossens, H.H.L.M., Van Opstal, A.J., 1999. Influence of head position on the. spatial representation of acoustic targets. J. Neurophysiol. 81, 2720–2736.

Guthrie, B.L., Porter, J.D., Sparks, D.L., 1983. Corollary discharge provides accurate eye-position information to the oculomotor system. Science 221, 1193–1195.

Hallett, P.E., Lightstone, A.D., 1976. Saccadic eye movements toward stimuli triggered by prior saccades. Vision Res. 16, 99–106.

Jürgens, R, Becker, W., Kornhuber, H.H., 1981. Natural and drug-induced variations of velocity and duration of human saccadic eye movements: evidence for a control of the neural pulse generator by local feedback. Biol. Cybern. 39, 87–96.

Mays, L.E., Sparks, D.L., 1980. Saccades are spatially, not retinocentrically, coded. Science 208, 1163–1165.

Parise, C.V., Knorre, K., Ernst, M.O., 2014. Natural auditory scene statistics shapes human spatial hearing. Proc. Natl. Acad. Sci. 111, 6104–6108.

Robinson, D.A., 1975. Oculomotor control signals. In: Lennerstrand, G., Bach- y-Rita, P. (Eds.), Basic Mechanisms of Ocular Motility and Their Clinical Implications. Pergamon Press, Oxford, pp. 337–374.

Scudder, C.A., 1988. A new local feedback model of the saccadic burst generator. J. Neurophysiol. 59, 1455–1475.

Sommer, M.A., Wurtz, R.H., 2002. A pathway in primate brain for internal monitoring of movements. Science 296, 1480–1482.

Sommer, M.A., Wurtz, R.H., 2004. What the brainstem tells the frontal cortex. II. Role of the SC-MD-FEF pathway in corollary discharge. J. Neurophysiol. 91, 1403–1423.

Sparks, D.L., Mays, L.E., Porter, J.D., 1987. Eye movements induced by pontine stimulation: interaction with visually triggered saccades. J. Neurophysiol. 58, 300–318.

Umeno, M.M., Goldberg, M.E., 1997. Spatial processing in the monkey frontal eye field. I. Predictive visual responses. J. Neurophysiol. 78, 1373–1383.

Van Barneveld, D.C.P.B.M., Binkhorst, F., Van Opstal, A.J., 2011a. Failure to compensate for vestibularly-evoked passive head rotations in human sound localization? Eur. J. Neurosci. 34, 1149–1160.

Van Barneveld, D.C.P.B.M., Kiemeneij, A.C.M., Van Opstal, A.J., 2011b. Absence of spatial updating when the visuomotor system is unsure about stimulus motion. J. Neurosci. 31, 10558–10568.

Van Gisbergen, J.A.M., Robinson, D.A., Gielen, S., 1981. A quantitative analysis of generation of saccadic eye movements by burst neurons. J. Neurophysiol. 45, 417–442.

Van Grootel, T.J., Van Wanrooij, M.M., Van Opstal, A.J., 2011. Influence of static eye and head position on tone-evoked gaze shifts. J. Neurosci. 31, 17497–17504.

Van Grootel, T.J., Van der Willigen, R.F., Van Opstal, A.J., 2012. Experimental test of spatial updating models for monkey eye-head gaze shifts. PLoS One 7 (10), e:47606.

Van Opstal, A.J., Hepp, K., 1995. A novel interpretation for the collicular role in saccade generation. Biol. Cybern. 73, 431–445.

Van Opstal, A.J., Hepp, K., Suzuki, Y., Henn, V., 1995. Influence of eye position on activity in monkey superior colliculus. J. Neurophysiol. 74, 1593–1610.

Vliegen, J., Van Grootel, T.J., Van Opstal, A.J., 2004. Dynamic sound localization during rapid eye-head gaze shifts. J. Neurosci. 24, 9291–9302.

Vliegen, J., Van Grootel, T.J., Van Opstal, A.J., 2005. Gaze orienting in dynamic visual double steps. J. Neurophysiol. 94, 4300–4313.

Walker, M.F., Fitzgibbon, E.J., Goldberg, M.E., 1995. Neurons in the monkey superior colliculus predict the visual result of impending saccadic eye movements. J. Neurophysiol. 73, 1988–2003.

Walton, M.M., Bechara, B., Gandhi, N.J., 2007. Role of the primate superior colliculus in the control of head movements. J. Neurophysiol. 98, 2022–2037.

Wang, X., Zhang, M., Cohen, I.S., Goldberg, M.E., 2007. The proprioceptive representation of eye position in monkey primary somatosensory cortex. Nat. Neurosci. 10, 640–646.

Xu, B.Y., Karachi, C., Goldberg, M.E., 2012. The postsaccadic unreliability of gain fields precludes the motor system from using a simple gain-field algorithm to calculate target position in space. Neuron 76, 1201–1209.

Zipser, D., Andersen, R.A., 1988. A back-propagation programmed network that simulates response properties of a subset of posterior parietal neurons. Nature 331, 679–684.

Chapter 12

Sound Localization Plasticity

12.1 INTRODUCTION

To localize a sound the auditory system relies on craniocentric acoustic cues that result from the interaction of sound waves with the body, head, and pinnae. We have seen (chapter: Acoustic Localization Cues) that these interactions depend strongly on the geometry of the head and ears. As the head grows from childhood to adulthood both the interaural distance and acoustic head shadow effect will increase, which affects the interaural time delays and frequency-dependent level differences, respectively (Clifton et al., 1988). But also the pinnae gradually change in size and shape, which has potentially profound effects on the spectral elevation cues too (Otte et al., 2013).

To cope with these gradual acoustic changes the auditory system should appropriately adapt its mappings of these cues onto spatial locations. Since it is highly unlikely that the entire process of growth and cue mapping is encoded in one's DNA, the auditory system should possess adequate neural adaptive mechanisms that allow the system to learn and store new cue-location associations.

We here wish to make a distinction between *implicit sensory–motor learning* as a response to (potentially long-term) changes in the acoustic periphery, or in the central auditory pathways due to a lesion, or to a neural or sensory degenerative process, and *explicit perceptual learning*, in which a particular auditory performance skill improves by means of an explicitly defined training paradigm. Let's first briefly discuss the latter, before we proceed with sensory–motor learning.

Perceptual learning: In a perceptual learning paradigm, a listener is repeatedly confronted with a particular auditory stimulus, which she has to learn to discriminate or detect. The idea is that by rigorous training, in which often (but not always) feedback about performance is provided, the subject will improve on the task. The objective measure for perceptual improvement is determined by a significant change in the *d*-prime or threshold from the psychometric function (chapter: Assessing Auditory Spatial Performance). The process of perceptual learning involves *active attention* of the listener to the particular stimulus features that define the task. Furthermore, the task is repeatedly presented to the subject (*sensitization*; Amitay et al., 2006). Finally, if feedback about

The Auditory System and Human Sound-Localization Behavior. http://dx.doi.org/10.1016/B978-0-12-801529-2.00012-X

performance is provided, the type of feedback (positive: correct, or negative: failure), and feedback probability, may both affect the learning speed (Amitay et al., 2015).

For example, in a certain perceptual learning task, the subject could be required to discriminate the durations of two sounds. Before training, the minimum detectable duration difference is assessed in a 2AFC task, in which the subject indicates the stimulus interval containing the longer sound. During training, the subject will be exposed to sounds with duration differences close to, or even below, the detection threshold for hundreds of trials. After training, a new psychometric curve is measured, with the result that the subject can detect shorter duration differences than before training. Amitay et al. (2006) could demonstrate that listeners even improve their frequency-discrimination thresholds when trained with *identical* stimuli that lack any objective difference.

An important aspect of any learning paradigm is whether the improvement remains exclusive for the stimuli used in the training set, or whether the subject *generalizes* the improved skill to other stimuli too (Wright et al., 2010). For example, if training was done only with 1000 Hz pure tones, a possible result could be that the subject only improved for these tones (and perhaps for some neighboring frequencies), but failed to show any improvement for different frequencies, for broadband noise bursts, or for spectral–temporal modulated sounds. The question then of course is: what did the subject really learn to discriminate? Was it really sound *duration*, or some other aspect of 1000 Hz tones? Of course, it is still possible that this particular task induced a true *local* perceptual change within the auditory system's tonotopic representations. In that case, the paradigm addressed a particular (but perfectly valid) learning aspect of the (bottom-up) auditory pathways.

A second important aspect of the learning is its *persistence*: is the improvement still measurable long after the training? Does it require continuous training to be persistent? (Wright and Sabin, 2007). We return to these issues in later sections of this chapter.

Sensory–motor learning: In active sound-localization behavior, discussed extensively in the previous chapters, the auditory system is heavily integrated with the gaze-control system to redirect the eyes and head toward peripheral sound sources. Each time the audiomotor system generates an eye–head gaze shift to a sound, it automatically receives sensory and motor feedback about its success (eyes and head on or near the target, just as planned), or its failure. In case of a considerable motor error, the system has to program a new response to acquire the target. If localization errors are consistent, the audiomotor system has to figure out whether the problem is due to errors in the sensory-to-spatial mapping (leading to wrong goal representations), in the motor pathways (leading to wrongly directed movements), or in both. As each of these problems will produce particular error patterns, the system can in principle identify (and remedy) the source of the problem. Fig. 12.1 illustrates this concept with two schematic examples of auditory (A) and motor (B) mismatches.

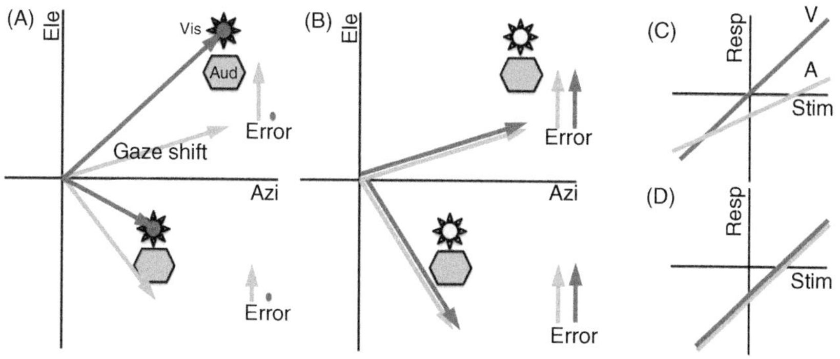

FIGURE 12.1 **Examples of error patterns for gaze shifts to visual and auditory targets.** (A) Error is due to a problem in the auditory system, since visual gaze shifts are accurate. (B) Errors show a constant downward bias for both visual and auditory gaze shifts, hinting at a motor problem. (C) Response patterns of (A) for many stimulus locations reveal that gain and offset of sound-elevation encoding are both wrong. (D) Downward bias for visual and auditory-evoked gaze shifts of (B) would be consistent with an error in the vertical pulse-step-generator (chapter: The Gaze-Orienting System).

In Fig. 12.1C,D the stimulus–response relations for auditory and visual stimuli are described by (elevation only, for simplicity):

$$\begin{array}{ll} \text{Responses} & \text{Error in trial } n \\ R^A = aT^A + b & E_n^A = (1-a)T_n^A - b \\ R^V = cT^A + d & E_n^V = (1-c)T_n^V - d \end{array} \tag{12.1}$$

For the examples in Fig. 12.1A the visual responses are assumed to be accurate: $c = 1, d = 0, E_n^V = 0\ \forall n$, but the auditory responses have systematic localization errors, as $a < 1$ and $b < 0$. The system could thus conclude that the errors are due to a problem in the auditory system. In Fig. 12.1B, however, the error patterns for auditory and visual stimuli are identical, so that the most likely cause for the errors would be in the gaze-control motor system. Since $a = c = 1$, and $b = \mathrm{d} < 0$ the errors are described by a constant downward bias, necessitating a fixed upward correction gaze shift that is independent of stimulus eccentricity and modality. The most likely candidate structure for producing such errors would be the vertical brainstem pulse–step generator (see chapter: The Gaze-Orienting System, where the PSG is introduced).

To learn the appropriate error corrections, the sound-localization system should acquire evidence across many trials from its own gaze-motor responses and from its own audio visual feedback evaluations, and combine the different sources of information to make adjustments in the system (which would be for A: $-E_n^A$, for V: $-E_n^V\ \forall n$). On the basis of these evaluations it should have to decide whether the problem relates exclusively to audition (and where), to gaze-control, or to some combination of the two (see Exercise 12.2).

The barn owl: This chapter deals with neural plasticity in the human audiomotor system, rather than with perceptual learning. We will first introduce the concepts of sound-localization plasticity by briefly referring to the seminal work of Eric Knudsen and coworkers, who have studied the barn owl's auditory system for almost four decades. The barn owl (*Tyto alba*) is a peculiar animal in the sense that its auditory system has a unique organization, not found in any mammalian species (nor in other birds, for that matter). The barn owl has its ear canals partially covered by earflaps that are oppositely directed on either side, such that the right ear canal receives the strongest acoustic input for sounds coming from above, while the left ear is more sensitive for sounds coming from below (Knudsen and Konishi, 1979; Fig. 12.2B, inset). This causes a systematic binaural level difference for sounds, which monotonically varies with the *elevation* angle. A second evolutionary adaptation in this animal is that phase locking of its auditory nerve fibers (chapter: The Cochlea) can follow frequencies up to 9–10 kHz. Thus, the barn owl still has ITD sensitivity above 8 kHz (Takahashi, 1989). As a result, this species localizes the azimuth and elevation coordinates of sounds on the basis of two, nearly perpendicular and monotonic localization cues (Fig. 12.2B).

A third important difference between barn owls and (most) mammals is that the owl's oculomotor range is extremely small (less than 2 degrees). As a result, the oculocentric and craniocentric reference frames in the barn owl are *always aligned*. This means that an orienting motor command to a light or a sound source can directly drive the head, as the eye will automatically move with the head (the vestibular ocular reflex is very limited in this species). Thus, for sensorimotor evoked orienting, the owl's head functionally operates as a giant eye.

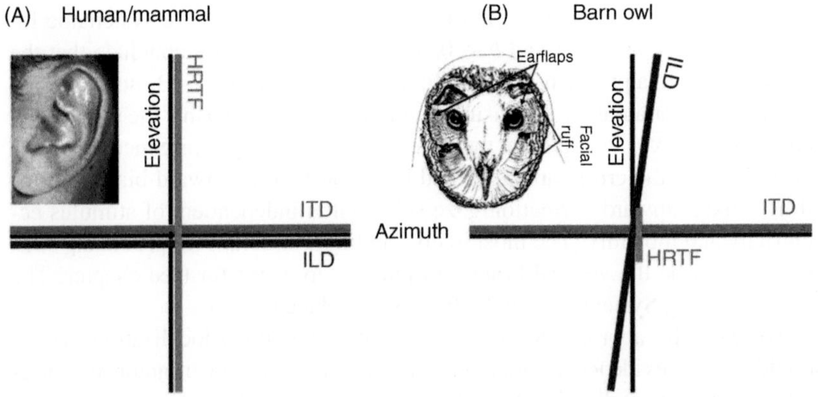

FIGURE 12.2 (A) In humans ILDs (blue) and ITDs (red) both encode azimuth for high and low frequencies, respectively, while HRTFs encode elevation angles. (B) In barn owls the asymmetric ears ensure that ILDs encode elevation, and ITDs azimuth, both for relatively high frequencies. The owl's HRTFs vary little with sound location. Note that the owl's ILDs also vary somewhat with azimuth, because of the head shadow.

Note, however, that although the owl does not have to solve the dynamic coordinate transformation problems for the eyes and head described in chapter: Coordinate Transformations, also this animal will have to compensate for intervening head movements in static and dynamic double steps (although, so far, whether owls do this, has not been verified experimentally).

Taken together, the barn owl's auditory system possesses a relatively simple, monotonic neural encoding of the azimuth and elevation coordinates by two independent acoustic variables (time/phase and level, respectively). In addition, the processing pathways for the ILDs and ITDs already diverge at the level of nucleus angularis and magnocellularis (the bird's homolog of the CN). Thus, the ILD and ITD tunings can be directly mapped to the posterior part of the brainstem ventral lemniscus and the nucleus laminaris (the bird's homologs of LSO and MSO, respectively). The topographically organized ILD and ITD pathways eventually converge in the external nucleus of the owl's inferior colliculus (ICx), where they cause its cells to have topographically organized 2D *spatial* tuning sensitivity. The owl's ICx thus contains an explicit map of *craniocentric auditory space* (Knudsen and Konishi, 1978).

Plasticity in the barn owl: When a young owl is raised with a unilateral plug, say in the left ear, the animal will initially make large, mainly upward, sound-localization errors, to which the otherwise normal animal adapts within a week. At that point, the owl correctly localizes sounds despite the presence of a plug. When the plug is removed, however, the owl will display a strong *aftereffect*: now its localization errors will be in the *opposite* direction (Fig. 12.3A, day 1). Knudsen and Knudsen (1985) tested owls under different adaptation conditions upon the removal of a unilateral plug to which they had already adapted: normal vision (Fig. 12.3A), occluded eyes (Fig. 12.3B), and shifting Fresnel prisms (Fig. 12.3C).

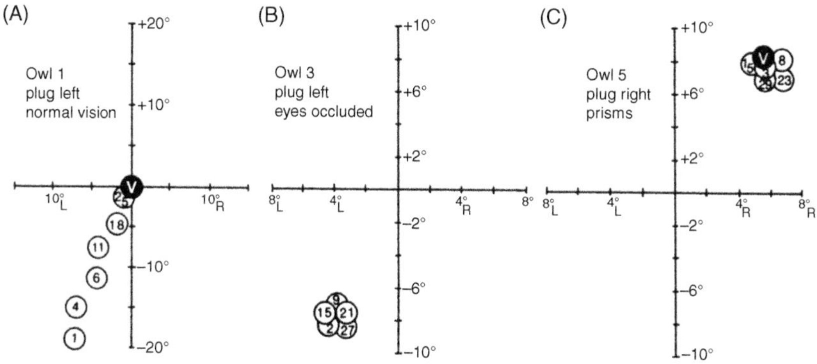

FIGURE 12.3 Sound-localization plasticity in the barn owl depends on the integrity of vision. (A) An owl with normal vision adapts to the aftereffect of a plug in the left ear, removed on day 1. The large downward localization errors are virtually gone after three weeks. (B) When the eyes are occluded the owl does not adapt to the plug's removal. (C) When the owl is also equipped with Fresnel prisms that shift its visual input by the same amount as the plug's aftereffect, the owl does not adapt either. *(Source: After Knudsen and Knudsen, 1985, with kind permission.)*

Only under normal vision would the owls readapt to regain normal sound-localization performance levels. When the eyes were occluded, the aftereffect remained, and when the prisms shifted their vision to the same location as the plug's aftereffect, there was no adaptation either. When the prisms shifted vision to a different location than the aftereffect, the owl would map its sound localization estimates to the location imposed by the prisms. These experiments thus clearly demonstrated that the auditory system (at least of barn owls) requires normal vision to calibrate the acoustic localization cues.

Cue perturbation in humans: The human audiomotor system differs in some profound ways from the barn owl's auditory system: first, humans need three independent acoustic cues to localize sound sources (Fig. 12.2A). Second, although the binaural difference cues can be monotonically mapped onto azimuth, the mapping of the HRTFs to elevation angle is highly nonlinear. Third, humans make large eye movements, which misalign the eyes and head under normal orienting conditions. This poses additional coordination problems for eye–head orienting toward sounds (chapter: Coordinate Transformations). Fourth, there is no evidence for a 2D craniocentric auditory map in the mammalian brain. Instead, sound locations appear to be mapped onto an *oculocentric* reference frame at the level of the motor map in the SC. It is therefore not a priori obvious that similar learning rules as in the barn owl would also apply to the human auditory system.

In the following sections we discuss the response of the human sound localization system to short- and long-term perturbations of the acoustic cues. These will mainly be manipulations to normal-hearing human subjects with unilateral plugs, or with monaural and binaurally applied molds to the pinnae.

Acute effects of a unilateral plug: Fig. 12.4 shows the acute effect of a unilateral plug in the left ear on the sound-localization responses (in this case, rapid head movements) of a human subject in azimuth and elevation. Sounds were high-pass filtered Gaussian white noises (>3 kHz) of 150 ms duration that covered a range of intensities between 30–60 dB A-weighted. The sound levels were randomized across trials in order to uncover the listener's use of the acoustic head shadow as an (ambiguous) localization cue (Van Wanrooij and Van Opstal, 2004; chapters: Acoustic Localization Cues; Impaired Hearing and Sound Localization, where we describe this phenomenon in detail for hearing-impaired listeners).

The immediate and most obvious effect of a unilateral plug is a large contralateral shift in the azimuth responses, which is accompanied by a strong reduction in the response gain. For the example listener of Fig. 12.4C, who was equipped with a plug in the left ear that attenuated high frequencies by about 25 dB, the bias was 45 degrees to the right, and the gain dropped from about 1.1 in the binaural control condition (both ears free) to only 0.18 with the plug. The strong effect on the bias is readily explained by the large ILD that favors the hearing ear. However, note that since the maximum ILD for a human head is about 20 dB (see chapter: Acoustic Localization Cues), a 25 dB attenuation would point to an azimuth of nearly 90 degrees. Thus, other cues (notably: spectral cues, discussed later) must play a role in this response behavior too.

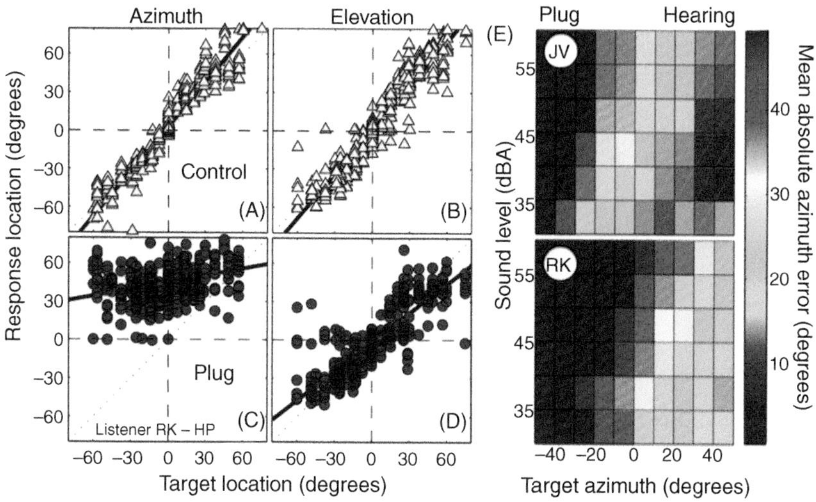

FIGURE 12.4 **Head-movement responses to HP sounds in the 2D frontal hemifield with a unilateral plug in the left ear.** Sound levels between 30 and 60 dBA. (A,B) Normal binaural hearing condition; localization performance is accurate. (C,D) Unilateral plug attenuated the high frequencies in this listener >25 dB. Note the large rightward bias (45 degrees) in the azimuth responses, and a very low gain (0.18). Data pooled across sound levels. Note also the decrease in gain for the elevation responses. (E) Azimuth localization errors (red = high, blue = low) depend systematically on sound level and azimuth. The lowest sound levels yielded the smallest errors. Data from two listeners. *(Source: After Van Wanrooij and Van Opstal, 2007.)*

An explanation for the effects on the azimuth *gain*, however, is less obvious. Because the head movements were not directed to the head–motor limits (which is well above 100 degrees to either side), this effect is not due to motor saturation. As will be discussed in chapter: Multisensory Integration, the response gain could be determined in part by the precision (inverse of the variance) of the percepts. Indeed, the response gain and the overall correlation between stimulus and responses covaried too: the lower the gain, the higher the variance in the data, and hence the lower the correlation.

An acute unilateral plug further reveals some subtle localization effects that are nonetheless quite consistent among listeners. First, the plug also affects the gain and variance of the *elevation* response components (compare panels B and D, where the gain dropped from 1.1 to 0.75). This hints at an interesting interaction between the initially independent cue-extraction pathways. We already alluded to such interactions in chapter: Assessing Auditory Spatial Performance, where we described the acute effects of unilateral and bilateral molds in the pinnae on 2D sound localization (Fig. 8.18).

Second, the azimuth localization gain and bias both depend systematically on the absolute intensity of the stimulus. Loud sounds are persistently localized with a larger contralateral bias (more into the hearing side) and a lower gain, than soft sounds (Fig. 12.5). The explanation for this phenomenon is that the

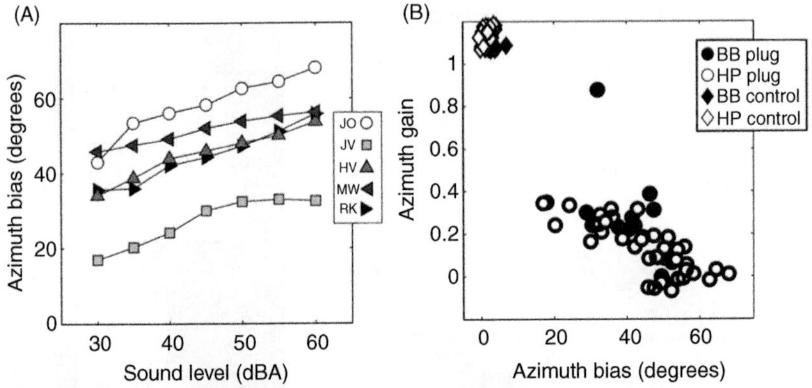

FIGURE 12.5 Head–shadow effect on sound localization responses with a unilateral plug. (A) Azimuth bias increases systematically with sound level. (B) Azimuth gain and bias for the plug data (circles) are negatively correlated. Diamonds in the top-left corner are the binaural control data (biases close to zero, gains close to one). *(Source: After Van Wanrooij and Van Opstal, 2007.)*

perceived intensity at the hearing ear [the proximal sound level, see Eq. (7.16)] for a sound at a fixed azimuth, say straight ahead, varies systematically with the sound's absolute level. A soft sound will therefore produce a lower perceived level (and hence a lower ILD dominance for the hearing ear) than a loud sound. Since in the acute plug condition the otherwise normal-hearing subject has no other cues to rely on than the perceived ILD, the perceived azimuth location will shift with sound level too.

Third, the localization *errors* in azimuth (and hence the gain and bias of the responses) depend in another interesting way on sound level (Fig. 12.4E): for the lowest sound levels (here, 30 dBA) the azimuth errors on both the hearing side and the plugged side were considerably *lower* than for high sound levels. This is immediately visible in the color-coded plots (two listeners), where lines of constant error run diagonally across the azimuth versus sound-level matrix. This effect disappears when the *spectral cues* of the hearing ear are removed by inserting a mold in that ear. Similar effects are observed for the elevation responses: errors are smallest for soft sounds presented at the hearing side.

Since the latter is an acute effect, measured immediately after plug insertion, and since subjects are tested under open-loop conditions with fully randomized trials, it cannot be due to learning. Rather, it demonstrates that the auditory system adopts the interesting strategy to increase the weight of subtle azimuth-related spectral pinna cues as soon as the binaural difference cues become unreliable. This happens in normal-hearing conditions, for example, for very soft sounds, when the contralateral ear doesn't receive enough acoustic power to unambiguously encode an ILD. In chapter: Impaired Hearing and Sound Localization we will see that also hearing-impaired patients often use this strategy.

Taken together, the independent cue-extraction pathways in the auditory brainstem appear to interact in an interesting way when cues become unreliable. After

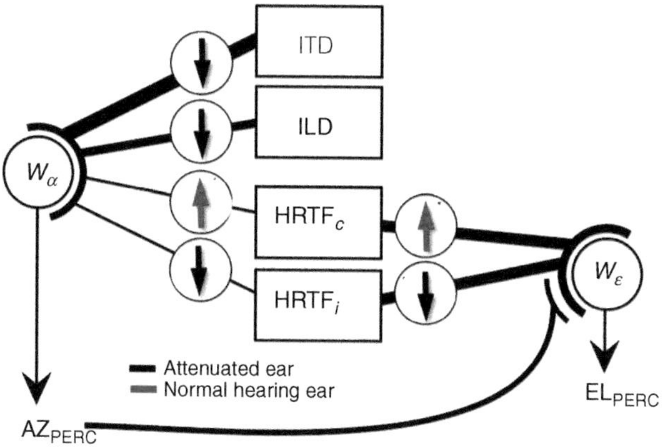

FIGURE 12.6 **Schematic model for the changes (arrows) in the different cue weightings upon unilateral plugging that determine perceived azimuth and elevation.** *(Source: After Van Wanrooij and Van Opstal, 2007.)*

unilateral plugging (to simulate an acute conductive hearing loss), the binaural difference cues suddenly change in a complex way. As a result, the weighting of the pinna cues of the contralateral ear becomes stronger. Fig. 12.6 shows a conceptual model for adaptive acoustic cue-weighting (increase/decrease) for the perceived azimuth and elevation coordinates upon unilateral plugging. At present it is not clear where in the auditory pathway the integration occurs, but the IC (chapter: The Midbrain Colliculus) and Auditory cortex are likely candidate structures.

12.2 LEARNING BINAURAL CUES

To learn new associations between the binaural cues and actual stimulus locations, different techniques have been employed. For example, Shinn-Cunningham et al. (1998) used headphone simulation of HRTF cues in the horizontal plane to distort the cue–azimuth relations. In this virtual reality simulation they imposed a sigmoid relation on the azimuth cues that had a slope > 1 around straight-ahead, and zero for the lateral positions ($\alpha = \pm 90$ degrees). This distortion magnified the cue differences in the frontal field ("supernatural" cues), while compressing these differences for more lateral positions. For example, subjects would initially perceive a sound as coming from an azimuth of 20 degrees, which was supposed to come from 10 degrees. They would receive visual feedback about the intended (10 degrees) location. Within one feedback session subjects would change their responses in the appropriate direction. However, listeners only learned to reduce the bias in their responses, not the response sensitivity (gain). Possibly, the training sessions were not long enough to reach full adaptation. Alternatively, the nonlinearity in the required response-gain reduction proved to be too difficult for the system to learn. Another important question, however, is whether the response changes were indicative of adaptation in the *auditory* system, or whether subjects learned a particular "mapping trick" that reassigned spatial percepts to new (visually defined) locations. Since the number of potential stimuli was restricted to 13 locations in the frontal horizontal plane, the inclusion of cognitive and memory factors cannot be excluded either.

Consistent simulation of altered azimuth cues for all sound directions and stimuli can also be achieved, for example, with binaural hearing aids. Javer and Schwarz (1995) applied this technique by artificially changing the ITDs. They added a constant binaural delay between 171 and 684 μs, while subjects wore the devices for several days. The participants rapidly learned to reduce the initially large contralateral shift in their azimuth responses.

More recently Kumpik et al. (2010) tested unilaterally plugged subjects in a free-field localization experiment. To attenuate the acoustic input, they used classic foam ear plugs. Stimulus levels were roved over a large range to avoid the use of head shadow as a valid cue (Van Wanrooij and Van Opstal, 2004). Also in this study, potential locations were confined to the horizontal plane ($n = 12$, distributed over the full 360 degrees), but now subjects wore the plug for a week, during which they were tested daily. Three groups of subjects underwent different test and training procedures: (1) flat noise/daily training, (2) flat noise/single-day training, and (3) random amplitude spectra/daily training.

Only the group of subjects that received daily training and flat noise spectra learned to improve localization of the stimuli. The group that only received a single (long) training session, and the group that had to localize randomized spectra, did not learn to compensate for the plug-induced perturbations (Fig. 12.7). Thus, in line with the results of acute plugging, shown in Fig. 12.4,

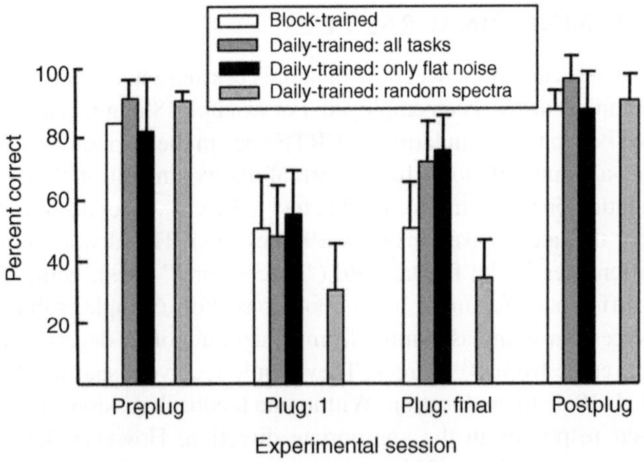

FIGURE 12.7 Learning to localize broadband sounds after unilateral plugging (Kumpik et al., 2010). White, gray and black bars: groups exposed to flat noise. Only the group that received daily training (gray and black bars) improved performance. The group that received all training on one day (white bar) did not learn. The group that was trained on randomized amplitude spectra (pink bars) did not improve either.

the availability of reliable *spectral* cues may be crucial to relearn localization with a single ear.

Although subjects could localize the random spectra in the azimuth direction without the plug (Fig. 12.7B), it is unclear whether these stimuli could also be localized across the full azimuth and elevation range. For example, Hofman and Van Opstal (2002) showed that strongly randomized spectra typically induce severe illusions in the elevation direction. As a result, these stimuli may be regarded by the auditory system as unreliable for localization, and hence not worth the effort for learning, as their spectra do not consistently single out a unique, true elevation angle (see chapter: Acoustic Localization Cues, on the spectral correlation model).

The question is whether specific training and explicit feedback to induce *perceptual learning* is required to regain, or speed-up, (near) normal localization performance after perturbing the localization cues (Shinn-Cunningham et al., 1998; Kumpik et al., 2010; Amitay et al., 2006, 2015; Irving and Moore, 2011). A potential problem with explicit training procedures applied under laboratory conditions, and with testing and training of only a small set of stimulus locations, could be that these procedures could interfere with generalization of localization performance, which is implicitly trained in the subject's natural environment. Clearly, in natural listening environments subjects are typically exposed to a myriad of different sound-sources, potentially coming from any direction, eliciting the full eye–head–body orienting behaviors and their sensorimotor error feedback, which is virtually impossible to train explicitly in a laboratory. Moreover, explicit training conditions may not only miss, but potentially even hamper, the use of natural sensorimotor mechanisms that underlie learning within unknown, natural scenes.

Therefore, to test whether prolonged exposure to a unilateral custom-made plug (for 14 days, or longer), would lead to general localization improvements in two dimensions without any specific training or feedback, subjects were tested for their open loop–sound localization responses to broadband and high-pass filtered noises, presented over a large range of intensities (30 dB) within the frontal hemifield (60 degrees eccentricity in all directions). All stimuli were presented in randomized order, while subjects generated fast head-saccades to the perceived target locations. To verify that the plugs induced a stable perturbation of the binaural difference cues over time, the subject's ear-specific hearing thresholds with the plug in situ were regularly assessed for tones presented over headphones to either ear (0.5–11.3 kHz, spaced at 0.5 octaves).

Fig. 12.8 summarizes the results for HP stimuli of one listener. The left panel shows acute results of inserting the plug in the left ear. It nicely shows the intensity and azimuth dependence of localization accuracy (measured by the mean absolute error per bin), which is indicative of the subtle use of spectral cues already at the start of the experiment (see also Fig. 12.4). In the center panel, which shows the data taken at the end of the experiment (session 11), it can be seen that the MAE has decreased in a systematic way, as the bluish colors (smaller MAE) gradually invaded the plot from right to left for all intensities. To quantify the total data set

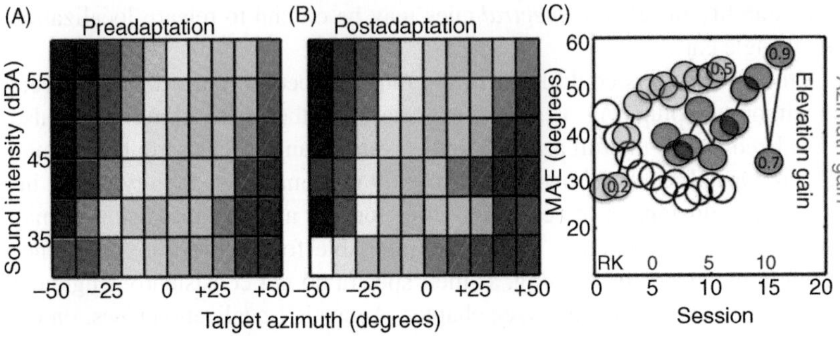

FIGURE 12.8 Adaptation of listener RK to a unilateral (left) plug. (A) (same format as Fig. 12.4E) shows acute mean absolute localization error in azimuth as function of target azimuth and sound level. (Dark red: MAE = 50 degrees, dark blue = 0 degrees). (B) Systematic improvements after 11 sessions (two weeks). (C) MAE (open circles), azimuth (gray), and elevation (blue) gains, as function of time (session nr). Elevation data are shifted, for clarity (min–max values inside symbols). *(Source: Van Wanrooij et al., unpublished results.)*

we determined the slope and offset of the azimuth–elevation stimulus–response relations for each session. The right-hand panel summarizes how the gains (gray circles) and mean absolute localization error in azimuth (open circles) gradually changed as function of time. Interestingly, also the elevation gain of the responses gradually increased over time (blue symbols).

In summary, it is possible to learn to use spectral cues of the hearing ear to improve monaural sound localization in 2D, without exposing listeners to explicit training procedures. The improvements, which can be quite substantial (the participant of Fig. 12.8 had an azimuth gain increase from 0.2 to 0.5, and an MAE decrease from 50 to 30 degrees), already start within the first days after insertion of the plug, and extend over the full 2D audiomotor and intensity range.

We hypothesize that the exposure to natural acoustic environments, in combination with cognitive and continuous sensorimotor feedback, which most of the time occurs unconsciously, may suffice to let the auditory system adapt to the cue changes. However, the following points are worth noticing.

1. When the plug blocks the ILDs only partially (say, by 20 dB, or less), the auditory system could still recalibrate the binaural cues for louder sounds (also see chapter: Impaired Hearing and Sound Localization on patients with conductive hearing loss).
2. When the binaural cues are totally absent [as in single-sided deafness (chapter: Impaired Hearing and Sound Localization), or for soft sounds, Fig. 12.4] the spectral cues from the hearing ear may help, but will be insufficient to regain normal localization performance across the full range of azimuth angles.
3. Learning can only take place when there exists a *unique* mapping between the newly available cues (no matter how weak they are), and the associated sound locations. As soon as inconsistent ambiguities arise (eg, front–back

confusions, or conflicts between remaining ITDs and frequency-dependent ILDs), it may be impossible for the system to cope with the problem, and therefore refrain from learning.

Concerning this latter point, Hofman et al. (2002) demonstrated that it is *impossible* to learn to cope with inverted ears, which was achieved by interconnecting a pair of in-ear-canal hearing aids. Although listeners indeed inverted their eye-movement azimuth localization responses in the laboratory with these devices, they did not learn to invert their acoustic cue mappings to achieve normal localization performance. The major reason for this lack of adaptation was that head movements (continuously made in daily life), or moving sound sources, invariantly produced conflicting front–back confusions that were inconsistent with the (static) reversed ILDs (Exercise 12.1).

12.3 LEARNING SPECTRAL CUES

The mapping from pinna cues to elevation angles cannot be described by a simple functional relationship (like the monotonic ILD–azimuth and ITD–azimuth relations), but rather seems to involve complex pattern associations, in which different frequency bands play a role, predominantly within the 3.5–12 kHz range (discussed in chapter: Acoustic Localization Cues; eg, Fig. 12.9B). Internal knowledge about these spectral patterns, which are highly idiosyncratic, must somehow be shaped by experience and sensorimotor feedback mechanisms, but until recently it was not clear whether and how the human auditory system would retain its ability to adapt to changes of the pinna cues.

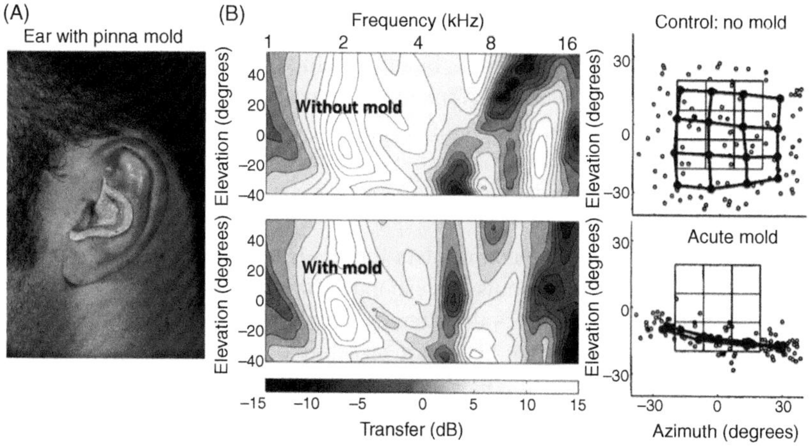

FIGURE 12.9 Effect of molds inserted in the concha of both pinnae (A) on the subject's HRTFs and localization responses. (B) The HRTFs (only the left ear is shown) are heavily perturbed by the molds. (C) Stimulus–response grids and individual responses (open dots). Localization of elevation is no longer possible immediately after inserting the mold. Azimuth responses are unaffected. *(Source: Adapted from Hofman et al., 1998.)*

The previous section hinted at the possibility that the weighting of spectral cues to the 2D sound-localization percept may be plastic, and dependent on the acoustic conditions (eg, soft sounds; Fig. 12.6). But what about the spectral pinna cues themselves? What happens if they are suddenly perturbed, or gradually change as a result of pinna growth? Can the auditory system adapt to such perturbations? And if it does, will it interfere with, or even remove, the representation of the ear's original spectral cues?

Hofman et al. (1998) set out to study this problem by introducing binaural silicon molds in the major pinna cavity, the *concha*, which listeners had to wear for 24 h/day (Fig. 12.9A). Their localization performance with the molds in situ was tested almost daily in the laboratory for broadband noises presented in all directions at random locations within the oculomotor range. The experiments were carried out in total darkness and did not provide the subject any feedback about performance. Fig. 12.9B (bottom panel) shows the acoustic effect of the left mold for one of the participants. It is clear that the small mold had a drastic effect on the spectral cues from about 4 kHz and higher, but the interesting question is, of course, whether the new cues still provide sufficient information for the listener to localize sounds in the elevation direction. Fig. 12.9C shows the immediate effects of the molds on 2D sound localization performance. The data are presented in a particular way to immediately visualize the spatial effects in 2D.

To generate these plots, stimulus locations were averaged within bins of 10 × 10 degrees with 5 degrees overlap. Each bin (containing 15–20 stimuli) thus yielded a single point. By connecting the averages of neighboring bins, one obtains the thin central squares shown in Fig. 12.9C. Since these squares are not distorted, they demonstrate that targets were presented homogeneously within the 2D oculomotor range [(−35, + 35) degrees in all directions]. For the responses one repeats this trick, binning them according to the stimuli that elicited them. If a subject would respond with a perfect gain of 1.0 for azimuth and elevation, and no bias, the average response grid (shown by the solid lines and large black dots) would coincide precisely with the stimulus grid. Any deviation can thus be interpreted as systematic and constant localization errors in the experiment.

The top panel shows the control data for this listener, prior to applying the molds. The small open symbols correspond to the individual responses. The response grid nicely resembles the stimulus grid, indicating that the listener's responses indeed captured the geometric structure of the stimulus distribution. Yet, some idiosyncratic errors may be noted too, exemplified by a down- and rightward bias, and an azimuth gain > 1.0. Also the gains for left and right hemifield elevation responses appear differ somewhat (also see chapter: Assessing Auditory Spatial Performance, Fig. 8.18, on the binaural interactions of pinna cues).

The bottom panel shows the response data immediately after inserting the molds. Clearly, the response pattern differs dramatically from the control

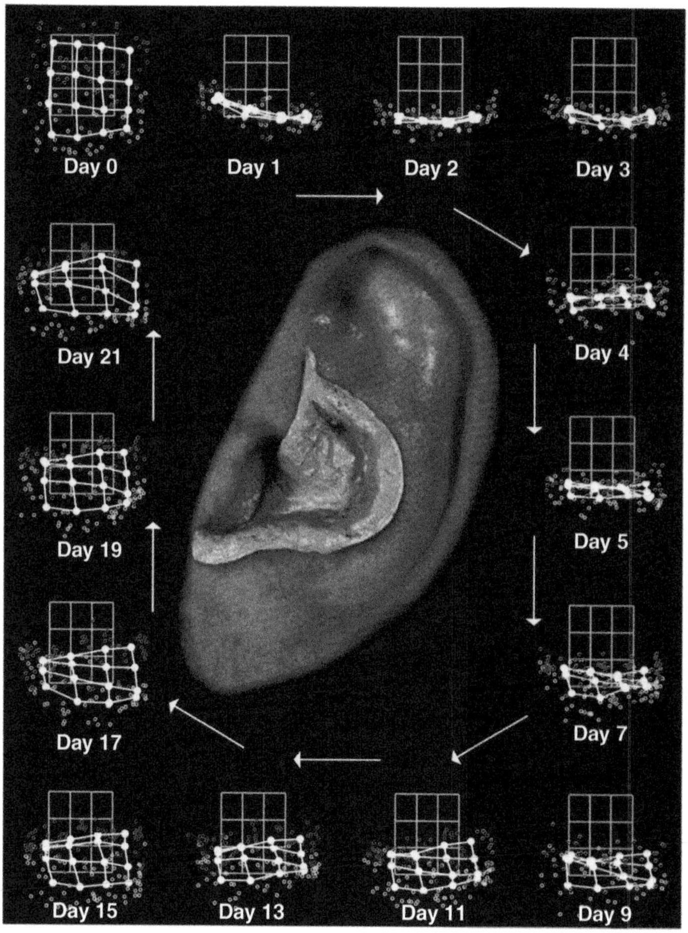

FIGURE 12.10 **Adaptation to binaurally perturbed spectral cues over the course of three weeks.** From day 4–5 onward, the elevation responses gradually recover, reaching near-normal behavior after about 15–17 days. Data from subject PH. Single dots are individual responses (indigo: downward target elevations; violet: upward elevations).

condition: the response grid has completely collapsed onto a straight line, indicating the impossibility for the subject to cope with the perturbed spectral cues. Note that if the subject had not been able to localize any of the stimuli at all (say, completely random behavior), the response grid would have collapsed into a single point. This is clearly not the case as to first order the azimuth responses appear unaffected. Yet, the effect of the molds on localization performance is quite dramatic and clearly demonstrates the importance of the pinna cues.

Fig. 12.10 shows what happens to the responses when the molds are kept permanently in place for about three weeks. During these three weeks, the subject just does his regular business: he goes to work, interacts with others, moves around in the city, at home, etc. but is never specifically instructed or trained to localize particular sounds. Perceptually, listeners report that the sound quality appears unaffected, and since the molds are very lightweight and fit exactly in the pinna cavities, subjects are hardly aware of particular acoustic problems, or even of the presence of the molds. During the first 2–3 days nothing much seems to happen, but after about 4 days it seems that the response-elevation range slightly expands in a systematic way, as the response grid shows a small separation between the lowest and highest stimulus elevations. This trend gradually expands over the next days. After about 15 days the upper and lower stimulus hemifield yield fully separated localization responses: the stimulus grid has now completely opened up. After 19–20 days, the responses seem to have stabilized, and do not recover further, at which point the adaptation experiment is terminated.

The experiments were conducted with four participants, and all gradually relearned to localize the elevation of sounds with their "new ears." Fig. 12.11 shows the regression results for the entire experiment. The final elevation gains strongly increased from nearly zero to about 0.6–0.8 for all subjects (where the preadaptation gains scattered around 1.0). The learning speeds, however, varied substantially. It is not clear whether the spectral quality of the molds (in terms of unambiguous, well separable elevation cues) differed critically among the eight ears, or whether perhaps learning speed is really idiosyncratic.

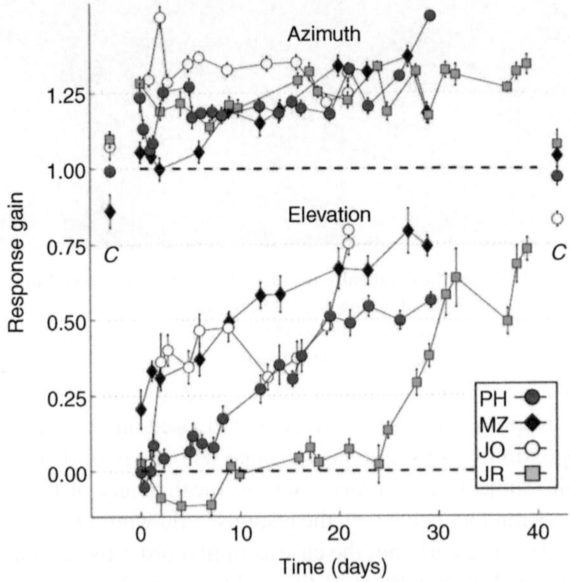

FIGURE 12.11 Regression results (gains) of azimuth (top curves) and elevation (bottom curves) for the four participants (colors) as a function of time. (C) control elevation gains without the molds. Note different learning speeds of the participants, and the absence of an aftereffect: immediately after removal of the molds, the elevation gains for the natural ears were back to normal (C on the right). *(Source: Adapted from Hofman et al., 1998.)*

No aftereffects: Quite remarkably, after removal of the molds, the subjects' localization behavior with their normal ears *immediately* returned to the pre-adapation control levels (see control data on the right-hand side of Fig. 12.11). In other words: there was *no aftereffect*! This important result has been replicated in a follow-up study with unilateral molds (Van Wanrooij and Van Opstal, 2005) and more recently in several studies from other laboratories too (Majdak et al., 2013; Carlile et al., 2013; Carlile and Blackman 2014).

It is important to note that the measurements of Hofman et al. (1998) were performed under *open-loop* conditions, in total darkness, and without any type of feedback. Thus, the subject had to rely entirely on the *acoustic* input to orient to the sound's location. Apparently, when the molds are removed the auditory system can immediately "switch" to the original HRTF representations, based on the sensory spectral input. Thus, it also appears that the learning process illustrated in Figs. 12.10 and 12.11 had not deteriorated the original "own-ear" representations. It even seems as if the auditory system, after learning, had simultaneous access to multiple (two? four?) HRTF representations, and that it can immediately decide which set (or even single HRTF?) is the most appropriate for a given input spectrum (Exercise 12.3).

These results differ profoundly from the adaptation experiments reported for the barn owl by Knudsen and coworkers. As can be seen in Fig. 12.3, the barn owl shows a large, oppositely directed aftereffect when the unilateral plug is removed, and it takes the animal almost two weeks under full visual feedback to get rid of this effect. In humans, these aftereffects are neither observed in the spectral perturbation experiments, nor in the unilateral plug experiments described above.

So why is this? Note, that humans *do* show strong aftereffects for certain *visual–motor* behaviors, like in vestibular adaptation. In such an experiment subjects wear for example, magnifying glasses that disrupt the relationship between an eye movement and the associated retinal slip. The normally functioning vestibular–ocular reflex (VOR) precisely compensates with an eye movement that has a gain of -1.0 with respect to the head movement, to stabilize the eye in space. With magnifying glasses (say with factor 1.5), however, the appropriate VOR gain should be -1.5. Indeed, the VOR can adapt to this new gain (see also Fig. 12.14C, for an example of gain reduction), but after removal of the glasses it will have to readapt to the original gain of -1.0. What could be the profound difference with spectral cue learning, or what may be the similarity with the barn owl's ILD learning?

We propose that the major difference lies in the nature of the stimulus–response relationships of the adaptation mechanism that is invoked for these different systems:

If the VOR requires a change of its responses to magnifying glasses, it could cope with the situation by tweaking relatively few parameters (gain, bias, phase, and sign) that determine the input–output relations of the whole system. If done correctly, the VOR will again be appropriate for the entire range of vestibular stimuli. The total system, however, has changed, and to undo the adaptation it will need to retune its parameters to their original values. This again will require sensory (motor) feedback, which becomes evident as an aftereffect. The same principle might hold for changing the ILDs in the barn owl with a unilateral plug.

However, this simple tuning strategy, cannot work for the spectral cues, because a parametric input–output relation between HRTFs and elevation angles (that would cover the entire elevation domain) does not exist. Instead, the system has to reconstruct a new representation (patterned map) that establishes such a relation, and to achieve this it will have to learn to make new spatial associations for each new spectral shape. Note that the resulting new mapping does not have to interfere with the original natural mapping, as long as the cue perturbations are strong enough.

Indeed, in a follow-up study with unilateral molds applied to 13 subjects, Van Wanrooij and Van Opstal (2005) showed that the fastest learning (within 6 days) was obtained for ears that received the *strongest* spectral perturbations. When the cue perturbations were relatively small (in that case, the subject could still localize stimuli reasonably accurately with the acute mold in place), learning did *not* occur, not even after 30 days.

Fast learning of the spectral cues hinges on two requirements:

1. Cues provided by the mold, do not correlate with the original cues.
2. Cues are unique for each elevation angle, and cover the full elevation range.

Conceptually, there may exist an analogy with learning a new language: also in this case learning doesn't lead to forgetting one's native language, and the learner is able to switch between languages without aftereffects. An important difference, of course, is that language acquisition is a truly cognitive process, whereas the learning of acoustic spectral cues, as reported here, occurs unconsciously, without cognitive involvement.

Just like for the unilateral plug experiments, we hypothesize that the auditory system learns to construct new spectral cue representations through active sensorimotor interactions with sounds in the natural environment, and that the use of eye (visual)- and head-movement (motor and acoustic feedback) information helps the system to gradually build a new set of HRTFs. Fig. 12.12 shows a simple computational neural-network model that learns a new set of HRTFs from scratch without interfering with the original ones.

In this model, the spectral input, $Y(\omega)$, for a target sound at ε_{Tar} leads to an elevation percept, $\varepsilon_{\text{Perc}}$, through an internal correlative analysis of the input spectrum with all stored HRTFs (see chapter: Assessing Auditory Spatial Performance and Fig. 8.16). However, whenever possible, it also acquires feedback information from the visual–motor (eye–head coordination) system about the actual source position, ε_{Vis}. If there is a consistent localization mismatch (and the error exceeds a certain criterion), the system will add the responsible sensory spectrum, $\eta \cdot Y(\omega)$, with η the learning rate (and determine a cumulative average)

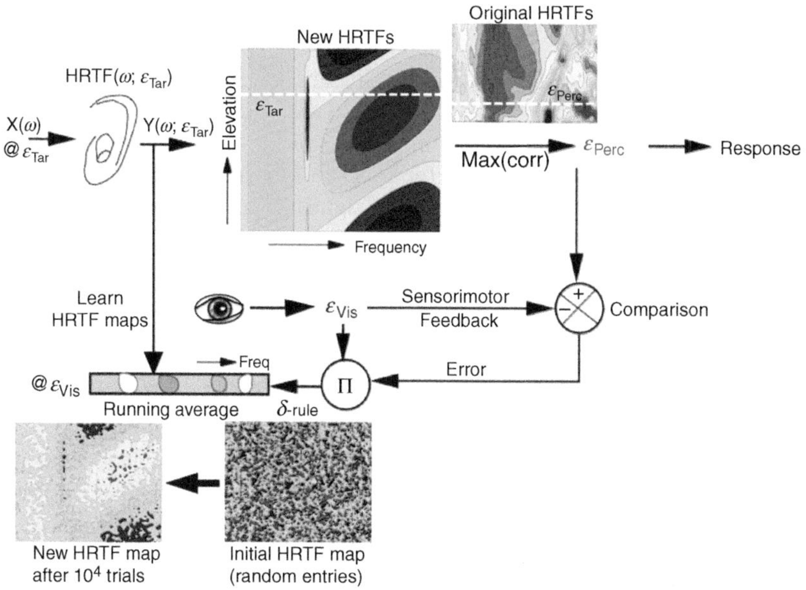

FIGURE 12.12 Computational model of how the auditory system could learn and store a new set of HRTFs. The perceived elevation (ε_{Perc} and hence the response) is determined by the maximum correlation between spectral input, *Y*, and all stored HRTFs. If the visual–motor system detects an error: $\Delta = \varepsilon_{Perc} - \varepsilon_{Vis}$, the input spectrum at ε_{Tar} is stored (as a long-term running average) in a new HRTF map at the visually perceived elevation. The new map does not erase the original map of stored HRTFs. A new set of HRTFs is thus acquired, and there will be no aftereffect.

in a new HRTF map, at the elevation that corresponds to the visual elevation. This new map is only formed when needed (ie, gated learning when errors are large enough), and it will not interfere with the original HRTFs. Strong evidence for a gated visual teaching signal (from the optic tectum, or SC) into the barn owl's auditory spatial map (in ICx) has been provided by Gutfreund et al. (2002).

This simple model explains how the auditory system can immediately use the new cues: for each sensory spectrum the perceived elevation will be determined by the maximum correlation *across the different maps*. The system doesn't even need to know from which map it derived its response! The model thus suggests that there may be many (uncorrelated) spectra that point to the same, single elevation angle (*many to one*). The reverse, however, is not true: a given spectrum should never point to more than one elevation angle (*ambiguity*).

12.4 VISUAL FACTORS IN LEARNING

Plasticity studies in the barn owl have unequivocally demonstrated that the auditory system of this species relies heavily on feedback from the visual system (Fig. 12.3; Knudsen and Knudsen, 1985; Gutfreund et al., 2002). We suggest that also in the human auditory system learning is, at least partly, driven by visual–motor feedback (Fig. 12.12). Here we discuss two lines of evidence that support the involvement of vision in training and calibrating the human auditory system: (1) localization performance of congenitally blind listeners, and (2) changes in the auditory system of normal-hearing listeners, induced by perturbed visual input.

Congenitally blind: Most studies with blind participants have been confined to the localization and discrimination of sound sources in azimuth only, and typically tested spatial hearing for sounds of a single intensity. In those situations, the blind may outperform the sighted (eg, Lessard et al., 1998; Röder et al., 1999). However, the use of a single intensity enables listeners to successfully employ the head–shadow effect, which for hearing conditions with a variety of sound levels proves to be an ambiguous localization cue (Van Wanrooij and Van Opstal, 2004). Furthermore, including the elevation direction greatly increases the number of potential sound source locations (thereby severely limiting the use of other cognitive factors, like memory), and directly tests the integrity and usefulness of the pinna-related spectral cues.

Zwiers et al. (2001a,b) thus tested congenitally blind listeners in a 2D head-pointing localization task, in which the target sound was hidden within a diffuse broadband noise background of 60 dB SPL. As an acoustic parameter they systematically varied the signal-to-noise (S/N) ratio from infinite (no background noise, serving as the control condition) to one of the following values, randomly presented in the experiment: $SNR = [0, -6, -12, -18, \text{and} -21]$ dB. In these acoustic conditions, also the normal-sighted will at a certain point no longer be able to localize the sound source. Indeed, the gains for both the azimuth and elevation response components vary in a systematic way with the SNR for both groups, and interestingly, the azimuth responses result to be much more robust against low SNR than the elevation components. This once again underlines the different neural mechanisms for the azimuth and elevation directions. By normalizing the gains for all participants in the following way:

$$G_E = 1 - \frac{G_{SNR}}{G_C} \tag{12.2}$$

with G_C the gain measured in the control condition (no background) and G_{SNR} the gain measured for the particular SNR, the gain values stay within 0 ($G_{SNR} = G_C$) and 1 (when $G_{SNR} = 0$). In this way, the idiosyncratic variability in absolute gain values is removed, and subject groups can be directly compared and averaged. Fig. 12.13B shows the result of this analysis for azimuth and elevation components of blind versus sighted as function of SNR.

The two major observations on these results are that (1) the sighted and blind are *equally sensitive, robust, and precise* for sound source changes in azimuth, as the normalized gains (left) and the stimulus–response correlations are indistinguishable between the groups. However, (2) the *sighted clearly outperform the blind* in elevation for the intermediate SNRs, where the curves deviate from each other by almost four standard deviations. Below an $SNR = -10$ dB the blind can no longer localize sound sources in elevation, whereas the sighted still

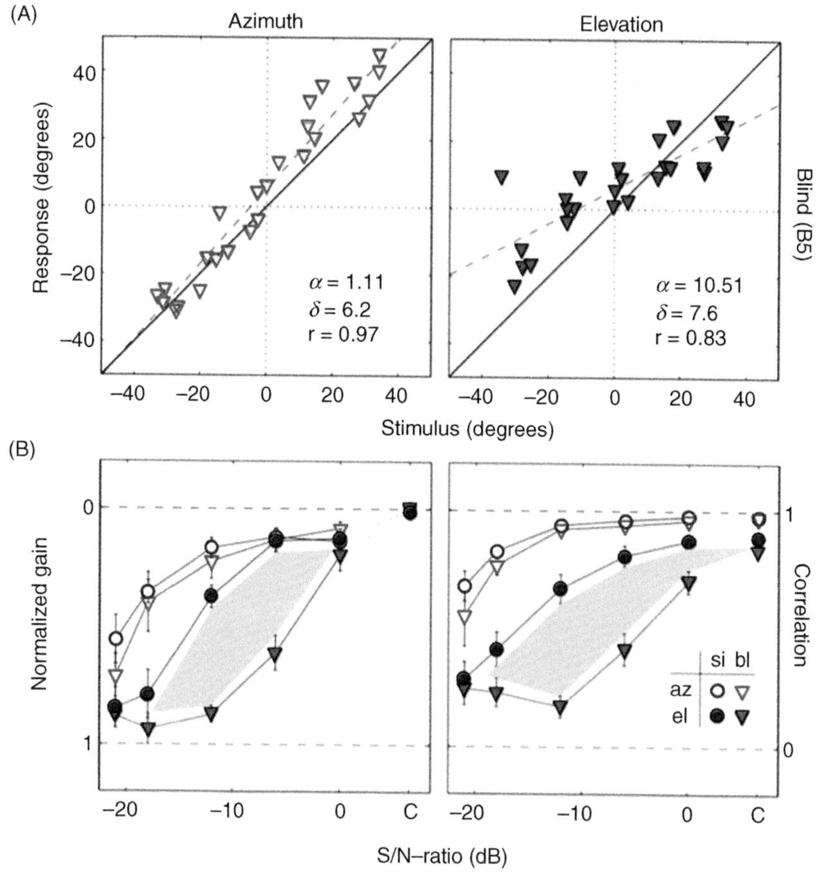

FIGURE 12.13 (A) Localization performance of a congenitally blind subject in azimuth and elevation to broadband noise. Azimuth performance is accurate; despite the low gain (α = 0.51), elevation responses correlate well with target elevation (r = 0.83). (B) Group comparison of sighted (si) and blind (bl) subjects for azimuth and elevation responses to single sounds presented within background noise at different signal-to-noise ratios. Left: normalized gains (Eq. 12.2). Right: correlation coefficients. Azimuth performance (open symbols) is indistinguishable between groups, but localization of elevation (closed symbols) of the blind is significantly worse for low S/N ratios (pink areas: at −6 and −12 dB, difference >4 sigma).

perform reasonably well at $SNR = -18$ dB. The correlation curves are shifted by about 12 dB SNR, which is quite substantial.

The conclusion that can be drawn from these results is that fine-tuning of the spectral cues requires visual feedback. Yet, other sources of information will have been used to generate at least a coarse spectral representation of the HRTF-to-elevation mapping in the blind. Here, one could think of head-movement feedback, or tactile feedback, which appear to be less precise than foveal vision. The coarse HRTF representation may thus miss the deep and sharp notches of the actual HRTFs, and therefore be more vulnerable to masking by background noise. Furthermore, the data also show that adequate performance in azimuth does not rely on intact vision. Possibly, the ILD and ITD cues are robust and simple enough to be trained by head movement and tactile feedback as well.

Perturbed vision: The barn owl adapts its auditory localization responses to a prism-induced displacement of the visual input, even when no plug had been applied in the ear canal. This seems a very strange response, since the normal acoustic cues are perfectly valid for catching its daily menu of 30 mice. The visually induced adjustments will in fact lead to large visual and sound-localization errors, and a very hungry owl! Yet, it clearly demonstrates the importance of vision as the major instructive signal to calibrate the acoustic cues in this species.

To test whether vision plays a similar role in the human auditory system, Zwiers et al. (2003) equipped nine listeners with minifying glasses (×0.5) to perturb their normal vision (Fig. 12.14A,B). The auditory system was left unperturbed. An additional aspect of the glasses was a strongly reduced field of view (FOV; Fig. 12.14B, as if looking through inverted binoculars), and a strong perturbation of vestibular–ocular reflex function, which led to an appropriately reduced VOR gain in all subjects within the first day (Fig. 12.14C).

Subjects responded with a laser pointer at the perceived sound location in two different ways: (1) keeping the eyes fixed on a central fixation spot

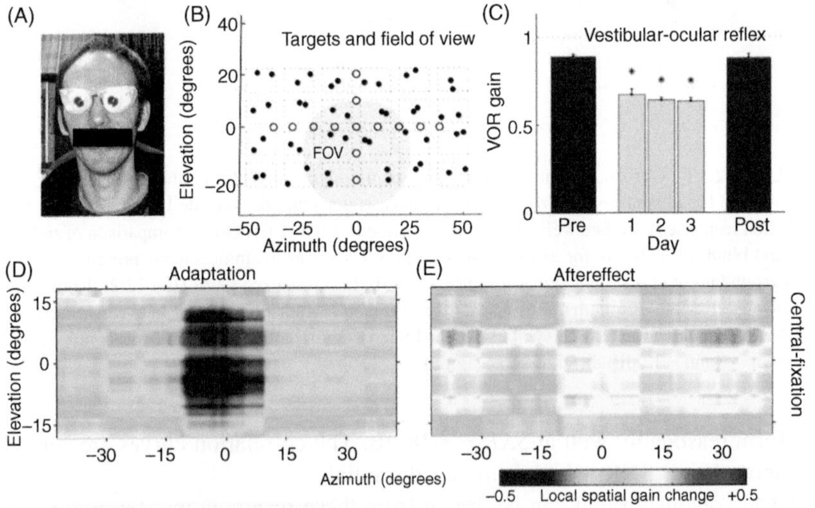

FIGURE 12.14 (A) Subject with minifying glasses, and a cover that blocks normal visual input from the sides. (B) Target distribution of the localization experiments and field of view of the glasses. (C) Vestibular ocular reflex gain before, during, and after the adaptation experiment. (D) Change in local spatial gain after three days: sound-localization gain changed by about −0.5 within the FOV, and close to zero outside the FOV. (E) Rebound aftereffect when the glasses were removed. *(Source: After Zwiers et al., 2003.)*

(ie, pointing with their peripheral retina, central fixation task), and (2) directing their fovea with an eye movement to the perceived location while pointing the laser (target fixation task). The difference in task (visual pointing vs. oculomotor pointing) should not matter if adaptation were to occur within the *auditory system*, rather than in the visual or oculomotor systems. This is indeed what happened: both tasks yielded identical results. Figure 12.4D shows the result of the central-fixation experiments, averaged across subjects. The data represent the local slope of the stimulus–response relation per azimuth-elevation bin [calculated over (Az × El) = (20 × 15)] degree bins, which were moved over the full target range in 1 degree steps). Note that the local gain changes are negative (blue, and approximately −0.5) within the central field of view of the glasses, but close to zero outside the FOV. Because of the restricted elevation range of the measurements, the boundaries could only be established for the azimuth components.

The results nicely show that, like the barn owl, also the human auditory system uses vision as an instructive signal to calibrate its sound-localization responses whenever there is a consistent mismatch between the auditory and visual percepts. This recalibration even occurs when the visual input is distorted, and actually wrong (eg, subjects could not correctly orient to, or grasp, visual targets in one go!).

Interestingly, in contrast to the spectral-cue learning paradigms described above, the visual adaptation experiment yielded a short-lived but consistent sound-localization *aftereffect* when the glasses were removed: the gain of the localization responses to the unperturbed vision was slightly higher (Fig. 12.14E; yellow colors) than before adaptation.

This aftereffect hints at a different plastic mechanism than for the spectral cues (also see Table 12.1). To appreciate the differences, it should be noted that the *local* change in *gain*, ΔG, shown in Fig. 12.14D, induced a pattern of *localization response changes*, ΔR, which extended over the *entire* azimuth domain in an interesting way (see Fig. 12.15A–C). In other words, the largest response changes (about 10 degrees inward) were obtained for far-lateral azimuth positions on either side, where the gain changes were zero! In contrast, no response change was obtained in the central FOV, where the gain change was largest.

TABLE 12.1 **Differences and Similarities Between Different Systems and Adaptation Mechanisms, and the Presence or Absence of an Aftereffect**

Systems	Relations	Adaptation mechanisms	Aftereffects
ILD owl	Monotonic	Gain, bias (or sign) change	Yes
VOR human	Monotonic	Gain, phase (or sign) change	Yes
ILD human	Monotonic	Spectral cues	No
HRTF human	Spectral pattern	Spectral cues	No

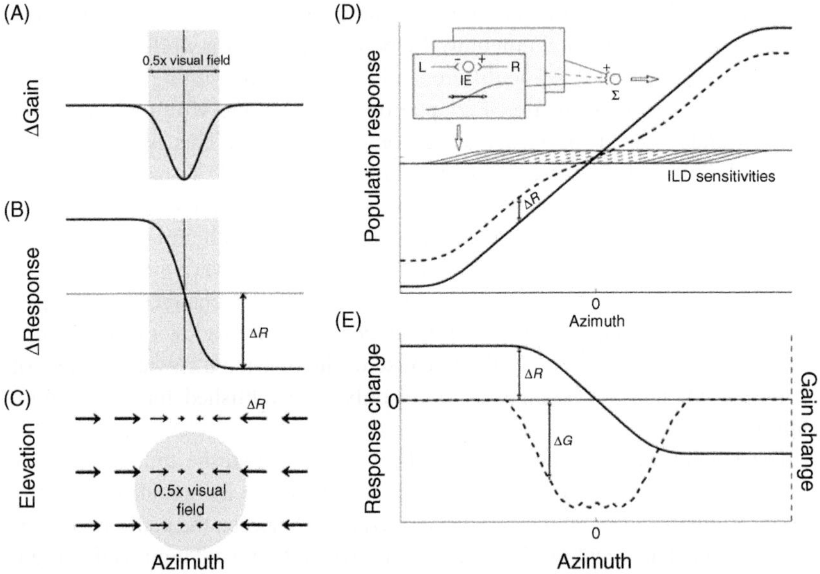

FIGURE 12.15 **Pattern of responses in the adaptation experiment.** (A) Gain changes are confined to the FOV. (B) Response changes, however, extend over the entire azimuth domain. (C) Pattern of response changes shown as vectors. (D) Population recruitment model of adaptable ILD cells can explain the patterns in panels (A–C). (E) Model response. *(Source: After Zwiers et al., 2003.)*

To explain this seemingly puzzling result, Zwiers et al. (2003) proposed a neural population model in which cells are sensitive to ILDs over restricted azimuth ranges, through sigmoid tuning curves with cell-specific recruitment thresholds (Fig. 12.15D). Together, all cells cover the entire azimuth range. The azimuth coordinate of the sound source is determined by summing the total right-left difference output of all cells. Clearly, the more a sound moves toward a lateral position, the more cells will be recruited. This leads to a total population response that varies in a monotonic way with source azimuth. During visual adaptation, only those cells with their thresholds within the FOV would locally change their gain (by lowering their ILD slope). This affects the total population output (response change and gain change) in a similar way as observed in the adaptation experiment (Fig. 12.15E).

12.5 LIMITS TO PLASTICITY?

In Section 12.2 we briefly mentioned some of the acoustic and procedural requirements for auditory spatial learning to occur. These requirements mainly concerned consistency and nonambiguity in the acoustic perturbations or experimental manipulations to allow the auditory system to establish new, unique acoustic-spatial relationships. One point we did not mention, however, is whether nonacoustic factors like age might also limit sensory-spatial learning. For example, the development of binocular vision in humans and animals is subjected to a critical period in the first few years (humans) or even weeks (eg, cats) of life. Auditory spatial tuning in the ferret's SC can only form within the first few weeks after birth, and also the barn owl's renowned plasticity (Fig. 12.3) seems to be limited to a critical period of a few months.

Remarkably, the human plasticity studies described in this chapter all involved adult participants (ages in the range of 20–55 years), who were all well able to adapt to the imposed acoustic perturbations. To what extent could age limit human sound-localization plasticity? Obviously, the integrity of the peripheral auditory system (cochlear function) should be sufficient to allow for the encoding of critical ITD, ILD, and spectral cue information. When outer hair cells give up, the frequencies to which they were tuned are no longer perceived, and cannot be used in the encoding of spatial location either. Normal aging, however, induces a high-frequency hearing loss (due to the loss of outer hair-cell function), which can be quite substantial: at the age of 60 (and above) the hearing thresholds for 4 kHz tones of unaided, good-hearing listeners are already 20 dB higher than for children and young adults (eg, Otte et al., 2013). At 8 kHz, this difference is already 35 dB, or more. What does this mean for sound-localization performance of the elderly? Otte et al. (2013) measured the sound-localization responses in 2D of elderly listeners (up to 80 years of age) and reported that *azimuth accuracy* and precision of the elderly is as good as that of young adults. In other words, the use of binaural difference cues for broadband sounds is still adequate in the elderly, even when there is substantial high-frequency hearing loss. But what about their elevation responses?

An interesting epiphenomenon in this regard is that the pinnae keep growing, throughout the human lifespan, at a speed of approximately 0.1 mm/year. Otte et al. (2013) recently quantified pinna growth for a group of 42 listeners between 7 and 80 years of age and observed that the pinna grows predominantly in concha *height*, not in concha width. The result is that the characteristic notches in the HRTFs systematically shift toward lower frequencies in the elderly, by a few kHz. As a gross simplification, one could describe the frequencies of the principle notch that runs between elevations of −40 and +40 degrees by a straight line in the elevation–frequency plot (eg, chapter: Acoustic Localization Cues, Fig. 7.15), according to:

$$f = f_{\text{min}} + \frac{E+40}{80} \cdot (f_{\text{max}} - f_{\text{min}}) \tag{12.3}$$

with f_{min} and f_{max} the minimum and maximum notch frequencies at $E = -40$ and $E = +40$ degrees, respectively. Otte et al. (2013) found that the minimum and maximum frequencies differ systematically for the different age groups (Table 12.2; Exercise 12.4).

Do elderly listeners benefit from these gradual frequency shifts? In other words, do they adapt to these changes throughout life? Otte et al. (2013) hypothesized that if the elderly would indeed use the lower-frequency spectral

TABLE 12.2 Estimated Locations of the First Notch for Two Elevation Angles Across the Three Different Age Groups

Ages	f_{min} (kHz) at $E = -40$	f_{max} (kHz) at $E = +40$
7–11	5.9–6.9	12.6–13.9
21–34	5.2–6.1	8.3–14.3
63–80	4.1–5.9	7.0–11.9

cues of their larger ears, they might actually *outperform younger listeners* in localizing elevation when the spectral bandwidth of stimuli would be reduced to, say, 7 kHz. As younger listeners require also higher frequencies to stimulate their HRTFs, they would fail to localize certain elevations, which could still be localized by the elderly. This is indeed what they observed for a few elderly listeners whose high-frequency hearing loss was limited to frequencies above 6–8 kHz. This unequivocally shows that spectral plasticity, at least in humans, is not limited by age, but by the integrity of the auditory system (see chapter: Impaired Hearing and Sound Localization).

Unfortunately, age-related hearing loss (above 3–4 kHz) usually far exceeds the relatively small benefit of having a larger ear (first notch around 4–5 kHz, but still running up to about 9 kHz). In other words, the ear growth itself does not appear to be an evolutionary determined compensatory response of the body to high-frequency hearing loss…

12.6 EXERCISES

Problem 12.1: Referring to Eq. 12.1, analyze (and graphically draw) the following error scenarios:

(a) A positive bias for auditory, a gain that is too high for visual, and motor undershoots.
(b) Auditory gain too low, and a negative bias on the motor responses.

Is it possible to identify the different sources of error from the stimulus-response relations? How?

Problem 12.2: Explain why head movements produce conflicting cue information in the situation of inverted ears. Make drawings to illustrate your arguments.

Problem 12.3: Apply the spectral correlation model of Chapter 8 to explain (i) the presence of multiple sets of HRTFs without interfering with localization accuracy, and (ii) the absence of after effects upon removal of molds.

Problem 14.4: From Table 12.2, estimate the concha height for the three different age groups.

REFERENCES

Amitay, S., Krwin, A., Moore, D.R., 2006. Discrimination learning induced by training with identical stimuli. Nat. Neurosci. 9, 1446–1448.

Amitay, S., Moore, D.R., Molloy, K., Halliday, L.F., 2015. Feedback valence affects auditory perceptual learning independently of feedback probability. PLoS One 10 (5), e0126412.

Carlile, S., Balachandar, K., Kelly, H., 2013. Accommodating to new ears: the effects of sensory-motor feedback. J. Acoust. Soc. Am. 135, 2002–2011.

Carlile, S., Blackman, T., 2014. Relearning auditory spectral cues for locations inside and outside the visual field. J. Assoc. Res. Otolaryngol. 15, 249–263.

Clifton, R.K., Gwiazda, J., Bauer, J.A., Clarkson, M.G., Held, R.M., 1988. Growth in head size during infancy: Implications for sound localization. Dev. Psychol. 24, 477–483.

Gutfreund, Y., Zheng, W., Knudsen, E.I., 2002. Gated visual input to the central auditory system. Science 297, 1556–1559.

Hofman, P.M., Van Riswick, J.G., Van Opstal, A.J., 1998. Relearning sound localization with new ears. Nat. Neurosci. 1, 417–421.

Hofman, P.M., Vlaming, M.S., Termeer, P.J., Van Opstal, A.J., 2002. A method to induce swapped binaural hearing. J. Neurosci. Meth. 113, 167–179.

Hofman, P.M., Van Opstal, A.J., 2002. Bayesian reconstruction of sound localization cues from responses to random spectra. Biol. Cybern. 86, 305–316.

Irving, S., Moore, D.R., 2011. Training sound localization in normal hearing listeners with and without a unilateral ear plug. Hear. Res. 280, 100–108.

Javer, A.R., Schwarz, D.W., 1995. Plasticity in human directional hearing. J. Otolaryngol. 24, 111–117.

King, A.J., 1999. Auditory perception: does practice make perfect? Curr. Biol. 9, R143–R146.

Knudsen, E.I., Knudsen, P.F., 1985. Vision guides the adjustment of auditory localization in young barn owls. Science 230, 545–548.

Knudsen, E.I., Konishi, M., 1978. A neural map of auditory space in the barn owl. Science 200, 795–797.

Knudsen, E.I., Konishi, M., 1979. Mechanisms of sound localization in the barn owl (Tyto alba). J. Comp. Physiol. 133, 13–21.

Kumpik, D.P., Kacelnik, O., King, A.J., 2010. Adaptive reweighting of auditory localization cues in response to chronic unilateral ear-plugging in humans. J. Neurosci. 30, 4883–4894.

Lessard, N., Paré, M., Lepore, F., Lassonde, M., 1998. Early-blind human subjects localize sound sources better than sighted subjects. Nature 395, 278–280.

Majdak, P., Walder, T., Laback, B., 2013. Effect of long-term training on sound localization performance with spectrally warped and band-limited head-related transfer functions. J. Acoust. Soc. Am. 134, 2148–2159.

Mendonça, C., 2014. A review on auditory space adaptations to altered head- related cues. Front. Neurosci. 8, 219.

Moore, D.R., Hine, J., Jiang, Z.D., Matsuda, H., Parsons, C.H., King, A.J., 1999. Conductive hearing loss produces a reversible binaural hearing impairment. J Neurosci. 19, 8704–8711.

Musicant, A.D., Butler, R.A., 1980. Monaural localization: An analysis of practice effects. Percept. Psychophys. 28, 236–240.

Otte, R.J., Agterberg, M.J.H., Van Wanrooij, M.M., Snik, A.F.M., Van Opstal, A.J., 2013. Age-related hearing loss and ear morphology affect vertical, but not horizontal, sound-localization performance. JARO 14, 261–273.

Röder, B., Teder-Sälejärvi, W., Sterr, A., Rösler, F., Hillyard, S.A., Neville, H.J., 1999. Improved auditory spatial tuning in blind humans. Nature 400, 162–166.

Shinn-Cunningham, B.G., Durlach, N.I., Held, R.M., 1998. Adapting to supernormal auditory localization cues. I. Bias and resolution. J. Acoust. Soc. Am. 103, 3656–3666.

Takahashi, T.T., 1989. The neural coding of auditory space. J. Exp. Biol. 106, 307–322.

Van Wanrooij, M.M., Van Opstal, A.J., 2004. Contribution of head shadow and pinna cues to chronic monaural sound localization. J. Neurosci. 24, 4163–4171.

Van Wanrooij, M.M., Van Opstal, A.J., 2005. Relearning sound localization with a new ear. J. Neurosci. 25, 5413–5424.

Van Wanrooij, M.M., Van Opstal, A.J., 2007. Sound localization under perturbed binaural hearing. J. Neurophysiol. 97, 715–726.

Wright, B.A., Sabin, A.T., 2007. Perceptual learning: how much daily training is enough? Exp. Brain Res. 180, 727–736.

Wright, B.A., Wilson, R.M., Sabin, A.T., 2010. Generalization lags behind learning on an auditory perceptual task. J. Neurosci. 30, 11635–11639.

Zwiers, M.P., Van Opstal, A.J., Cruysberg, J.R.M., 2001a. A spatial hearing deficit in early-blind humans. J. Neurosci. 21, 1–5, RC421.

Zwiers, M.P., Van Opstal, A.J., Cruysberg, J.R.M., 2001b. Two-dimensional sound-localization behavior of early-blind humans. Exp. Brain Res. 140, 206–222.

Zwiers, M.P., Van Opstal, A.J., Paige, G.D., 2003. Plasticity in human sound localization induced by compressed spatial vision. Nat. Neurosci. 6, 175–181.

Chapter 13

Multisensory Integration

13.1 INTRODUCTION

In the preceding chapters we made a case for the natural alliance and integration between the auditory and the visual–motor control systems. This holds true especially in foveate species that have to rely on fast, accurate, and precise saccadic gaze shifts to allow detailed visual inspection of a peripheral target as quickly as possible.

As our exteroceptive sensors, the visual and auditory systems identify, select, and localize sensory stimuli in the external environment (our smell does too, but is kept out of the picture here). Their receptive ranges have considerable overlap in the frontal hemifield, but their spatial and temporal resolutions are quite different (Fig. 1.5). Vision has a very high resolution in the fovea (in the order of arc minutes, or less, in the human), which rapidly decays (as $1/r^2$) toward the visual periphery (Fig 9.1). The auditory system has its highest spatial resolution around straight ahead [about 0.7 degrees (40') in azimuth, and 2.5 degrees (150') in elevation; chapter: Assessing Auditory Spatial Performance; Fig. 8.7], but the degradation toward the periphery is less dramatic than for vision. Moreover, the auditory system receives input from all directions on the hypothetical sphere surrounding the listener, so it can respond to acoustic events that occur way beyond the visual field (eg, at the rear), in total darkness, or from behind objects that block vision (eg, in the next room). Finally, audition has a high resolution for temporal amplitude modulations (>100 Hz), for which vision is relatively poor (<25 Hz).

Efficient audiovisual (AV) integration could help the observer to perform better (ie, be faster, more accurate, more reliable, and precise) than when depending on one sense only. For example, a simple AV strategy could be *complementary integration*, in which vision would fully dominate orienting and perceptual responses for AV stimuli within the parafoveal visual field (say, within a 20 degrees radius from the fixation point), while audition would completely take over beyond that range. Although in this way AV stimuli will evoke rapid responses across the entire hemisphere, this is not really what is meant by multisensory integration.

To identify a multisensory integration effect, the responses (eg, measured by reaction time, accuracy, precision, speed) to a combination of sensory inputs should differ significantly from the responses evoked by any of the sensory modalities presented in isolation, in a way that cannot be explained by mere statistical summation effects, or by increased vigilance. Note that according to this definition multisensory integration does not necessarily imply better (ie, faster,

The Auditory System and Human Sound-Localization Behavior. http://dx.doi.org/10.1016/B978-0-12-801529-2.00013-1

and more accurate) responses, but also allows for negative interactions (slower and more variable) in the responses.

The study of AV integration has a long tradition in the psychophysical literature, and many different models have been proposed. Some of these models will be briefly reviewed in the next section. Traditionally, one first measures the responses to the unimodal stimuli in isolation (visual-, and auditory-, or tactile-only, respectively), and then compares the results to the situation in which the different sensory modalities are presented together. The parameter that has been studied most in multisensory experiments is the response *reaction time*, but also the properties of saccadic eye movements (amplitude, direction, and variability) have been investigated for potential multisensory interaction effects.

For example, Frens et al. (1995) studied saccadic eye-movement trajectories to synchronous visual and auditory stimuli presented at any combination of four possible locations within the 2D frontal hemifield. The stimuli could be spatially aligned, or spatially incongruent in three ways. The task was to *ignore* the auditory accessory stimulus (it served as a distracter, not as a target), and to respond only to the visual target. As in most other studies, the sensory environment in this study was kept quite simple: the stimulus intensities were high, only two stimuli were present, and there were no additional distracters or noisy backgrounds. Yet, the spatial trajectories and saccade reaction times (SRTs) varied systematically with the spatial stimulus configurations, and thus provided clear support for real multisensory integration: when stimuli were spatially congruent, SRTs were about 50 ms shorter than for the fastest modality (in this case, auditory), and SRTs increased systematically with increasing distance between visual target and auditory distracter. Note that the latter phenomenon is the real indicator for AV integration, because a decrease in overall SRT could in principle be due to a nonspecific effect, such as increased vigilance, or attention: the SRT could be shorter just because multisensory inputs deliver more sensory energy to the brain than each of the unimodal inputs. However, for any of those effects the spatial configuration of the stimuli should not matter!

Because visual, auditory, and tactile stimuli evoke rapid gaze-orienting behaviors, an obvious candidate in the central nervous system to reveal a neural correlate for multisensory integration would be the midbrain superior colliculus (SC). Indeed, the neural mechanisms of multisensory interactions have been studied extensively in this structure, although many questions on orienting behavior still remain unanswered. Before going into that, let's briefly review the original neurophysiological findings, and identify some important remaining issues.

Superior Colliculus: Stein and coworkers were among the first to measure responses of cells in the SC of anesthetized cats to visual, auditory, and tactile stimuli presented across sensory space. They found that (1) the receptive fields of different modalities typically overlapped to a large extent, and (2) that a single modality was often ineffective in evoking a strong sensory response, even when the stimulus was presented in the center of the receptive field.

However, when stimulating an SC cell with multiple sensory modalities from the same location, the cell would often respond with a vigorous burst, which could well exceed the linear sum of the unimodal sensory responses (*nonlinear response enhancement*; Meredith and Stein, 1986a). The multisensory enhancement was strongest when unimodal stimuli were ineffective (ie, at low stimulus levels), a phenomenon that has become known as "the *principle of inverse effectiveness*." The following index quantifies multisensory integration effects on neural responses with respect to the linear response sum:

$$MI \equiv \left(\frac{R_{A+V+T}}{R_A + R_V + R_T} - 1 \right) 100\% \tag{13.1}$$

with R_{A+V+T} the multisensory response (say, for auditory, visual, and tactile inputs), and R_A, R_V, and R_T the respective unimodal responses. If, for example, the multisensory response equals the linear sum of unimodal responses, the index $MI = 0\%$. If it's less than the linear sum, the index is negative, and if the cell does not respond at all, $MI = -100\%$. There is no upper bound to the value of MI; Meredith and Stein even reported cells with $MI > 1000\%$.

A series of follow-up studies showed that the multisensory interactions depend in a systematic way on the spatial (Meredith and Stein, 1986a,b; 1996) and temporal (Meredith et al., 1987) congruence of the stimuli: strongest enhancements are obtained for stimuli in spatial–temporal alignment, while responses may become suppressed for spatial–temporal misalignments (Stein and Meredith, 1993).

Fig. 13.1 schematically summarizes the major findings of these SC recordings, with some example responses for high-intensity and low-intensity stimuli

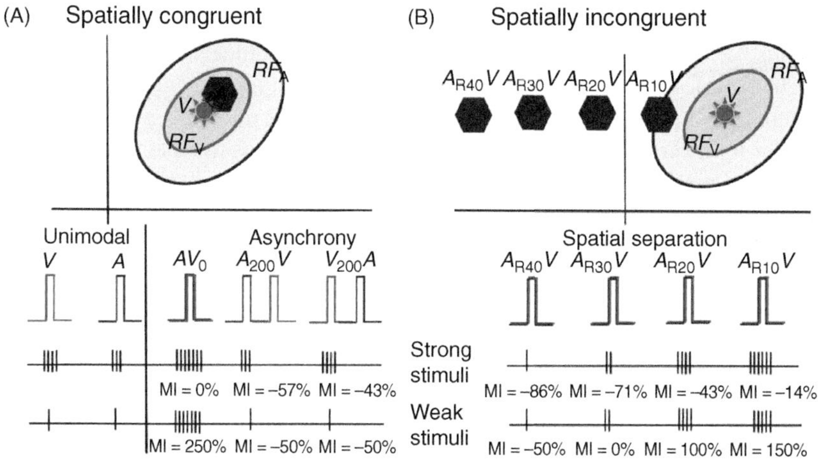

FIGURE 13.1 Summary of the major findings of sensory SC responses to multisensory stimuli by Stein and coworkers. RF_A, auditory receptive field; RF_V, visual receptive field. (A) AV stimuli are spatially aligned, but can be asynchronous. (B) AV stimuli are synchronous, but can be spatially unaligned. Responses to strong (top spike traces) and weak stimuli (bottom spike traces) exemplify the principle of inverse effectiveness. Note that the MI for weak stimuli is higher than for strong stimuli, even though the neural activity may be the same.

for different spatial–temporal configurations. The principle of inverse effectiveness is best seen in the AV_0 condition (spatially and temporally aligned): although the multisensory responses are equal (seven spikes), the *MI* differs considerably for the two cases, and is by far highest for the weak stimuli.

Note that these effects were measured for the *sensory* responses of SC cells only, and in (anesthetized) animals that could not benefit from the interactions to prepare or optimize a behavioral response. They nonetheless constitute strong evidence for the idea that the neural mechanisms that mediate multisensory integration assess whether signals arose from the *same object in space-time.* It thus serves the clear potential advantage of combining sensory evidence from multiple modalities to generate an accurate goal-directed response in the fastest possible way.

In later studies, other groups reported qualitatively similar results for SC responses in the *awake* cat (Peck, 1996), and for SC activity and saccades in the monkey (Van Opstal and Frens, 1996; Frens and Van Opstal, 1998; Bell et al., 2005). Also at the behavioral level (gaze orienting, food approaching, etc.) congruent multisensory stimuli prove to be more effective than incongruent or unimodal sensory stimuli, even under relatively simple sensory conditions like in Fig. 13.1 (Stein et al., 1989; Frens et al., 1995).

Integration Problem: But first let's identify an important problem, not immediately obvious from the examples in Fig. 13.1. AV integration is far from trivial when we consider the dynamic coordinate transformations in natural gaze control, discussed extensively in chapter: Coordinate Transformations. Indeed, integration should only occur when visual and acoustic signals arose from the *same* object, which means from the *same point in space and time.* But (1) because of the different reference frames for vision (oculocentric) and audition (craniocentric) that misalign with every eye and head movement, (2) given the differences in propagation speed of the sensory signals through space (infinitely fast for light versus 343 m/s for sound, taking about 3 ms/m), and (3) given the additional neural processing delays (at cortex, for vision ~70 ms, for audition ~10 ms), how can the system reliably decide whether a visual and auditory event both originated from the same space–time point? Bringing the auditory and visual signals into a common reference frame would certainly help, but whether and how this occurs has received relatively little attention in multisensory studies, and cannot be inferred from the electrophysiological recordings discussed previously.

Fig. 13.2 illustrates the problem for two simple examples. In both cases, eyes and head look at different spatial locations: the eye (*E*) 20 degrees leftward, while the head (*H*) points straight ahead. In panel A, a sound-source is located 20 degrees to the right, and a visual stimulus at straight ahead. Note that both sensory systems would measure a 20 degrees rightward stimulus location. However, it is clear (at least to us!) that the two stimuli belong to different objects in space. If the sensory–motor system would keep targets in their original reference frames, the motor SC would show one active population at 20 degrees

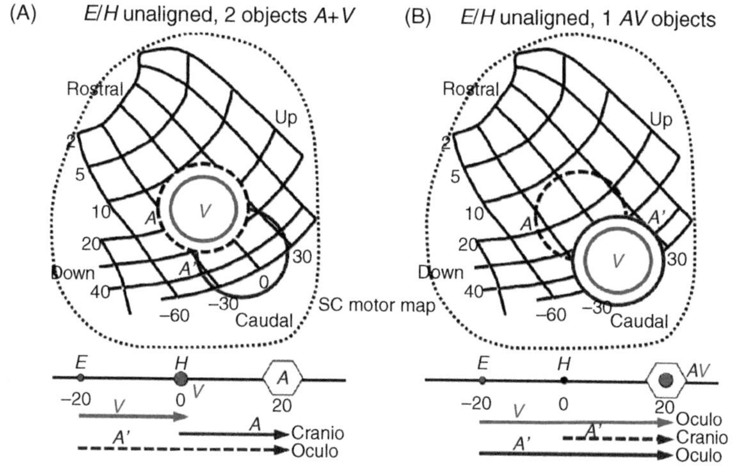

FIGURE 13.2 Reference frame problem for audio–visual integration. In this example, eyes and head are unaligned (eye looks 20 degree left, head straight ahead). (A) Visual and auditory coordinates are the same in their own reference frames. (B) Visual and auditory coordinates are the same in the oculocentric reference frame (solid arrows). In (A) AV integration should not occur as stimuli belong to two unrelated objects. In (B) AV integration should occur, provided that signals also have the appropriate timing alignment.

rightward. However, we have seen in chapter: The Midbrain Colliculus that the SC motor map represents *oculocentric* gaze–shifts to the selected target (Jay and Sparks, 1987). Therefore, in this case, *two sites* will be activated: one at 20 degrees (for the visual stimulus), and one at 40 degrees (for the auditory stimulus; the blue circle). *Clearly, this sensory configuration should not lead to AV integration.*

In panel B a single AV target is presented at 20 degrees right from the head. In this case, the sensory inputs encode different spatial signals, but the *derived* oculocentric (and also craniocentric, and world-centered) coordinates for both modalities are identical. As a result, there will be only one multisensory-evoked population in the SC motor map at 40 degrees rightward. *This is the sensory configuration that should induce AV integration!*

In complex environments and under natural orienting behaviors, the alignment of eyes and head and the potential sensory locations of auditory and visual stimuli will typically vary rapidly and independently. In chapter: Coordinate Transformations we proposed that the gaze-control system first builds a representation of targets in world-centric coordinates to cope with complex dynamic localization problems. In this chapter we propose to extend this concept to multisensory integration.

Inverse effectiveness? A second interesting point in multisensory integration is in part suggested by the example responses in Fig. 13.1A. Suppose spatial–temporal alignment (AV_0), but high-intensity stimuli. In that case a true effect

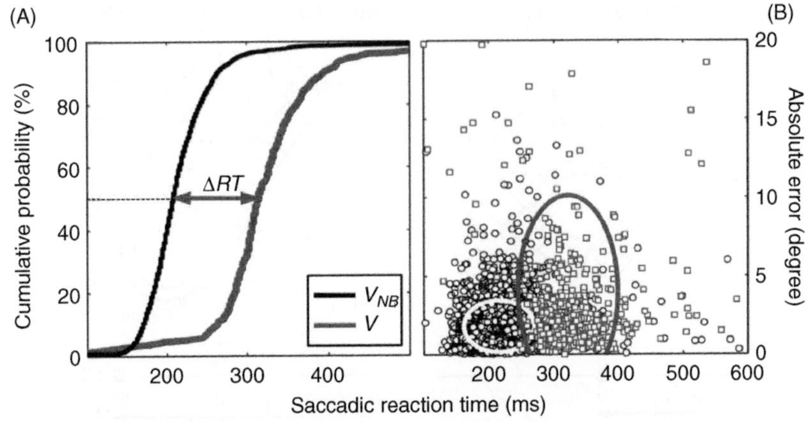

FIGURE 13.3 First-saccade visual responses with and without a cluttered background. The effect of a background (red) on first-saccade responses to a visual target is twofold: (A) increased reaction times (here, for 84 distractors, $\Delta RT \sim 110$ ms) when compared to no background (NB; black), and (B) increased absolute errors and response variability. *(Source: After Van Wanrooij et al., 2009.)*

of multisensory integration may not be visible, as the unimodal stimuli may already have led to ceiling performance, with little room for improvement. How could the principle of inverse effectiveness (with potentially the largest effects) be studied in a behavioral context? One possibility may be to present stimuli near the detection threshold, but such an approach may run the risk of inattentiveness and fatigue of the participant, which could strongly influence potential multisensory effects.

An alternative approach would be to increase task difficulty of localizing the target, by embedding target stimuli in a sea of visual and auditory distracters. In such cluttered AV environments, which arguably approaches more natural situations, visual localization of a target is typically slow (ie, long saccadic reaction times), and variable (initial saccades make large errors).

Fig. 13.3 shows an example of the cumulative distributions of first-SRT (panel A) and response accuracy (absolute localization error, panel B) to a dim red target presented in otherwise complete darkness (ie, without a background, black data), or when hidden among 84 green distracters (visual search task; red data). The difference in SRTs is more than 100 ms (the averages are 200 and 310 ms, respectively), and it is also quite clear that the first saccadic response in a cluttered visual environment has increased response variability (red ellipse) and larger mean absolute error.

A similar manipulation can be introduced for auditory saccades, by embedding the target sound (eg, a broad-band buzzer) in a diffuse Gaussian white noise background. Fig. 13.4 compares the results for four different signal-to-noise ratios [$SNR = (-6,-12,-18,-21)$ dB; see legend) versus the no-background

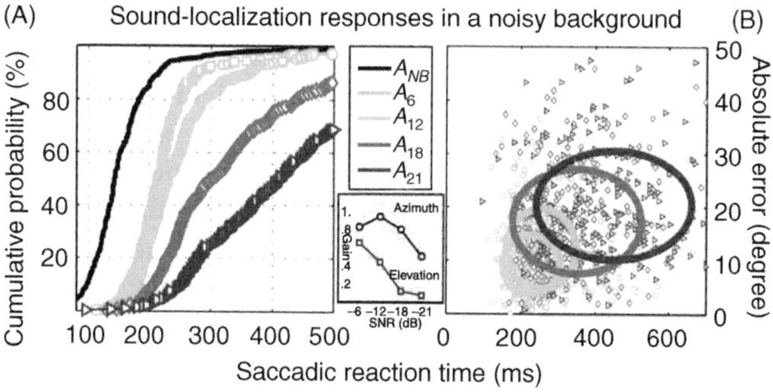

FIGURE 13.4 **The effect of diffuse broadband background noise (60 dB SPL) on first saccades to an auditory target at different signal-to-noise ratios.** (A) Reaction times of goal-directed responses systematically increase with decreasing SNR. (B) Absolute errors (predominantly in elevation) and response variability both increase with SNR. A_{NB}: control condition without background. Center inset: different azimuth and elevation gains as function of SNR. *(Source: After Van Wanrooij et al., 2009.)*

hearing condition (black curve). Again, without a background auditory saccades are faster (shorter reaction times), more accurate (smaller errors) and more precise (less variable) than with the background. Interestingly, the SNR nicely serves as an acoustic control parameter with which task difficulty can be systematically manipulated (see central inset).

Thus, eye-movement responses to unimodal visual or auditory targets hidden in a cluttered background demonstrate considerable task difficulty:

- Much longer reaction times, which for auditory stimuli depend on SNR.
- Larger absolute errors, which for sounds depend on SNR, and on azimuth versus elevation target coordinates.
- Larger variability, again for audition a function of SNR and azimuth/elevation components.

These conditions provide solid ground to reveal inverse effectiveness, and hence real multisensory interactions. These effects will be discussed in Sections 13.3 and 13.4.

Ventriloquism: So where does the *ventriloquist effect* fit into all this? As the ventriloquist moves the dummy's lips, the spectator cannot escape the impression that the dummy is really talking: "visual capture of audition". The idea is that the brain regards visual spatial information as much more reliable than auditory spatial input, and hence the spatial percept is completely dominated by vision. Is this multisensory integration in the sense as discussed previously? Clearly, in the ventriloquist problem we deal with a situation of incongruent sensory inputs, because of the spatial disparity between the AV stimuli. Hence, according to the rules of multisensory integration, these signals should not be integrated, as they constitute different objects. However, when the dummy's lip movements are approximately synchronized with the ventriloquist's voice, the natural independence of unrelated sensory objects is clearly violated. So the ventriloquist scene introduces a real *ambiguous problem* to the brain. Note, however, that the real ventriloquist effect refers to a highly cognitive phenomenon, in which complex visual patterns (moving lips, interpretation of the scene) and complex synchronized sounds (voices) are merged into a single audio–visual percept. Yet, despite the reported percept of the dummy talking, it is not entirely clear where exactly the auditory source is perceived: is it really at the location of the dummy's lips, or perhaps at some intermediate location between dummy and ventriloquist? To study the major aspects of this effect under more abstract and controllable laboratory situations, the stimuli are typically reduced to an abstract audio–visual pair (a dot, or a visual blob, and a noise burst, or a click sound) with an added spatial disparity (eg, Alais and Burr, 2004). In Sections 13.3 and 13.6 we will return to this issue.

13.2 MODELS

To identify an effect of multisensory integration on the reduction of reaction times, it is important to discard two potential alternative explanations: (1) increased vigilance, versus (2) statistical facilitation.

The former relates to the possibility that the increased sensory energy of two stimuli already speeds up the sensory processing time, regardless of the sensory modalities and their configuration. It's nonspecific (may only depend on stimulus levels) and therefore is not caused by true multisensory integration.

Statistical facilitation is a more subtle effect on multisensory reaction times, and usually serves as a quantitative benchmark against true multisensory integration. Conceptually, statistical facilitation is better known as *race model*, which holds that the reaction time to the multisensory stimulus is determined by which sensory input signal reached the trigger for eliciting a response first:

$$SRT(AV) = \begin{cases} SRT(A) & \text{if } t_A < t_V \\ SRT(V) & \text{if } t_V < t_A \end{cases} \tag{13.2}$$

Note that this definition does not assume any interaction between the stimuli, as it contains a simple detection threshold that is not influenced by the stimulus configuration.

In the exact version of the race model (Gielen et al., 1983) the distribution of AV reaction times is calculated as the *distribution of minimum reaction times* for any pair of A versus V responses. Because the reaction times, τ, for auditory- and visual-evoked responses are described by overlapping probability distributions, $A(\tau)$ and $V(\tau)$, respectively, the race distribution of minima of all possible pairs (Exercise 13.1) is given by:

$$R_{AV}(\tau) = A(\tau)\int_{\tau}^{\infty} V(s)ds + V(\tau)\int_{\tau}^{\infty} A(s)ds \tag{13.3}$$

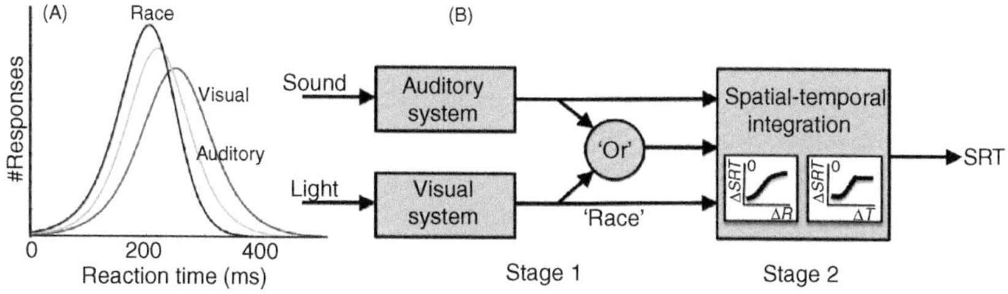

FIGURE 13.5 A two-stage model after Colonius and Arndt (2001) that could explain the properties of AV reaction times. The first stage implements a race model, which assumes a "first-come-first-serve" mechanism. This leads to statistical facilitation: the distribution of minimal reaction times is shifted to shorter reaction times than either unimodal distribution [left graphs; Eq. (13.3)]. In the second stage, real AV integration takes place where an additional change in reaction time depends on the spatial–temporal configuration of the stimuli.

Note that the race distribution is shifted toward *shorter reaction times* than either the auditory or the visual responses (Fig. 13.5), but this is merely due to a *statistical* property of the distribution of minima, not to an effect of multisensory integration.

A more conservative version of the race model prescribes that the boundary for the race distribution is simply determined by the sum of the cumulative auditory and visual reaction-time distributions:

$$\mathrm{Cum}\left[R^*_{AV}(\tau)\right] = \int_{-\infty}^{\tau} A(s)ds + \int_{-\infty}^{\tau} V(s)ds \tag{13.4}$$

(in which it should be noted that the cumulative values of the race prediction do not add up to one, but to two).

To account for the true effect of multisensory integration on reaction times, a model should include the contribution of statistical facilitation and vigilance (which add up to about 30 ms in studies using suprathreshold stimuli), and a true interaction effect of the multisensory inputs that depends on the spatial–temporal stimulus configuration. Colonius and Arndt (2001) proposed a two-stage model, which separates the nonspecific factors (statistical facilitation, vigilance, auditory warning effect, or auditory search) from true spatial–temporal factors in two consecutive neural stages (Fig. 13.5).

The two-stage model of Fig. 13.5 accounts for multisensory effects on response reaction times (both their reduction and lengthening), but does not include spatial and temporal effects on the trajectories of saccadic gaze shifts or manual responses (eg, Frens et al., 1995; Corneil and Munoz, 1996). For example, the endpoints of *AV* saccades often land in between the visual (*V*) and auditory (*A*) stimulus locations, resembling a weighted average response (also see chapter: The Gaze-Orienting System):

$$\vec{S}_{AV} = a\vec{S}_V + b\vec{S}_A \tag{13.5}$$

In the case of pure visual dominance (as described for the ventriloquist scene), $a = 1$ and $b = 0$; for total auditory dominance (say, for far eccentric, or rear stimulus locations), $a = 0$ and $b = 1$. However, for stimuli within the frontal hemifield responses may be neither purely visual, nor purely auditory, in which case $a > 0$ and $b > 0$, with $a + b = 1$.

Spatial averaging of saccadic gaze shifts has been described (and modeled) extensively for neural responses in the midbrain SC (see chapter: The Midbrain Colliculus; Van Opstal and Van Gisbergen, 1989, 1990). The work of Munoz and coworkers proposes that the SC also provides a neural signal for saccade *initiation* (the SRT). Their spike–train analyses revealed that whenever the prelude build-up activity of the SC population surpasses a fixed criterion, the saccade is triggered. Interestingly, the rise in build-up activity (Bell et al., 2005) depends on the stimulus configuration. The two mechanisms, a cumulative saccade trigger, and spatial averaging in the SC motor map, may thus provide a neural correlate for the spatial–temporal implementations of the second stage in Fig. 13.5 (the trigger mechanism for multisensory integration), as well as its influence on the saccade trajectories.

Fig. 13.6 gives a schematic account of how the SC motor map could be involved in multisensory integration. The population activity for the gaze shift depends on the sensory conditions: firing rates for saccades to auditory stimuli are typically lower than for visual saccades (eg, Frens and Van Opstal, 1998), and the auditory-evoked population (blue) is more diffuse in the elevation direction than for azimuth. Yet, the total number of spikes of the A population is the same as for the V population (red). The SC output provides two types of signals: the instantaneous cumulative number of spikes encodes the saccade trajectory (where) and its velocity (how) by the instantaneous population firing rates (Eq. 10.16). This signal drives the pulse generator (see chapter: The Midbrain Colliculus). Depending on the spatial–temporal distribution of the spikes, the trajectory can be fast and straight, or slow and even curved. The cumulative

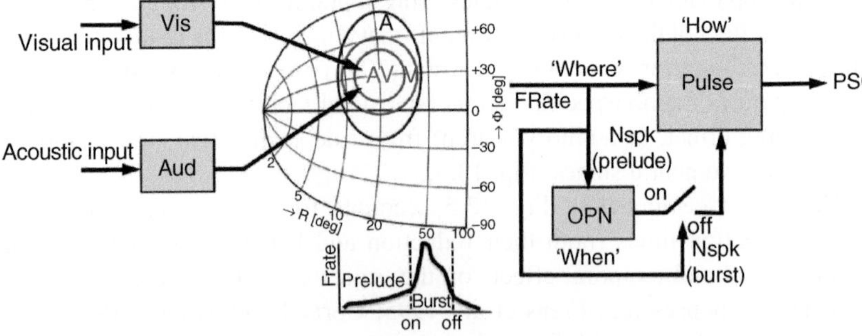

FIGURE 13.6 The saccade trigger (OPNs) and its trajectory (pulse generator) are both controlled by the SC population. Visual-only (*V*) stimuli yield a classical population and normal saccades. Auditory-only (*A*) yields slower saccades, because of lower firing rates in a diffuser population. Audio–visual (*AV*) stimuli induce spatial–temporal nonlinear interactions that create a smaller population with higher firing rates and better synchronization.

number of spikes of the *prelude* (the SC activity *prior* to saccade initiation; lower inset) determines the trigger moment for opening the pulse generator circuit (ON). The OPN gate is closed, and hence the saccade ends (OFF), when the number of spikes of the SC saccade-related burst reaches a fixed offset criterion (Eq. 10.26). This latter mechanism can account for response averaging, in cases where auditory- and visual-evoked populations are not centered at the same point (see chapter: The Midbrain Colliculus). A congruent AV event causes nonlinear interactions in the SC motor map, which are thought to result in higher firing rates and stronger synchronization within a smaller population.

Up to this point, we have treated AV integration in a mechanistic-descriptive way, without wondering why the integration might indeed be beneficial for the organism, beyond yielding shorter reaction times through statistical facilitation, and a wider sensory measurement range through complementary integration. This point becomes especially interesting in the case of inverse effectiveness, where the reliability of the unimodal sensory input signals is typically low. The next section introduces the Bayesian framework that provides a statistical–theoretical account for the benefits of multisensory integration (first coined by Anastasio et al. (2000), to model the MI effects on SC cells).

13.3 BAYESIAN INFERENCE

The response to a sensory stimulus can be regarded as a random variable, as it will scatter around a mean with certain variability (see eg, Figs. 13.3 and 13.4). The uncertainty in the responses is due to noisy processes in the sensory-to-neural transformations, and in the transformation from sensory-evoked response to perceptual or motor output. If the response is an overt gaze shift (here we assume that it reflects the brain's internal estimate of the sensory stimulus) and the stimulus is a sound repeatedly presented at a given location, identical stimulus repetitions will lead to a distribution of internal sensory (and hence gaze-shift) responses that we can denote by the conditional probability of a response, given stimulus S:

$$p(R|S) \tag{13.6}$$

This conditional probability is also called the likelihood function, and is usually described by $L(S;R)$ (note that it's a function of the stimulus, S). It describes the observer's belief about the stimulus, based on the sensory input only, and without any prior information. What one really would like to know, though, is the environmental state (that is, the stimuli) that produced a certain perceptual response, and this will likewise be denoted by:

$$p(S|R) \tag{13.7}$$

In statistics, these conceptually quite different conditional probabilities are related to each other by Bayes' rule:

$$p(S|R)p(R) = p(R|S)p(S) \tag{13.8}$$

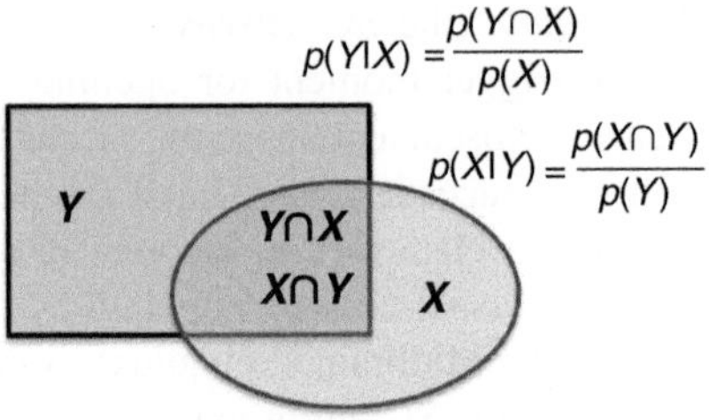

FIGURE 13.7 Random variables *Y* and *X* have a certain overlap, given by the intersection between the two distributions. Bayes' theorem follows directly from this overlap.

The rule can be readily derived from the Venn diagram of Fig. 13.7.

In perceptual and sensory–motor theories, Bayes' theorem is formulated as follows:

$$p(S|R) = \frac{p(R|S)p(S)}{p(R)} \propto p(R|S)p(S) = L(S;R)p(S) \tag{13.9}$$

where the distribution over all responses in the experiments for all sensory conditions [$p(R)$, the *marginal distribution*] is a normalization factor for $p(S|R)$:

$$p(R) = \int p(R|S)p(S)dS \tag{13.10}$$

Eq. (13.9) gives a statistical estimate of the sensory events that may have caused the measured response. This conditional distribution is called the *posterior distribution.* As such, Eq. (13.9) is a theory of the brain. The Bayesian theory then states that the posterior estimate (which in our case makes an *inference* about the underlying sensory causes of an overt motor response) is the result of two factors: the *likelihood* (or belief) that a certain stimulus leads to this response, and assumptions about the probability that this stimulus indeed occurs. The latter is called the *prior probability* of stimulus events, and contains all the knowledge of the subject about the environment, outside the stimulus: her expectations, experiences, preknowledge, motivation, etc.

Bayes' rule is always true, regardless the shape of the underlying probability distributions, but if we assume that the likelihood function and prior are both normally distributed in stimulus space:

$$p(R|S) \sim \exp\left(-\frac{(x_S - \mu_S)^2}{2\sigma_S^2}\right)$$
$$p(S) \sim \exp\left(-\frac{(x_S - \mu_P)^2}{2\sigma_P^2}\right)$$

it can be readily shown (Exercise 13.2) that the posterior distribution is also a Gaussian in response space:

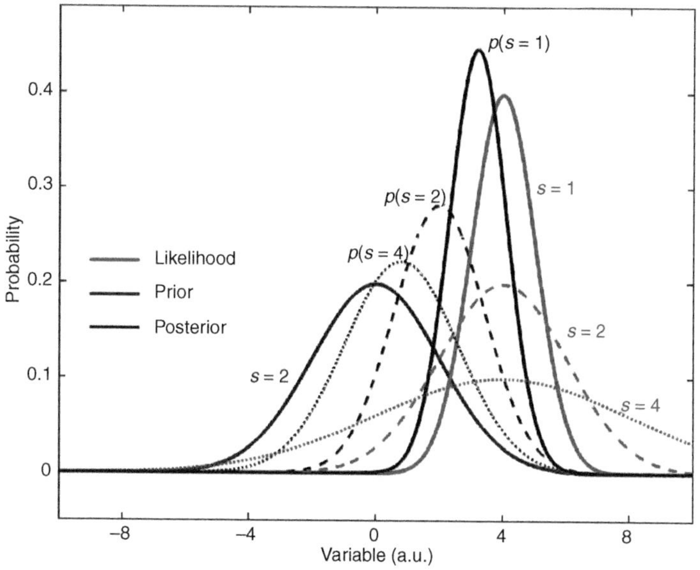

FIGURE 13.8 **Bayes' rule [Eq. (13.9)].** The three likelihood functions (red), centered at $x = 4$ and standard deviations = 1, 2, and 4, respectively, are combined with a fixed prior at zero and std = 2 (blue), to produce different posteriors (black). Note that the mean of the posterior shifts toward the prior, with increasing std of the likelihood.

$$p(S|R) \sim \exp\left[-\frac{(x_R-\mu_R)^2}{2\sigma_R^2}\right]$$
$$\text{with } \mu_R = \frac{\sigma_P^2}{\sigma_S^2+\sigma_P^2}\mu_S + \frac{\sigma_S^2}{\sigma_S^2+\sigma_P^2}\mu_P \text{ and } \sigma_R^2 = \frac{\sigma_S^2\sigma_P^2}{\sigma_S^2+\sigma_P^2} \qquad (13.11)$$

This important result, illustrated in Fig. 13.8, states that

1. The mean of the posterior is a *weighted average* of the means of the likelihood function and prior distribution, in which their variances determine the weights. The combined mean lies between the means of the prior and likelihood function.
2. The variance of the posterior is *smaller* than either variance of the likelihood function and prior.

In case of a congruent AV stimulus, in which the likelihood of the auditory and visual events can be regarded as *independent* signals, Eqs. (13.9) and (13.11) can be expanded to:

$$p(A,V|R) = \frac{p(R|V)p(R|A)p(A,V)}{p(R)} \propto p(R|V)p(R|A)p(A,V) \qquad (13.12)$$

Assuming a prior distribution of all potential stimuli around straight ahead, ie, $\mu_P = 0$, the mean and variance of the AV posterior distribution are given by (Exercise 13.3):

$$\mu_{R,AV} = S_{R,AV}\left(\frac{\mu_A}{S_V}+\frac{\mu_V}{S_A}\right) \text{ and } \frac{1}{S_{R,AV}} = \frac{1}{S_A}+\frac{1}{S_V}+\frac{1}{S_P} \tag{13.13}$$

where $S_X \equiv \sigma_X^2$ the variance of signal X.

In Bayesian inference, the final step in the process is application of a *decision rule*, which ultimately selects the response from the posterior distribution. Often, the optimal decision rule is the *maximum-aposteriori (or MAP)* rule, which simply selects the maximum (mode) of the posterior distribution (in case of Gaussians, this would be the mean of the posterior).

It is important to realize that because of the noise in the sensory processing, each experimental trial (stimulus presentation) yields a new likelihood function, and hence a new posterior. Thus, picking the MAP value is a stochastic process too, and the inferred distribution of response estimates according to the MAP rule has a (Gaussian) probability distribution with

$$\mu_{R,\text{Map}} = \mu_{R,\text{Post}} \text{ and } \sigma^2_{R,\text{Map}} = \frac{\sigma_P^2}{\left[1+(\sigma_P^2/\sigma_S^2)\right]^2} \tag{13.14}$$

The MAP decision rule is *optimal* in the sense that it:

- Minimizes the absolute mean-squared errors of the responses for Gaussian distributions.
- Minimizes the response variance.

We will now discuss three illustrative examples from psychophysics that apply the Bayesian model.

1. Ventriloquism. A simple example, often used in Bayesian applications of perceptual data, is a uniform (flat) prior: $p(S) = \text{constant}$. In that case, Eq. (13.9) reduces to

$$p(S|R) \propto p(R|S) \tag{13.15}$$

so that the posterior distribution is the same as the likelihood function, and the MAP rule becomes a maximum-likelihood estimation (MLE) problem.

Applying a flat prior to bimodal integration, leads to the following result (Ernst and Banks, 2002; Alais and Burr, 2004):

$$p(A,V|R) \propto p(R|V)p(R|A) \tag{13.16}$$

for which mean and variance of the AV posterior are determined as

$$\mu_{R,AV} = S_{R,AV}\left(\frac{\mu_A}{S_V}+\frac{\mu_V}{S_A}\right) \text{ with } \frac{1}{S_{R,AV}} = \frac{1}{S_A}+\frac{1}{S_V} \tag{13.17}$$

Note that the predicted mean AV estimate of the posterior is a *weighted average* of the visual and auditory percepts, in which the variances of the unimodal sensory responses act as weighting factors. It is readily verified that the weights add to one [compare this important result to Eq. (13.5)]:

$$w_A = \frac{S_V}{S_A + S_V} \text{ and } w_V = \frac{S_A}{S_A + S_V} \tag{13.18}$$

When the prior is flat, the MAP estimate will have the *same variance* as the posterior. If subjects were to rely on the MLE of Eq. (13.17), AV responses are expected to be *more precise* than visual- and auditory-only responses, and the endpoints of pointing would be a weighted average of the unimodal responses (Fig. 13.8). Note that the MLE model is *parameter free*: the unimodal responses to visual and auditory stimuli completely determine the (Gaussian) distribution of congruent AV responses by Eq. (13.17), without fitting any parameters to the data.

Alais and Burr (2004) applied this simple model to test whether subjects would respond in an optimal way to incongruent AV stimuli, such as in the ventriloquist effect. To that end they blurred the visual input to Gaussian blobs of different sizes that make visual localization increasingly noisy, while keeping the auditory accessory stimulus (a brief click) constant. Fig. 13.9A shows the psychometric

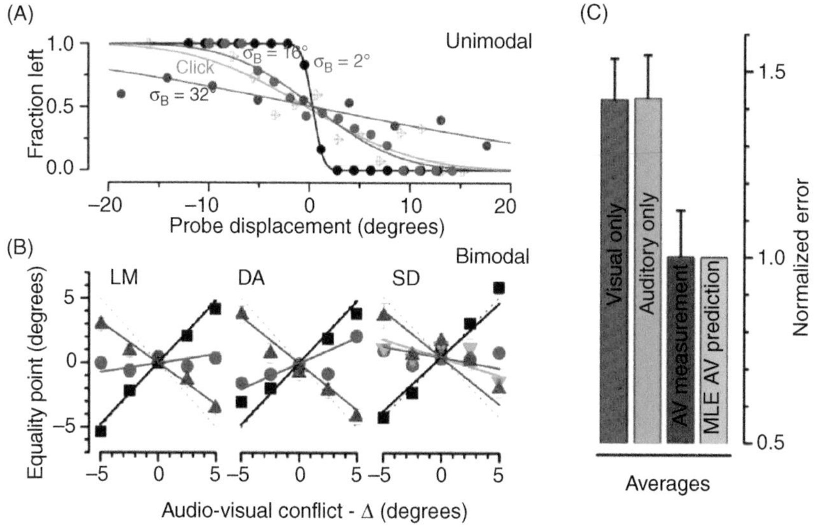

FIGURE 13.9 **The ventriloquist effect follows the near-optimal MLE model [Eqs. (13.17) and (13.18)].** (A) Unimodal psychometrics for auditory and visual location judgments to various Gaussian blob sizes (σ_B). (B) Sensitivity of AV alignment for the different blobs. Solid lines: Eq. (13.17). Dotted lines: purely visual (black), or purely auditory (blue) responses. (C) Model comparison of data variance [light blue; Eq. (13.18)] for the blob size that yielded similar psychometrics as the auditory stimulus. *(Source: Alais and Burr, 2004, with kind permission.)*

curves for the unimodal stimuli in which subjects had to indicate whether the stimulus was seen (or heard) left or right from a center probe. With increasingly blurred visual blobs the subjects had more problems in judging the difference, which is seen as a widening of the response variance (flatter curves). Note that the visual spatial judgments are inferior to the auditory percepts for the largest blob size (blue), and that they are similar for the intermediate blob ($\sigma_B = 16$ degrees).

In bimodal trials subjects had to judge when an AV displaced pair differed from an aligned pair in the center. Results for three of their subjects are shown in panel B, which shows that responses were fully guided by vision for the smallest blob (width of 2 degrees; black). In contrast, they relied on the auditory stimulus for the largest blob (width 32 degrees; blue), and interestingly, their responses were a weighted average for the intermediate blob sizes [solid lines are predicted from Eq. (13.17)]. The responses followed the model prediction quite well (Fig. 13.9C).

2. Sound-localization response gain. The Bayesian formalism can also be applied to answer the question why the elevation gain of auditory-evoked saccades is typically lower than the azimuth gain. To understand this problem we refer to Fig. 13.8, which shows how the weighted average shifts toward, or away from the prior with increasing, or decreasing, variance of the sensory input, respectively (the likelihood function). Recall the data of Fig. 13.4 (central inset), which demonstrate that the azimuth response components are much more robust to background noise than the elevation components. In Bayesian terms, the auditory system should regard its elevation estimate as less reliable than azimuth, and as a result, the relative weight of the prior in Eq. (13.11) should be higher for elevation than for azimuth. We can calculate some estimates of how this would work.

Suppose that the variances for the azimuth and elevation components of sound-sources are fixed (we know they are not, but let's suppose this, for simplicity) at S_{Az} and S_{El}, respectively, in which $S_{El} > S_{Az}$. Let's further assume that the auditory system adopts a straight-ahead, circular prior ($\mu_P = 0$, and S_P), then the MAP estimates for the azimuth and elevation components are determined by

$$\begin{aligned} \mu_{R,Az} &= \frac{S_P}{S_P + S_{Az}} \mu_{Az} \equiv G_{Az}\mu_{Az} \quad & S_{R,Az} &= \frac{S_P}{\left[1+(S_P/S_{Az})\right]^2} \\ \mu_{R,El} &= \frac{S_P}{S_P + S_{El}} \mu_{El} \equiv G_{El}\mu_{El} \quad & S_{R,El} &= \frac{S_P}{\left[1+S_P/S_{El}\right]^2} \end{aligned} \tag{13.19}$$

Thus, the comparison yields for their ratios:

$$\frac{G_{El}}{G_{Az}} = \frac{S_P + S_{Az}}{S_P + S_{El}} < 1 \quad \text{and} \quad \frac{S_{R,El}}{S_{R,Az}} = \frac{\left(1+S_P/S_{Az}\right)^2}{\left(1+S_P/S_{El}\right)^2} > 1 \tag{13.20}$$

If we take the straight-ahead resolutions (the MAA; chapter: Assessing Auditory Spatial Performance) in azimuth (0.7 degrees) and elevation (2.5 degrees) as an educated guess for the relative ratio of their standard deviations (ie, a factor 3.5), and a prior std of 10 degrees, the ratio of the response gains yields about 0.94 (which is close to human data), but for the response variances the ratio would yield a value of 12.0, which seems far too large. Note that the MLE model (a uniform prior) cannot account for a lower gain of the elevation responses, as Eq. (13.19) predicts a gain of 1.0 for both components ($S_P \rightarrow \infty$).

Probability matching: Instead of taking the optimal MAP estimate from the posterior distribution function, an alternative strategy could be to randomly select a sample from the posterior distribution as the response estimate for the target location in any given trial. This decision rule has become known as *probability matching*, PM, and would lead to responses with the same distribution as the posterior (eg, Mamassian et al., 2003). Because the posterior variance is larger than the MAP variance, responses would not be optimal. However, this rule may have an important advantage above the MAP decision rule: in the absence of sensory evidence (say, in total darkness and silence) Bayes' rule predicts that the posterior distribution equals the prior distribution. The MAP decision rule then selects the maximum of the prior, which would always result in the same response, or no response at all when the prior is centered at straight ahead.

In contrast, probability matching of the posterior would lead to a "blind" exploration of the environment according to the samples taken from the prior distribution. Indeed, eye movements (gaze shifts) follow an active explorative strategy under such conditions. For the analysis of Eq. (13.20) it would mean that a probability-matching model would yield a ratio for the variances of about 3.5 (Exercise 13.4), which seems more in line with experimental data than the MAP estimate. The reader may derive the following interesting relationship between the response gain and response variance for the PM rule, in Exercise 13.5:

$$G_{R,S} = 1 - \frac{\sigma_{R,S}^2}{\sigma_P^2} = 1 - \frac{S_{R,S}}{S_P} \tag{13.21}$$

This equation states that the standard deviation of the responses and the response gain are related by a parabolic relation with an intercept, $G = 0$, at $\sigma = \sigma_P$. Thus, in the absence of sensory evidence (eg, when the SNR in a sound-localization experiment is so low that the response gain drops to zero), the response variance will equal the variance of the straight-ahead (Gaussian) prior. In Exercise 13.6 the reader may check the potential validity of this relationship for the raw auditory data of Fig. 13.4.

3. Spectral Pinna Cues. As a third example of the use of Bayesian inference, we return to the spectral correlation model described in chapter: Assessing Auditory Spatial Performance, for estimating the elevation of a sound-source (Hofman and Van Opstal, 1998, 2002). The central idea is that the elevation estimate of the subject (c.q. her eye-movement response) results from finding

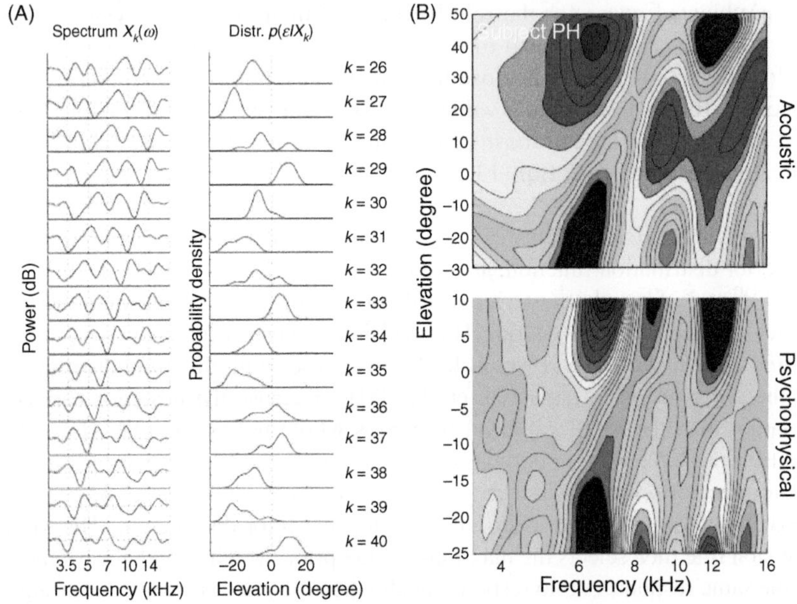

FIGURE 13.10 Psychophysical reconstruction (B) of the HRTFs based on MLE of saccadic eye-movement response distributions (A, right) to random broadband amplitude spectra (A, left). *(Source: After Hofman and Van Opstal, 2002, with kind permission).*

the maximum correlation of the sensory spectrum within the library of stored HRTFs. For spectra that are relatively flat the model guarantees an accurate estimate of the true elevation. This is not necessarily the case for spectra with strong spectral amplitude modulations. In that case, the elevation estimate may point to a phantom location instead of the physical source direction. Hofman and Van Opstal (2002) exploited this idea to test whether subjects would use a Bayesian strategy in their spectral-to-elevation mapping.

The idea runs as follows: the experiment presented 180 different broadband random spectra to the subject that had similar amplitude statistics as real HRTFs. The left-hand side of Fig. 13.10A shows 15 example spectra. All sounds were played from a speaker that was positioned at a fixed, straight-ahead location, and were presented in random order between 10 and 20 times. Subjects responded with an eye movement to the perceived location of the stimulus, which in general was not to the straight-ahead speaker location, but varied from stimulus to stimulus around different perceived elevation angles. Importantly, the responses were quite consistent, indicating a true spatial percept of the listener (see the right-hand column of Fig. 13.10A). The resulting eye-movement distributions in elevation can then be described as conditional probabilities (Likelihood functions):

$$\text{for stimulus k:} \quad p[\varepsilon|X_k(\omega)] = L[X_k(\omega);\varepsilon] \qquad (13.22)$$

Applying Bayes' rule to calculate the posterior distribution function:

$$\begin{aligned} p[X_k(\omega)|\varepsilon] &= \frac{p[\varepsilon \mid X_k(\omega)]p[X_k(\omega)]}{p(\varepsilon)} \\ &= \frac{p[\varepsilon \mid X_k(\omega)]p[X_k(\omega)]}{\sum_{m=1}^{180} p[\varepsilon \mid X_m(\omega)]p[X_m(\omega)]} \end{aligned} \tag{13.23}$$

The posterior describes which stimulus spectrum underlies a given elevation response. Assuming a uniform prior, $p(X_k) = 1/180$, reduces Eq. (13.23) to MLE:

$$p[X_k(\omega)|\varepsilon] \propto p[\varepsilon \mid X_k(\omega)] \tag{13.24}$$

Eq. (13.24) states how much a given spectral stimulus, X_k, contributes to the perceived elevation. If the stimulus spectrum has no features that belong to the subject's HRTF at elevation ε, the posterior (and likelihood) will be low. If the spectrum contains many elevation-specific features, it will be high. To estimate the total stimulus spectrum (ie, the subject's HRTF) that would explain the elevation response, we can regard the posterior for each stimulus that generated responses to that elevation as a weighting factor for that stimulus, and add all weighted stimuli, according to:

$$H_{\text{psych}}(\omega;\varepsilon_0) = \sum_{k=1}^{180} p[X_k(\omega)|\varepsilon_0]X_k(\omega) \tag{13.25}$$

Fig. 13.10 (lower-right panel) shows the result of applying this simple MLE model to the psychophysical results. Two points are worth noticing: first, the overall pattern of spectral features in the psychophysical reconstruction have a remarkable resemblance to the acoustic HRTFs, and second, the range of elevation angles covered by the subject's responses (from −25 to +10 degrees), is substantially compressed (about a factor 2) with respect to the true elevation range (from −30 to +50 degrees). A possible explanation for this latter discrepancy may be the assumption of a uniform prior. If the auditory system would instead rely on a prior for sound spectra that corresponds, for example, to its own stored HRTFs for elevations weighted around straight ahead in a Gaussian way, say:

$$p(X_k) \sim \sum_k \exp\left(-\frac{\varepsilon_k^2}{2\sigma_P^2}\right) \text{Corr}\left[HRTF(\omega;\varepsilon_k), X_k(\omega)\right] \tag{13.26}$$

the predicted elevation responses of Eq. (13.23) are expected to be weighted averages between the true elevation and the straight-ahead direction.

Recently, Parise et al. (2014) argued that the particular spectral shape of the HRTFs may have a clear correspondence to the statistics of natural sounds. For example, humans consistently tend to associate high-frequency tones to high elevations and low frequencies to low elevations in absolute world coordinates.

Interestingly, high-pitch natural sounds tend to originate from high elevations, and low-pitch sounds from low elevations. Thus, the perceptual mapping and complex ear geometry both seem to match the natural sound statistics.

13.4 AV CONGRUENT

We have so far described AV integration for relatively simple situations, in which a subject was confronted with only two stimuli, either congruent in space and time, or incongruent. Even in those situations clear effects of multisensory integration can be observed, and the models described in Section 13.2 can account for these effects. The Bayesian framework provides an abstract mathematical description for the spatial response data (not for the reaction times, and not the underlying neurobiological mechanisms) under more general conditions that may include arbitrary levels of uncertainty in the sensory signals. The Bayesian model is therefore well suited to deal with the phenomenon of inverse effectiveness. Section 13.1 presented a possible scenario in which multisensory integration might become truly beneficial: when visual and auditory performances become highly degraded due to interfering AV backgrounds (Figs. 13.3 and 13.4).

Let's see what happens under such localization conditions to saccadic eye movements. Fig. 13.11A shows three eye-movement trajectories for trials in which the subject was searching for the target, which could be visual (red square), auditory (blue star), or audiovisual (all sensory conditions were randomly mixed in the experiment; Corneil et al., 2002; Van Opstal and Munoz, 2004). A search trial could last 3–4 s, and during all that time the stimuli were on. The trigger to indicate stimulus presentation was provided by a central fixation

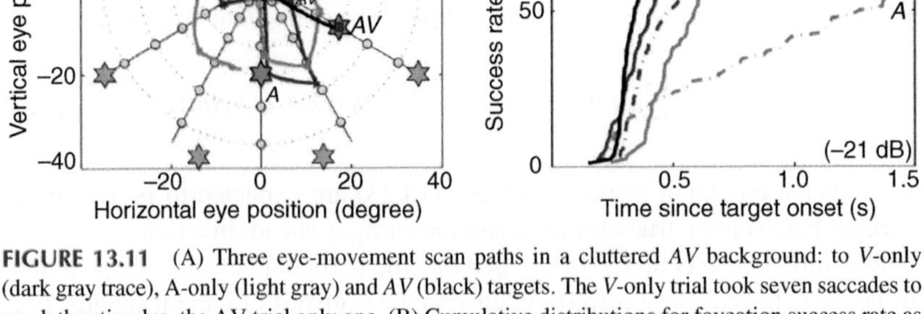

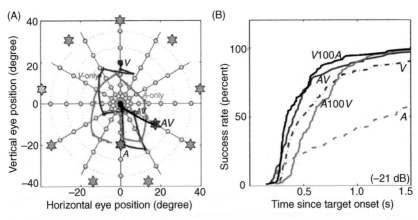

FIGURE 13.11 (A) Three eye-movement scan paths in a cluttered *AV* background: to *V*-only (dark gray trace), A-only (light gray) and *AV* (black) targets. The *V*-only trial took seven saccades to reach the stimulus, the AV trial only one. (B) Cumulative distributions for foveation success rate as function of time for five different stimulus conditions: A-only, V-only, A100V, AV, and V100A. The latter two multimodal stimuli are the most successful.

spot. Since AV stimuli were always spatially aligned, both sensory modalities served as a target. The auditory target (a broad-band buzzer) was hidden in diffuse background GWN (produced by the peripheral speakers, gray stars) of 60 dBA, and the SNR in the A and AV-trials was −21 dB (also see Fig. 13.4); the visual target (red LED) was hidden among 84 green distracters. Note that in this particular V-only trial it took the subject seven saccades to foveate the upward target. The A-only scan path (light gray) took the eye close to the (downward) target in the first saccade, after which the subject started to search further for a potential visual stimulus (which clearly wasn't there...). Interestingly, the AV trial brought the subject immediately on target in one saccade. This behavior was typical for this experiment. Fig. 13.11B summarizes all data of this subject for the A = −21 dB sounds, as cumulative distributions of success rate versus time (a success was recorded as soon as the subject foveated the target within a few degrees accuracy). Clearly, the A-only saccades were least successful, as even after 1.5 s only 50% of the trials resulted in target fixation, but also the V-only trials were only 50% correct after 500 ms (ie, 2–3 saccades).

The AV trials in the experiment also included three temporal manipulations: in A100V trials, the sound preceded the visual stimulus by 100 ms, AV means that they were synchronous, and in V100A trials the visual target came on first. This additional manipulation had a clear effect on the success rate: the most successful congruent AV combination is the one in which the visual stimulus slightly preceded the auditory target. Indeed, for the AV and V100A stimuli the success rates were already close to 90% within two saccades (ie, within 500 ms). Given the processing delays within the visual (70 ms) and auditory (10 ms) systems (the travel time of sound in the lab adds only a few milliseconds), the V100A condition induces nearly synchronous activation of multisensory neurons in the brain.

Interestingly, although the A100V condition would lead to the very earliest neural response of the three AV conditions, the success rate is very low (still less than 50% after 500 ms), indicating that a race mechanism may not be of help in complex AV search tasks. Clearly, all three AV conditions show signs of multisensory integration, leading to enhanced (V100A, AV) or suppressed (A100V) performance in accuracy and time, relative to A- or V-only stimuli.

Fig. 13.12 summarizes how multisensory integration on the first saccadic response to congruent AV stimuli in a complex AV environment depends on stimulus synchrony and task difficulty. The experiment employed 12 different multisensory conditions: four auditory SNRs times three temporal manipulations; Corneil et al., 2002).

In Fig. 13.12A it can be seen that the large variability of saccade accuracy and reaction time to the unimodal stimuli is strongly reduced for synchronous AV stimuli. Interestingly, the AV responses seem to be generated by a mechanism that yields *better than the best of both worlds* performance: the variability in the accuracy of the AV responses is smaller than for either A- or V-only responses. This property is nicely in line with a Bayesian integration mechanism, as shown also by Alais and Burr (2004) for the (incongruent) ventriloquist

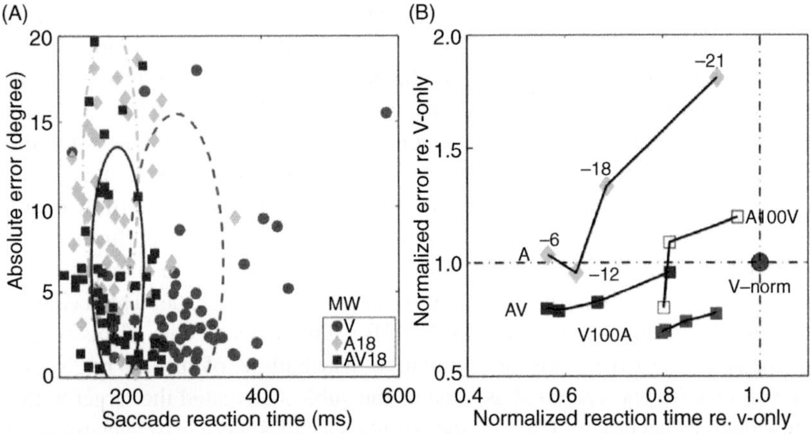

FIGURE 13.12 AV integration on first-saccade responses in a cluttered AV environment. (A) AV responses follow a "better than best of both worlds" principle: they are more accurate and precise than *V*-only, but at auditory reaction times (here: −18 dB re. GWN background). (B) AV integration for all 12 SNR-timing conditions. Data averaged across five subjects and normalized to their *V*-only performance. All AV conditions induce multisensory integration; effects vary systematically with SNR (−21, −18, −12, and −6 dB) and stimulus synchrony (colors). *(Source: After Corneil et al., 2002.)*

effect with blurred visual stimuli (Fig. 13.9). Furthermore, the AV reaction times are at least as short as, but less variable than the reaction times to A-only stimuli.

Panel B shows the pooled (averaged) data, from which we can verify the principle of *inverse effectiveness* on the saccades. The results for the four A-only and 12 AV stimulus types are plotted with reference to the V-only condition (ie, the slowest, but most accurate unimodal responses). A normalized value <1.0 indicates that performance is better (faster or more accurate) than visual performance. It is immediately clear from this plot that all auditory conditions yielded shorter reaction times than visual. However, the −18 and −21 dB conditions also yielded highly inaccurate responses (principally due to increased errors in elevation).

So let's compare the reaction times and errors for the most difficult ($SNR = -21$ dB) and easiest ($SNR = -6$ dB) auditory localization conditions (Table 13.1).

TABLE 13.1 Comparison of MI Effects on the Normalized Reaction Times (RT) and Mean Absolute Errors (ERR) for the Two Extreme Auditory SNRs in Fig. 13.12

	A-only		AV		V100A	
SNR	ERR	RT	ERR	RT	ERR	RT
−6 dB	1.05	0.55	0.8	0.55	0.7	0.8
−21 dB	1.8	0.9	0.95	0.8	0.8	0.93

Note that to quantify the strength of inverse effectiveness from these data we cannot simply use the definition of Eq. (13.1), which was designed for multisensory effects on neural responses. Instead, we quantify the effect by taking the difference between the normalized A-only and AV responses:

$$MI^{X}_{\text{Behavior}} = \Delta SNR_X 100\% \text{ with } \Delta SNR_X \equiv R_{A_X} - R_{AV_X} \quad (13.27)$$

This simple definition leads to the following multisensory effects for the AV and V100A stimuli:

	Reaction times	Errors (%)
AV_6	$MI_{\text{Behavior}} = 0$	$MI_{\text{Behavior}} = +25$
$V100A_6$	$MI_{\text{Behavior}} = -25\%$	$MI_{\text{Behavior}} = +35$
AV_{21}	$MI_{\text{Behavior}} = +10\%$	$MI_{\text{Behavior}} = +85$
$V100A_{21}$	$MI_{\text{Behavior}} = -3\%$	$MI_{\text{Behavior}} = +100$

By comparing the −6 and −21 dB conditions, the largest positive effects of multisensory integration are obtained for the weakest auditory stimuli: this is a clear fingerprint for inverse effectiveness, and the effect results to be quite strong. Interestingly, it holds for the change in reaction time, as well as for the response accuracy. In Exercise 13.7 the reader calculates the indices for the other SNRs too, to check for a trend as function of SNR.

13.5 AV INCONGRUENT

We have seen that in relatively simple "ventriloquist-like" situations with only two stimuli the strength of multisensory integration varies systematically with the spatial disparity between the stimuli (Frens et al., 1995; Alais and Burr, 2004; Meredith et al., 1987; Frens and Van Opstal, 1998). In addition, even in spatially congruent conditions there is a strong effect of the temporal order and synchrony of the stimuli. In the previous section we described a myriad of interaction effects in a cluttered environment for spatially congruent stimuli, but in that experiment subjects were specifically instructed that both stimuli served as a target for localization. However, in a real world this is not the default situation. When searching for a particular object, it is not known a priori whether the visual event will or will not be accompanied by an auditory event. However, if so, the auditory event should be in spatial–temporal congruency with the visual event to induce integration. But how does the perceptual system decide whether there is spatial–temporal congruency between AV events (eg, Körding et al., 2007)? Figure 13.2 explained why this is not a trivial problem when the eyes (and head) move around during stimulus presentation. So what happens with multisensory integration in more complex environments, like the one described previously?

Van Wanrooij et al. (2009) studied this problem in the same AV environment as in Fig. 13.11A, but now subjects were instructed to *ignore* the auditory accessory (presented at $SNR = -12$ or -18 dB re. the 60 dBA background) as it provided no clue for the location of the visual target. Still, in about 16% of

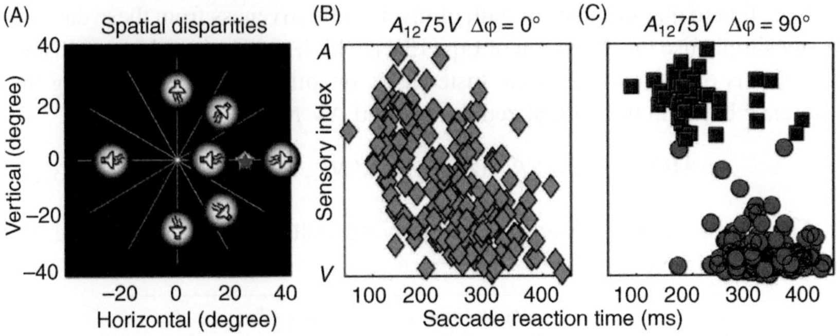

FIGURE 13.13 (A) One of eight AV spatial disparities could be selected for each of the 24 visual targets (red star). (B) Sensory index [Eq. (13.30)] as function of reaction time for spatially aligned stimuli, and (C) for stimuli at the same eccentricity but at a 90 degrees direction difference. For aligned stimuli SI gradually varies from A-evoked (short SRT) to *V*-evoked (late SRT). For unaligned stimuli responses become bimodal (blue and red symbols were separated by k-means cluster analysis). *(Source: After Van Wanrooij et al., 2009.)*

the trials stimuli were spatially congruent. Also, in this study the authors used a temporal manipulation by presenting A75V (auditory leads visual by 75 ms) and V75A (visual leads) stimuli at equal probability. The spatial disparity of the stimuli was chosen from eight locations around the visual target, either at different eccentricities along the same target direction, or at the same eccentricity as the target, but in different directions (Fig. 13.13, left). The visual target was randomly selected from one of 24 possible locations within the oculomotor range ($R = 20$ or 27 degrees, at 12 different directions).

In correct trials the saccade is directed toward the visual target. However, under certain conditions it may occur that the subject cannot ignore the auditory accessory, and her first saccade may be primarily guided to the perceived location of the auditory distracter. How do we quantify that? Here it is important to recognize that the *perceived* auditory location may differ substantially from the *physical* stimulus location, because of the considerable differences in gain for azimuth versus elevation response components at low SNRs. To account for this, the perceived sound location was estimated from the responses to the A-only targets, performed in a separate control experiment.

The physical and perceptual spatial disparities between auditory and visual stimuli in a given AV pair are calculated as the vectorial distances between the actual stimulus locations (SD), and between the measured response locations (PD), respectively. The latter are derived from linear regression fits on the azimuth/elevation components of unimodal auditory- and visual-evoked responses (eg, $\alpha_{R,A} = a + b\alpha_{T,A}$):

$$\begin{aligned} SD &\equiv \sqrt{(\alpha_{T,V} - \alpha_{T,A})^2 + (\varepsilon_{T,V} - \varepsilon_{T,A})^2} \\ PD &\equiv \sqrt{(\alpha_{R,V} - \alpha_{R,A})^2 + (\varepsilon_{R,V} - \varepsilon_{R,A})^2} \end{aligned} \tag{13.28}$$

To quantify to what extent AV responses are driven by the auditory versus the visual modality, or by both (ie, integration), we introduce the sensory index (SI).

The SI is constructed as follows: calculate the *perceptual* localization error of the AV response with respect to the visual and auditory unimodal percepts by

$$PE_V \equiv \sqrt{(\alpha_{R,V} - \alpha_{R,AV})^2 + (\varepsilon_{R,V} - \varepsilon_{R,AV})^2}$$
$$PE_A \equiv \sqrt{(\alpha_{R,A} - \alpha_{R,AV})^2 + (\varepsilon_{R,A} - \varepsilon_{R,A}V)^2} \tag{13.29}$$

The SI is then defined by

$$SI = \frac{PE_V - PE_A}{PE_V + PE_A} \tag{13.30}$$

The $SI = -1$ if the AV response lies very close to the A-only percept (ie, $PE_A \gg PE_V$), and $SI = +1$ if the response is close to the V-only percept ($PE_V \gg PE_A$). If $SI \sim 0$ the AV response is a weighted average of the A- and V-percepts and reflects clear multisensory integration.

Figure 13.3B,C shows the SI-index as a function of the saccade-reaction time for two stimulus configurations: spatially aligned (congruent) $A_{12}75V$ stimuli, versus the same stimuli with a separation angle of 90 degrees. For the aligned as well as unaligned stimuli the earliest responses are mainly auditory driven (note that although AV targets are spatially aligned, $SD \neq PD$!), and the late responses are mainly directed to the visual target (correct trials). However, for the aligned stimuli the SI index gradually evolves from -1 to $+1$ for the intermediate reaction times, whereas for the unaligned stimuli, a k-means cluster analysis reveals that the SI index reflects bistable behavior: responses are either auditory driven (blue cluster), or visually driven (red cluster). Still, even for unaligned stimuli Van Wanrooij et al. (2009) showed that the responses were less variable, were more accurate, and had shorter reaction times than their unimodal counterparts.

Interestingly, the sensory–motor system seems to "recognize," which stimuli to spatially integrate (when aligned), and which not (when unaligned), but the intermediate reaction times suggest that it costs some time to suppress the dominance of earlier acoustic input. This interesting result is further quantified for all spatial stimulus configurations in Fig. 13.14. It can be seen that when responses are ordered according to the *perceived AV* distance [*PD*, Eq. (13.28)], instead of the *physical AV* distance (*SD*), integration only occurs for the lowest *PD*s (<15 degrees). For larger perceived stimulus separations, responses become bistable: early eye movements are invariantly driven by the auditory stimulus (incorrect responses), late saccades by the visual target (correct). Note that this behavior does not depend strongly on the temporal order of the two stimuli. Finally, the results indicate that the range of reaction times for multisensory integration is limited to about 150–350 ms.

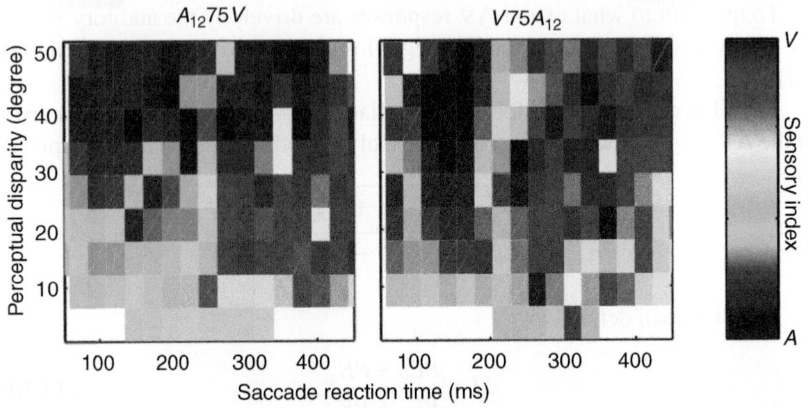

FIGURE 13.14 Sensory index (colors) as function of perceived *AV* spatial disparity (*PD*) for auditory first (left) and visual first (right) stimuli. Note that integration only occurs for the smallest perceived disparities. Above *PD* ~ 10–15 degrees the responses become bistable. *(Source: After Van Wanrooij et al., 2009.)*

AV trial history. SRTs become shorter when AV stimuli are integrated, than when they are not integrated. Clearly, if the sensory–motor system would know beforehand that AV stimuli are always congruent, fast integration can be the default response mode (Section 13.4). But when the stimuli have only a low probability for spatial–temporal alignment, the system needs to be more cautious. Indeed, Fig. 13.14 shows that in those cases the assessment of spatial congruency requires some time. Does the sensory–motor system build a representation of the environmental statistics, even when events are completely randomized, like in the laboratory tests?

To test for this possibility, Van Wanrooij et al. (2010) designed an experiment in which the probabilities for congruent/incongruent AV stimuli in different blocks of trials could be either 100/0, 50/50, or 10/90%, respectively. Interestingly, the SRT reflected a small, but very systematic influence of the *stimulus history*. For example, if the current trial was congruent (C) and the previous trial also (a CC sequence), then the reaction time was shorter than for a DC sequence in which the previous trial was incongruent (disparate). Interestingly, even the reaction times for CDC sequences differed from those in DCC series (Fig. 13.15A,B). By calculating the current probability of stimulus alignment in the experiment (ie, the likelihood of congruence for the next trial) on the basis of a simple memory-decay model:

$$p(C;n=0)=\sum_{n=0}^{N} w_0 \exp\left(-\frac{n}{\tau}\right) p(C|C,D) \tag{13.31}$$

where $p(C|C)=1$ and $p(C|D)=0$, and by assuming that the SRT is modulated by $p(C;\, n=0)$, the observed changes in the reaction times for the different

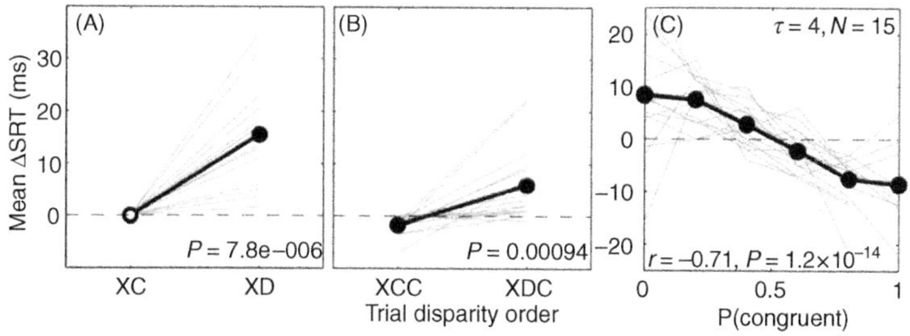

FIGURE 13.15 **Effects of trial history on the SRT.** (A) If the current trial (right-most letter) is congruent, reaction times are always faster than when it is disparate, regardless the history (*X*). (B) If the previous trial was also C, reaction times are shorter than when it was D. This effect disappears if the current trial is D (not shown here). (C) A memory model with decay [Eq. (13.31); solid line]) explains the observed changes of reaction times (thin lines) as function of the current probability of congruence. *(Source: After Van Wanrooij et al., 2010.)*

possible trial sequences could be described quite accurately with a memory of $N = 15$ trials, and a decay time constant of $\tau = 4$ trials (Fig. 13.15C).

13.6 AUDIO-VESTIBULAR INTEGRATION

When head orientation changes with respect to a world-fixed sound source, the auditory system should incorporate this change in order to recalculate the stimulus coordinates (see chapter: Coordinate Transformations). To that end, three sources of information relate to the head movement: (1) an efference copy signal that represents the planned motor program for the change in head orientation, (2) a proprioceptive signal that transmits information about the current status of the neck muscles, and (3) vestibular signals from the semicircular canals and otoliths that measure the change in head orientation through space as 3D rotations and translations, and as tilts with respect to gravity.

The brain will use all possible sources of sensory and nonsensory information to arrive at the best estimate of the sensory environment, and the Bayesian framework described in Section 13.3 provides a mathematical–statistical theory on how this integration could be achieved in an optimal way.

In the context of this chapter, also the combination of acoustic input with proprioceptive and vestibular information should be regarded as multisensory integration, in this case of interoceptive and exteroceptive sensory signals. The Bayesian theory predicts that when acoustic cues become increasingly unreliable, the weights for other sources of information will have to increase. A particular spatial range for which the acoustic cues become more unreliable than elsewhere in terms of their spatial resolution is around the auditory zenith (straight above the head when sitting upright).

Van Barneveld et al. (2011) used a psychometric 2AFC task to study how well the auditory zenith was defined under different head tilts within the frontal

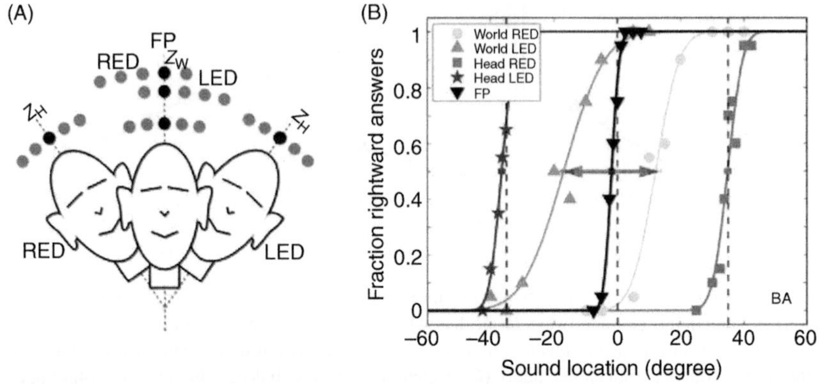

FIGURE 13.16 Effect of static head tilts on the perceived sound–source zenith. (A) Experimental paradigm. (B) Psychophysical results for a representative subject. *(Source: After Van Barneveld et al., 2011.)*

plane (FP) with respect to gravity: straight up (FP), and tilted by approximately 35 degrees with the left and right ear down (LED and RED, respectively; Fig. 13.16A). Subjects had to indicate whether a free-field broadband sound was presented either left or right from the head-centered (head task) or world-centered (world task) zenith.

The head-centered task could be performed quite well (Fig. 13.16B, head RED, FP, and LED data), as the psychometric curves were shifted by the same amount as the head tilt, without a significant increase in response variability. Thus, subjects had an accurate representation of their actual head orientation, which could be due to reliable information from their vestibular and proprioceptive neck–muscle systems. Their estimates of the world-centered zenith, however, were severely distorted. If responses had been correct, the world RED/LED data (gray triangles and circles) would have coincided with the frontal-plane (FP) data. Instead, their psychometric curves were shifted in the direction of head tilt by about 15–20 degrees (red and blue arrows), and the variability in their responses (slopes of the curves) had also increased significantly. Thus, in the world task, the sound-localization system made systematic errors (Fig. 13.16).

How could these errors arise when the information about head orientation is accurate (and precise), and also the acoustic input is accurate? The Bayesian framework can account for these error patterns in the world task by assuming an accurate estimate of the sound's head-centered location (with considerable variability around the true mean) in combination with a uniform prior for sound-source directions. The errors arise through the use of an upright (around FP) prior for head orientation in the world task. This leads to a biased estimate in the posterior for head orientation, as the noise in the system's otolith signal (oto) is comparable to the prior's variance:

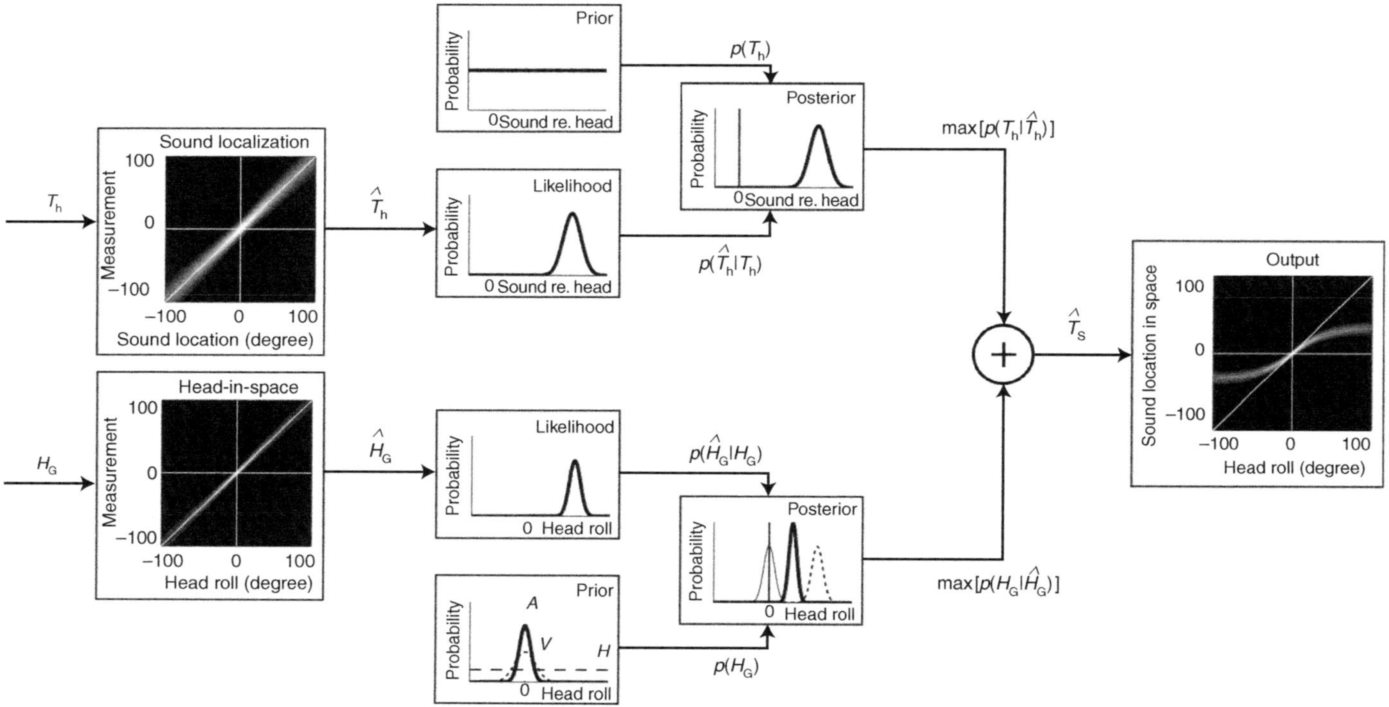

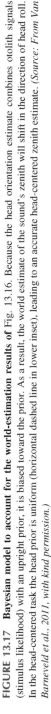

FIGURE 13.17 **Bayesian model to account for the world-estimation results of** Fig. 13.16. Because the head orientation estimate combines otolith signals (stimulus likelihood) with an upright prior, it is biased toward the prior. As a result, the world estimate of the sound's zenith will shift in the direction of head roll. In the head-centered task the head prior is uniform (horizontal dashed line in lower inset), leading to an accurate head-centered zenith estimate. *(Source: From Van Barneveld et al., 2011, with kind permission.)*

$$\hat{H}_{\text{post},W} = \frac{S_{p,W}}{S_{p,W} + S_{\text{oto}}} \hat{H}_{\text{oto}} + \frac{S_{\text{oto}}}{S_{p,W} + S_{\text{oto}}} \hat{H}_p \tag{13.32}$$

The proprioceptive neck–muscle signals may be reliable initially (upon assuming the new head orientation), but because of co-contraction (which is not constant) during the trials, which last a couple of seconds, might become less reliable than the otoliths. The mean and variance of the head-centered prior in Eq. (13.32) could be task dependent. Thus, when making a head-centered judgment, the prior is taken to be uniform, leading to an accurate estimate of head orientation:

$$\hat{H}_{\text{post},H} = \hat{H}_{\text{oto}} \tag{13.33}$$

Fig. 13.17 shows the different components of the Bayesian model for the world-task. The world-centered estimate of the sound's zenith is determined by the Bayesian MAP estimates for the head-centered sound location and head-orientation with respect to gravity (upright):

$$\hat{T}_S = \mathbf{Max}[p(T_h \mid \hat{T}_h)] + \text{Max}[p(H_G \mid \hat{H}_G)] \tag{13.34}$$

In summary, by including task-dependent priors (uniform head postures for head-centered judgments that do not require an estimate of gravity, but upright for world-centered judgments of head posture, which do require an estimate of gravity) the model can successfully account for the results of Fig. 3.16.

13.7 EXERCISES

Problem 13.1: Derive the race model distribution of minimum reaction times [Eq. (13.3)], and identify each of the two terms.

Problem 13.2: Derive expressions for the mean and variance for the multiplication of two Gaussians, $N(\mu_1,\sigma_1)$, and $N(\mu_2,\sigma_2)$, respectively.

Problem 13.3: Derive expressions for the mean and variance for the multiplication of three Gaussians, $N(\mu_1,\sigma_1)$, $N(\mu_2,\sigma_2)$, and $N(\mu_3,\sigma_3)$, respectively.

Problem 13.4: Derive expressions for the gain and variance ratios in case the responses are described by mean and variance of the posterior distribution.

Problem 13.5: Derive Eq. (13.21).

Problem 13.6: Apply 13.21 to the data of Fig. 13.4, and estimate the prior variance, S_P, for gaze shifts. Assume that the largest variance in the data is due to scatter in the estimates for elevation. However, from the gain plots (center insets in Fig. 13.4) you may now also estimate the azimuth scatter on the basis of Eq. (13.21)!

Problem 13.7: Calculate the multisensory enhancement index (*MI*) for the −12 dB and the −18 dB results, plot the results for all SNRs, and verify whether there is indeed a trend for "inverse effectiveness" of multisensory integration as a function of the sound's SNR.

REFERENCES

Alais, D., Burr, D., 2004. The ventriloquist effect results from near-optimal bimodal integration. Curr. Biol. 14, 257–262.

Anastasio, T.J., Patton, P.E., Belkacem-Boussaid, K., 2000. Using Bayes' rule to model multisensory enhancement in the superior colliculus. Neural. Comput. 12, 1165–1187.

Bell, A.H., Meredith, M.A., Van Opstal, A.J., Munoz, D.P., 2005. Crossmodal integration in the primate superior colliculus underlying the preparation and initiation of saccadic eye movements. J. Neurophysiol. 93, 3659–3673.

Colonius, H., Arndt, P., 2001. A two-stage model for visual–auditory interaction in saccadic latencies. Percept. Psychophys. 63, 126–147.

Corneil, B.D., Munoz, D.P., 1996. The influence of auditory and visual distractors on human orienting gaze shifts. J. Neurosci. 16, 8193–8207.

Corneil, B.D., Van Wanrooij, M., Munoz, D.P., Van Opstal, A.J., 2002. Auditory–visual interactions subserving goal-directed saccades in a complex scene. J. Neurophysiol. 88, 438–454.

Ernst, M.O., Banks, M.S., 2002. Humans integrate visual and haptic information in a statistically optimal fashion. Nature 415, 429–433.

Frens, M.A., Van Opstal, A.J., 1998. Visual–auditory interactions modulate saccade-related activity in monkey superior colliculus. Brain Res. Bull. 46, 211–224.

Frens, M.A., Van Opstal, A.J., Van der Willigen, R.F., 1995. Spatial and temporal factors determine auditory–visual interactions in human saccadic eye movements. Percept. Psychophys. 57, 802–816.

Gielen, C.C.A.M., Schmidt, R.A., Van den Heuvel, P.J.M., 1983. On the nature of intersensory facilitation of reaction time. Percept. Psychophys. 34, 161–168.

Hofman, P.M., Van Opstal, A.J., 1998. Spectro-temporal factors in two-dimensional human sound localization. J. Acoust., Soc., Am. 103, 2634–2648.

Hofman, P.M., Van Opstal, A.J., 2002. Bayesian reconstruction of sound localization cues from responses to random spectra. Biol. Cybernet. 86, 305–316.

Jay, M.F., Sparks, D.L., 1987. Sensorimotor integration in the primate superior colliculus. I. motor convergence. J. Neurophysiol. 57, 22–34.

Körding, K.P., Beierholm, U., Ma, W.J., Quartz, S., Tenenbaum, J.B., Shams, L., 2007. Causal inference in multisensory perception. PLoS ONE 2, e943.

Mamassian, P., Landy, M., Maloney, L.T., 2003. Bayesian modeling of visual perception. In: Rao, R.P.N., Olshausen, B.A., Lewicki, M.S. (Eds.), Probabilistic Models of the Brain: Perception and Neural Function. MIT Press, Cambridge, pp. 13–36.

Meredith, M.A., Stein, B.E., 1986a. Visual, auditory, and somatosensory convergence on cells in superior colliculus results in multisensory integration. J. Neurophysiol. 56, 640–662.

Meredith, M.A., Stein, B.E., 1986b. Spatial factors determine the activity of multisensory neurons in cat superior colliculus. Brain Res. 365, 350–354.

Meredith, M.A., Stein, B.E., 1996. Spatial determinants of multisensory integration in cat superior colliculus neurons. J. Neurophysiol. 75, 1843–1857.

Meredith, M.A., Nemitz, J.W., Stein, B.E., 1987. Determinants of multisensory integration in superior colliculus neurons. I. temporal factors. J. Neurosci. 7, 3215–3229.

Parise, C.V., Knorre, K., Ernst, M.O., 2014. Natural auditory scene statistics shapes human spatial hearing. PNAS 111, 6104–6108.

Peck, C.K., 1996. Visual-auditory integration in cat superior colliculus: implications for neuronal control of the orienting response. Prog. Brain Res. 112, 167–177.

Stein, B.E., Meredith, M.A., 1993. The Merging of the Senses. MIT Press, Cambridge, MA.

Stein, B.E., Meredith, M.A., Huneycutt, W.S., McDade, L., 1989. Behavioral indices of multisensory integration: orientation to visual cues is affected by auditory stimuli. J. Cog. Neurosci. 1, 12–24.

Van Barneveld, D.C.P.B.M., Van Grootel, T.J., Alberts, B., van Opstal, A.J., 2011. The effect of head roll on perceived auditory zenith. Exp. Brain Res. 213, 235–243.

Van Opstal, A.J., Frens, M.A., 1996. Task-dependence of saccade-related activity in monkey superior colliculus: implications for models of the saccadic system. Prog. Brain Res. 112, 179–194.

Van Opstal, A.J., Munoz, D.P., 2004. Auditory–visual interactions subserving primate gaze orienting. In: Calvert, G., Spence, C., Stein, B. (Eds.), The Handbook of Multisensory Processes. MIT Press, MA, USA, pp. 373–394 (Chapter 23).

Van Opstal, A.J., Van Gisbergen, J.A.M., 1990. Role of monkey superior colliculus in saccade averaging. Exp. Brain Res. 79, 143–149.

Van Opstal, A.J., Van Gisbergen, J.A.M., 1989. A nonlinear model for collicular spatial interactions underlying the metrical properties of electrically elicited saccades. Biol. Cybernet. 60, 171–183.

Van Wanrooij, M.M., Bremen, P., Van Opstal, A.J., 2010. Acquired prior knowledge modulates audiovisual integration. Eur. J. Neurosci. 31, 1763–1771.

Van Wanrooij, M.M., Bell, A.H., Munoz, D.P., Van Opstal, A.J., 2009. The effect of spatial-temporal audiovisual disparities on saccades in a complex scene. Exp. Brain Res. 198, 425–437.

Chapter 14

Impaired Hearing and Sound Localization

14.1 INTRODUCTION

Given the tedious microhydrodynamics of the inner ear, its hair cells and the organ of Corti, and the micromechanics of the tympanic membrane and middle ear ossicles (see chapter: The Cochlea), it should come as no surprise that disorders of the peripheral auditory system are quite prevalent: about 10% of the population suffers from mild to severe hearing loss, with about 0.3% being deaf. Severe bilateral hearing loss can lead to serious problems, like social isolation, degraded communicative and language skills, and significant delays in intellectual development; the associated costs for society are high and expected to increase steeply with the aging population in the modern world.

Hearing loss gets manifested in different forms. It can be unilateral (affecting only one ear) or bilateral (symmetrical or asymmetrical); the hearing loss can be congenital, or acquired later in life (due to an infectious disease, a trauma, or to a genetic disorder, like type III Usher's syndrome). Furthermore, the hearing loss can be *complete*, inducing total deafness of the ear, or *partial*, affecting only the perceptual quality of a restricted range of frequencies.

Clearly, hearing loss affects the binaural processing of spectral–temporal acoustic input and as a consequence also influences the listener's capacity to localize and segregate sound sources in the environment.

The causes of impaired hearing are divided into three major classes:

1. *Conductive* hearing loss concerns an impaired transmission of acoustic energy from air to cochlea, which can be due to relatively simple causes, like excessive earwax, or an outer ear infection, or to a real disease condition, such as otosclerosis (stiffening of the ossicles), or atresia (a closed ear canal, which can be a congenital condition).
2. *Sensorineural* hearing loss has its origin in the inner ear (most commonly the outer hair cells), the eighth nerve (vestibular–cochlear nerve), or in the pathways and nuclei of the central nervous system. *Sensory* hearing loss concerns damage to the (outer) hair cells, for example, due to sudden trauma, caused by extremely loud noise blasts (from guns and exploding bombs), or to prolonged loud (>80 dB SPL) headphone exposure, which leads to

The Auditory System and Human Sound-Localization Behavior. http://dx.doi.org/10.1016/B978-0-12-801529-2.00014-3

impaired hearing around the mid-to-high frequency range (near 4 kHz). Certain antibiotic drugs, like Gentamicin, may result in ototoxic side effects that lead to sensory hearing loss. Neural hearing loss could result from brain tumors, intracranial hemorrhages, head trauma, etc. Also Menière's disease (an excess of fluid in the inner ear, which is also associated with attacks of vertigo) may cause permanent sensorineural hearing loss in the low frequency range (<1 kHz).

3. *Mixed* hearing loss is a combination of conductive and sensorineural hearing loss.

In the clinic, two quick qualitative screening tests may differentiate conductive from sensorineural hearing loss:

1. The *Weber test* can detect a unilateral conductive hearing loss (UHCL). Suppose that the patient complains about impaired hearing on the left side. A tuning fork (512 Hz) is placed on the center of the skull and the patient is asked to indicate which side is heard louder. If she hears the fork louder on the affected side, the impairment indicates a conductive hearing loss. If it's heard louder on the better ear, however, the impairment refers to a sensorineural hearing loss.
2. The *Rinne test* can be used when the patient is unaware of any hearing loss. The vibrating 512 Hz fork is now placed on the mastoid behind the ear, until the vibration is no longer heard. Then, the (still vibrating) fork is immediately placed close to the ear canal and the patient indicates whether or not the tone is still heard. A positive Rinne test (normal hearing) results when the air-borne sound is still perceived, while the bone-conducted sound is imperceptible: $AC > BC$. In conductive hearing loss, $BC > AC$. Also in sensorineural hearing loss, however, $AC > BC$, but the perceived sounds at the mastoid and ear canal will last for much lesser time than in normal hearing.

As both tests only use one particular frequency, they cannot lead to any general conclusion about the quality of hearing. To obtain a quantitative assessment of the hearing loss, more precise and systematic audiometric measurements are required.

Air–bone gap: In quantitative audiometry, the listener indicates whether she perceives a pure tone presented at different frequencies and intensities (typically at 0.125, 0.25, 0.5, 1, 2, 4, and 8 kHz). Two tests measure the subject's frequency perception thresholds: (1) stimulation over headphones through the external acoustic pathways determines the (normal) *air-conduction* hearing thresholds, which include transmission through the ear canal, eardrum, and middle-ear bones, to the cochlea and brain. These thresholds can be quantitatively compared to the normal-hearing average human (defined as 0 dB); (2) stimulation of the skull determines the *bone-conduction* hearing thresholds, bypassing the ear canal–middle-ear chain through direct stimulation of the cochlea-to-brain pathway. If the bone-conduction thresholds are lower than the air-conduction

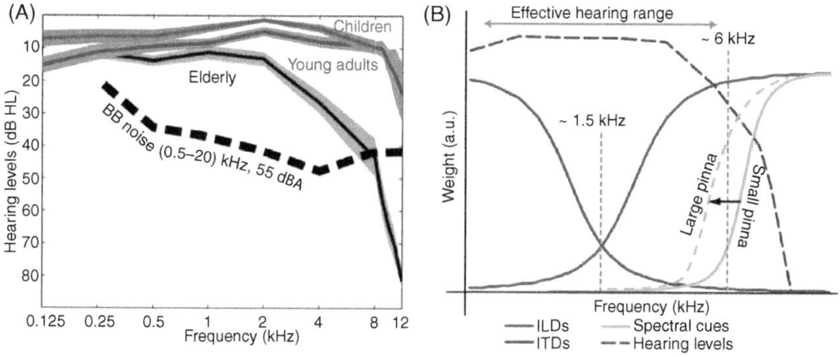

FIGURE 14.1 (A) Presbycusis (gray curve; listeners > 63 years) demonstrates a significant hearing loss already above 2–3 kHz. Dashed line: a flat-spectrum broadband noise sound presented at 55 dBA, is replotted on the same hearing-level scale. The elderly no longer perceive this sound for frequencies > 8 kHz (B) Available localization cues for the elderly. A larger pinna could shift a significant part of the spectral cues into the perceptual hearing range < 6 kHz. Also see Section 14.6 and chapter: Sound Localization Plasticity. *(Source: After Otte et al., 2013, with kind permission.)*

thresholds, there is a conductive hearing loss, which is quantified by the difference between the air- and bone-conduction thresholds: the so-called *air–bone gap*. If the air–bone gap is zero, the hearing impairment has a sensory–neural origin.

In severe cases of conductive hearing loss (eg, loss of the middle-ear bones, or a complete ossification of the ear canal), the maximum air–bone gap is about 60 dB; yet, for high sound levels in the free field the cochlea will still be stimulated through bone conduction. In single-sided *deafness* (SSD), however, the sensory hearing loss is complete (>70–90 dB), and cannot be circumvented through bone stimulation. Note that since at high sound levels bone conduction will also stimulate the hearing ear, a masking noise at the hearing side is used to dissociate SSD from severe conductive hearing loss.

Presbycusis: As we get older especially the high-frequency outer hair cells start to lose their functionality, leading to a gradual high-frequency sensory hearing loss (*presbycusis*; Fig. 14.1A). When also the middle ear loses its mechanical flexibility the hearing loss may affect the lower frequencies too. In the US, presbycusis affects about 30% of the elderly between 65 and 74 years of age, and 50% of 75 years and older.

Fig. 14.1 illustrates the average hearing thresholds of normal-hearing children (ages 7–11), young adults (21–34), and elderly listeners (63–80). While children and young adults have comparable hearing across the frequency range from 0.125 to 12 kHz, the elderly already start to deviate significantly above 2–3 kHz. As a result, the elderly no longer perceive frequencies >8 kHz when presented at a moderate intensity of 55 dBA.

In principle, the gradual (usually symmetric) loss of high-frequency perception need not dramatically interfere with the binaural difference cues (ITDs and

ILDs) for the localization of the sound's azimuth angle, although it has been suggested that also the central temporal processing capacities (and hence ITD processing) deteriorate with age (Dobreva et al., 2011). In any case, the loss of higher frequencies will surely affect the processing of the HRTF spectral cues, which typically extend beyond 8–10 kHz (Fig. 14.1B, solid green curve).

As mentioned in chapter: Sound Localization Plasticity, the pinnae of elderly people (in particular, the concha height; Fig. 7.14) are on average larger than of young adults and children. Interestingly, this particular mode of pinna growth induces a systematic shift of elevation-related notches in the HRTFs toward *lower* frequencies (Table 12.2; Otte et al., 2013). It is conceivable that for some elderly, a significant part of these notches may enter the perceptual hearing range (say < 6 kHz), and thus become useful for sound localization in the elevation direction, provided their auditory system adapts to these changes (Fig. 14.1B, dashed green curve). This point is discussed further in Section 14.6.

Single-sided deafness: SSD patients suffer from a severe sensory–neural hearing loss on the affected side that exceeds 60–80 dB over the entire frequency range. As a result, they have no access at all to binaural difference cues. Potentially, however, they could still use the spectral localization cues of their hearing ear. However, these spectral cues provide at best a weak signal for sound–source azimuth, and will be limited to locations around the midsagittal plane, as the acoustic head–shadow will mask the high-frequency cues for more lateral locations on the deaf side (see chapter: Assessing Auditory Spatial Performance). Since the acoustic sensory input is inherently unreliable, these listeners will have to rely on other strategies too to cope with their handicap.

One other potential cue for sound–source location in SSD is the acoustic *head–shadow*: as the source moves in the horizontal plane, its perceived intensity (the proximal sound level) varies systematically with its azimuth. However, as argued in chapter: Acoustic Localization Cues, this cue is ambiguous since it is always confounded with the absolute intensity of the sound source. Since the latter is unknown to the listener the auditory system of SSD patients has access to one "equation" (the monaural acoustic input) with two unknowns (azimuth and intensity), and thus faces an ill-posed localization problem. Yet, if listeners could make an adequate estimate (ie, *construct a prior,* chapter: Multisensory Integration) of the absolute source intensity, the head–shadow could become their most reliable acoustic localization cue.

In chapter: Multisensory Integration, we described how according to the Bayesian framework, the best possible strategy in a statistical sense would be to reweigh the different sources of evidence, depending on their reliabilities: as the acoustic input becomes less reliable, the weight of prior information should increase accordingly [weighted averaging, Eq. (13.11)]. Whether SSD listeners may indeed use such a strategy is discussed in Section 14.3.

Unilateral partial hearing loss: With a partial unilateral hearing loss, listeners do in principle have access to perturbed binaural difference cues, and to the spectral cues of the pinna from the intact hearing side. Over the course of time these listeners may therefore have developed strategies to cope with the perturbation, and have recalibrated and the remaining localization cues to veridical spatial coordinates. In chapter: Sound Localization Plasticity, we described some potential localization strategies in response to *acute* binaural perturbations (monaural plugging) in normal-hearing subjects. In Section 14.4, we describe and discuss sound-localization performance of listeners with a unilateral conductive hearing loss (UHCL), with and without the use of a bone conduction hearing aid.

14.2 RESTORATIVE HEARING TECHNOLOGIES

Hearing aid (HA): The different hearing disorders described previously require different technologies to restore hearing function. If the disorder is a mild to moderate hearing impairment, and the middle-ear transmission is functional, air-mediated acoustic transmission to the auditory system may be possible through an acoustic hearing aid. Currently, many different types of HA are on the market, covering a large variety of amplification algorithms (band-specific, flat, compressive, and directional), filters, and personalized programming capabilities. Fig. 14.2A displays different HA models, varying from attachment of the device's microphone behind the ear (pinna), to full insertion of both microphone and acoustic transmitter within the ear canal. Besides the obvious cosmetic differences (visible vs. nearly invisible), and size, the major functional

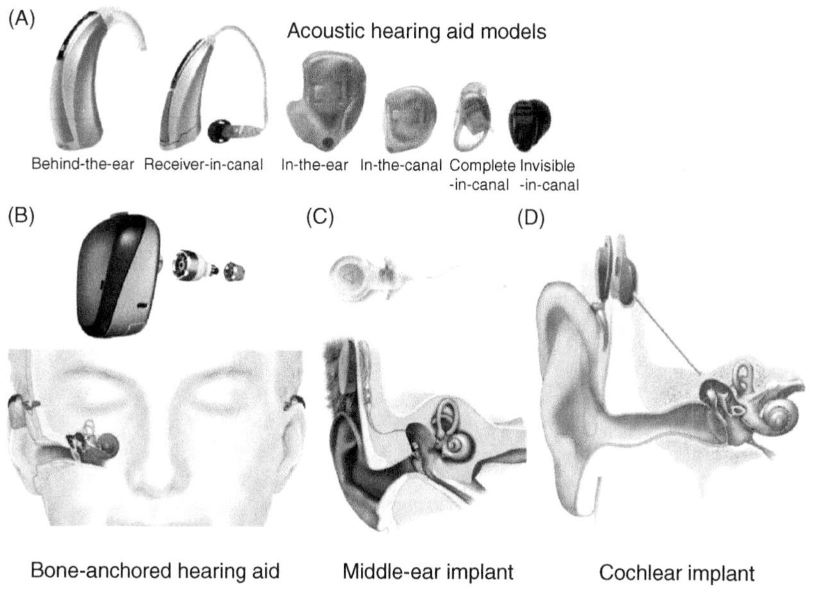

FIGURE 14.2 (A) Different types of hearing aids. (B) BCD (bone-anchored hearing aid, courtesy Cochlear). (C) Middle-ear implant (vibrant SoundBridge, courtesy Medel), (D) CI. *(Source: NIH, from Wikipedia, public domain.)*

difference between behind-the-ear and in-the-ear models is the potential use of the spectral pinna cues in the latter technology. Indeed, if the HA could restore hearing for frequencies up to about 7–8 kHz, a significant portion of the spectral localization cues could in principle be preserved (Figs. 7.15 and 14.1) and used for front-back and up-down sound localization.

Bone-Conduction Device (BCD): If the conductive hearing loss is more severe, or when wearing a HA is not possible (eg, in the case of atresia, chronic ear canal infections, absence of the pinna, poor function of the middle-ear bones), acoustic stimulation may only be achieved through bone conduction (Von Békésy, 1948; Zurek, 1986; Stenfelt et al., 2000) or via a middle-ear implant (for the latter, see further sections). Modern BCDs consist of a titanium screw implanted in the mastoid, which is set into acoustic vibration by a speech processor (Fig. 14.2B). The acoustic signal travels through the skull at a propagation velocity of 570 m/s and can directly stimulate the ipsilateral cochlea, thereby bypassing the ear canal to middle-ear acoustic path. Note, however, that the same acoustic signal also reaches the *contralateral* cochlea after a short transmission delay, thereby inducing so-called *cross hearing*. With a normally functioning contralateral ear, the direct (air-transduced) and indirect (bone-conducted) signals at the normal ear could interfere, thereby potentially deteriorating binaural integration. Section 14.4 will dive into this topic in more detail.

A BCD may also be implanted at the side of the impaired ear in case of SSD (total sensory hearing loss). In this way, the head–shadow at the deaf side will be compensated, which significantly expands the spatial range of acoustic input for the normal-hearing ear through cross hearing. Although it does not restore binaural hearing, it will provide a benefit for speech reception and sound detection across all spatial locations on the deaf side. However, note that the (delayed) acoustic input arising from the deaf side could deteriorate the processing of normal air-conducted acoustic input at the hearing ear (Lin et al., 2006; Wazen et al., 2003).

Middle-ear implant: If hearing loss is due to a malfunctioning middle ear, a middle-ear implant may replace the air-mediated acoustic pathway from ear canal to stapes (Fig. 14.2C). As the implant directly drives the stapes of the ipsilateral cochlea, this method does not suffer from potential cross-hearing interference like the BCD, for which it could serve as an alternative option (Agterberg et al., 2014). The quality of transmitted acoustic energy is usually excellent (distortions <1%, as compared to hearing aids of about 5%), and the signal bandwidth nearly spans the entire audible frequency range (0.1–10 kHz). Note that since the acoustic receiver is placed behind the ear it does not incorporate the spectral-shape HRTF information from the pinna.

Cochlear implant: In the case of total bilateral sensory hearing loss, but with the auditory nerve still intact, currently the only option to restore hearing function is the cochlear implant (CI; Fig. 14.2D). An electrode with up to 24 contacts is positioned inside the scala tympani of one ear, close to the auditory nerve, through an opening in the round window. The flexible electrode can usually be inserted up into the second winding of the cochlea (from about 8 kHz down to 1 kHz), but usually fails to access the low-frequency (<1 kHz) representation in the third winding. An external speech processor performs an ongoing Fourier analysis on the acoustic input, and distributes the instantaneous amplitude of the 24 different spectral components over the corresponding electrodes as a series of rapid electric pulses that together encode the sound's temporal envelope. The CI thus makes clever use of the cochlea's tonotopic organization, meanwhile aiming to preserve acoustic spectral–temporal patterns.

Note that although the CI can replicate the subtle changes in proximal sound level, this mode of stimulation does not preserve the temporal fine structure of the acoustic signal. Since the microphone of the speech processor is placed near the mastoid behind the ear, spectral-shape information of the pinna is not available either. Moreover, the spatial–temporal electric fields of the 24 stimulating electrodes overlap considerably, leading to cross-stimulation and interference of adjacent electrodes. Yet, despite these potentially severe encoding and decoding problems, many patients learn to use the CI to achieve excellent speech comprehension, and sometimes they even learn to appreciate music. The benefits of a CI are especially high for prelingually deaf children when implantation is performed at a young age. These children may develop normal verbal communication skills, lead normal social lives, and can often even pursue a normal academic career. Since the technology has considerably matured over the years, implanting a second CI has now become an option in more and more countries. This could in principle restore binaural integration, provided the devices capture precise timing and intensity information to the brain in overlapping frequency bands from either ear.

In an increasing number of countries a CI can be implanted when there is still some (typically low-frequency) rest hearing in the nonimplanted ear, which is equipped with a hearing aid. In such cases, the auditory system has to cope with *bimodal acoustic inputs*: relatively coarse electrical pulsatile stimulation from the CI for the 1–8 kHz range versus amplified analog acoustic stimulation from the HA at the low frequencies. Although the spectral overlap between the two modes of stimulation may be quite limited, there exists a possibility for binaural integration that could help the patient to localize sounds in the horizontal plane (see Section 14.5). Current developments in restorative hearing therapies aim to restore binaural integration for these challenging sensory conditions, by designing strategies that fine-tune device settings and integrate the stimulation and encoding algorithms for both devices.

14.3 SINGLE-SIDED DEAFNESS

Complete sensory loss in the cochlea leads to total deafness in the affected ear, which can only be restored with a CI. In many countries, however, SSD is still not an indication for CI surgery. Clearly, SSD patients lack any binaural processing, and can rely only on spectral cue information from the unaffected ear to localize sounds in both azimuth and elevation. In chapters: Assessing Auditory

Spatial Performance and Sound Localization Plasticity, we saw that the weighting of spectral cues for azimuth localization is relatively low in normal-hearing listeners, but perhaps these weights may have increased over time in SSD patients.

A second potential cue for sound localization in the horizontal plane is the head–shadow effect (HSE), which was introduced in chapter: Acoustic Localization Cues. We argued that the HSE is an *ambiguous* spatial cue, because it always consists of a mixture of azimuth-related information (due to the acoustic shadow) and absolute intensity of the sound source, which are both unknown to the auditory system. Although unimportant for binaural listeners, who can extract the ILDs (illustrated in Fig. 14.3A–C for a representative subject), the HSE could still be useful for SSD listeners, provided they can acquire reliable prior knowledge regarding the absolute intensity of the sound source.

Fig. 14.3C,D shows the head-movement responses of a congenital SSD listener (left ear deaf) to HP noise bursts, pooled across sound levels between 30–60 dB SPL that were randomly interleaved in the localization experiment. The listener is unable to localize these sound sources, as the pooled responses are virtually unrelated to the actual stimulus locations. Yet, when only the responses for the 45 dB stimuli in this experiment are selected, a reasonable stimulus–response relation seems to emerge, both for azimuth (Fig. 14.3C) and

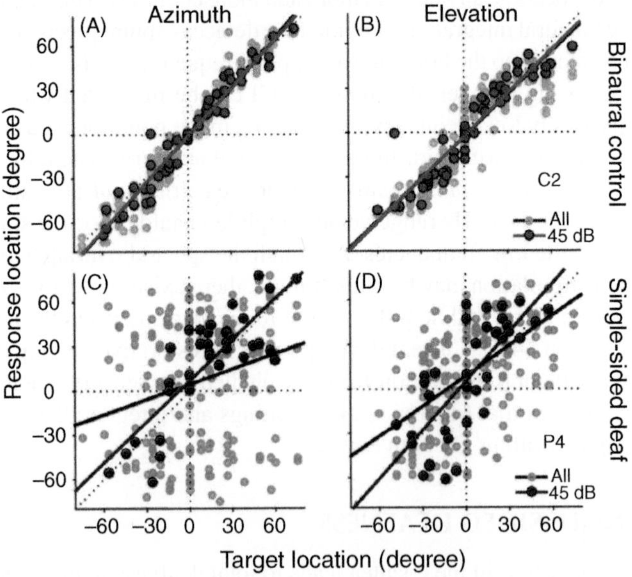

FIGURE 14.3 (A,B) Localization of HP noise sounds by control listener C2. Data pooled for sound levels. Azimuth responses for the 45 dB HP sounds are highlighted and fitted separately. Response accuracy and precision are independent of sound level. (C,D) Data taken from single-sided (left ear) deaf listener P4. (D,E) For the pooled intensities a stimulus–response relation is virtually absent. However, responses to the 45 dB stimuli correlate well with stimulus location.

for elevation (Fig. 14.3D). This interesting observation suggests that the listener indeed used the HSE for sound localization, and that the internal representation (the *prior*) of the absolute sound level was around 45 dB SPL (which corresponded to the midrange of the employed sound levels in the experiment). As a result, stimuli presented at lower intensities were predominantly localized at the deaf side (large negative bias), whereas the louder stimuli were all localized at the hearing side (large positive bias). To quantify the contribution of the HSE to horizontal localization in listeners the following regression model was used for the azimuth response components:

$$\hat{\alpha}_{RESP} = p\hat{\alpha}_{TAR} + q\hat{L}_{PROX} \tag{14.1}$$

where the dimensionless *z*-scores for variable *x* are defined by (see chapter: Assessing Auditory Spatial Performance):

$$\hat{x} \equiv \frac{x - \mu_x}{\sigma_x}$$

and *p* and *q* are the partial correlation coefficients for the target's azimuth and its proximal sound level, $\hat{L}_{PROX}$ [Eq. (7.16) and Fig. 7.9], respectively. For a normal-hearing binaural listener one expects $p \sim 1$ and $q \sim 0$.

Fig. 14.4 compares the results of this multiple regression analysis for nine SSD and six binaural control listeners. The responses of normal-hearing subjects indeed do not rely at all on the proximal sound level, as the regression coefficients lie close to the ideal values of $(p,q) = (1,0)$. The SSD listeners, however, show nearly opposite results, as they yield a high contribution of the HSE to their responses, in combination with low azimuth gains [for three listeners the coefficients are even close to zero, and $(p,q) \sim (0,1)$].

For 6/9 SSD patients, however, the contribution of target azimuth to their spatial percept is small but significant, and is not explained by the HSE. To check whether the azimuth contribution is indeed due to spectral cues from the hearing ear, Fig. 14.5 compares the partial correlations for the azimuth and elevation response components of all listeners. To that end, the following multiple regression model was used to describe the elevation response components:

$$\hat{\varepsilon}_{RESP} = r\hat{\varepsilon}_{TAR} + m\hat{\alpha}_{TAR} + n\hat{L}_{PROX} \tag{14.2}$$

For normal-hearing control listeners one expects $(r,m,n) \sim (1,0,0)$. In Fig. 14.5A the elevation coefficients, *r*, of Eq. (14.2), are plotted against the azimuth coefficients, *p*, of Eq. (14.1). Indeed, the binaural control listeners yield high partial correlations for elevation that do not relate to the azimuth coefficients, which again underlines the notion of independent binaural and monaural neural pathways to extract sound–source azimuth and elevation, respectively.

The SSD listeners, however, produce an essentially different result, as there is a highly significant correlation between the elevation and azimuth coefficients

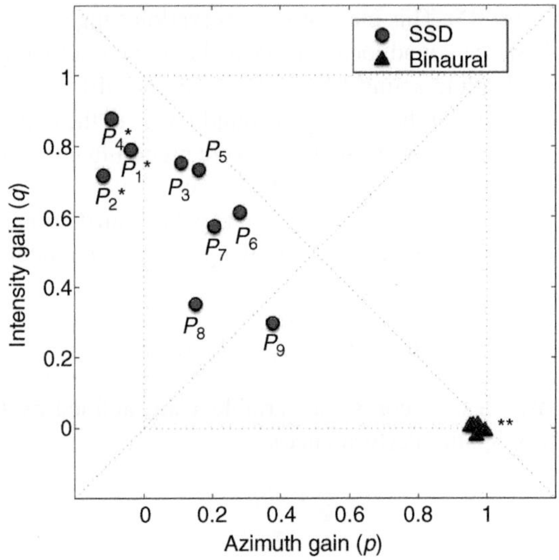

FIGURE 14.4 Results of multiple linear regression [Eq. (14.1)] on the azimuth response components to HP noises of varying intensities. The SSD listeners depend heavily on the proximal sound level, and hence the HSE. Binaural control listeners rely entirely on their veridical binaural difference cues (* and **: p and q do not differ significantly from zero). *(Source: After Van Wanrooij and Van Opstal, 2004, with kind permission.)*

for this group of listeners. This finding nicely demonstrates that those SSD listeners who show a significant gain for azimuth can do so by relying on their spectral cues. Fig. 14.5B shows that the SSD listeners indeed use spectral cues

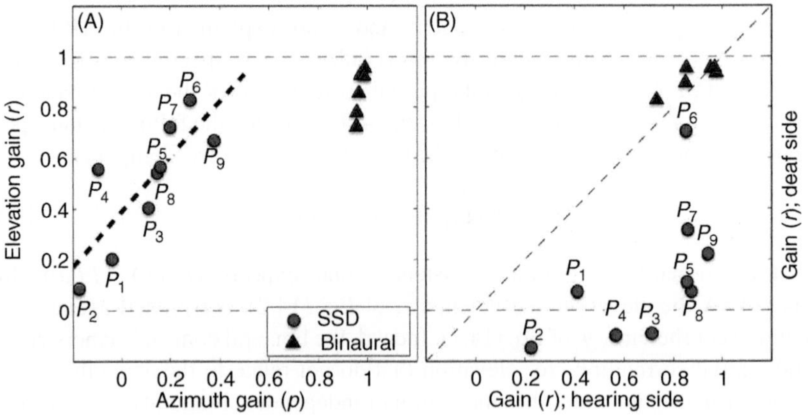

FIGURE 14.5 SSD listeners with a significant gain for azimuth (all but P_1, P_2 and P_4; see Fig. 14.4) rely on the spectral cues of the hearing ear for localization in the horizontal plane. *(Source: After Van Wanrooij and Van Opstal, 2004, with kind permission.)*

from their hearing ear, since elevation performance on their hearing side far exceeds that of the deaf side.

In the aforementioned examples, the hearing ear of the SSD patients was within the normal hearing range (ie, hearing loss < 20 dB). However, SSD listeners may also suffer from conductive or sensory–neural hearing loss in that ear. For example, as a result of aging the high-frequency sensitivity of the good ear may be impaired. How tolerant is the auditory system until SSD listeners start to lose the capacity to use spectral cue information from that ear? Recently Agterberg et al. (2014) conducted an experiment with a group of SSD listeners with ages varying from young adult to elderly. Their results indicated that as long as the hearing loss below 8 kHz is less than 40 dB, listeners can still use the spectral cues of their hearing ear to localize the azimuth of sound-sources. Indeed, for those listeners there is a strong correlation between the azimuth and elevation response gains too (Fig. 14.6A, filled symbols). However, unlike the data of Fig. 14.5A, this correlation was absent for SSD listeners with a high-frequency hearing loss that exceeded 40 dB at 8 kHz (open symbols).

Fig. 14.6B shows what happens when the spectral cues are removed by a mold inserted in the pinna of the hearing ear. In that case both the elevation gain and the azimuth gain of good-hearing SSD listeners decreased substantially (filled symbols). This decrease in gain was confined to elevation for the control listeners with binaural molds, as they relied entirely on binaural difference cues to localize azimuth.

In conclusion, SSD listeners weigh the acoustic evidence from the HSE and spectral cues from their hearing ear, provided the latter are indeed available (hearing loss below 40 dB at about 8 kHz). According to the Bayesian

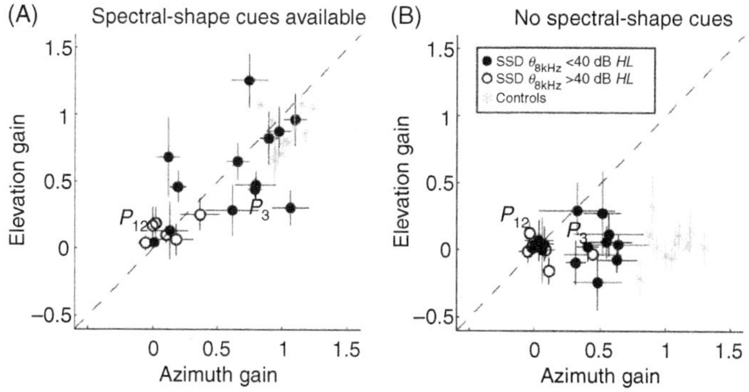

FIGURE 14.6 (A) The correlation between azimuth and elevation gains breaks down for SSD listeners with a hearing loss at the hearing ear of 40 dB or more at 8 kHz (open symbols). (B) When the spectral cues are perturbed by a mold both the elevation and azimuth gains decrease for SSD listeners with hearing loss <40 dB at 8 kHz (filled symbols). Gray asterisks: normal-hearing listeners without (A) and with (B) binaural molds. *(Source: After Agterberg et al., 2014.)*

framework of chapter: Multisensory Integration this is an optimal strategy, given the weak and noisy evidence from the available localization cues. Hence, listeners will rely more and more on their experience with (familiar) sound-sources and adopt a prior for the expected absolute sound level in their daily environment. Whether they rely on different priors for different types of sounds and environments is an interesting question that will require further study.

14.4 UNILATERAL CONDUCTIVE HEARING LOSS

Unless a SSD listener receives a CI, it is not possible to restore binaural hearing in these patients. In case of a conductive hearing loss, however, it should be possible to regain adequate binaural integration, even though the impairment and the restorative hearing device may heavily perturb the natural difference cues for sound localization (Priwin et al. 2007; Agterberg et al., 2011, 2012; Sparreboom et al. 2015). This expectation is based on the existing lifelong plasticity of the human sound-localization system (illustrated extensively in chapter: Sound Localization Plasticity).

To test whether listeners with a severe UCHL are indeed able to integrate binaural acoustic inputs when aided with a BCD, they can be subjected to similar (eye-) head-orienting experiments as normal-hearing control listeners, such as described in chapter: Assessing Auditory Spatial Performance. As the BCD can be turned "off" (monaural localization) or "on" (binaural localization), any improvement in localization performance can be attributed to the device. Further, by using stimuli with different bandwidths [low-pass noise (<1.5 kHz), high-pass noise (>3 kHz), and broadband noise (0.5–20 kHz)] and intensities (30–60 dB SPL) the potential contributions of ITD versus ILD cues and the HSE in these listeners can be assessed too. Fig. 14.7 summarizes the results of such an experiment.

The BCD of the listener in Fig. 14.7A,B provides a clear binaural integration benefit; the localization responses in the azimuth direction to broadband noise improve from practically no relationship at all (panel A) to near-normal behavior (in B). An improvement in localization performance is typical for the majority of patients (panels C,D), although in a fraction of listeners (about 30%) the improvement seems only modest. This considerable variability in performance may be due to several causes.

One potential cause could be that UCHL listeners employ different weightings on the available cues for sound localization. Here, the following five factors come to mind: *remnant ITDs and ILDs*, *spectral cues of the intact ear*, *the HSE*, and *priors*. To get a handle on this, Fig. 14.8A,B quantifies the relative contributions of the four acoustic factors as partial correlation coefficients: proximal sound level, q, versus the veridical sound location, p (regardless whether ITD, ILD, or HRTF cues were used to extract source location). The results for UCHL patients (Fig. 14.8A,B) are compared to the responses of normal-hearing listeners with and without a unilateral acute plug (Fig. 14.8C,D).

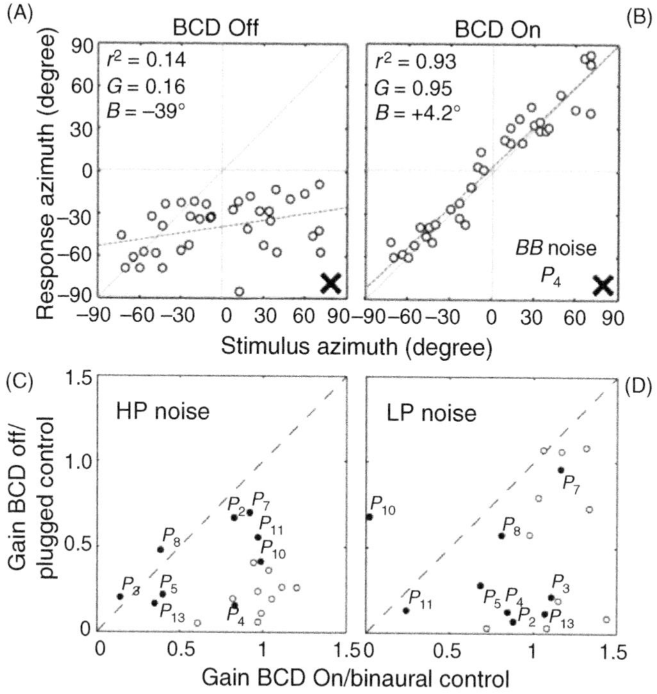

FIGURE 14.7 (A,B) Localization responses to broadband noise (pooled intensities) of a UCHL listener (P_4) equipped with a BCD on the right side (cross). With the BCD off (left) the patient localizes all stimuli on the hearing side, quantified by a large leftward bias of 39 degrees. With the BCD on, responses are accurate (high gain) and precise (high coefficient of variance). (C,D) Stimulus–response gains for HP and LP noise with BCD on versus off (solid dots). Control listeners (open symbols) wore a plug in one ear, to simulate the BCD off condition. The gains with BCD-on/binaural hearing are systematically higher than for BCD-off/plug (most data points lie below the main diagonal). Pooled intensities. *(Source: After Agterberg et al., 2012, with kind permission.)*

Clearly, the binaural control group is expected to put a heavy weight on the veridical azimuth cue(s): $p \gg q$. Indeed, without the plug (open symbols), all responses are fully determined by the sound's azimuth, and sound intensity plays no role whatsoever. In the case of an acute monaural plug, it is possible that low frequencies can leak (albeit attenuated) through the plug, thus still allowing for relatively intact ITDs. Indeed, the intensity coefficients tend to be very low for localization responses to BB noise, which contains low frequencies (Fig. 14.8C). For the HP noise, however, the HSE also contributes considerably in most of the acutely plugged control listeners (Fig. 14.8D).

For the UCHL listeners the picture is quite different (Fig. 14.8A,B): with the BCD off they invariably rely strongly on the HSE, for BB as well as for HP noise, although the coefficient for the HSE varies considerably from listener to listener. But even with the BCD on, some patients still base their responses to a large extent on the HSE. Perhaps these listeners put stronger weighting on prior

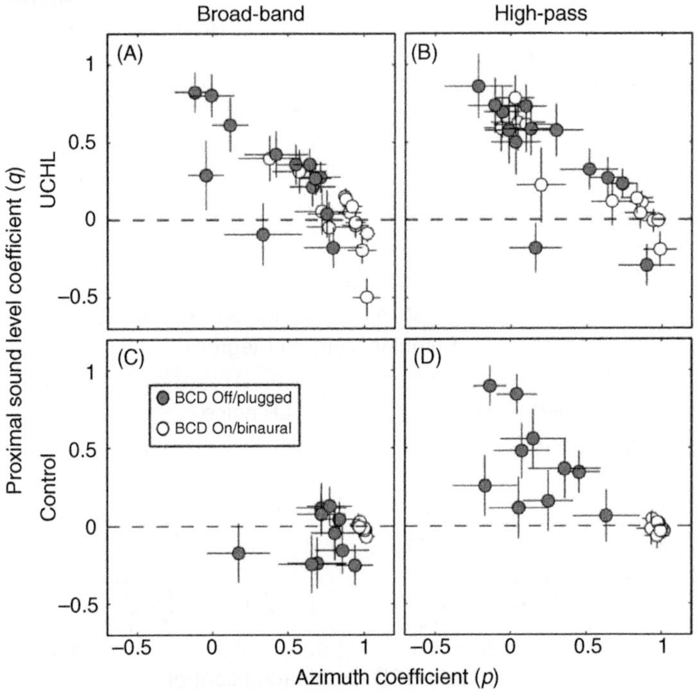

FIGURE 14.8 Contribution of proximal sound level (acoustic head–shadow) and target azimuth to perceived sound–source azimuth in control listeners (C,D) and BCD users (A,B). Two conditions are shown: device on (open symbols) or off (gray symbols) for the BCD users, and normal binaural hearing (open) versus monaural plugging (gray) in the controls. Further, responses are shown for two stimulus types: broadband noise (left) versus high-pass filtered noise (>3 kHz). Note that the BCD-on condition shifts responses toward the normal localization behavior (at azimuth: 1.0, sound level: 0.0). *(Source: From Agterberg et al., 2012, with kind permission.)*

information regarding familiar sound-sources, which would allow them to use the HSE as a more or less reliable localization cue. Due to this, however, their localization accuracy in the laboratory remains relatively poor when compared to listeners who have learned to ignore the HSE with their BCD on.

A second potential factor to explain variability among UCHL listeners may relate to the reasonably good localization capacity in some patients, even with the BCD off: in that case, little improvement from the device may be expected, and their auditory system may simply have decided not to reweigh the cues arising from the BCD. Good sound localization with the BCD off could have two potential causes, which are not mutually exclusive: (1) the hearing loss is moderate, allowing for some acoustic power to be directly transmitted to the cochlea without active bone vibration from the device; and (2) effective use of spectral cues from the intact ear. To dissociate these different mechanisms, Fig. 14.9 shows the result of one UCHL listener with a significant localization performance with the BCD off (Fig. 14.9A). Note that the localization performance

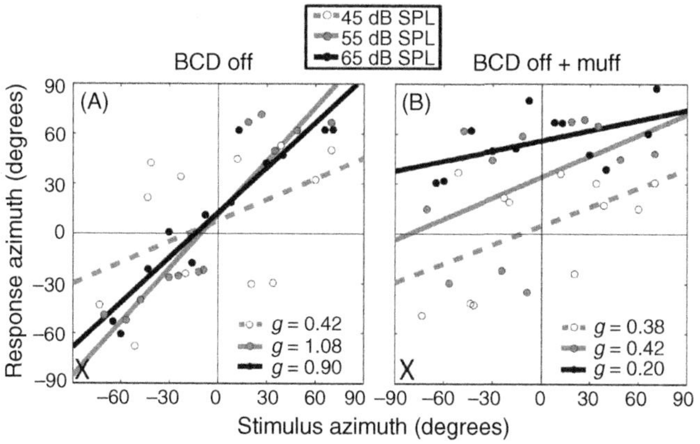

FIGURE 14.9 (A) A UCHL listener with some remnant hearing uses air-conducted acoustic input for the louder sounds to localize sound-sources. (B) Additional attenuation of the impaired ear with a muff abolishes this capacity. *(Source: From Agterberg et al., 2012, with kind permission.)*

depends strongly on sound level: the weakest stimuli are localized poorest, whereas the loud sounds (>50 dB SPL) are localized quite well. This suggests a potential transmission of air-conducted acoustic power to the affected cochlea. Indeed, when a muff attenuates the *impaired ear* even further, the high-intensity responses break down: they are now best described by an intensity-dependent bias (ie, a pure HSE response). Note that the low-intensity responses are hardly affected by the muff: they tend to rely on spectral cues from the intact ear (also see chapter: Sound Localization Plasticity, for normal-hearing plugged listeners, who use the same strategy).

This surprising behavior nicely demonstrates the existence of flexible and seemingly automatic strategies adopted by the auditory system of hearing-impaired listeners, which depend on the acoustic conditions. As soon as cues become available they contribute to the spatial percept. Within the Bayesian framework, the weights should depend on the reliability of cues with respect to each other and to the subject's prior expectations.

14.5 BIMODAL HEARING: CI–HA

When a patient received a CI in one ear, and a HA for the other ear, the auditory system faces *bimodal* (electric and acoustic) inputs. It is far from trivial how to present the auditory inputs to the two ears such that they can be integrated into a single binaural percept. To achieve this, many hurdles need to be overcome: the frequency bands of the *CI* (1–8 kHz) and *HA* (<3 kHz) may overlap poorly, and their spectral resolutions may be quite different too; the perceptual loudness from the two devices, as well as their sound images (timbre, chromo, pitch) may differ, and additional time delays may be introduced from the speech-processing unit in the *CI*; amplifier compression strategies may differ between devices because of differences in technology, algorithms, and hearing loss on the two sides, etc. Despite this, most patients seem to benefit from bimodal stimulation, as they yield higher scores on speech comprehension; sometimes patients can even localize sounds in the horizontal plane (Ching et al. 2007; Mok et al., 2006; Morera et al. 2005). However, whether benefits are due to real binaural integration, or to mere binaural summation, remains unclear.

One potential problem with the two hearing devices is that they operate in different frequency bands and at different dynamic ranges. The *CI* covers the broad range of relatively high frequencies from about 1 up to 8 kHz, whereas the *HA* complements the CI by covering mainly the lower end of the spectrum. The CI signal is directly transmitted to the auditory nerve and therefore is not confined to the limited dynamic range for hair cell stimulation (nonlinear compression). Together, however, the devices may augment the perceived chroma of complex sounds, and hence increase listening comfort.

To integrate sound into a binaural localization percept requires: (1) sufficient frequency overlap; (2) that the ILDs derived from the devices lie in the physiological range, and vary monotonically (and consistently) with azimuth for all frequencies; (3) ITD (envelope) cues should vary in a similar way as the ILDs. If these conditions are not met, the cues become ambiguous and cannot (will not) be used for localization.

However, the amount of frequency overlap and frequency-dependent compression will vary from patient to patient, which could necessitate tedious individualized settings for the two devices. Fig. 14.10 illustrates a simple psychophysical strategy to determine whether binaural integration and localization are at all possible in the patient, given the current settings of the device. Fig. 14.10A–E sketches different stimulus-response relationships for a hypothetical patient with a *HA* on the left side and a *CI* on the right. From left to right the sound's bandwidth is systematically increased: in Fig. 14.10A sounds contain only low frequencies, which will be processed by the *HA* only. As a result the sound-localization percept will be fully dominated by the *HA* (left ear), and the listener will lateralize all sounds on the far left. In Fig. 14.10E we see the reverse: the bandwidth is at its broadest, and now the *CI* will fully dominate the percept. Hence, all sounds will be lateralized on the *CI* side (far right). The idea is, that there can be a gradual transition from left monaural lateralization to right monaural lateralization, via real *binaural integration* for stimuli with intermediate bandwidths. Fig. 14.10B–D sketch this potential transition scenario.

Fig. 14.10F–I shows data from a bimodal listener (*CI* on the right, *HA* on the left), subjected to four different sound bandwidths. When the frequencies are low-pass filtered at 1.5 kHz (Fig. 14.10F) responses are directed to the side of the hearing aid. As the bandwidth is gradually increased to 2.0, 2.5, and 3.0 kHz the responses of the listener move from *HA* dominated toward the *CI* side. Note, however, that the slope to the best-fit regression line changes quite abruptly in Fig. 14.10G; this suggests that bimodal integration may indeed take place in this patient when stimuli have a bandwidth of ~2 kHz.

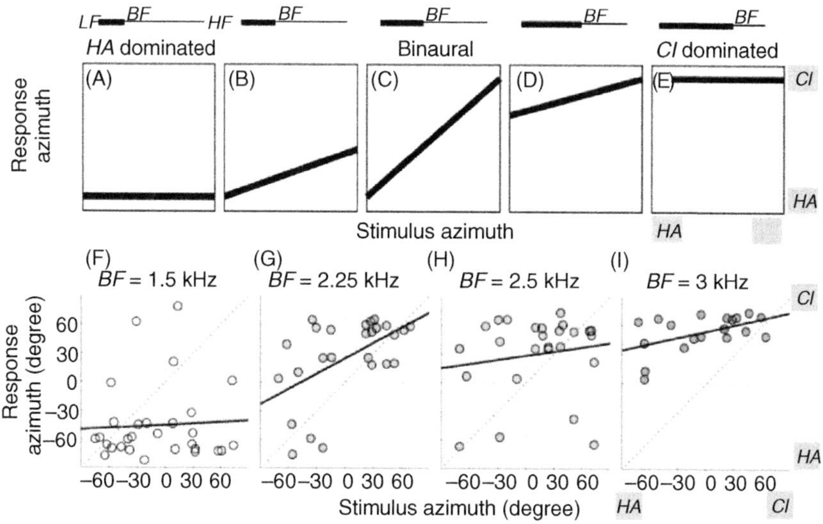

FIGURE 14.10 (A–E) Hypothetical sound-localization scenario for a bimodal listener to sounds with different frequency bands from *LF* = 0.25 kHz to BF (indicated at the top). When BF is low (A), the HA dominates responses (left lateralization); when BF is high (E) the *CI* dominates (right lateralization). For intermediate BF there may be binaural integration (B–D). (F–I) Results for bimodal patient P7 suggest that binaural integration takes place for *BF* ~ 2 kHz (G). *(Source: From Veugen et al., unpublished results.)*

Clearly, for patients such as these the requirements for optimal sound localization performance versus optimal speech comprehension and sound appreciation seem to impose conflicting constraints on the stimulus properties. As the former is optimal for a highly restricted bandwidth (2 kHz), the latter thrives best with a bandwidth that should be as broad as possible. However, the data also clearly indicate that in those cases the CI would fully dominate the percept, and there will be no binaural benefit for such patients. To truly integrate and benefit from bimodal stimulation for all hearing tasks (localization, target-background segregation, as well as identification and appreciation), the amount of frequency overlap should be maximized and perceptually comparable for the two devices. This may be impossible when the hearing ear is severely impaired, as is the case for the example listener in Fig. 14.10F–I. It may be expected that the better residual hearing in the hearing ear, the higher the chances for true bimodal integration over a broad frequency range.

14.6 PRESBYCUSIS: SOMETIMES SUPERIOR PERFORMANCE?

In age-related high-frequency hearing loss, such as illustrated in Fig. 14.1A, the perception of high-frequency spectral localization cues will deteriorate. However, otherwise normal-hearing elderly will keep sufficient frequency sensitivity

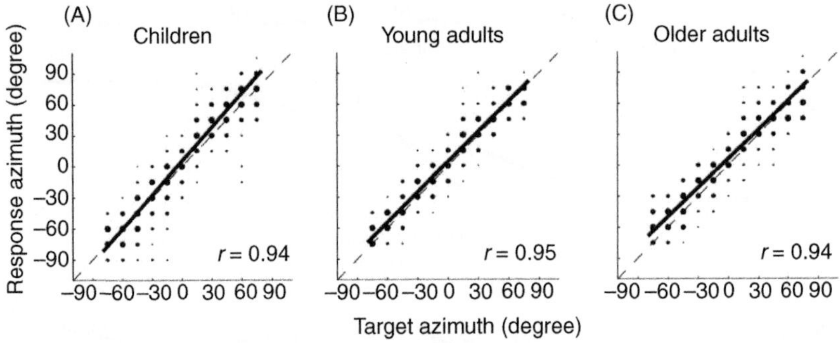

FIGURE 14.11 Azimuth localization performance of children (A), young adults (B), and the elderly (C) is indistinguishable across the three groups. Responses pooled for stimuli with bandwidths (from 0.5 to 5, 7, 11, and 20 kHz, respectively). Solid lines: regression through the data. Data were binned for visual purposes only; symbol size is proportional to the number of responses in each bin. *(Source: From Otte et al., 2013, with kind permission.)*

to potentially allow for the use of binaural difference cues (ITDs and ILDs) over a frequency range of 4–6 kHz (Fig. 14.1B). Recent evidence has suggested, however, that the elderly might lack sufficient *temporal* modulation sensitivity around 1–2 kHz, which would suggest that their use of ITDs for localization in the horizontal plane might be hampered too (Dobreva et al., 2011).

When comparing the head-movement sound-localization responses of children (7–11), young adults (20–34) and elderly (63–80) listeners to a variety of broadband sounds with different bandwidths, presented across the frontal hemifield (−75 to +75 degrees in azimuth and −55 to +55 degrees in elevation), the azimuth response components for the three groups result to be virtually indistinguishable (Fig. 14.11). Thus, although the elderly may poorly localize specific narrow-band sounds (Dobreva et al., 2011), overall performance in the horizontal plane seems to be unaffected by age.

The vertical extent of the pinna's concha of the elderly is, on average, larger than for young adults and children, because of slow pinna growth (about 0.1 mm/year) during life. The interesting acoustic result of this directed pinna growth is that the characteristic elevation-dependent notches in the HRTFs shift systematically toward lower frequencies (see chapter: Sound Localization Plasticity). Wouldn't it be nice if this shift would compensate for the high-frequency hearing loss, and bring the notches back into the perceptual hearing range (Fig. 14.1B, dashed green curve)? The effective elevation-encoding frequency range for the HRTFs of the elderly (say, between 3.5 and 7 kHz) is restricted to lower frequencies than for young adults and children (say, from 5 to 12 kHz). A similar difference holds for men (larger ears) versus women (smaller ears). Indeed, if the elderly would have adapted to the shifts of their spectral notches, and if they still have good hearing at the higher frequencies, they could in principle outperform younger listeners in localizing sounds that have limited bandwidths

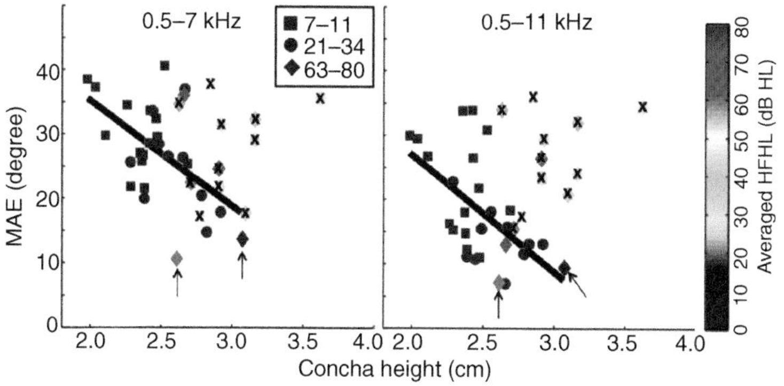

FIGURE 14.12 **Mean absolute localization error in elevation as a function of concha height for two band-limited noise bursts.** Age groups are indicated by the different symbols (inset), and only listeners with limited high-frequency hearing loss (color bar on the right) were included in the regression analysis. Arrows: two of the elderly outperformed everybody else.

in elevation. To test for this interesting possibility Otte et al. (2013) analyzed the elevation responses of the three age groups shown in Fig. 14.11 separately for the four different broadband stimuli (0.5–5k, 0.5–7k, 0.5–11, 0.5–20 kHz). The elderly with excellent hearing (<25 dB *HL*; blue symbols in Fig. 14.12) were then compared with the young adults and children groups.

Fig. 14.12 shows a clear negative correlation between concha height and the mean absolute localization error in elevation for those stimuli with limited bandwidths that potentially contain information of elevation (cut-off at 7 and 11 kHz). As expected, for the 5 and 20 kHz stimuli this correlation was absent. Interestingly, however, two of the elderly listeners with limited high-frequency hearing loss (the *Blue Diamonds*, indicated by the arrows) yielded the best performance of all 42 subjects who participated in this experiment. Yes, there is hope: not all our capacities have to disappear with age: thanks to its lifelong plasticity, our sound-localization system keeps on fighting until the very end!

REFERENCES

Agterberg, M.J.H., Snik, A.F.M., Hol, M.K.S., TVan, Esch.E.M., Cremers, C.W.R.J., Van Wanrooij, M.M., Van Opstal, A.J., 2011. Improved horizontal directional hearing in Baha users with acquired unilateral conductive hearing loss. J. Assoc. Res. Otolaryngol. 12, 1–11.

Agterberg, M.J.H., Snik, A.F.M., Hol, M.K.S., Van Wanrooij, M.M., Van Opstal, A.J., 2012. Contribution of monaural and binaural cues to sound localization in patients with unilateral conductive hearing loss: improved directional hearing with a bone-conduction device. Hearing Res. 286, 9–18.

Agterberg, M.J.H., Hol, M.K.S., Van Wanrooij, M.M., Van Opstal, A.J., Snik, A.F.M., 2014. Single-sided deafness and directional hearing: contribution of spectral cues and high-frequency hearing loss in the hearing ear. Front. Neurosci. 8, 188.

Ching, T.Y., Van Wanrooy, E., Dillon, H., 2007. Binaural–bimodal fitting or bilateral implantation for managing severe to profound deafness: a review. Trends Amplific. 11, 161–192.

Dobreva, M.S., O'Neill, W.E., Paige, G.D., 2011. The influence of aging on human sound localization. J. Neurophysiol. 105, 2471–2486.

Lin, L.M., Bowditch, S., Anderson, M.J., May, B., Cox, K.M., Niparko, J.K., 2006. Amplification in the rehabilitation of unilateral deafness: speech in noise and directional hearing effects with bone-anchored hearing and contralateral routing of signal amplification. Otol. Neurotol. 27, 172–182.

Mok, M., Grayden, D., Dowell, R.C., Lawrence, D., 2006. Speech perception for adults who use hearing aids in conjunction with cochlear implants in opposite ears. J. Speech Lang. Hear. Res. 49, 338–351.

Morera, C., et al.,2005. Advantages of binaural hearing provided through bimodal stimulation via a cochlear implant and a conventional hearing aid: A 6-month comparative study. Acta Oto-Laryngol. 125, 596–606.

Otte, R.J., Agterberg, M.J.H., Van Wanrooij, M.M., Snik, A.F.M., Van Opstal, A.J., 2013. Age-related hearing loss and ear morphology affect vertical, but not horizontal, sound-localization performance. J. Assoc. Res. Otolaryngol. 14, 261–273.

Priwin, C., Jönsson, R., Hultcranz, M., Granström, G., 2007. Baha in children and adolescents with unilateral or bilateral conductive hearing loss: a study of outcome. Int. J. Pediat. Otorhinilaryngol. 71, 135–145.

Sparreboom, M., Langereis, M.C., Snik, A.F., Mylanus, E.A., 2015. Long-term outcomes on spatial hearing, speech recognition and receptive vocabulary after sequential bilateral cochlear implantation in children. Res. Dev. Disabil. 36, 328–337.

Stenfelt, S., Håkansson, B., Tjellström, A., 2000. Vibration characteristics of bone conducted sound in vitro. J. Acoust. Soc. Am. 107, 422–431.

Van Wanrooij, M.M., Van Opstal, A.J., 2004. Contribution of head shadow and pinna cues to chronic monaural sound localization. J. Neurosci. 24, 4163–4171.

Veugen, L.C.E., Chalupper, J., Snik, A.F.M., Van Opstal, A.J., Mens, L.H.M.: Matching automatic gain control across devices in bimodal cochlear implant users. Ear Hearing, in press.

Von Békésy, G.V., 1948. Vibration of the head in a sound field and its role in hearing by bone conduction. J. Acoust. Soc. Am. 20, 749–760.

Wazen, J.J., Spitzer, J.B., Ghossaini, S.N., Fayad, J.N., Niparko, J.K., Cox, K., Brackmann, D.E., Soli, S.D., 2003. Transcranial contralateral cochlear stimulation in unilateral deafness. Otolaryngol. Head Neck Surg. 129, 248–254.

Zurek, P.M., 1986. Consequences of conductive auditory impairment for binaural hearing. J. Acoust. Soc. Am. 80, 466–472.

Subject Index

A

D

E

F

G

H

M

Q

R

S

T

U

Printed and bound by CPI Group (UK) Ltd, Croydon, CR0 4YY
08/05/2025
01865022-0004